PIGMENTS OF ENGLISH MEDIEVAL WALL PAINTING

Pigments of English Medieval Wall Painting

HELEN HOWARD

Archetype Publications

First published 2003 by Archetype Publications Ltd.

Archetype Publications Ltd.
6 Fitzroy Square
London W1T 5HJ

www.archetype.co.uk

Tel: 44(207) 380 0800
Fax: 44(207) 380 0500

ISBN 1-873132-48-4

British Library Cataloguing in Publication Data
A catalogue record for this book is available from the British Library.

Printed on acid-free paper.

Typeset by Kate Williams, Abergavenny
Printed and bound in Great Britain by Henry Ling Limited, The Dorset Press, DT1 1HD

For Christopher, Alexander and Ursula

Contents

Acknowledgements ix
List of tables x
List of colour plates xi
Abstract xxxi

1 **Introduction** **1**
Background to the research 1
Overview of English medieval wall painting techniques 3
The focus of this book 8
Sources of information 9
Chronological and geographical scope 10
Initial summary of the results 11

2 **Methodology** **15**
Introduction 15
Primary sources: medieval treatises, recipe books and accounts 15
Secondary literature: antiquarian literature and more recent published sources 21
The author's work and unpublished analytical sources 23
Methods of scientific examination employed 24

3 **Blue pigments** **27**
Ultramarine 27
Vivianite 35
Egyptian blue 39
Azurite 40
Synthetic copper blues 50
Indigo 57

4	**Green pigments**	**65**
	Green earth	65
	Salt green	72
	Malachite	78
	Verdigris and copper resinate	85
5	**Red pigments**	**97**
	Vermilion	97
	Red lake pigments	111
	Red lead	125
6	**Red and yellow earth pigments**	**141**
7	**Yellow pigments**	**153**
	Orpiment	153
	Lead-tin yellow	159
8	**White pigments**	**165**
	Lime white	165
	Chalk	171
	Calcium sulphate whites	173
	Lead white	176
	Bone white	185
9	**Carbon black pigments**	**187**
	Charcoal black	189
	Bone black	193
	Coal black	193
	Black earth pigments	194
10	**Discussion**	**195**
	Pigments previously unidentified in English medieval wall painting	195
	Use of pigments and the changing nature of English medieval wall painting	197
	Pigment alteration in English medieval wall painting	202
	Relationships with continental wall painting, polychrome sculpture and panel painting: implications for workshop practice	205
	Conservation implications of the present research	208
	Avenues for further research	211
	Appendix 1: Glossary	213
	Appendix 2: Examination proforma for use *in situ*	215
	Appendix 3: Tables	223
	Appendix 4: Abbreviations used in the tables	287
	References	289
	Index	307

Acknowledgements

I am very grateful to Jim Black of Archetype Publications for publishing this work. Particular thanks go to the Paul Mellon Centre for Studies in British Art whose generous funding made possible the inclusion of the many essential colour plates.

The Leverhulme Trust generously funded the five-year study (1991–96) of English medieval wall painting techniques which provided much of the groundwork for the PhD thesis on which this book is based. I would also like to thank: English Heritage for funding access to FTIR facilities at Imperial College and for providing access to SEM/EDX facilities at the Ancient Monuments Laboratory; Dr Ashok Roy, Raymond White, Dr Jenny Pilc, Marika Spring and Catherine Higgitt of the Scientific Department of the National Gallery for advice and help with analysis (XRD, GC-MS and FTIR); Dr Gordon Cressy of the Natural History Museum for XRD analysis; Dr Robert Withnall of University College, London for Raman spectroscopy; Dr Sophie Stos-Gale of the Isotrace Laboratory, Oxford for lead isotope analysis; Dr Arie Wallert of the Rijksmuseum for fluorescence spectroscopy and direct temperature-resolved mass-spectrometry; and Dr M. P. Colombini of Pisa University for GC-MS analysis. I am also grateful to Jo Kirby of the National Gallery for HPLC analysis of lake pigments and for her detailed comments on the chapters on red pigments, and to Caroline Villers of the Courtauld Institute for enthusiastic and professional support including helpful comments on large parts of the text. The late Professor Stephen Rees Jones kindly discussed his experience in developing the field of painting examination and analysis in the 1940s.

I am indebted to David Park and Sharon Cather of the Courtauld Institute without whom this work could not have been done. They won funding for the initial Leverhulme study and invited me to take up the resultant post. They provided a supportive professional environment in which the significance of interdisciplinary evaluation of the technical results, in terms of art history, condition and environment of the paintings, was paramount. David Park generously provided access to his (largely unpublished) survey of over two thousand English medieval wall paintings, a resource which was fundamental to the work. As supervisor to my PhD, he painstakingly and patiently commented on every section of the text, and for this and all his help I am profoundly grateful.

I thank many friends for their help and forbearance, in particular John Stewart. Above all, I thank Christopher Griffiths who has given untold support, enthusiastic encouragement and assistance at every turn. Without his help this work would never have been completed.

List of tables

Table 1 Selected analytical data relating to the use of blue pigments in medieval polychromy.
Table 2 Selected analytical data relating to the use of green pigments in medieval polychromy.
Table 3 Selected analytical data relating to the use of red pigments in medieval polychromy.
Table 4 Selected analytical data relating to the use of yellow pigments in medieval polychromy.
Table 5 Selected analytical data relating to the use of white pigments in medieval polychromy.
Table 6 Selected analytical data relating to the use of black pigments in medieval polychromy.
Table 7 Selected analytical data relating to the use of oil in medieval polychromy.
Table 8 Selected analytical data relating to the use of resin and oil/resin in medieval polychromy.
Table 9 Selected analytical data relating to the use of protein in medieval polychromy.

List of colour plates

Plate 1 St. Gabriel's Chapel, Canterbury Cathedral. Detail showing the *Naming of St. John* on the north side of the apse. Natural ultramarine was extensively employed for the background areas of this scheme of *c*1130. (© Crown copyright. NMR)

Plate 2 Detail of Plate 1 photographed in raking light showing the undulating surface of the plaster and the rich blue of the ultramarine background. (© Courtauld Institute, Conservation of Wall Painting Department)

Plate 3 St. Gabriel's Chapel, Canterbury Cathedral. Thin section of a sample taken from the blue robe of the *Christ in Majesty*, stained for protein with Amido black (Sol.1); photographed in transmitted light with crossed polars, original magnification 200×. Analysis of the original binding media confirmed the presence of protein in both the blue and red layers in this sample, while linseed oil was detected in a sample taken from the blue background of the scheme. (Photomicrograph: H. Howard; © Courtauld Institute, Conservation of Wall Painting Department)

Plates 4 and 5 Details of the scheme of *c*1130 at Idensen (Germany), where natural ultramarine is used extensively for the panelled backgrounds and other details. The use of charcoal black mixed with white as an underpaint for the ultramarine mirrors that in the approximately contemporary paintings in St. Gabriel's Chapel at Canterbury and in the scheme of *c*1170 at Cormac's Chapel, Cashel (Ireland). (Plate 4 © Courtauld Institute, Conservation of Wall Painting Department; Plate 5 © Dr Rolf-Jürgen Grote)

Plates 6 and 6a Details of the *Crucifixion* above the north nave arcade at Ickleton (Cambridgeshire). Natural ultramarine was employed to pick out details of the hair and beard of Christ in each scene of this scheme of *c*1125. (Photographs: H. Howard; © Courtauld Institute, Conservation of Wall Painting Department)

Plate 7 Ickleton (Cambridgeshire). Cross-section of a sample taken from Christ's hair in the *Betrayal*, original magnification 200×. Ultramarine was combined with lime white and applied in a single layer approximately 150μm thick directly over the coarse underlying plaster. The blue pigment is of exceptionally high quality, consisting of extremely small particles only 5 × 3μm. (Photomicrograph: H. Howard; © Courtauld Institute, Conservation of Wall Painting Department)

Plate 8 Detail of the scheme of *c*1170 at Cormac's Chapel, Cashel (Ireland), where natural ultramarine was employed extensively for the backgrounds and other details. (Photograph: H. Howard; © Courtauld Institute, Conservation of Wall Painting Department)

Plate 9 Cormac's Chapel, Cashel (Ireland). Cross-section of a sample from the west side of the chancel arch, original magnification 400×. Natural ultramarine has been applied over a dark purple underpaint of charcoal black combined with red earth, presumably to produce a purple tint. (Photomicrograph: H. Howard; © Courtauld Institute, Conservation of Wall Painting Department)

Plate 10 Detail of the *Feeding of the Five Thousand* from the scheme of *c*1165 at Pürgg (Austria). Here ultramarine was also applied over coloured underpaint: over dark red where a purple effect was sought, and over grey where a dark blue colour was intended. (Photograph: H. Howard; © Courtauld Institute, Conservation of Wall Painting Department)

Plate 11 Pürgg (Austria). Cross-section of a sample from Cain's drapery, original magnification 422×. Charcoal black and red earth were combined to produce the dark red colour of the drapery, while decorative elements were added in natural ultramarine. Traces of limewash are visible on the surface of the sample. (Photomicrograph: H. Howard; © Courtauld Institute, Conservation of Wall Painting Department)

Plate 12 Detail of the 'Losinga' cycle of *c*1190 at Norwich Cathedral showing Bishop Losinga and the intense ultramarine blue of the background. (Photograph: H. Howard; © Courtauld Institute, Conservation of Wall Painting Department)

Plates 13 and 14 'Losinga' cycle, Norwich Cathedral. Cross-section of a sample from the hem of the standing figure on the west segment of the vault, original magnification 340×. The drawing of the drapery was undertaken in yellow earth, and over this is a layer of ultramarine including other siliceous minerals and some unusual columnar-shaped inclusions. In UV light, these inclusions – among others in the paint layer – fluoresce a brilliant orange colour (Plate 14). Ground lapis lazuli is known to display a spotty fluorescence in UV of long wavelengths, but the level of fluorescence here is exceptional. In this instance, it is possible that a red lake pigment, which may also fluoresce in UV, was originally combined with the blue pigment, perhaps to produce a more purple tint. Although any red lake pigment has clearly faded substantially, close examination of one of the samples revealed the slightest trace of a pink hue at the base of the ultramarine layer. (Photomicrographs: H. Howard; © Courtauld Institute, Conservation of Wall Painting Department)

Plates 15 and 16 Details of the scheme of *c*1230 on the vault of the Guardian Angels' Chapel, Winchester Cathedral. (©Crown copyright. NMR) In UV light (Plate 16), the scrollwork, which was added in *c*1280, is clearly visible. (© Courtauld Institute, Conservation of Wall Painting Department)

Plate 17 Guardian Angels' Chapel, Winchester Cathedral. Cross-section of a sample from the lowest angel roundel on the west segment of the vault, original magnification 394×. Here, ultramarine combined with azurite was applied over azurite in a white matrix on a local yellow ground. Ultramarine was also identified in samples taken from a flower in the late 13th-century scrollwork. (Photomicrograph: H. Howard; © Courtauld Institute, Conservation of Wall Painting Department)

Plate 18 Detail of the scheme of *c*1250 on the eastern arch of the Ante-Reliquary Chapel, Norwich Cathedral. Ultramarine, applied over a thin skim of plaster on the stone support, was used for the background of the bishop's head to the south of the arch. In raking light the incised details of the bishop's mitre and face are clearly visible. (Photograph: H. Howard; © Courtauld Institute, Conservation of Wall Painting Department)

Plate 19 Natural ultramarine continued to be employed in high-quality schemes of wall painting in the 14th century as in the important mid-14th-century paintings in St. Stephen's Chapel, Westminster Palace. (© Copyright The British Museum)

Plate 20 The extensive use of natural ultramarine employed in Romanesque wall painting is paralleled in Romanesque sculpture as on the limestone figure of Moses (*c*1180–85) from St. Mary's Abbey, York. (© The Yorkshire Museum, York)

Plates 21 and 22 Ultramarine was also employed on the early Gothic south portal of Lausanne Cathedral in *c*1216–20. (©André Held). Ultramarine was commonly used in panel painting and notable examples include the Adisham reredos of *c*1180, the ceiling paterae from the Painted Chamber of Westminster Palace of *c*1263–66, and the Westminster Retable of *c*1270 (Plate 22). (© Dean and Chapter of Westminster)

Plate 23 Detail of the *Traditio Legis* of *c*1100 on the east wall of the nave at Coombes (West Sussex). In this plate the 'false blue' fictive marbling at dado level is clearly visible. (© Courtauld Institute, Conservation of Wall Painting Department)

Plates 24 and 25 Detail of the fictive marbling shown in Plate 23. Modelling effects have been created by varying the amounts of lime white combined with charcoal black, and a cross-section taken from a dark shadow clearly shows a higher proportion of charcoal black in the upper layer (Plate 25), original magnification 200×. The use of this pigment mixture is particularly prevalent in Romanesque wall painting and a much cheaper alternative to the costly mineral blues then available. (Photograph and photomicrograph: H. Howard; © Courtauld Institute, Conservation of Wall Painting Department)

Plate 26 'False blue' continued to be employed in the Gothic period, and a blue-black mixture made from charcoal black combined with lime white was also employed for the archbishop's shoes on the painted coffin lid of Archbishop Walter de Gray at York Minster of *c*1255. (© Crown copyright. NMR)

Plate 27 'False blue' was also used to produce a blue-grey colour on a 13th-century polychromed stone column in Lichfield Cathedral. The cross-section, original magnification 400×, shows particles of charcoal black combined with lime white and a little red earth. (Photomicrograph: H. Howard; © Courtauld Institute, Conservation of Wall Painting Department)

Plate 28 Winchester Cathedral, Holy Sepulchre Chapel: detail of the head of Joseph of Arimathea in the *Deposition* on the east wall of the chapel (*c*1175). Vivianite was extensively employed for the blue portion of the panelled backgrounds and for numerous details such as the vair lining of Nicodemus' cloak. (© Courtauld Institute, Conservation of Wall Painting Department)

Plate 29 Winchester Cathedral, Holy Sepulchre Chapel. Cross-section of a sample from the background of the *Deposition*, now an olive-green colour, original magnification 500×. A thin layer of natural ultramarine is visible over the grey plaster substrate, while the present green appearance is due to a layer of vivianite on the surface of the sample. Originally blue in colour, some of the particles of this vivianite have altered and now appear yellow, producing an overall green effect. (Photomicrograph: H. Howard; © Courtauld Institute, Conservation of Wall Painting Department)

Plates 30 and 31 Initial identification of vivianite at Winchester was made by polarised light microscopy, since the particles display a characteristic blue-yellow pleochroism. The thin section shown in these plates has been photographed in plane-polarised transmitted light and is shown in varying alignments as the microscope stage was rotated, original magnification 450×. The blue-yellow pleochroism is particularly visible in the large elongated particle which appears blue at the centre of Plate 30, while it is yellow in Plate 31. XRD provided further confirmation of the presence of vivianite. (Photomicrographs: H. Howard; © Courtauld Institute, Conservation of Wall Painting Department)

Plate 32 Holy Sepulchre Chapel, Winchester Cathedral: detail of the *Deposition* of *c*1220 now transferred to the west end of the chapel. Vivianite was employed for the blue panelled backgrounds of this later scheme in the chapel and these areas now appear green. (© Courtauld Institute, Conservation of Wall Painting Department)

Plate 33 Cross-section of a sample from the 'green' background shown in Plate 32, original magnification 470×. Vivianite was applied over grey underpaint of charcoal black combined with lime white which increased its covering power. The majority of the blue particles have now altered to yellow although the blue centres of the larger particles are still clearly visible. (Photomicrograph: H. Howard; © Courtauld Institute, Conservation of Wall Painting Department)

Plate 34 Examination of polished cross-sections of the partially altered vivianite at high magnification indicates that it is the largest particles (19 × 7μm) which have remained blue in colour. Although yellow particles of similar dimensions are present, they show clear divisions along cleavage planes allowing for increased oxidation and therefore colour change. (Photomicrograph: H. Howard; © Courtauld Institute, Conservation of Wall Painting Department)

Plate 35 Detail of the eastern vault of the Holy Sepulchre Chapel, Winchester Cathedral. Vivianite has retained its blue colour only where is has been protected from subsequent conservation treatments; for example, in the paint layer beneath the white dots on the background to the *Pantocrator*. (© Courtauld Institute, Conservation of Wall Painting Department)

Plate 36 A cross-section from one of the dots in Plate 35 shows the dark 'indigo' blue of vivianite beneath the white impasto layer, original magnification 500×. (Photomicrograph: H. Howard; © Courtauld Institute, Conservation of Wall Painting Department)

Plate 37 Horsham St. Faith Priory (Norfolk): detail of the monumental figure of St. Faith on the east wall of the refectory (*c*1250). Although azurite has been identified in English wall painting as early as *c*1130, it was not commonly employed until the 13th century, as here at Horsham St. Faith. (© Crown copyright. NMR)

Plate 38 Westminster Abbey, chapter house, *Apocalypse* cycle of *c*1375–1404: detail of *Banishment to Patmos*. Azurite combined with lead white was employed for St. John's blue drapery. (Photograph: R. Gowing; © Courtauld Institute, Conservation of Wall Painting Department)

Plate 39 Horsham St. Faith Priory (Norfolk): detail of the *Foundation Narrative* on the east wall of the refectory. (© Crown copyright. NMR)

Plates 40 and 41 Horsham St. Faith Priory (Norfolk). Cross-section of a sample taken from the first scene in the *Foundation Narrative*, original magnification 500×. Natural azurite, adulterated with its synthetic equivalent, has been applied over a layer of verdigris on a lead white ground. A dispersion of the blue pigment photographed in transmitted light (Plate 41) with crossed polars shows the angular particles of natural azurite; these display striations indicating their mineral origin. The synthetic copper carbonate has a combined aggregate structure seen at the top centre of the illustration, original magnification 400×. (Photomicrographs: H. Howard; © Courtauld Institute, Conservation of Wall Painting Department)

Plate 42 Ante-Reliquary Chapel, Norwich Cathedral: detail of the scheme of *c*1300 on the west segment of the vault. (© Crown copyright. NMR)

Plate 43 Ante-Reliquary Chapel, Norwich Cathedral. Detail of the elaborate blue vinescroll surrounding the figures on the vault. (Photograph: H. Howard; © Courtauld Institute, Conservation of Wall Painting Department)

Plate 44 Ante-Reliquary Chapel, Norwich Cathedral. Cross-section of a sample from the vinescroll showing the use of pure high-quality natural azurite. The largest particles, of an intense blue, measure some 75 × 50μm. Overall the pigment would have produced a beautiful surface texture, sparkling as the crystals were caught by the light. (Photomicrograph: H. Howard; © Courtauld Institute, Conservation of Wall Painting Department)

Plate 45 St. Albans Cathedral: detail of St. William of York on the north-east pier of the feretory (*c*1302–08). (Photograph: H. Howard; © Courtauld Institute, Conservation of Wall Painting Department)

Plate 46 St. Albans Cathedral. Cross-section of a sample from St. William's blue dalmatic, original magnification 400×. Natural azurite has been applied over red underpaint, presumably to provide a richer, more purple hue. Later in the medieval period, the sparkling blue of this high-quality azurite was covered with a duller, darker layer of indigo. (Photomicrograph: H. Howard; © Courtauld Institute, Conservation of Wall Painting Department)

Plate 47 St. Albans Cathedral: detail of an unidentified archbishop on the south-east pier of the feretory. (Photograph: H. Howard; © Courtauld Institute, Conservation of Wall Painting Department)

Plate 48 Cross-section of a sample taken from the purple chasuble of the archbishop shown in Plate 47, original magnification 400×. The soft purple colour was achieved by combining azurite with a red lake pigment and a little lead white, and applying the mixture in a single layer over a lead white ground. (Photomicrograph: H. Howard; © Courtauld Institute, Conservation of Wall Painting Department)

Plates 49 and 50 Westminster Abbey, chapter house, *Apocalypse* cycle of *c*1375–1404: detail of *St. John the Evangelist before Domitian*. Both azurite and indigo have been identified in this cycle of wall paintings, with azurite employed extensively for areas such as St. John's blue drapery. In UV light (Plate 50), the modern wax coating – which is particularly thick in hollows in the paint layer and areas such as the eyes which have been gouged – appears a milky white colour. (Photographs: Howard/Kakoulli; © Courtauld Institute, Conservation of Wall Painting Department)

Plates 51 and 52 Westminster Abbey, chapter house. Cross-section of a sample taken from another scene in the *Apocalypse* cycle, where azurite has been combined with red lake and lead white to produce a purple colour. This mixture was applied over a layer of tin leaf, presumably to provide additional luminosity to the purple colour. In UV light (Plate 52), the organic layer beneath the tin leaf (which appears as a dark zone in this photomicrograph) fluoresces an orange/yellow colour. The waxy surface coating fluoresces a milky blue colour, original magnification 382×. (Photomicrographs: H. Howard; © Courtauld Institute, Conservation of Wall Painting Department)

Plate 53 Ingham (Norfolk): tomb of Sir Roger and Lady Margaret de Bois (*c*1380). (Photograph: H. Howard; © Courtauld Institute, Conservation of Wall Painting Department)

Plate 54 Ingham (Norfolk), de Bois tomb. Despite severe damage to the polychromy, traces remain in the carved detail as on the cushion below Lady Margaret's head. (© John Burbidge)

Plate 55 Cross-section of a sample from the red lozenge pattern shown in Plate 54, original magnification 500×. Here, perhaps with the intention of increasing its luminosity and brilliance, azurite was applied over silver leaf. Red impasto linear details were then added in vermilion and red lead. (Photomicrograph: H. Howard; © Courtauld Institute, Conservation of Wall Painting Department)

Plate 56 Dispersion of azurite particles from the cross-section shown in Plate 55, photographed in plane-polarised transmitted light. The large particles, typically 60 × 50μm, attest to the high quality of the mineral blue pigment. The now-yellow binding medium is particularly apparent at the right side of the plate. Analysis by FTIR established that the original binding medium contained both oil and a proteinaceous component. This may indicate that the medium is egg yolk, or that the blue particles were ground in protein before being bound in oil (as found in the wall paintings of *c*1270–80 in the choir of Angers Cathedral). (Photomicrograph: H. Howard; © Courtauld Institute, Conservation of Wall Painting Department)

Plate 57 Azurite was also employed in Gothic panel painting. In the ceiling paterae of *c*1263–66 from the Painted Chamber, Westminster Palace, azurite was combined with black and white and used to outline the figures. (© Copyright The British Museum)

Plate 58 Peterborough Cathedral, nave ceiling (*c*1220): detail of St. Peter's head, showing the crude repainting and severe cupping and lifting of the paint layers before the recent phase of conservation by the Perry Lithgow Partnership. Although very few original paint layers remain on the ceiling, natural azurite combined with a little yellow earth and lead white were combined to produce effects of modelling in the flesh tones. (Photograph: Howard/Heritage; © Courtauld Institute, Conservation of Wall Painting Department)

Plate 59 Peterborough Cathedral, nave ceiling. Cross-section of a sample from St. Peter's left temple, original magnification 500×. A layer of dark green underpainting, produced by mixing natural azurite, yellow earth and lead white, was employed to produce the tonal modelling for the flesh tones of vermilion combined with lead white. (Photomicrograph: H. Howard; © Courtauld Institute, Conservation of Wall Painting Department)

Plate 60 Oyster-shell palette of *c*1300 excavated from the Greyfriars, Norwich, which may have been used for large-scale painting. (Photograph: H. Howard; © Courtauld Institute, Conservation of Wall Painting Department)

Plate 61 Oyster-shell palette from Greyfriars, Norwich. The azurite particles in the Norwich palette are of average size with a few larger inclusions. This natural azurite particle has been photographed in transmitted light with crossed polars, original magnification 1500×. Examination suggests that the mineral blue pigment here was bulked out with its synthetic equivalent. This practice has also been found in the azurite employed for the mid-13th-century scheme at Horsham St. Faith Priory, Norfolk (Plates 37, 39, 40 and 41). (Photomicrograph: H. Howard; © Courtauld Institute, Conservation of Wall Painting Department)

Plate 62 Natural azurite also occurs in an oyster-shell palette (of between the mid-14th and mid-16th century), excavated from the Carmelite Friary, Coventry. Here the particles are slightly larger than average size (the majority are 30 × 15μm, with others of approximately 50 × 30μm). These particles have been photographed in transmitted light with crossed polars, original magnification 1380×. (Photomicrograph: H. Howard; © Courtauld Institute, Conservation of Wall Painting Department)

Plate 63 St. Albans Cathedral, feretory: Shrine of St. Alban (*c*1302–08). (© The Cathedral and Abbey Church of St. Alban)

Plate 64 St. Albans Cathedral, feretory: Shrine of St. Alban. The alteration of azurite to form a copper-chloride alteration product has been identified in a number of instances, notably in samples from the shrine. Cross-sections clearly demonstrate the blue-green alteration signalling the transformation, which in this case progresses further to a dark material visible at the centre of the particles. A single blue particle with a green halo is visible at the centre of the cross-section (original magnification 400×), while green, copper chloride-rich particles with dark centres are also visible. (Photomicrograph: H. Howard; © Courtauld Institute, Conservation of Wall Painting Department)

Plate 65 SEM micrograph showing a backscatter image of the sample in Plate 64. Details of the particles at top left of this image are shown in the subsequent plates. (SEM micrograph: H. Howard; © Courtauld Institute, Conservation of Wall Painting Department)

Plate 66 Detail of the cross-section in Plate 64, original magnification 1000×. The subsequent four plates relate to the particle at the top left of this image. (Photomicrograph: H. Howard; © Courtauld Institute, Conservation of Wall Painting Department)

Plate 67 SEM micrograph showing a backscatter image of the green particle with a dark centre at the top left of Plate 66. The disrupted structure of the centre of the particle is clearly visible. (SEM micrograph: H. Howard; © Courtauld Institute, Conservation of Wall Painting Department)

Plate 68 X-ray map showing the distribution of chlorine which is clearly present around the green perimeter of the particle, although not in its dark centre. (SEM micrograph: H. Howard; © Courtauld Institute, Conservation of Wall Painting Department)

Plate 69 SEM micrograph with X-ray map for copper showing that its distribution is fairly consistent throughout the green and dark brown areas of the particle. (SEM micrograph: H. Howard; © Courtauld Institute, Conservation of Wall Painting Department)

Plate 70 SEM micrograph with X-ray map for oxygen showing a slightly increased concentration in the centre of the particle. (SEM micrograph: H. Howard; © Courtauld Institute, Conservation of Wall Painting Department)

Plate 71 SEM backscatter image of the particle at top left of Plate 65, with an X-ray line scan for chlorine. The line scan confirms the increased concentration of chlorine around the green perimeter of the particle. (SEM micrograph: H. Howard; © Courtauld Institute, Conservation of Wall Painting Department)

Plate 72 SEM backscatter image of the same particle as in Plate 71, but here with an X-ray line scan for oxygen. The line scan confirms the increased concentration of oxygen in the dark brown centre of the particle. The analysis suggests that while the green alteration product of azurite particles on the Shrine of St. Alban consists of a copper chloride-based material, the dark centre of the particle is principally copper oxide. (SEM micrograph: H. Howard; © Courtauld Institute, Conservation of Wall Painting Department)

Plate 73 Little Wenham (Suffolk): detail of the early 14th-century wall paintings on the chancel east wall, showing the turquoise blue drapery and darkened flesh tones which are such features of the scheme. (© Courtauld Institute, Conservation of Wall Painting Department)

Plate 74 Little Wenham (Suffolk). Cross-section of a sample from the drapery of one of the female saints on the east wall of the chancel, original magnification 1000×. At low magnification the synthetic copper blue pigment appears to have no particulate structure, appearing almost as a stain on the white lime matrix. However, at high magnification, as here, tiny blue particles with slightly rounded edges are apparent in a few places. The blue paint layer has been applied directly over a lime ground and details in charcoal black applied on the surface. (Photomicrograph: H. Howard; © Courtauld Institute, Conservation of Wall Painting Department)

Plate 75 Little Wenham (Suffolk). A sample from the blue border at the south end of the chancel east wall indicates the presence of the same synthetic copper blue pigment. A few tiny rounded particles are just visible, original magnification 500×. (Photomicrograph: H. Howard; © Courtauld Institute, Conservation of Wall Painting Department)

Plate 76 Brent Eleigh (Suffolk): detail of the fictive reredos of *c*1335 on the east wall of the chancel. The background appears bright turquoise blue with tiny rosettes painted in pale blue on the surface. In a few places, particularly at either side of the reredos and beneath some of the rosettes, a grey colour is visible. (© Courtauld Institute, Conservation of Wall Painting Department)

Plate 77 Brent Eleigh (Suffolk). Examination of a cross-section from the blue background indicated the presence of a synthetic copper blue pigment, similar to that found at neighbouring Little Wenham. The sample, original magnification 700×, also includes grey-brown particles. Analysis of these areas suggested the presence of copper oxide, indicating an alteration of the synthetic blue pigment. (Photomicrograph: H. Howard; © Courtauld Institute, Conservation of Wall Painting Department)

Plate 78 A similar 'stained blue' pigment occurs in the 14th-century paintings in the north aisle of Great Hockham Church (Norfolk). The cross-section shows that the blue-green paint layer (75μm thick) was applied over a lime white ground. The dark particles are inclusions of charcoal black. (Photomicrograph: H. Howard; © Courtauld Institute, Conservation of Wall Painting Department)

Plate 79 Brent Eleigh (Suffolk), chancel east wall: detail of a censing angel set against a now dark background scattered with stars. This painting forms part of the same scheme as the fictive reredos illustrated in Plate 76. (© Crown copyright. NMR)

Plate 80 Brent Eleigh (Suffolk), chancel east wall. Cross-section of a sample from a pale blue area on the dark background. Here verdigris has been applied over the white ground and the pale blue layer apparent at the surface appears to be an alteration product of this. In this instance, limewash applied over the surface of the painting has probably initiated a reaction resulting in the formation of the turquoise blue alteration product of verdigris in some areas. (Photomicrograph: H. Howard; © Courtauld Institute, Conservation of Wall Painting Department)

Plate 81 Westminster Abbey, chapter house, east bay: detail of *Seraph* (*c*1375–1404). (Photograph: Howard/Kakoulli; © Courtauld Institute, Conservation of Wall Painting Department)

Plate 82 Westminster Abbey, chapter house. Detail of the angels' nimbi in the first arch of the east bay. (Photograph: Howard/Kakoulli; © Courtauld Institute, Conservation of Wall Painting Department)

Plate 83 Cross-section of a sample from the nimbus on the right side of Plate 82, original magnification 180×. Red earth was applied over the plaster support, and over this a layer of indigo combined with lead white. A thin layer of red 'bole'– comprising red earth combined with red lead – has then been applied in preparation for the gilding. Thick layers of wax and other surface accretions are evident on the surface of the sample. Indigo was employed quite extensively throughout the *Majesty* and the *Apocalypse* cycle in the chapter house. (Photomicrograph: H. Howard; © Courtauld Institute, Conservation of Wall Painting Department)

Plate 84 Durham Deanery, detail of the paintings of *c*1470 featuring the Virgin: the *Resurrection*. (© Crown copyright. NMR)

Plate 85 Durham Deanery. For the armour of one of the soldiers in the *Resurrection*, indigo was combined with lead white and applied in a layer 60μm thick over a warm orange-yellow layer, which in turn was applied over a fine layer of bone black. The resultant blue-grey colour of this layer has an almost metallic sheen, original magnification 400×. (Photomicrograph: H. Howard; © Courtauld Institute, Conservation of Wall Painting Department)

Plate 86 The Lord Mayor's Chapel (St. Mark's), Bristol: detail of the paintings of *c*1510, showing the *Noli me Tangere*. (Photograph: H. Howard; © Courtauld Institute, Conservation of Wall Painting Department)

Plate 87 The Lord Mayor's Chapel (St. Mark's), Bristol. Cross-section of a sample from the Virgin's drapery in the *Nativity*, where thick black underpaint was applied beneath a layer of indigo combined with lead white. This increased the intensity of the blue colour by reducing the light scatter from the lead white ground below, and also made the dye less susceptible to fading. Similarly, grey underpaint was employed beneath the pale purple (indigo combined with vermilion and lead white) of the veil of the female figure behind Christ in the *Noli me Tangere*. (Photomicrograph: H. Howard; © Courtauld Institute, Conservation of Wall Painting Department)

Plate 88 Eton College Chapel: detail of the scheme of *c*1479–87. Although the presence of indigo has yet to be confirmed in these paintings, it is interesting to speculate whether the trace of blue – apparently much faded – that has been observed in the scheme may indeed be indigo. The blue colour was only detected on the sword of the figure farthest west on the south wall, an area largely protected from direct light. (© Crown copyright. NMR)

Plate 89 Confirmed examples of the use of green earth in English medieval wall painting fall principally in the Romanesque period. The pigment was employed, for example, in the scheme of *c*1175 in the Holy Sepulchre Chapel of Winchester Cathedral, for flesh tones (Plate 28) and other areas such as drapery. This cross-section (original magnification 410×) of a sample taken from a shadow in Christ's flesh in the *Entombment* shows a layer of green earth applied over lime white and red and yellow earth pigments. (Photomicrograph: H. Howard; © Courtauld Institute, Conservation of Wall Painting Department)

Plate 90 Lambach (Austria): scheme of *c*1089 in the west choir, where a mixture of green earth and malachite has been identified. The malachite may have been added to intensify the colour of the green earth pigment. (Photograph: A. Heritage; © Courtauld Institute, Conservation of Wall Painting Department)

Plate 91 Hardham (West Sussex): detail of the *Annunciation* on the east wall of the nave (*c*1100), showing the use of a copper chloride green in Gabriel's halo and for details on his drapery. (Photograph: Sawdy/Heritage; © Courtauld Institute, Conservation of Wall Painting Department)

Plate 92 Hardham (West Sussex). Cross-section (original magnification 500×) of a sample from Gabriel's halo, showing the copper chloride green in a lime white matrix. The copper chloride green has undergone partial alteration to form what appears to be dark copper oxide. (Photomicrograph: H. Howard; © Courtauld Institute, Conservation of Wall Painting Department)

Plate 93 Hardham (West Sussex). Cross-section of a sample from the drapery of the *Virgin* in the *Flight into Egypt* on the nave north wall. Here, copper chloride green combined with lime white was applied over very pale grey underpaint of charcoal black combined with lime white on a pink preparation. The grey underpaint would have improved the covering power of the green pigment and thereby enhanced its appearance. (Photomicrograph: H. Howard; © Courtauld Institute, Conservation of Wall Painting Department)

Plate 94 Backscatter SEM micrograph of the sample shown in Plate 92 with X-ray map for copper. The copper is clearly associated with the green and brown (partially altered) particles. (SEM micrograph: H. Howard; © Courtauld Institute, Conservation of Wall Painting Department)

Plate 95 Backscatter SEM micrograph of the same sample with X-ray map for chlorine. The chlorine is associated with the copper-rich particles. (SEM micrograph: H. Howard; © Courtauld Institute, Conservation of Wall Painting Department)

Plate 96 Hardham (West Sussex): detail of the *Agnus Dei* on the east wall of the nave (*c*1100), showing the use of a copper chloride green for the nimbus and cross-staff. (Photograph: Sawdy/Heritage; © Courtauld Institute, Conservation of Wall Painting Department)

Plate 97 Hardham (West Sussex). Cross-section (original magnification 426×) of a sample from the nimbus of the *Agnus Dei*, showing a copper chloride green in a lime white matrix applied over a local red earth ground. (Photomicrograph: H. Howard; © Courtauld Institute, Conservation of Wall Painting Department)

Plate 98 Hardham (West Sussex). Cross-section (original magnification 466×) of a sample from the nimbus of the southern angel censing the *Agnus Dei*, where yellow earth was employed beneath the green pigment. The variety of coloured underpaints used beneath the copper chloride green suggests that particular coloristic effects were sought in each instance. In this sample the green pigment has undergone partial transformation to form an alteration product presumed to be tenorite. (Photomicrograph: H. Howard; © Courtauld Institute, Conservation of Wall Painting Department)

Plate 99 Norwich Cathedral: detail of the 'Losinga' cycle of paintings (*c*1190), showing the green mantle of the western figure on the west segment of the vault. A copper chloride green pigment, combined with lime white and applied over a lime ground, was employed in this case. (Photograph: H. Howard; © Courtauld Institute, Conservation of Wall Painting Department)

Plate 100 Norwich Cathedral, 'Losinga' cycle. Cross-section of a sample from the brilliant green roof of the cathedral in the lowest roundel on the soffit of the arch, showing the copper chloride green applied over dark grey underpaint of charcoal black combined with lime white. The grey underpaint would have increased the covering power of the copper chloride green and thereby enhanced its green appearance, original magnification 250×. (Photomicrograph: H. Howard; © Courtauld Institute, Conservation of Wall Painting Department)

Plate 101 Malachite has been identified by instrumental analysis in only one instance in English Romanesque wall painting – in the excavated fragments of *c*1180 from the chapter house of Sherborne Abbey (Dorset). Similarly, in the Gothic period, identification has usually only been made by polarised light microscopy. This thin section, photographed in transmitted light with semi-crossed polars, original magnification 200×, shows the green copper-containing pigment – probably malachite – employed for the original background of St. William of York (*c*1302–08) in the feretory of St. Albans Cathedral (Plate 45). The original green background was overpainted later in the medieval period with verdigris. A thick wax coating is visible on top of the paint sample. (Photomicrograph: H. Howard; © Courtauld Institute, Conservation of Wall Painting Department)

Plate 102 Detail of the *Nativity* in the scheme of *c*1165 in the Johanneskapelle, Pürgg (Austria). Both malachite and a green made from combining charcoal black and yellow earth have been identified in this scheme. (Photograph: H. Howard; © Courtauld Institute, Conservation of Wall Painting Department)

Plate 103 Horsham St. Faith Priory (Norfolk): detail of the *Return to England* from the *Foundation Narrative* on the east wall of the refectory (*c*1250, but repainted in the 15th century). The presence of verdigris combined with prepolymerised linseed oil was confirmed in a sample taken from the repainting in the area above the boat. (© Crown copyright. NMR)

Plate 104 Rochester Cathedral: detail of the painting of *c*1340 on the back wall of the tomb of William of Perth. Analysis of the vine leaves confirmed the presence of verdigris bound with walnut oil and subsequently glazed with copper resinate. (© Courtauld Institute, Conservation of Wall Painting Department)

Plate 105 Canterbury Cathedral, Chapel of Our Lady Undercroft: detail of the central canopy of the reredos (*c*1370). (© Canterbury Cathedral Wall Painting Workshop)

Plate 106 Cross-section of a sample from one of the raised decorative elements on the canopy shown in Plate 105, original magnification 400×. A luminous effect was produced by applying verdigris in an oil medium over a layer of silver leaf. (Photomicrograph: H. Howard; © Courtauld Institute, Conservation of Wall Painting Department)

Plate 107 Windsor Castle, King John's Tower: detail of the recently discovered 14th-century wall paintings. Accounts for 1365–66 list 12lb of verdigris and other pigments including 7lb *asure de bys* at 3s per lb, purchased for these paintings. Preliminary analysis of the scheme indicated the presence of both verdigris and red lead. (© English Heritage Photographic Library)

Plate 108 Farleigh Hungerford Castle, St. Leonard's Chapel: detail of the early 15th-century *St. George* on the east wall. (© Courtauld Institute, Conservation of Wall Painting Department)

Plate 109 Farleigh Hungerford Castle, St. Leonard's Chapel. Cross-section of a sample from the green lining of the saint's cloak, original magnification 320×, showing a layer of verdigris combined with lead white and charcoal black applied in an oil medium over lead white. The particles of verdigris show signs of alteration to a pale blue copper chloride-rich product. (Photomicrograph: H. Howard; © Courtauld Institute, Conservation of Wall Painting Department)

Plate 110 Farleigh Hungerford Castle, St. Leonard's Chapel. Cross-section of a sample from the contemporary *Kneeling Knight* on the south wall of the chapel, original magnification 240×. Here multiple layers of verdigris – combined with varying quantities of lead white – have been applied over the red preparatory layers. A thin layer of vermilion is also visible on the surface of the sample. The uppermost verdigris layer has also partially altered to form a blue copper chloride-rich product. (Photomicrograph: H. Howard; © Courtauld Institute, Conservation of Wall Painting Department)

Plate 111 Backscatter SEM micrograph of the sample shown in Plate 109. The line scan for chlorine indicates a higher concentration of the element in the partially altered verdigris particle and at the surface of the sample. (SEM micrograph: H. Howard; © Courtauld Institute, Conservation of Wall Painting Department)

Plate 112 Backscatter SEM micrograph of the sample shown in Plate 109. The line scan for copper indicates a higher concentration of the element in the blue particle and at the surface of the sample, though copper is present throughout the green paint layer. Significantly, the transformation of verdigris to form a copper chloride-rich alteration product seems to be associated with contact between the pigment and the overlying plaster. It appears to occur initially at the interface between the two layers, or where cracks in the green paint have provided a passage for further penetration of the lime-based plaster. (SEM micrograph: H. Howard; © Courtauld Institute, Conservation of Wall Painting Department)

Plates 113 to 116 Ingham (Norfolk), tomb of Sir Oliver de Ingham (*c*1344): cross-section (original magnification 324×) of a sample from the now blue polychromy on the bascinet. A pale blue alteration of verdigris has formed at the interface of this pigment and the coloured limewash. Where a crack exists in the verdigris layer, allowing increased contact with the alkaline limewash layer (Plate 114), the alteration has progressed to a greater depth in the green paint layer. In other areas where the alteration is more advanced, only a tiny trace of the green colour visible at the base of the pale blue layer (Plate 115) provides an indication of the original colour and nature of the paint layer. The same alteration phenomenon occurs on the de Bois tomb (*c*1380) at Ingham (Plate 116). This sample, taken from one of the weepers on the tomb chest, shows the formation of the pale blue alteration product of verdigris at the surface of the paint layer (from which the covering limewash has now been removed). (Photomicrographs: H. Howard; © Courtauld Institute, Conservation of Wall Painting Department)

Plate 117 The alteration of verdigris has also been noted in several Gothic wall paintings as here in the mid-14th-century scheme over the chancel arch at Ickleton (Cambridgeshire). The alteration product has formed in a layer 15μm thick over the original layer of verdigris combined with lead white. It is perhaps significant that the transformation product is not apparent where an original dark green glaze over the verdigris layer remains intact. (Photomicrograph: H. Howard; © Courtauld Institute, Conservation of Wall Painting Department)

Plate 118 Müstair (Switzerland): detail of the late 12th-century scheme showing the extensive use of various red pigments: vermilion, red lead and red earth. Vermilion was employed extensively in cathedral schemes in England in the Romanesque period, though there is little evidence for its use in parish churches at this time. (© Colorphoto Hans Hinz SWB)

Plate 119 Winchester Cathedral, Holy Sepulchre Chapel: detail of the *Deposition* (*c*1175) on the east wall of the chapel. Vermilion was used for numerous details such as the panelled background and drapery of St. John. (© Courtauld Institute, Conservation of Wall Painting Department)

Plate 120 Vermilion has sometimes altered, or partially altered, to its black form, metacinnabar. In the 13th-century scheme in the Guardian Angels' Chapel, Winchester Cathedral, the once brilliant red paint layers now appear almost black as in this rosette on the west segment of the vault. (© Courtauld Institute, Conservation of Wall Painting Department)

Plate 121 Cross-section (original magnification 400×) of a sample from the rosette shown in Plate 120. Here the particles of darkened vermilion can be seen at the top of the otherwise brilliant red paint layer. A sulphated crust is visible on the surface of the paint layer. (Photomicrograph: H. Howard; © Courtauld Institute, Conservation of Wall Painting Department)

Plate 122 Norwich Cathedral, Ante-Reliquary Chapel: detail of St. Paul's red drapery (*c*1300). This has a rather patchy red-brown appearance, again due to the partial alteration of vermilion. (Photograph: H. Howard; © Courtauld Institute, Conservation of Wall Painting Department)

Plate 123 Norwich Cathedral, Ante-Reliquary Chapel. Cross-section of a sample taken from St. Paul's drapery, original magnification 775×. Red particles of vermilion are mixed throughout the paint layer with those which have darkened. (Photomicrograph: H. Howard; © Courtauld Institute, Conservation of Wall Painting Department)

Plate 124 Canterbury Cathedral, Chapel of Our Lady Undercroft: detail of the painting on the eastern bay of the vault before cleaning (*c*1370). (© Canterbury Cathedral Wall Painting Workshop)

Plate 125 Canterbury Cathedral, Chapel of Our Lady Undercroft. Detail of one of the attached suns which once had a reflective centre and gilded rays. (© Canterbury Cathedral Wall Painting Workshop)

Plate 126 Canterbury Cathedral, Chapel of Our Lady Undercroft. Cross-section (original magnification 400×) of a sample from the dark background of the vault before cleaning. Although it was thought that the background was originally blue, examination of the cross-section confirmed the presence of vermilion applied over a layer of red lead to produce a rich orange-red effect. (Photomicrograph: H. Howard; © Courtauld Institute, Conservation of Wall Painting Department)

Plate 127 Canterbury Cathedral, Chapel of Our Lady Undercroft. Cross-section of a sample taken from the base of one of the attached stars. A layer of partially oxidised tin leaf is present at the base of the sample. This was originally gilded and then the orange-red background was painted in the areas between the raised rays. The layers of vermilion over red lead which formed the red background are again clearly visible. (Photomicrograph: H. Howard; © Courtauld Institute, Conservation of Wall Painting Department)

Plate 128 Westminster Abbey, chapter house: detail of Christ's red drapery in the *Majesty*, showing the attached 'brooch' and linear details in gold. (Photograph: Howard/Kakoulli; © Courtauld Institute, Conservation of Wall Painting Department)

Plate 129 Westminster Abbey, chapter house. Cross-section of a sample taken from Christ's brilliant red drapery, original magnification 430×. The modelling of the drapery has been built up with numerous layers of vermilion combined with varying quantities of lead white. (Photomicrograph: H. Howard; © Courtauld Institute, Conservation of Wall Painting Department)

Plate 130 Westminster Abbey, chapter house: detail of one of the angels in the *Judgement*, showing the brilliant red flesh tone which consists of vermilion applied in a thin layer (10μm) over red earth. (Photograph: Howard/Kakoulli; © Courtauld Institute, Conservation of Wall Painting Department)

Plate 131 Westminster Abbey, chapter house: detail of the nimbus of one of the angels in the *Judgement*. Lines used to set out the design were incised into the red preparatory layer. The fine black linear detail was applied with a brush on the surface of the gold leaf. (Videomicroscopy: Heritage/Howard; © Courtauld Institute, Conservation of Wall Painting Department)

Plate 132 Coventry, Holy Trinity: detail of the *Last Judgement* (*c*1435–60) over the chancel arch. (Photograph: H. Howard; © Courtauld Institute, Conservation of Wall Painting Department)

Plate 133 Coventry, Holy Trinity. Cross-section of a sample from one of the large drips of blood on Christ's foot, original magnification 338×. For the rendering of fresh blood a combination of vermilion with red lead and red lake has been applied over a thin layer of carbon black on the flesh tone. (Photomicrograph: H. Howard; © Courtauld Institute, Conservation of Wall Painting Department)

Plate 134 Bristol, St. Mark's: cross-section (original magnification 500×) of a sample from the angel's wing in the *Nativity* (*c*1510) (Plates 86 and 87), showing vermilion applied over a layer of lead white in oil. Beneath the lead white ground, an oily sealant layer on a layer of oil bulked with chalk and calcium sulphate is clearly visible. (Photomicrograph: H. Howard; © Courtauld Institute, Conservation of Wall Painting Department)

Plate 135 Vermilion has been identified in external polychromy as on the west front of Wells Cathedral where the pigment was applied in an egg tempera medium over a lead white ground. This plate shows details of a roof delineated in vermilion in the *Naming of St. John* on the north side of the apse of St. Gabriel's Chapel, Canterbury Cathedral. (© Courtauld Institute, Conservation of Wall Painting Department)

Plate 136 Canterbury Cathedral, cloister: detail showing part of the *Martyrdom Doorway* of *c*1236. Here vermilion has been identified, combined with a little red lead and applied over a lead white ground. (© Courtauld Institute, Conservation of Wall Painting Department)

Plate 137 Witley (Surrey): detail of the *Feast of Dives* on the south wall of the nave (*c*1110). The presence of an altered lead pigment in this scheme was signalled by the dark zones which occur in the collar and cuffs of Dives, and in the brooch and collar of two other figures. (© Courtauld Institute, Conservation of Wall Painting Department)

Plate 138 Witley (Surrey). Detail of the figure of Dives showing the alternating red and white ornament on his collar, with grey-black areas characteristic of altered lead. (© Courtauld Institute, Conservation of Wall Painting Department)

Plate 139 Witley (Surrey). Cross-section of a sample taken from Dives' collar, showing the abraded paint layer surviving in the interstices of the uneven plaster. SEM/EDX analysis confirmed the presence of lead in the paint layer. (Photomicrograph: H. Howard; © Courtauld Institute, Conservation of Wall Painting Department)

Plate 140 Canterbury Cathedral, St. Gabriel's Chapel: detail of the *Birth of the Baptist* on the north side of the apse (*c*1130). The grey 'crust' over the red drapery occurs across many areas of the paintings and is associated with the alteration of red lead. (© Courtauld Institute, Conservation of Wall Painting Department)

Plates 141 to 144 Cross-section of a sample (141) from the red drapery shown in Plate 140, original magnification 1000×. Particles of cerussite – converted from the red lead layer beneath – can be seen bursting through the upper layer of vermilion and forming a grey crust on the surface. Similarly, in a sample from the tree on the north buttress (Plate 142), the white alteration product of red lead has pushed through the vermilion layer and formed a crust on the surface. In another sample from the apse paintings, a layer of cerussite has formed on the surface of the original layer of red lead combined with a little vermilion (Plate 143). In this sample (Plate 144) from a detail in red lead over the vermilion drapery, the crust forms a particularly uneven layer on the surface. (Photomicrographs: H. Howard; © Courtauld Institute, Conservation of Wall Painting Department)

Plate 145 Winchester Cathedral, Holy Sepulchre Chapel: detail of the paintings of *c*1175 on the east wall, showing the border dividing the narrative tiers. The geometric elements within the border were initially incised into the wet plaster with the aid of a compass; when the border was moved to a slightly lower position new incisions were made, but a different quality of line is evident since the plaster was clearly no longer fresh. Details added in impasto in red lead are visible at the right. (© Courtauld Institute, Conservation of Wall Painting Department)

Plate 146 Cross-section (original magnification 302×) of a sample from the raised detail in red lead shown in Plate 145. The red lead is combined with lead white and calcium carbonate, and analysis of the binding medium suggested the presence of a proteinaceous material. (Photomicrograph: H. Howard; © Courtauld Institute, Conservation of Wall Painting Department)

Plate 147 SEM micrograph of the sample in Plate 146, showing what appear to be hyphae within the paint layer. The presence of microbiological growth may herald an alteration in the lead-based paint layer, since humidity has been implicated in the alteration of lead pigments. (SEM micrograph: H. Howard; © Courtauld Institute, Conservation of Wall Painting Department)

Plate 148 Glasgow Cathedral: detail of a fragment of wall painting excavated from the crypt (*c*1190). (Photograph: H. Howard; © Courtauld Institute, Conservation of Wall Painting Department)

Plate 149 Cross-section of a sample taken from the shield in Plate 148, original magnification 446×. For the orange colour at the tip of the shield, red lead was applied over a local lime white ground on a layer of charcoal black combined with lime white and a little red lead. (Photomicrograph: H. Howard; © Courtauld Institute, Conservation of Wall Painting Department)

Plate 150 Norwich Cathedral, Ante-Reliquary Chapel: detail of an angel beside the eastern arch (*c*1250). (Photograph: H. Howard; © Courtauld Institute, Conservation of Wall Painting Department)

Plate 151 Norwich Cathedral, Ante-Reliquary Chapel. Cross-section of a sample from the angel's collar, showing a layer of red lead applied over lead white on the lime plaster, original magnification 500×. (Photomicrograph: H. Howard; © Courtauld Institute, Conservation of Wall Painting Department)

Plate 152 St. Albans Cathedral, feretory: detail of the unidentified archbishop on the south-east pier (*c*1302–08), after recent cleaning. (© The Perry Lithgow Partnership)

Plate 153 St. Albans Cathedral, feretory. Cross-section of a sample from the archbishop's orange maniple, original magnification 400×. A lead white ground covers the lime substrate used to level the tooled ashlar. The paint layer consists of red lead combined with charcoal black and is covered with a wax coating which was removed during recent conservation work. (Photomicrograph: H. Howard; © Courtauld Institute, Conservation of Wall Painting Department)

Plate 154 Westminster Abbey, chapter house: detail of *Babylon is Fallen* from the *Apocalypse* cycle of *c*1375–1404. The tool marks of the ashlar support are clearly visible. The red background of the scene consists of red lead applied over a chalk ground. (Photograph: Howard/Kakoulli; © Courtauld Institute, Conservation of Wall Painting Department)

Plate 155 Westminster Abbey, chapter house. Cross-section (original magnification 422×) of a sample from arch 3 of the south bay of the *Apocalypse* cycle. The upper orange-red paint layer consists of red lead with white, lead-containing inclusions which represent an alteration of the red pigment. Below this is a thin dark red layer of red earth which was applied over a third red layer consisting of vermilion combined with red lead and large lead-containing inclusions, which are also due to the alteration of red lead. A yellow zone – probably due to incomplete alteration – is visible at the centre of some of the white particles. Of particular interest is the volume change associated with this transformation and the consequent disruption of the paint layers. (Photomicrograph: H. Howard; © Courtauld Institute, Conservation of Wall Painting Department)

Plate 156 Westminster Abbey, chapter house. This sample from the south-east bay shows a similar alteration of the red lead layer as in Plate 155. In UV light it produces a slightly yellow fluorescence, while the coating on the surface has a milky appearance. (Photomicrograph: H. Howard; © Courtauld Institute, Conservation of Wall Painting Department)

Plate 157 SEM micrograph showing one of the particles of altered red lead from the sample in Plate 156. (SEM micrograph: H. Howard; © Courtauld Institute, Conservation of Wall Painting Department)

Plate 158 Backscatter SEM micrograph of the particle shown in Plate 157. The backscatter image provides information relating to atomic number – the higher the atomic number the lighter the portion of the image. Thus it is clear that the centre of the altered particle (which appears yellow in normal light, as in Plate 155), is of higher atomic number (or more dense, or both) than the rather disrupted outer portion. Elemental analysis by SEM/EDX confirmed the presence of only lead and oxygen in both of these zones. The disrupted portion of the particle shows a pattern of radial cracks, and perpendicular to these, slightly wavy striations. The transformation of red lead here is due to a reaction between the pigment and an oil medium. (SEM micrograph: H. Howard; © Courtauld Institute, Conservation of Wall Painting Department)

Plate 159 Red lead was employed extensively in the mid-14th-century wall paintings of St. Stephen's Chapel, Westminster Palace. The red lead also employed for the raised appliqué decoration is clearly visible where the uppermost paint layers have been lost. (© Copyright The British Museum)

Plate 160 Winchester Cathedral, Holy Sepulchre Chapel: detail of the *Deposition* on the east wall (*c*1175), showing Nicodemus' pale pink robe. (© Courtauld Institute, Conservation of Wall Painting Department)

Plates 161 and 162 Winchester Cathedral, Holy Sepulchre Chapel. Cross-section (original magnification 266×) of a sample from Nicodemus' robe, showing how the unusual pink colour was achieved by combining a red lake pigment with a tiny trace of ultramarine in a calcium carbonate matrix. (Photomicrograph: H. Howard; © Courtauld Institute, Conservation of Wall Painting Department) This is the earliest identification of red lake in English medieval wall painting, although at Müstair (Switzerland) it was employed in the Carolingian scheme of *c*800, of which a detail is shown in Plate 162. (© Colorphoto Hans Hinz SWB)

Plate 163 Glasgow Cathedral: detail of a fragment of wall painting (*c*1190) excavated from the crypt, showing a male head. (Photograph: H. Howard; © Courtauld Institute, Conservation of Wall Painting Department)

Plate 164 Tiny traces of a red lake pigment were detected in another of the fragments from Glasgow Cathedral. The cross-section (original magnification 500×) shows that it was applied over a layer of vermilion, presumably to saturate its appearance. This is the first identified instance of the use of red lake as a glaze in Romanesque wall painting in Britain. (Photomicrograph: H. Howard; © Courtauld Institute, Conservation of Wall Painting Department)

Plate 165 Norwich Cathedral, Ante-Reliquary Chapel: detail of the paintings on the western arch (*c*1300). (© Crown copyright. NMR)

Plate 166 Norwich Cathedral, Ante-Reliquary Chapel. Cross-section of a sample from the drapery of one of the apostles on the arch, original magnification 500×. Red lake was combined with natural azurite and applied over a lead white ground to produce the rich purple colour. A combination of azurite and red lake was also employed for the purple draperies of St. Peter and St. Andrew on the vault of the chapel. Analysis by fluorescence spectroscopy of the dyestuff employed in this scheme confirmed the presence of a madder lake. (Photomicrograph: H. Howard; © Courtauld Institute, Conservation of Wall Painting Department)

Plate 167 Tower of London, Byward Tower: detail of St. Michael (*c*1395). (© Courtauld Institute, Conservation of Wall Painting Department)

Plate 168 Tower of London, Byward Tower. Cross-section of a sample from St. Michael's robe, showing the mixture of red lake, azurite and lead white employed to achieve the soft purple colour. The purple robe was painted over the green background of the scheme, for which verdigris was combined with lead white. (© Courtauld Institute, Conservation of Wall Painting Department)

Plate 169 St. Albans Cathedral, feretory: detail of St. William of York on the north-east pier (*c*1302–08). The highlights of the chasuble show considerable signs of fading. (Photograph: H. Howard; © Courtauld Institute, Conservation of Wall Painting Department)

Plates 170 and 171 Cross-section of a sample from the faded area of drapery shown in Plate 169. Here the red lake has been applied in a matrix of lime white and lead white over the lime ground, making it more susceptible to internal reflection. In these areas it has faded almost entirely. By contrast, where it has been applied over coloured underpaint and mixed with charcoal black for an area of deep pink shading (Plate 171) there is less internal reflection and consequently the colorant shows less sign of fading. Madder lake was identified in a sample taken from the corresponding archbishop on the south-east pier. (Photomicrographs: H. Howard; © Courtauld Institute, Conservation of Wall Painting Department)

Plates 172–174 Red lakes have frequently been identified in English medieval polychrome sculpture as on the tomb of Sir Oliver de Ingham (*c*1344) at Ingham, Norfolk. This sample (172), taken from a fragment probably from the tomb canopy, shows a layer of red lake applied as a glaze over vermilion. The same sample, stained for protein with FITC and viewed in UV light (Plate 173), indicates that while the vermilion layer was applied in a proteinaceous binding medium, the red lake layer is in an oil medium. In a sample from the appliqué decoration on Sir Oliver's camail (Plate 174), a thick red lake glaze was applied over gold leaf. Instrumental analysis of the constituent dyestuff confirmed the presence of madder lake. (Photomicrographs: H. Howard; © Courtauld Institute, Conservation of Wall Painting Department)

Plate 175 Detail of an oyster-shell palette, dating from between the mid-14th and the mid-16th centuries, excavated from the Carmelite Friary at Coventry. (Photograph: H. Howard; © Courtauld Institute, Conservation of Wall Painting Department)

Plate 176 Analysis of the pink paint around the perimeter of the shell palette shown in Plate 175 confirmed the presence of a red lake pigment. The photomicrograph shows particles of the lake pigment in plane-polarised transmitted light, original magnification 900×. The blue pigment in the centre of the palette is natural azurite. (Photomicrograph: H. Howard; © Courtauld Institute, Conservation of Wall Painting Department)

Plate 177 Winchester Cathedral, Holy Sepulchre Chapel: detail of the *sinopia* for three scenes set beneath architectural canopies (*c*1175). While red earth is typically employed for *sinopie*, vermilion combined with red earth was used in this case. (© Courtauld Institute, Conservation of Wall Painting Department)

Plate 178 Glasgow Cathedral: detail of a fragment of wall painting (*c*1190) excavated from the crypt. Extensive areas of red paint are present on the penultimate plaster layer of this fragment – visible at lower left centre – which may represent either a *sinopia* or the presence of two distinct decorative schemes. (Photograph: H. Howard; © Courtauld Institute, Conservation of Wall Painting Department)

Plate 179 Witley (Surrey): detail of the paintings on the nave south wall (*c*1110). Preliminary drawing in red earth was employed to place the main pictorial elements within the visual field. Preliminary drawing for figure subjects is particularly conspicuous in the *Miraculous Draught of Fishes* shown here. (© Courtauld Institute, Conservation of Wall Painting Department)

Plate 180 Canterbury Cathedral, St. Gabriel's Chapel: detail of the *Birth of the Baptist* (*c*1130) on the north side of the apse. Here, preliminary drawing in yellow earth has been strengthened and refined in red earth. Pentimenti clearly show that the drawing was modified for a third time during the painting phase, as can be seen for the setting out of the scroll at the centre of the plate. (© Courtauld Institute, Conservation of Wall Painting Department)

Plate 181 Witley (Surrey): detail of the *Harrowing of Hell* on the west wall of the nave (*c*1110). Yellow earth, red earth and lime white are the dominant colours employed in this scheme. For areas of flesh painting, a pink colour – red earth combined with lime white – was applied directly to the plaster, then linear details in red or yellow earth combined with lime white were applied in impasto. Highlights in lime white were added last. (© Courtauld Institute, Conservation of Wall Painting Department)

Plate 182 Witley (Surrey). Cross-section of a sample from the flesh tone of one of the figures in the *Harrowing of Hell*, original magnification 340×. Red and yellow earth have been combined with quartz particles and lime white to produce a yellow highlight painted over a thin layer of red earth on the coarse lime plaster. (Photomicrograph: H. Howard; © Courtauld Institute, Conservation of Wall Painting Department)

Plate 183 Farleigh Hungerford Castle (Somerset) St. Leonard's Chapel: cross-section from the early 15th-century wall paintings, original magnification 500×. This sample, taken from St. George's helmet, shows the red preparatory layers for two layers of silver leaf. In preparation for the figures of St. George and the kneeling knight, a layer of red earth combined with calcium carbonate was applied in large rectangular zones in order to provide particular emphasis within the overall decorative scheme. Over this dark red layer – and throughout the remainder of the scheme – another red ground comprising red lead combined with oil and a little red iron oxide was applied to provide a unifying substructure throughout. This layer would also have acted as a sealant, ensuring that the translucency of the polychromy would not be compromised by leaching of the paint medium into the plaster support. (Photomicrograph: H. Howard; © Courtauld Institute, Conservation of Wall Painting Department)

Plates 184 and 185 Backscatter SEM micrograph with an X-ray line scan for silver confirming the presence of the two layers of the metal leaf in the cross-section shown in Plate 183. A gold effect was achieved by applying a yellow glaze over the lower layer of silver leaf, and further embellishment provided by slightly raised curvilinear details in silver leaf highlighted with a green glaze. Backscatter SEM micrograph (Plate 185) of the upper part of the sample shown in Plate 183. (SEM micrographs: H. Howard; © Courtauld Institute, Conservation of Wall Painting Department)

Plate 186 Tingelstad I altarpiece of *c*1260–1300 (Norway). Orpiment occurs on all three of the altarpieces from Tingelstad. In this instance particularly large particles of the pigment were detected, some measuring 5µm in length. (Photograph: Jacques Lation; © University Museum of Cultural Heritage, University of Oslo)

Plates 187 and 188 Orpiment has only been detected in six English wall paintings from the Gothic period including the *Apocalypse* cycle of *c*1375–1404 in the chapter house of Westminster Abbey. This cross-section, taken from the window tracery in *The Seventh Vial*, shows particles of orpiment combined with lead white, and red and black inclusions in the mordant for the thick layer of tin leaf above. In UV light (Plate 188) the mordant fluoresces a warm yellow colour. (Photomicrographs: H. Howard; © Courtauld Institute, Conservation of Wall Painting Department)

Plate 189 Orpiment also occurs in the early 15th-century scheme at Farleigh Hungerford Castle (Somerset). This sample, from the shield above St. George (Plate 108), shows a thick layer of red lake applied over the pigment (at lower left of the sample). The lake shows considerable evidence of fading in the uppermost portion of the paint layer. (Photomicrograph: H. Howard; © Courtauld Institute, Conservation of Wall Painting Department)

Plate 190 Lead-tin yellow has only been identified in a few English wall paintings including those of *c*1510 in St. Mark's, Bristol. This cross-section shows the pigment applied over a complex system of pale pink, white and grey underpaints for the halo of the Virgin Mary (?) in the *Noli Me Tangere* (Plate 86). (Photomicrograph: H. Howard; © Courtauld Institute, Conservation of Wall Painting Department)

Plate 191 Lime white was often used to particularly good effect for highlights of flesh painting and other details such as 'pearling' in fictive jewelled borders. The lime white used for highlights of the flesh in the apse paintings of St. Gabriel's Chapel, Canterbury Cathedral (*c*1130), shows up especially well in UV light. (© Courtauld Institute, Conservation of Wall Painting Department)

Plate 192 Lime white highlights in considerable impasto were similarly employed for the 'Losinga' scheme of *c*1190 in Norwich Cathedral. Some details, such as the collars and clasps of the standing figures on the vault, are now difficult to detect and are only clearly visible in UV light. (Photograph: H. Howard; © Courtauld Institute, Conservation of Wall Painting Department)

Plate 193 Lime white reacts with casein to form a stable calcium caseinate, and this combination has been identified, for example, in the scheme of *c*1240 at St. Georgen-bei-Judenburg, Austria, a detail of which is shown here. (Photograph: H. Howard; © Courtauld Institute, Conservation of Wall Painting Department)

Plate 194 Canterbury, Eastbridge Hospital: detail of the paintings of *c*1220, showing the angel symbol of St. Matthew. (Photograph: H. Howard; © Courtauld Institute, Conservation of Wall Painting Department)

Plate 195 Canterbury, Eastbridge Hospital. Cross-section of a sample from the white inscription 'MATHEVS', original magnification 228×. A layer of lead white has been applied in impasto over verdigris on the lime ground. Analysis by FTIR confirmed that both the lead white and verdigris layers have an oil medium. Yellow-brown surface coatings are evident at the top of the sample. (Photomicrograph: H. Howard; © Courtauld Institute, Conservation of Wall Painting Department)

Plate 196 Lead white was extensively employed in both the *Last Judgement* and the *Apocalypse* paintings in the chapter house of Westminster Abbey. This detail of the seraph in arch 2 of the east bay shows the subtle modelling of the flesh tones. To produce the pale flesh tones in the *Majesty*, layers of lead white were combined with varying quantities of vermilion and applied over a grey-green underpaint of charcoal black and yellow in a white matrix. (Photograph: Howard/Kakoulli; © Courtauld Institute, Conservation of Wall Painting Department)

Plate 197 Westminster Abbey, chapter house. *The Majesty and the Twenty-four Elders* in the north-west bay: for green details on the border of the mandorla, verdigris in a lead white matrix was applied over a secondary ground of lead white over tin leaf. (Photograph: Howard/Kakoulli; © Courtauld Institute, Conservation of Wall Painting Department)

Plate 198 Farleigh Hungerford Castle (Somerset), St. Leonard's Chapel: detail of the south wall of the nave showing the kneeling knight (early 15th-century). (© English Heritage Photographic Library)

Plate 199 Farleigh Hungerford Castle (Somerset), St. Leonard's Chapel. Detail of the impasto rendering of the mail around the neck of the knight. (© Courtauld Institute, Conservation of Wall Painting Department)

Plate 200 Farleigh Hungerford Castle (Somerset), St. Leonard's Chapel. Cross-section of a sample from the mail, original magnification 500×. A thick paste of lead white in an oil medium was applied and the mail modelled *in situ*. The undulating surface was then covered with silver leaf, and although this would have been glazed or varnished to prevent tarnishing, no trace of this remains on the surface of the cross-section. (Photomicrograph: H. Howard; © Courtauld Institute, Conservation of Wall Painting Department)

Plate 201 Ingham (Norfolk), tomb of Sir Roger and Lady Margaret de Bois (*c*1380): detail of Sir Roger's helmet showing the pink 'jewels' which have been applied in considerable impasto on the surface. (© John Burbidge)

Plate 202 Ingham (Norfolk), de Bois tomb. Cross-section of a sample from one of the fictive jewels, original magnification 250×. The impasto paint layer consists of lead white combined with red lead, red earth, calcium carbonate and azurite in an oil medium, applied over a layer of silver leaf. The lead and copper pigments here would have functioned as driers for the oil. GC-MS analysis confirmed the presence of linseed oil which was partially heat-bodied to promote drying. (Photomicrograph: H. Howard; © Courtauld Institute, Conservation of Wall Painting Department)

Plate 203 Longthorpe Tower (Cambridgeshire): detail of the *Three Living and the Three Dead* in the early 14th-century cycle in the first floor solar. The darkening of the flesh tones in these and other paintings has been the subject of a detailed study (Welford 1991) which demonstrated that the phenomenon was due to the alteration of lead white to form dark lead dioxide (plattnerite). (© Courtauld Institute, Conservation of Wall Painting Department)

Plates 204 and 205 Longthorpe Tower (Cambridgeshire). Cross-section (original magnification 268×) of a sample of darkened lead white from the flesh painting. This section (204) shows the curiously patchy nature of the darkening phenomenon, with zones of unaffected lead white juxtaposed with completely darkened areas. A similar partial alteration of lead white appears in a sample taken from the fictive tapestry on the south wall (Plate 205) where lead white was applied over vermilion. (Photomicrographs: H. Howard; © Courtauld Institute, Conservation of Wall Painting Department)

Plate 206 Winchester Cathedral, Holy Sepulchre Chapel: detail of the *Deposition* of *c*1220. Alteration of a lead pigment is evident here in Nicodemus' hat which has a dark spotty appearance. (© Courtauld Institute, Conservation of Wall Painting Department)

Plate 207 Norwich Cathedral, Ante-Reliquary Chapel: detail of the *Virgin and Child* on the vault (*c*1300). Although dark flesh tones are frequently due to the alteration of lead white, XRD analysis of samples taken from the dark flesh areas in this scheme produced no evidence of altered lead pigments. (Photograph: H. Howard; © Courtauld Institute, Conservation of Wall Painting Department)

Plate 208 Norwich Cathedral, Ante-Reliquary Chapel. The flesh tone of lead white combined with vermilion remains unaltered in this sample from the Norwich paintings, although wax and surface accretions are visible on the surface, original magnification 314×. Here, the patchy grey appearance of the flesh is simply due to the partial removal of past surface coatings and accretions. (Photomicrograph: H. Howard; © Courtauld Institute, Conservation of Wall Painting Department)

Plate 209 Norwich Cathedral, Ante-Reliquary Chapel. This cross-section from the Virgin's robe also shows the effect of partial cleaning. A thin layer of wax remains on the surface of the cross-section, and below 19th-century resin varnish penetrates the azurite layers. The coatings have been removed from the high points in the paint layer, but have been left in the hollows producing the patchy blue-green-brown colour of the drapery. (Photomicrograph: H. Howard; © Courtauld Institute, Conservation of Wall Painting Department)

Plate 210 Longthorpe Tower (Cambridgeshire): detail of the *Three Living and the Three Dead* (early 14th century). (© Courtauld Institute, Conservation of Wall Painting Department)

Plate 211 Longthorpe Tower (Cambridgeshire). Dispersion of a sample from a white stencilled motif on one of the king's mantle, original magnification 940×. Tiny nanofossils known as coccoliths display characteristic crosses in transmitted light with crossed polars, indicating the presence of chalk. (Photomicrograph: H. Howard; © Courtauld Institute, Conservation of Wall Painting Department)

Plate 212 Winchester Cathedral, Holy Sepulchre Chapel: detail of the censing angel in the *Entombment* on the east wall (*c*1175). A tiny trace of the original gilding remains on the angel's nimbus. (© Courtauld Institute, Conservation of Wall Painting Department)

Plate 213 Winchester Cathedral, Holy Sepulchre Chapel. Cross-section of a sample from the gilding on the angel's nimbus, original magnification 500×. The oil-based mordant for the gold leaf has been bulked out with calcium carbonate, which appears to be in the form of chalk. (Photomicrograph: H. Howard; © Courtauld Institute, Conservation of Wall Painting Department)

Plates 214 and 215 Cross-section of a sample from a wall painting fragment of *c*1180 excavated from the chapter house of Sherborne Abbey, original magnification 241×. Calcium carbonate – probably in the form of chalk – has been identified as a component of the oil-based mordant for the gold leaf. Multiple layers of gold leaf have been applied to the surface of the fragment, and these are particularly clearly visible in the SEM micrograph in Plate 215. (Photomicrograph and SEM micrograph: H. Howard; © Courtauld Institute, Conservation of Wall Painting Department)

Plate 216 SEM micrograph of the sample in Plate 214, showing the coarse aggregate of the plaster support and the finer layers applied in preparation for gilding. (SEM micrograph: H. Howard; © Courtauld Institute, Conservation of Wall Painting Department)

Plate 217 Thin section of another sample from the Sherborne fragments, photographed in plane-polarised transmitted light, original magnification 90×. The crushed fossil-rich limestone aggregate of the lime-based plaster is clearly visible. Since the Sherborne region is rich in Inferior Oolitic limestone, it is probable that this material is of local origin. (Photomicrograph: H. Howard; © Courtauld Institute, Conservation of Wall Painting Department)

Plate 218 Westminster Abbey, chapter house: detail of the *Celestial Host* in the *Apocalypse* cycle of *c*1374–1404. (Photograph: Howard/Kakoulli; © Courtauld Institute, Conservation of Wall Painting Department)

Plate 219 Westminster Abbey, chapter house. Cross-section (original magnification 376×) of a sample from the nimbus of St. John in the *Celestial Host*. The chalk ground is visible at the base of the sample and over this a secondary ground of lead white with a single azurite inclusion has been applied. The gold leaf has been adhered with an oil-based mordant. (Photomicrograph: H. Howard; © Courtauld Institute, Conservation of Wall Painting Department)

Plate 220 Backscatter SEM micrograph of the sample shown in Plate 219. The coccoliths in the chalk ground are clearly visible at high magnification. (SEM micrograph: H. Howard; © Courtauld Institute, Conservation of Wall Painting Department)

Plate 221 Westminster Abbey, chapter house: detail of the *Last Judgement* (*c*1374–1404), showing the raised crown held by the seraph in arch 2 of the east bay in raking light. (Photograph: H. Howard; © Courtauld Institute, Conservation of Wall Painting Department)

Plates 222 and 223 Backscatter SEM micrograph of a sample from the raised crown in Plate 221, original magnification 532000×, with an X-ray map for phosphorus (in orange). The dark particles in the cross-section clearly show the presence of phosphorus. Backscatter SEM micrograph (Plate 223) as above, but with X-ray dot maps for phosphorus (magenta), calcium (yellow), lead (cyan) and silica (orange). The dark particles in the cross-section are rich in calcium and phosphorus indicating the presence of bone white, while the matrix is lead-rich. (SEM micrographs: H. Howard; © Courtauld Institute, Conservation of Wall Painting Department)

Abstract

This research constitutes the first comprehensive study of the pigments employed in English medieval wall painting based on both scientific examination and documentary sources. Results of the scientific examination of wall paintings and related material are discussed in the context of wall painting, sculptural polychromy and panel painting techniques in the medieval period in England and elsewhere in northern Europe.

Principal findings are as follows. First, thirty pigments have been identified in total, establishing the sophistication of the English medieval wall painting palette. Second, four pigments have been detected which were previously unknown in the context of English medieval wall painting, namely vivianite, salt green, kermes lake and madder lake. Third, pigments such as red lakes whose earliest known use was in the Gothic period have now been identified in Romanesque wall painting. Fourth, the Romanesque period saw extensive use of lead pigments, presupposing partial use of organic binding media. Fifth, application methodology is more complex than previously supposed, with different binding media sometimes employed for specific layers in the stratigraphy of paintings, and for particular pigments in a single layer. Finally, three pigment alterations previously unknown in English medieval wall painting have been detected: the lightening of red lead; alteration of vivianite to a yellow form; and the transformation of verdigris to a blue copper chloride-based alteration product.

In some cases close correspondences in painting technology have been detected between English wall paintings and others elsewhere in northern Europe. More similarities also exist than hitherto thought in the techniques of wall painting and sculptural polychromy. The painting technology of the medieval period is enormously complex with variation between schemes of similar date and even within the same scheme. Assumptions about technique without supporting analysis are often ill-founded. The profound implications these findings have for the conservation of these paintings are discussed.

❦ ONE

Introduction

Background to the research

The first published account of the analysis of any pigment was that by J. Haslam (1807) on the splendid mid-14th-century wall paintings from St. Stephen's Chapel, Westminster Palace (London). Haslam not only carried out chemical analysis of the painting materials employed, but also demonstrated a close correlation between the medieval documentary evidence and his analytical results. Only a few years later, Jean Chaptal undertook analysis of pigments from a colour merchant's shop in Pompeii, while Sir Humphrey Davy examined wall paintings and pigment samples from Rome.[1] After this promising start, however, the analysis of easel painting techniques has in general taken precedence over that of wall paintings.

In 1914, A. P. Laurie published his technique for making cross-sections of samples from easel paintings by embedding them in wax (Laurie 1914). In the 1930s, various laboratories dedicated to the examination of painting materials – again principally of easel paintings – were established: at the Louvre in Paris (1931); the Forbes Laboratory at the Fogg Museum, USA (1932); the National Gallery (1934) and the Courtauld Institute (1934) in London; the Laboratoire Central des Musées de Belgique, Brussels (1934); the Doerner Institut, Munich (1937); and the Istituto Centrale del Restauro, Rome (1939). Since their inception, these institutions have provided significant contributions to the analysis of materials from works of art. Of particular note is the seminal work of Joyce Plesters (National Gallery, London) who developed a methodology for the examination and analysis of paint samples which still forms the basis of much of the work undertaken today (Plesters 1956).[2] Detailed optical examination of cross-sections followed by elemental analytical confirmation of observations is today almost standard procedure, but was an approach refined by Joyce Plesters.

1. For a discussion of the work of Haslam, Chaptal and Davy, see Rees-Jones 1990.
2. A detailed discussion of the history of the investigation of works of art was given by Dr Ashok Roy in his Forbes Prize Lecture at the 1998 IIC Congress in Dublin; see Roy 1998.

Although many of the methods developed for the analysis of easel paintings are directly transferable to the investigation of wall paintings, until recently very little work had been undertaken in this field in England. By contrast, pioneering analytical work on wall painting techniques has been undertaken elsewhere in Europe. The conservation and analytical procedures developed in Florence in the 1960s resulted in the discovery of considerable evidence regarding the technology of Italian wall painting (Arts Council 1969; Borsook 1980). More recently, detailed analyses have been undertaken by Mauro Matteini and Archangelo Moles of the Opificio delle Pietre Dure, Florence. Although their work primarily addresses Italian wall painting technology, they have also examined schemes more directly relevant to this book such as the 12th-century paintings at Idensen, Germany (Matteini and Moles 1994). Elsewhere in northern Europe many detailed studies have also been carried out, for example the investigation in the 1980s (Mairinger and Schreiner 1986: Emmenegger 1986) of the Carolingian and Romanesque wall paintings at Müstair (Switzerland) and of the Romanesque scheme of paintings at Mårslet, Denmark (Graebe *et al.* 1986). In recent years examination and analysis of other major north European schemes have been published; the paintings of *c*1270–80 in the choir of Angers Cathedral (France) (Subes-Picot 1992; Demailly *et al.* 1998), and the scheme of *c*1332–49 on the inner face of the choir screen of Cologne Cathedral (Jägers and Schulze-Senger 1989; Maul 1992). In general, such investigations have involved detailed examination of the technology of a particular scheme; and though McGrath (1987) and Nord and Tronner (2000) provide studies of the painting materials employed for small groups of medieval wall paintings in Ireland and Sweden, no up-to-date overview exists of painting technology of medieval wall paintings in northern Europe.[3]

It is extraordinary that after Haslam's pioneering analysis of the St. Stephen's wall paintings, further investigation of the technique of English medieval wall painting was neglected until the work of Professor Stephen Rees Jones at the Courtauld Institute in the 1940s and 1950s. Several of his most important results are published by Professor Tristram in his seminal volumes on English medieval wall painting, together with detailed discussions of techniques by Tristram (1944, 1950, 1955) based on the medieval documentary evidence and *in situ* observations.[4] The analytical methods employed by Professor Rees Jones at this time were principally wet chemistry, together with chromatography (for the analysis of proteins) and spectrography.[5] Thus, samples from 12th-century paintings in the apse of St. Gabriel's Chapel, Canterbury Cathedral, were examined by these methods, and pigments including vermilion, green earth, yellow ochre, carbon black, and calcium carbonate were identified (Tristram 1944: 104).

The National Survey of English Medieval Wall Painting, begun in 1980 as a joint project of the Courtauld Institute and the Royal Commission on Historical Monuments, highlighted both the importance and extent of surviving wall painting, and led directly to the establishment in 1985 of the Conservation of Wall Painting Department at the Courtauld. At that time knowledge of the techniques of English medieval painting was

3. Mora, Mora and Philippot's chapter on western medieval painting is still the best available general coverage of this subject, and though brief, provides good illustrations of aspects of technology such as preparatory techniques and paint application; see Mora *et al.* 1984:117–37.
4. Other investigations undertaken by Professor Rees Jones are archived in a card index at the Courtauld Institute of Art.
5. The late Professor S. Rees Jones kindly discussed his early analytical work at the Courtauld Institute with me.

still woefully inadequate, and the department therefore began to undertake studies of particular aspects of wall painting technology.[6] These initial studies provided the basis – in terms both of data and methodology – for a five-year Leverhulme Research Fellowship undertaken by the author in 1991–96. This constituted the first comprehensive study of English medieval wall painting techniques based on both scientific examination and documentary sources, and, for the first time, permitted a detailed understanding of how English medieval wall paintings were executed.

Based on initial historical research, fifteen major wall painting schemes were chosen for scientific examination during the Leverhulme study. The selection was specifically chosen to provide a broad chronological and geographical distribution, and to reflect different painting types from major cathedral schemes to those in parish churches. As well as including some of the most important surviving paintings, such as examples in Winchester and Canterbury Cathedrals, the opportunity was also taken to examine other paintings undergoing conservation, to which access would otherwise have been unavailable, and where the research had a direct impact on the conservation treatments adopted. In addition, some other closely related polychromy was studied, namely two limestone tombs, one early panel painting, and three schemes of wall painting from elsewhere in northern Europe. Previous analytical work and relevant samples available in the department were also utilised allowing the project to draw on evidence from approximately a hundred sites. All aspects of the technique of the paintings selected for the Leverhulme research project were studied in detail. For each site a detailed report or article was produced, and these have been collated in a separate volume (Howard 1996e). In many instances the information was published jointly with colleagues from the Conservation of Wall Painting Department, underlining the significance of interdisciplinary evaluation of the technical results, both art historically and in the context of the condition and environmental issues relating to the painting (Cather and Howard 1994; Park and Howard 1996; Gill and Howard 1997).

Overview of English medieval wall painting techniques

Our understanding of the techniques of English medieval wall painting has developed considerably since 1985, and it is now clear that they are often much more complex than previously supposed. A brief overview of these techniques is therefore provided here before addressing more specifically the pigments that form the subject of this book.

English Romanesque wall painting technique

Many English Romanesque wall paintings have been presumed in the past to have been painted almost exclusively in a *buon fresco* technique.[7] Recent investigations, however,

---

6. For example, a study of blue pigments in English medieval wall painting (Howard 1988), research on red pigments in English medieval wall painting (Hluvko 1991), and the phenomenon of darkened flesh tones (altered lead pigments) in English medieval wall painting (Welford 1991).
7. In *buon fresco*, that is, true *fresco* painting, pigments are combined with water and applied to fresh plaster. Fresh plaster includes slaked lime – calcium hydroxide – and this sets on contact with air ($Ca(OH)_2 + CO_2 \rightarrow CaCO_3 + H_2O$). Setting – that is, carbonation – is an irreversible chemical process that binds pigments to the surface of the plaster.

indicate that relatively few were executed wholly in this technique. It is now clear that, while the techniques of the Romanesque period were indeed rooted in the *fresco* tradition, they were much more complex than this, with one (or more) significant variation from established procedures apparent in every example studied. Thus, to classify the wall paintings of this period simply as *fresco* constitutes a gross over-simplification.

The presence of a *sinopia* and a thin surface layer of plaster normally suggests the use of *fresco* technique. Further evidence for the use of *fresco* may be provided when the final plaster layer is applied in fairly small areas which can be painted during the carbonation period. These are distinguishable by the overlapping of adjacent subsequently applied sections of plaster. Such evidence is clearly visible in the early Romanesque paintings of *c*1100 at Coombes, Clayton and Hardham (West Sussex).

In the Hardham scheme, the application of the *intonaco* in small overlapping patches and indications of the presence of a *sinopia* suggest that the intention was to paint in a *buon fresco* technique, with the primary mechanism for binding the pigments being the carbonation of calcium hydroxide from the lime plaster. Local lime grounds, applied in one or two layers, were also employed in some areas.[8] Furthermore, lime white was also frequently combined with pigments here, not only to produce tonal variation but also to provide additional binding capacity.[9] The use of this white pigment – which retains a high proportion of unreacted calcium hydroxide – is consistent with the *fresco* process, and indeed is a crucial element of the technique. *Impasto* highlights applied in lime white are also clearly visible at Coombes, as are the impressions of snapped cord lines for setting out the scheme in the fresh plaster. The latter evidence indicates that the painting is coeval with the plastering and provides further evidence of a *buon fresco* technique (Cather *et al.* 1990: xi–xvi). As such, these examples represent very much the exception in Romanesque wall painting technique.

The presence of a *sinopia* and the application of the uppermost plaster layer in small overlapping patches indicates that the technique of the late Romanesque scheme in the Holy Sepulchre Chapel, Winchester Cathedral (*c*1175) is also rooted in the *fresco* process. However, while analysis suggests that the carbonation of calcium hydroxide from the lime plaster is the primary binding mechanism, at least for the preliminary drawing and initial pigment layers, the extensive use of lead pigments and the sweeping changes made to the composition at an advanced stage of the painting process indicates that a significant proportion was completed in *secco* (Howard 1995a).

In a number of other instances – at Houghton-on-the-Hill (*c*1090) (Rickerby 1996), at Witley (*c*1110) (Howard 1992a), Kempley (*c*1130) (Rickerby 1990a: 252), and St. Gabriel's Chapel in Canterbury Cathedral (*c*1130) (Cather and Howard 1994) – the uppermost plaster layer was applied in large horizontal bands relating to scaffolding lifts. Indeed, as evidence for plastering in relation to wall painting accumulates, it would appear that application in wide horizontal bands was typical of English Romanesque wall painting.[10] In these examples, although it appears that the preliminary drawing and some

8. For pigment analysis by Howard, see Sawdy *et al.* 1998: 34–5.
9. For lime white (*bianco di San Giovanni*), see Denninger 1974.
10. Mora *et al.* (1984: 120) consider that working in *pontate* with overlapping joins was the most common method of painting in the Romanesque period. Winfield (1968: 71) also believes that technical investigation of Romanesque painting in Europe will mainly reveal plaster joins coincident with borders of compositions and with levels at which scaffolding was erected.

of the initial paint layers were applied while the plaster was still fresh and were therefore bound by its carbonation, the extremely large areas of plaster – at Witley up to 15m^2 – certainly preclude describing the overall technique as *buon fresco*.

Furthermore, while the paintings of *c*1130 in St. Gabriel's Chapel, Canterbury Cathedral, have been regarded as typical of Romanesque *fresco* technique, recent analysis shows a complex stratigraphy with lead white underpainting used in some areas. Significantly, the palette includes pigments such as red lead and lead white which are incompatible with painting in a *fresco* technique. In addition, instrumental analysis of the binding media confirmed that both protein and linseed oil were used as part of the original technique of the paintings, pushing back the earliest confirmed use of oil in English medieval wall painting by over a century from the previously documented and analytically substantiated date of *c*1236 when oil was employed for painting in the cloister at Canterbury (Tristram 1950: 519; Hluvko 1991: 41–2).

The St. Gabriel's Chapel paintings illustrate one of the most important recent findings regarding Romanesque wall painting technology: the incorporation of organic binding materials in the overall binding mechanism of wall paintings as early as *c*1130. Analysis of the paintings of *c*1175 in the Holy Sepulchre Chapel at Winchester Cathedral has also shown that, although the carbonation of lime is the principal binding mechanism, an organic component is also present in some paint layers. Protein has been identified in some samples, and instrumental analysis has also confirmed the presence of linseed oil – which had been partially heat-bodied to promote drying – in the mordant below the gold leaf of the angel's halo in the *Entombment* (White 1998: 1). The presence of partially pre-polymerised linseed oil in this scheme is significant since it indicates considerable sophistication in the use of this medium as early as the Romanesque period.

These findings concerning the incorporation of organic binding media in Romanesque wall paintings relate not only to major schemes. In the paintings of *c*1110 in the parish church of Witley (Surrey), the extensive use of lime-sensitive lead pigments suggests that an organic binding medium would have been employed.

The available data for contemporary English polychromed sculpture show similar results. A recent examination of the traces of original polychromy on the Lincoln frieze of *c*1140 has shown that linseed oil was used (Larsen 1992: 188). Similarly, analysis of the late 12th-century polychromed sculpture from the west portal of St. Mary's Abbey, York indicated that egg tempera and glue were employed.[11]

Taken together, this evidence suggests that the findings at Canterbury, Winchester and Witley do not represent an exception. Rather, that from at least the beginning of the Romanesque period, organic binding media were commonly used as a component of the original binding mechanism, in conjunction with lime – the major constituent – in some form, in both complex cathedral schemes and more modest parish church wall paintings in England.[12]

The use of oil and protein as binding media in wall paintings has also been confirmed in the scheme of *c*800 at Müstair (Switzerland) (Mairinger and Schreiner 1986: 196). The wall paintings of *c*1130 at Idensen (Germany) provide closely comparable results to

11. For the medium analysis by R. White, see Brodrick 1993: 26.
12. Analysis of excavated plaster fragments from the Anglo-Saxon monastery at Monkwearmouth has also indicated a trace of oil in the paint layer (Cramp and Cronyn 1990: 22), suggesting a continuity of technical practice from the Anglo-Saxon to the Romanesque period.

those obtained at Canterbury, since not only a drying oil – probably linseed oil – but also an egg protein have been identified (Matteini and Moles 1994: 85–6). Continental examples also provide evidence for the use of oil in polychrome sculpture at this date, for example analysis of Romanesque stone sculpture in S. Maria im Capitol, Cologne suggested that oil is both the sealant for the stone and the binding medium of the pigments.[13] Mixed media were employed for the 12th-century polychromy of the *Montvianex Madonna*. Here a drying oil was used for flesh tones, a casein/egg mixture for relief decoration, and a proteinaceous binding medium containing egg for other paint layers (Kargère 2002: 509).

In summary, recent investigation of English Romanesque wall paintings suggests that the technology employed is rooted in the *fresco* tradition and that the preliminary drawing and initial paint layers are bound by the carbonation of lime from the uppermost plaster layer. However, there are generally one or more substantial variations from established procedure which preclude describing the technique as *buon fresco*, and in the majority of cases mixed media are employed.

English Gothic wall painting technique

By contrast with the opaque and predominantly primary colours of the Romanesque palette, Gothic wall painting technique after about 1300 is governed by concepts of luminosity and translucency, perhaps influenced by the increasingly large areas dedicated to stained glass windows. Evidence for the extensive use in Romanesque wall painting of organic binding media, alkali-sensitive pigments and lake pigments (all of which have been associated with Gothic painting), suggest that the considerable developments in Gothic wall painting techniques were driven by artistic intent rather than any sudden introduction of new materials.

These developments may be summarised as: a move towards the use of lead white grounds; an increasing use of organic binding media, particularly linseed oil; increasingly complex combinations of media, often for different paint layers; increasingly complex layering of colours to produce sumptuous effects; increased use of translucent glazes to provide luminosity and translucency; and the more frequent use of relief ornament.

Clearly the changes in technique required to realise the ambitions of the Gothic wall painter took some time to achieve, and the period from about 1200 to 1300 may be regarded as one of transition. Initially, the carbonation of lime was still the principal mechanism for binding pigments, although organic binding media became an increasingly important component. In the early to mid-13th century, lime-based grounds still predominated, as in the scheme of *c*1220–30 in St. George's Chapel, Ely Cathedral (Cambridgeshire) and that of *c*1250 at Horsham St. Faith Priory (Norfolk). By contrast, lead white grounds became a feature of Gothic wall painting from *c*1300. A period of transition in the types of ground employed also appears to occur elsewhere in northern Europe, as exemplified perhaps by the scheme of *c*1270 in the choir of Angers Cathedral where both lime and lead white grounds have been detected (Demailly *et al.* 1998: 11).

The two schemes of wall painting in the Ante-Reliquary Chapel of Norwich Cathedral provide good examples of the changes in technique from this 'transitional' period to the 14th century (Park and Howard 1996; Howard 1995c). On the eastern arch of the

13. E. Jägers, pers. comm.

chapel are fragmentary paintings dated on stylistic grounds to *c*1250; analysis has shown that these are characterised essentially by dense opaque unmixed pigments and a primary palette. By contrast, for the paintings of *c*1300 on the vault and western arch of the chapel, a lead white ground was employed. This overall ground was clearly intended to interact optically with often translucent paint layers applied subsequently. Its introduction marks a sweeping change in both the artist's intent and technology, and is a characteristic of the majority of high-quality wall paintings from *c*1300.

In order to achieve the translucent effects which are central to Gothic wall painting, an increasingly large proportion of organic binding media – principally linseed oil (see Appendix 3, Table 7) – was employed. Also, characteristic of sophisticated Gothic wall paintings are complex combinations of media, often used for different paint layers or for pigments in a particular layer. The mid-13th-century wall paintings at Horsham St. Faith Priory (Norfolk) are important in this respect, since they demonstrate that this level of sophistication was sometimes employed in high-quality schemes of the 'transitional' period. Here, rabbit-skin glue was used to bind azurite in the first Foundation scene of the narrative cycle, while egg tempera was employed as the adhesive for gold leaf on St. Faith's crown. In addition, linseed oil was found combined with green and reddish/brown paint (White 1989). For the 15th-century repainting of this scheme, heat-bodied (pre-polymerised) linseed oil was employed.

The frequent use of a proteinaceous medium for copper-containing blue and green pigments in England during this period is mirrored by similar findings in wall paintings elsewhere in northern Europe. For example, this combination was identified both in the paintings of *c*1300 in the Ante-Reliquary Chapel at Norwich (Howard 1995c), and in the late 13th-century scheme in the choir of Angers Cathedral. Similarly, both protein and oil have been identified as constituents of the binding media of azurite-containing paint layers in the paintings of *c*1332–49 on the inner face of the choir screen of Cologne Cathedral.[14]

The extensive palette of particularly sophisticated Gothic wall paintings is exemplified by the early 15th-century scheme at Farleigh Hungerford (Somerset). Here the rich palette – which includes gold and silver leaf, a red lake pigment and orpiment – was extended further by mixing and layering pigments to produce particularly sumptuous effects. Especially remarkable is the extensive use of translucent glazes over gold and silver leaf. For example, a gold effect was achieved for St. George's helm by a yellow glaze over silver leaf, and further embellishment provided by raised curvilinear details made from a second layer of silver leaf highlighted with a green glaze (Howard *et al.* 1998: 60). These paintings, until recently virtually unknown, are perhaps even more complex technically than the renowned late fifteenth-century scheme of Eton College Chapel. In the latter paintings, linseed oil has been identified along with a palette including lake pigments and lead-tin yellow (Plummer 1986: 36).

Although sometimes used previously, relief ornament becomes increasingly common in Gothic wall painting (Nadolny 2001). A particularly interesting example is again provided by the early 15th-century scheme at Farleigh Hungerford, where the mail armour around the neck of the kneeling knight was modelled *in situ* from a paste of lead white and oil. The undulating surface was then covered in silver leaf and would have been glazed to prevent tarnishing, although none of this glaze layer now remains (Howard *et al.* 1998: 62). More commonly, relief ornament was manufactured in tin-lined moulds

14. Analysis by Jägers in Maul 1992; see also Heritage 1990: 8.

and applied subsequently in sections. This technique was used for borders and other elements in the royal wall paintings of *c*1352–63 in St. Stephen's Chapel, Westminster Palace (van Geersdaele and Goldsworthy 1978:9–12), and on the effigy of *c*1380 of Sir Roger de Bois at Ingham, Norfolk (Howard 1993d: 5). The relatively little-known decoration of the Chapel of Our Lady Undercroft in Canterbury Cathedral is perhaps the most sumptuous example of the use of relief ornament. Here the eastern vault of the chapel is covered with suns and stars, some of which originally contained mirrors to enhance the candlelight which illuminated the chapel. The stone screens enclosing the chapel are also encrusted with tiny relief appliqué designs, and the central gilded niche in the eastern screen is studded with imitation jewels. In addition, for raised decorative elements of the canopy, silver leaf was glazed with verdigris in an oil medium to produce a rich *translucida* effect (Howard 1993b).

To summarise, in the Romanesque schemes discussed above, calcium hydroxide – in a variety of forms – is the principal binding agent. In almost all cases this basic technique is elaborated to a greater or lesser extent, precluding description of the technique as simply *buon fresco*. Furthermore, the common presumption that lime-based binders are the sole binding agents of English medieval wall painting has led to a wholesale misunderstanding of the complexity of their technology, in turn spawning widespread use of inappropriate conservation treatments. It is now clear that organic binding media were commonly used as a component of the original binding media of English Romanesque wall painting. Characteristic of the Gothic period is an increase in the quantity, complexity and sophistication in the use of these additional binding materials, a development that was central to achieving the effects of luminosity and translucency which were so highly valued at that time.

The focus of this book

The focus of this book is the analysis of the pigments of English medieval wall painting. It demonstrates that an understanding of the pigments and their application methodology is pivotal both to an understanding of the original techniques of individual paintings, and to the overall development of such techniques in England and elsewhere in northern Europe during the medieval period. The alteration of many of the pigments detected highlights the effect of previous conservation treatments, and perhaps most importantly, underscores the importance of an understanding of original materials in the planning of future conservation treatments. Comprehensive discussion of all aspects of medieval wall painting technology is beyond the scope of this work, but discussion of topics such as binding media and grounds is included where relevant.

This work draws not only on the findings of the author's Leverhulme study, but extends this with the results of the examination of a range of paintings undertaken subsequently. The report to the Leverhulme Trust was presented as a series of detailed studies, each addressing a separate painting. A key feature of this book, therefore, is that all relevant work is drawn together to provide for the first time a comprehensive overview of the pigments of English medieval wall painting. Each pigment identified is dealt with systematically, and the results of the analysis of samples from English medieval wall paintings are discussed in the context of the pigments identified elsewhere in northern Europe, both in wall painting and in other media.

Sources of information

This book is based primarily on the analytical work undertaken during the Leverhulme Fellowship, together with results from the scientific examination of wall paintings and related material subsequently undertaken by the author. It also includes analytical information from the examination of wall paintings and other media in England and elsewhere in northern Europe obtained from published sources. Selected unpublished analytical reports by others (for example, by J. Darrah, E. Sinclair, T. Curteis and UCL Painting Analysis) have also been incorporated in an effort to provide as comprehensive and reliable a coverage as possible of current analytical results.

This information has been compiled in tabulated form in Appendix 3, which provides details of how the analysis was undertaken, by whom, and where (if at all) the results have been published. Since reliability is clearly increased where observations made by polarised light microscopy and microchemical tests have been confirmed by instrumental analysis, or the examination undertaken by a researcher highly experienced in painting analysis, the tabulated information should help the reader to assess the accuracy, reliability and significance of the reported results.

In addition to the *in situ* investigations and subsequent examination and analysis of paint samples, an integral part of the overall research involved a review of the relevant literature. Published primary sources utilised included medieval treatises – which provided recipes, details about the pigments and media employed, and technical instruction for their preparation and application – and financial accounts that detailed the materials purchased and employed in specific contexts (for detailed descriptions of these sources see pp. 15–21).

The medieval accounts comprise both ecclesiastic and royal records, and are particularly detailed for England in the 13th and 14th centuries; examples include the cathedral fabric rolls for Exeter Cathedral, Norwich Cathedral Priory and the accounts for Westminster Palace. These accounts often contain detailed information on the materials purchased, but in only a very few cases can these be linked directly to a particular surviving scheme of wall painting.

Many of the medieval treatises may, at first sight, have much less relevance to English wall painting technology; for example, they may describe the southern European tradition where *buon fresco* painting predominates. In other instances, they may deal primarily with panel painting and manuscript illumination. However, since the documentary evidence shows that artists worked in various media, the artificial distinctions made between painting on panels, walls or sculpture may be unhelpful overall, and it may be more useful to look at how similar goals were accomplished in these varying contexts.

Secondary sources of information examined included conservation reports, antiquarian literature, and other studies of wall painting and related art forms. Much of this information has been assembled in the National Survey of Medieval Wall Painting, undertaken by David Park since 1980. The Survey archive incorporates records of more than two thousand paintings dating from the Anglo-Saxon period to the 16th century. The present study has drawn heavily on this invaluable resource, and I am profoundly grateful to David Park for his generosity in providing access to this wealth of, as yet, largely unpublished information (for detailed descriptions of these sources see pp. 21–4).

Chronological and geographical scope

For the purposes of this book, the medieval period is taken as covering the years from the beginning of the Romanesque period in England (*c*1066) to the Reformation in the 16th century. The earliest wall paintings examined are excavated fragments of *c*1090 from Lewes Priory (Sussex), while the latest is the scheme of *c*1510 at St. Mark's (The Lord Mayor's Chapel), Bristol.

Sites examined during the Leverhulme research were selected to cover a broad geographical spread throughout England during the medieval period. Most of the schemes of wall painting examined subsequently are also English. However, two sites outside England have been included in the overall discussions of English medieval wall painting technique: fragments of wall painting of *c*1190 excavated from the crypt of Glasgow Cathedral (Scotland); and the wall paintings of *c*1170 from Cormac's Chapel, Cashel (Ireland).

The Glasgow Cathedral fragments have recently been analysed both technically and art historically (Park 1998; Howard 1998a; Park and Howard forthcoming). Park (1998: 39) concludes that stylistically they can be compared with wall paintings from northern England, and indeed that they may have been painted by an English workshop. For this reason, and due to the close technical comparisons which can be drawn with English examples, the discussion of the pigments employed at Glasgow is included with the English examples in the relevant pigment chapters. Similarly, samples taken from the scheme of *c*1170 from Cormac's Chapel are also discussed with the English material since they share a particular technical affinity with the approximately contemporary scheme in the apse of St. Gabriel's Chapel, Canterbury Cathedral, and again, may well have been painted by an English workshop (McGrath 1987: 101–2). In the tables (Appendix 3), however, these sites are listed with other examples of wall painting elsewhere in northern Europe.

The relationship of English medieval wall painting to northern European wall painting and painting in other media

Comparison of the findings from England are made with those from wall paintings elsewhere in northern Europe in an attempt to establish how typical – or untypical – the English traditions may have been. Broad patterns in the selection and application of pigments have emerged, such as the primacy of ultramarine as the mineral blue of choice in the Romanesque period (Howard 1988). In addition, a degree of similarity has been established in the selection and application of pigments from the Romanesque to the Gothic periods in England and elsewhere in northern Europe.

Very few panel paintings survive from the medieval period, and although the evidence for sculptural polychromy is more considerable, thorough study has so far been limited. Nevertheless, findings in these closely related fields throw some light on technical developments in wall painting. For example, media examination of the 14th-century tombs at Ingham (Norfolk), and of the shrine of St. Alban at St. Albans Cathedral (Hertfordshire), suggests that their technology is closely comparable to that of wall painting. This is not surprising since medieval documentary evidence indicates that artists worked in various media, making distinctions between painting on panels, walls or sculpture to some extent artificial.

Initial summary of the results

This is the first comprehensive study of the pigments employed in medieval wall painting to have been undertaken anywhere in northern Europe. For the first time, it permits a detailed understanding of the nature and extent of the palette employed in English wall painting, and the sophistication with which the pigments were applied to achieve a wide variety of effects. Several pigments have been identified in English medieval wall painting for the first time: vivianite, kermes lake and madder lake.

Other results will be considered in more detail in the Discussion (pp. 195–210), but may be summarised as follows. First, the research has revealed broad chronological developments in the choice, fabrication and application of materials, linked to changes in artistic intent, technology, and workshop practice. Thus, by the end of the 13th century, there is a clear increase in the use of alkaline-sensitive pigments such as lead white and red lead, and verdigris. These are ideally suited to the Gothic palette and painting technology, and are instrumental in achieving translucent effects. This development in the use of pigments was made possible by correspondingly greater use of organic binding media.

Secondly, chronological developments in the use of blue pigments are particularly significant, with a transition during the first half of the 13th century from natural ultramarine to azurite as the mineral pigment normally employed. In relatively modest schemes, a range of alternatives such as synthetic copper blues and indigo become more widespread in the Gothic period.

Thirdly, a red lake pigment has been detected in the paintings of *c*1175 in the Holy Sepulchre Chapel at Winchester Cathedral (Hampshire), and constitutes the earliest identified example of a lake pigment in English wall painting. Analysis of the dyestuff employed to make the lake pigments in the important late 14th-century scheme in the chapter house of Westminster Abbey has confirmed the presence of kermes.[15] This is the only instance in which this very expensive organic colorant made from a scale insect has been detected in English medieval wall painting, highlighting the importance and quality of the Westminster paintings. Madder lake has been detected in the only other two instances where analysis of the dyestuff has been possible.

Fourthly, analysis of the Romanesque scheme in the Holy Sepulchre Chapel at Winchester, and of two other 13th-century schemes in the same cathedral, revealed that vivianite – a blue iron-phosphate pigment – was employed. These are the first instances in which vivianite has been identified in English medieval wall painting. Moreover, an understanding of the partial alteration of this pigment to form an overall green colour in areas which were intended to be blue is crucial to our understanding of the original appearance of these paintings.

Art historical implications

In several instances, the technical investigation has provided crucial evidence for the redating of wall paintings. For example, evidence from the 12th-century scheme in the apse of St. Gabriel's Chapel, Canterbury Cathedral, showed that the paintings were executed some 25 years earlier than previously thought. Scientific examination indicated

15. I am grateful to J. Kirby of the National Gallery for undertaking this analysis by high performance liquid chromatography (HPLC), see Kirby 1999.

that the preparatory drawing and initial paint layers were executed while the plaster was still fresh, confirming that plastering and painting were coeval. In addition, the examination suggested that the plaster was applied in a single layer over the rubble and ashlar construction. There are three roughly horizontal plaster joins dividing the two narrative registers into four bands. None of these plaster bands conform to the divisions of the composition, and indeed they arbitrarily cut through them. Particularly important is the fact that the lowest plaster join corresponds to the change in construction material of the apse, from ashlar masonry of the wall to the rubble of the curved vault. These paintings have been thought to date from as late as *c*1155, but the chapel itself was completed in *c*1110; there is only one layer of plaster, and there is no doubt that the painting is coeval with it. Since it is inconceivable that the rubble vault would have been left exposed for 40 or 50 years, the paintings have now been dated much earlier in the 12th century, and this also agrees with stylistic comparisons that can be made with Canterbury manuscripts (Cather and Howard 1994: 141–2).

Another example where technical examination has assisted dating is in the Ante-Reliquary Chapel, Norwich Cathedral. Although an early 14th-century date has been generally accepted, there has been some speculation that the vault paintings are slightly more advanced stylistically, perhaps *c*1330 as opposed to *c*1310 for those on the western arch. On the eastern arch of the chapel are some more fragmentary paintings which have been dated on stylistic grounds to *c*1250. The analysis has shown that these eastern arch paintings are indeed earlier than the rest. It is also clear from both the similarity in preparatory layers and the identical stratigraphy and materials of the paint layers that the 14th-century paintings on the vault and western arch form a single scheme now dated to *c*1300 (Park and Howard 1996: 397).

Finally, important findings with respect to workshop practice and dating resulted from examination of two early 14th-century paintings of saints in St. Albans Cathedral. These are situated at either end of the east side of the feretory, the area around the shrine of St. Alban. Although they are certainly of the same date, analysis has established significant technical differences between the two stylistically similar paintings; they are clearly by two different artists. Significantly, the examination revealed that, although these painting techniques are so different, aspects of both are found in the polychromy of the shrine itself – also recently conserved and analysed. Although the wall paintings, which are technically and stylistically very similar to paintings of this period at Westminster, have previously been dated to *c*1330, they can now be seen as part of a single, coherent overall scheme which presumably must have been executed by a single workshop at the time the shrine was created, known to have been between 1302 and 1308 (Howard 1993a).

Comparison with wall painting technology elsewhere in northern Europe

Results of the study suggest a close link between the technology of English medieval wall painting and that of elsewhere in northern Europe. Comparison can be drawn, for example, between the scheme of *c*1130 in St. Gabriel's Chapel, Canterbury Cathedral and the materials and techniques employed in the approximately contemporary scheme at Idensen (Germany). Likewise, close parallels can be drawn between the Canterbury scheme and that of *c*1170 in Cormac's Chapel, Cashel, with the palette and application methodology very similar in both schemes. Moreover, the richness of the palette at Cashel is at odds with the limited range of earth pigments identified in other medieval wall paintings in Ireland, adding further evidence to the view that English painters may have undertaken the commission at Cashel (McGrath 1987: 102). Furthermore, examination of wall painting fragments of *c*1190 from the crypt of Glasgow Cathedral revealed a broad palette and complex painting technology. This, together with striking stylistic parallels which have been drawn with other Romanesque schemes in northern England, again suggests that English artists may have been involved in the execution of these paintings (Park 1998).

Although a comprehensive study of medieval wall painting techniques elsewhere in northern Europe has yet to be undertaken, from the site-specific studies which have been carried out, many of the overall patterns of development observed in English wall painting may be traced across northern Europe; for example, the limited but significant use of organic binding media and alkaline-sensitive pigments in the Romanesque period, and a desire to achieve more translucent effects in the Gothic period.

Implications for conservation

The research has shown that many of the widely held presumptions regarding the technology of English medieval wall painting do not withstand technical scrutiny. For example, Laurie (1967: 224–5) asserted that English wall paintings were generally executed in oil and varnish, producing a shiny surface; Waller (in Keyser 1883: xc, 40, 63) concluded that the medium was invariably size; and Tristram described the scheme in the apse of St. Gabriel's Chapel, Canterbury Cathedral as the 'finest example of northern European fresco technique' (James and Tristram 1928–9: 40). These presumptions regarding techniques have frequently influenced the types of materials used in conservation treatments and the methodology for their application (see Cather and Howard 1986).

One of the key findings of the present study is that English Romanesque wall paintings were rarely painted entirely in a *buon fresco* technique as has commonly been supposed, and that, from at least the beginning of the Romanesque period, an organic component of the original binding medium is frequently present. The presumption that wall paintings have a lime-based binding medium alone has led to the use of inappropriate conservation treatments, such as lime-watering, a practice still in use today. This type of treatment – which may involve multiple soakings of the entire stratigraphy of the painting – has been implicated in the alteration of original materials, and in particular in the transformation of lead pigments and verdigris.

Although the study has established some general patterns regarding original technology, it has also indicated that it is inadvisable to extrapolate information from one scheme to another. It is now clear that the original technique and the causes of deterioration should be rigorously studied in each case prior to the commencement of any conservation programme. For example, although the phenomenon of dark flesh painting in English wall painting has generally been shown to be due to the alteration of a lead pigment to form lead dioxide – as in the paintings of *c*1330 at Longthorpe Tower (Cambridgeshire) (Welford 1991) – this is not universally the case. Analysis of samples taken from the scheme of *c*1300 on the vault of the Ante-Reliquary Chapel (Norwich Cathedral) produced no evidence of altered lead pigments in the flesh tones. Rather, the patchy grey appearance of the flesh is, as in many other areas of the painting, due to the partial removal of past surface coatings and accretions. The surface accretions remaining on the flesh tones now give them a dark appearance, while the backgrounds and draperies have been more rigorously cleaned.

Likewise, study of the paintings in the feretory of St. Albans Cathedral has established that even with two stylistically identical paintings in a single scheme there may be fundamental differences in technology. This has clear implications not only for workshop practice but also for the conservation of the overall scheme, since a single conservation procedure is unlikely to be suitable in all areas.

The extent and variety of the pigment alterations observed in the study is significant, particularly since in many cases the mechanisms of these transformations – and how these may relate to conservation treatments and environmental conditions – are not fully understood. Such alterations, and their implications for conservation treatments, are discussed in more detail in the Discussion (pp. 202–5), and will be the subject of further research in the future by the author.

The alteration of vivianite to a yellow form, the transformation of red lead to white cerussite, and the alteration of verdigris to a blue copper-chloride containing product, have been confirmed for the first time in English medieval wall painting. Alkalinity and high humidity have been implicated in the alteration of lead pigments, verdigris and other copper pigments. The presence of these pigments would similarly make lime-watering an inadvisable and potentially damaging conservation treatment.

Finally, the study has highlighted a number of instances in which organic colorants such as red lake pigments and indigo have been employed in wall painting throughout the medieval period. Since these materials are subject to fading, the need for adequate preventive care has been highlighted.[16] The key factor in the fading of lake pigments is exposure to light, although other factors may affect the rate of colour loss (Saunders and Kirby 1994: 91–2). For example, lake pigments when mixed with white and unprotected by a glaze are particularly vulnerable to internal reflection and consequently fade more quickly (Johnston-Feller and Bailie 1982; Johnston-Feller 1986). The type of substrate, pigment particle size, and the type and quantity of binding medium may also affect the rate of fading of fugitive pigments (Saunders and Kirby 1994: 86–93). All these factors have implications in relation to the cleaning of such paintings since all commonly used solvents cause leaching of the binding media, and thus make the lake pigment particles increasingly susceptible to fading (Stolow 1957).

In summary, to undertake conservation treatment of an English medieval wall painting without first being fully informed of its technology represents an unacceptable risk.

16. To protect lake pigments from fading Saunders and Kirby (1994: 93) recommend limiting overall light exposure and eliminating UV light where possible.

❦ TWO

Methodology

Introduction

In conjunction with the *in situ* investigation and subsequent analysis of paint samples, an integral part of the present study involved a review of the relevant literature. This chapter, therefore, provides details of this literature followed by a description of the methods of investigation, examination and analysis employed.

The literature is considered under three headings: primary sources; secondary literature; and the author's own and other relevant unpublished analyses. Primary sources comprise medieval treatises, recipe books and financial accounts. The treatises and recipe books usually include detailed information about pigments and media, and technical instruction for their application. Financial accounts, both English and continental, detail the materials purchased for specific contexts and wages paid to craftsmen. Secondary literature comprises antiquarian literature and more recent published information – in particular studies of wall paintings and related material elsewhere in northern Europe. Finally, the author's work (published and unpublished reports), together with selected unpublished analytical reports by others, are reviewed to provide a comprehensive coverage of current analytical results.

Primary sources: medieval treatises, recipe books and accounts

Treatises and recipe books

Numerous early manuscripts contain technical information relating to the preparation and use of artists' materials. Clarke's (2001) introduction and catalogue of more than four hundred manuscripts to AD 1500 is an invaluable guide to understanding this diverse material which includes medical and alchemical texts, a small number of artists' treatises written by practitioners from first-hand knowledge, and numerous compilations of extracts from earlier material.

The alchemical and medical treatises, which ultimately derive from ancient Classical Greek, Middle Eastern and Indian sources, were translated into Latin and spread across

Europe from *c*1200 introducing methods of purification and preparation of materials. Other compilations such as the *Mappae Clavicula* and Lucca manuscript have their roots in Arabic-Greek sources, though the main body of the text was clearly compiled around 800 (Clarke 2001: fig. 1).

Manuscripts were transmitted through the laborious process of hand copies. The copyist would copy, add to, or amend the model text in the light of other related texts, contemporary knowledge, occasionally also drawing on personal experience. Indeed many of the early medieval artists' recipe books may be regarded as compilations of compilations which have undergone alteration and addition over a long period. Thus the date of the manuscript may not be the date of a recipe copied from a precursor, and many recipes will predate the date of a particular compilation. However, other manuscripts contain large amounts of almost completely original material (such as the mid-12th-century treatise of Theophilus), or were supplemented by new additions.

Since the majority of these manuscripts are, at least in part, copies from one another, they contain very much the same material. Most of the technical information included relates to the preparation of pigments for manuscript illumination with a few details regarding application, and very limited information, if any, regarding panel painting and other polychromy. However, one French source *Liber diversarum arcium* is of particular relevance here since it seems to refer, at least in part, to sculptural polychromy rather than MS illumination.[1] This manuscript, which may date from the late 14th or early 15th century, includes much from Theophilus, some from Eraclius, as well as other material including the character of some pigments in an oil medium.[2]

A number of the medieval manuscripts are English, for example, MS Cotton Titus D.XXIV, a late 12th-century compilation on manuscript illumination made at the Cistercian Abbey of Rufford (Nottinghamshire; Petzold 1995), and MS Brogyntyn 2.1, a 15th-century manuscript containing illuminators' recipes (Clarke 2001: 60; Halliwell 1848, 1855). Some of the Brogyntyn recipes also appear in the slightly earlier BL Sloane 2584 (Clarke 2001: 89), though there is some discrepancy in spelling and wording.[3]

The majority of the medieval treatises may, at first sight, seem to have limited relevance to northern European wall painting technology; for example they may describe the southern European tradition where *buon fresco* painting predominated (Cennini). In other instances, as discussed above, they deal primarily with manuscript illumination, and include only a very few details relating to panel painting and other polychromy.

Since the texts copy from one another, the material is often very much the same in many of the manuscripts. For this reason, the texts referred to here are principally the more obvious and easily accessible published sources. Those most relevant to the present study are: the *Mappae Clavicula*; Eraclius' *de Coloribus et Artibus Romanorum*; Theophilus' *De Diversis Artibus*; the manuscript of Peter of St. Omer; and the Strasbourg manuscript. Although Cennino Cennini's late 14th-century *Il Libro dell'arte* primarily

1. '*Liber diversarum arcium*' [Montpellier, Ecole de Médecine, MS 277, 14th–15th century, text no. 17 ff. 81^{v}–100^{v}]: *Catalogue général des manuscrits des bibliothèques publiques des départements,* Vol. 1, Paris (1849) 394–9, 739–811, 788–92. For a full discussion of this and other technical documentary sources, see Campbell *et al.* 1997: 14–16.
2. Also dealing with polychrome sculpture is the rare 14th-century Icelandic manuscript *Likneskjusmith* 3025, for which see Clarke 2001: 20.
3. I am grateful to J. Kirby for discussing the early technical documentary sources with me. Clarke (2001) lists a number of other medieval manuscripts in English.

describes southern European wall painting technology, it is also frequently cited in this work since, in many instances, it is one of the only sources to describe a particular material or wall painting technique in any detail. Where a more appropriate source for northern painting technique is available, it is cited in preference to Cennini.

1. *The* Mappae Clavicula

Mappae Clavicula quite literally means 'a little key to the chart'. This literal translation has been elaborated by Smith and Hawthorne in order to reflect more closely the contents of the manuscript which is now often referred to as 'A little key to the world of medieval techniques' (*Mappae Clavicula* 1974: 14). The *Mappae Clavicula* may be considered as a compilation of compilations since it includes transcriptions of recipes on various topics including pigments, painting methods, metals and miscellaneous chemical technology.

The edition of the *Mappae Clavicula* referred to in this text is that of Smith and Hawthorne.[4] This is a translation and collation of two of the most complete early copies of the *Mappae Clavicula*: the Sélestat manuscript and Phillipps-Corning MS 3715. The Sélestat manuscript, which dates to the 9th century, is thought to originate from the scriptorium of St. Amand, near Tours. It is designated as MS 17 in the Library of Sélestat (France) (*Mappae Clavicula* 1974: 4–5). Phillipps-Corning MS 3715 (in the Corning Museum of Glass, New York) is the most comprehensive copy of the *Mappae Clavicula* and is dated to the 12th century (*Mappae Clavicula* 1974: 5–7).

The earliest dated evidence (821–2) to a manuscript entitled *Mappae Clavicula* is in the catalogue of the library of the Benedictine monastery of Reichenau (*Mappae Clavicula* 1974: 4). Although this manuscript no longer survives, other early fragments are extant: the 9th-century Sélestat manuscript and a slightly more complete 10th-century version. These earliest fragments of the *Mappae Clavicula* are contemporary with the Lucca manuscript. The late 8th- or early 9th-century manuscript at Lucca is a Latin manuscript associated with the *Liber Pontificalis* in Codex Lucensis 490. Also known as *Compositiones Variae*, the Lucca manuscript is a compilation that originates from south of the Alps and includes recipes for pigments and dyes (*Mappae Clavicula* 1974: 3–4). As a result of palaeographic and stylistic study of the Phillipps-Corning manuscript, Smith and Hawthorne conclude that, by contrast, the *Mappae Clavicula* originated from north of the Alps, perhaps in northern France or the bordering areas of Germany. The *Mappae Clavicula* is a compilation which includes everything in the Lucca manuscript but with a considerable amount of additional information, some of which they suggest is from an English source which was later subject to German influence (*Mappae Clavicula* 1974: 7).

2. de Coloribus et Artibus Romanorum

The *de Coloribus et Artibus Romanorum* is attributed to the Italian writer Eraclius, who, in Merrifield's opinion, originated from the Lombard dukedom of Benevento (Merrifield 1967: 1:171). Eraclius' original work consists of two books written in verse. Numerous additions to these have been made in prose at various times, and these constitute the third 'book'. The compilation includes instructions for gilding, glass manufacture, engraving glass and precious stones, pigment preparation – including lake pigments and salt green –

4. See *Mappae Clavicula* 1974 for the Smith and Hawthorne edition.

and the preparation of wooden and stone surfaces for painting. Although the dating of this work still remains uncertain, the first two books have been attributed to the 10th century. Dodwell, however, considers that this dating is incorrect and suggests a 12th-century date for the original work.[5]

The two oldest known copies of the manuscript are Egerton MS 840, A, dated to the second half of the 13th century, and that transcribed in Jehan Le Bègue's 15th-century technological miscellany. The edition used here is that by Merrifield.[6]

3. De diversis artibus

By contrast with the *Mappae Clavicula, De diversis artibus* by Theophilus is not a compilation but a treatise written by a practitioner from first-hand knowledge.

Unfortunately the original manuscript does not survive, but the earliest copy – written in a German hand – is dated to the mid-12th century. An attribution written in the 17th century on an added title page to the 12th-century manuscript now in the Bibliothek National, Vienna (MS 2527) was the basis on which Ilg (1874) suggested that Theophilus was the metalworker Roger of Helmarshausen, active around 1100. This suggestion was challenged by Degering (1928) who, for a number of reasons, considered Theophilus to be a Benedictine monk working in Cologne in the mid-10th century. Dodwell, however, concludes that the evidence points to the manuscript being written by the Benedictine monk and metalworker Roger of Helmarshausen between 1110 and 1140.[7] The treatise consists of three books covering painting, glass and metalwork. The Hawthorne and Smith edition, *On Divers Arts*, was used for this work since one of the editors is a metallurgist and therefore able to pay particular attention to the detailed technology described by Theophilus.[8]

4. The manuscript of Peter of St. Omer (Liber Magistri Petri de Sancto Audemaro de coloribus faciendis)

It is now generally thought that this manuscript was compiled by Master Peter – a painter and cleric – at St. Omer in northern France in *c*1300.[9] The compilation covers painting on panel, wall and wood and gives the media to be used in conjunction with some pigments on walls; for example, azurite with water and egg, and minium with gum. Some recipes included appear to have been copied from English sources – for example from Adelard of Bath (*c*1120) (van Acker 1972: 155) – while others are found in the *Mappae Clavicula*, and in the contemporary Sloane MS 1754.[10] The only known copy of the work

5. For Dodwell's discussion of the dating, see Theophilus 1986: xiv, and Dodwell 1993: 375. Merrifield (1967: 176, 180) dates the third book to the 12th or 13th century.
6. Eraclius 1967. Garzyo Romano's 1996 Italian edition of Eraclius's compilation is also referred to as appropriate, see Eraclius 1996.
7. Dodwell challenged Degering's dating on a number of grounds: first, the earliest manuscript of *De diversis artibus* is dated to the middle of the 12th century; secondly, the supposedly incorrect spelling, *Rogkerus*, of Roger's name on the 17th-century frontispiece is not crucial since spelling of proper names varied considerably at this date; thirdly, Degering's belief that many of the words employed in the manuscript were not in use after the 11th century is incorrect; and finally Degering suggested that the author was not a practitioner of metalwork, which seems very unlikely from a reading of book three (see Theophilus 1986: xxxiii–xliv).
8. See Theophilus 1979 for the Hawthorne and Smith edition.
9. For the dating of this manuscript, see Morgan 1982:38, n.72.
10. Merrifield 1967, Vol. I: 112. For the contemporary Sloane MS 1754, see Thompson 1926. Van Acker (1972: 151–4) concludes that the St. Omer manuscript and Sloane MS 1754 are independently derived from the same sort of sources and are not by the same author or workshop.

is found in Jehan Le Bègue's 15th-century technological miscellany. Evidence to support the northern French origin of the text includes the use of the Latin name for Rouen in the recipe for Rothomagesian green (Merrifield 1967: 1: 112).

Alternatively, it has been also suggested that the author of this manuscript was the famous writer, painter and canon Peter Pictor – a contemporary of Adelard of Bath and also probably of Theophilus – who was born and lived in St. Omer in *c*1100, approximately two hundred years before Peter of St. Omer.[11] However, van Acker concludes that the author of this manuscript could not have been Peter Pictor.[12] First, much of the Latin is rather vulgar and contains several errors, and so is unlikely to have been written by Peter Pictor who styled himself as a poet and master of grammar at a time when correct grammar and use of language was highly valued (van Acker 1972). Second, the best of the Latin is in sections which appear to have been copied directly from Theophilus' treatise. For example, chapters 31 and 32 from Theophilus' first book appear to have been directly copied in the St. Omer manuscript (see Merrifield 1967: 1: 156).

The edition used here is that by Merrifield (1967: 1: 116–65), with note taken of Thompson's comments on Merrifield's translation.[13]

5. Il Libro dell'arte

Il Libro dell'arte, probably written in Padua in the 1390s by Cennino Cennini – a Tuscan painter and pupil of Agnolo Gaddi – describes 14th-century Italian painting practice (Bomford *et al.* 1990: 2). The treatise covers all aspects of painting technology from the preparatory stages – including the manufacture of pigments, and grinding them with the appropriate binding medium – through to the final paint and varnish layers. The treatise, which reads rather like a recipe book, describes the master–pupil tradition in which Cennini both trained and practised.

The 1960 reprint of Thompson's 1933 edition is used here (Cennini 1960). Some earlier editions date the treatise to 1437 on the basis of an inscription added by a transcriber at the end of the text. Other English editions – by Merrifield and Herringham – use this as the date of the work, although Thompson determined that Cennini describes *trecento* painting technique (Cennini 1960).

6. *The Strasbourg manuscript*

This German text, known only from a transcript, probably dated from the 15th century.[14] The original manuscript perished in the fire of 1870 in the Strasbourg City Library, although an incomplete translation of the transcript was incorporated by Eastlake in his *Materials for a History of Oil Painting* (1847). The only complete English translation, used here, is that of Viola and Rosamund Borradaile.[15] The handbook provides valuable

11. See discussion of de Smet's hypothesis in van Acker 1972. Mora *et al.* (1984: 119) also incorrectly attribute this compilation to the 12th century, and cite it as evidence for the use of oil and tempera in the 12th century.
12. D. V. Thompson also argues for a dating later than the 12th century, but confirms only the manuscript's *terminus ante quem* (1431), the year of le Bègue's transcription; Thompson 1935d.
13. Regarding the dating of the manuscript, Thompson simply provides a *terminus ante quem* of 1431, the date of Le Bègue's transcription; Thompson 1935d.
14. Ernst Berger dates the manuscript to the 15th century, while Eastlake proposes a date of no later than the first quarter of the 14th century. The Borradailes also suggest a 15th-century date; see Strasbourg MS 1966: 5 and 9.
15. See Strasbourg MS 1966 for the Borradaile edition.

information on late medieval northern painting technology, covering painters' methods and materials from the preparation of oils and varnishes and the grinding of pigments to application methodologies. Although the author of the manuscript is not known, it seems likely that it was written by a practising artist, perhaps to aid in the instruction of his pupils (Strasbourg MS 1966: 13–15).

Accounts

The medieval accounts reviewed for this work comprise both ecclesiastic and royal records. These are particularly detailed for England in the 13th and 14th centuries and include the fabric rolls for Exeter Cathedral (see Erskine 1981, 1983), Norwich Cathedral Priory (Saunders 1930), Ely Cathedral (Chapman 1907), and the accounts for Westminster Palace.[16] Accounts often include detailed information about materials purchased; in some cases they can be linked directly to a particular scheme of wall painting, such as the 14th-century decoration of St. Stephen's Chapel, Westminster Palace (Haslam 1807), and the scheme of *c*1365–66 in the Rose Tower at Windsor Castle (Brindle and Kerr 1997). From 1200 onwards the scale and quality of documentary evidence in England – particularly for the royal works – is very rich with many aspects rigorously recorded. Fewer records survive for France, although the early 14th-century accounts of the Countess of Artois provide valuable information on materials purchased for painting in north-east France (Richard 1887). Records such as painters' contracts and accounts submitted to the Treasury provide details regarding Scottish painters' wages and the cost of materials from the early 14th century onwards.[17]

The accounts of Norwich Cathedral Priory are particularly explicit. For example, the Sacrist's rolls provide information on painting within the cathedral itself, and include details regarding the quantities of pigments purchased, the painters' wages, and sums for their board and lodging. The Obedientiary rolls provide references to painting in the cathedral priory and describe the purchase of pigments, pens, oil, jars for pigments, metal leaves such as gold and silver, and in some instances specify the particular paintings for which these are to be used. For example, painting materials including orpiment and lead white, gold leaf and oil were purchased for the Shrine of St. William in 1304–05.[18] The painters themselves are frequently named and some, such as Edmund de Hemenhall, probably lived locally. In some instances, the accounts also provide details of where the pigments were purchased. Thus in 1329–30, various colours were purchased from near London and others from near Norwich.[19]

The fabric accounts for Exeter Cathedral are similarly detailed and provide important information relating to the building for the years 1279–1514, including details of the cost of the painters' materials and their wages. Erskine's translation of the rolls for the period 1279–1353 is an indispensable aid to the study of these documents (Erskine 1981, 1983). Much of the information relating to painting refers to sculptural polychromy, and a detailed analysis of how the documentary evidence supports analytical results for the examination of specific items such as the roof bosses is provided by Hulbert (1991, 1998).

16. See, for example, Colvin 1963 and 1971.
17. The sources include the Treasury Accounts, the Works Accounts and the Edinburgh Burgh Accounts; see Apted and Hannabuss 1978.
18. NRO DCN 1/4/6; Park and Howard 1996: 401.
19. NRO DNC 1/1/29; Park and Howard 1996: 403.

The royal accounts are particularly detailed for the reigns of Henry III and Edward III.[20] A great majority of the royal castles and palaces underwent structural change under Henry III, but his greatest achievement was perhaps the new Westminster Abbey built to house the shrine of Edward the Confessor. Among the many paintings recorded in his reign are those in Westminster Palace, where accounts record that 117s 10d was paid to Odo the Goldsmith for colours, oil and varnish for painting the queen's chamber (Tristram 1950: 574), and in the Painted Chamber – the king's bedchamber – which was extensively redecorated following the fire of 1263. The accounts record that more than £53 was spent on the wall paintings around the king's bed in the Painted Chamber (Binski 1986: 17–18). The decoration of St. Stephen's Chapel, Westminster Palace, was undertaken under Edward III and the accounts list the painters and the materials provided for these important paintings, which still partially survive. The price of materials such as brushes, shears to cut tin leaf, pointed brushes and the pigments purchased are recorded. Thus, in 1351–4, 1lb of *cynople* was priced at 30s, 2lb *cynopre of Montpellier* (de Monte Pessulano) was 16s, and 2lb of vermilion cost 3s 4d.[21] Many thousands of gold leaves were also purchased for the decoration, an entire year's supply of 2,300 leaves costing £4 12s (Smith 1807: 221). In 1351, Master Hugh of St. Alban was working on the paintings at the daily rate of one shilling, while painters including Richard Norwich and John Oxford were paid nine pence per day. Others priming walls, making prints, painting on columns or gilding were paid between four and sixpence a day (Tristram 1955: 282–7; van Geersdaele and Goldsworthy 1978).

Secondary literature: antiquarian literature and more recent published sources

Antiquarian literature

The antiquarian literature used for this study comprises 18th- and 19th-century studies of wall paintings. These sometimes include detailed observations regarding the appearance and extent of wall paintings, often immediately after their uncovering from beneath layers of limewash or panelling.[22] This may be particularly important in relation to pigment alterations, since the present appearance may be correlated with these early records. For example, Canon Scott Robertson's detailed description (1880) of the important 12th-century scheme in the apse of St. Gabriel's Chapel, Canterbury Cathedral suggests that the red lead had not significantly altered at that time.[23] Similarly

20. The painting materials recorded in the royal accounts are discussed by several authors. For example, Tristram (1950 and 1955) transcribes items from the accounts where they relate to wall painting, and Colvin (1963 and 1971) and Salzman (1952) provide excerpts arranged both chronologically and thematically. Extracts from the accounts for the materials purchased for St. Stephen's Chapel are also published by Smith (1807).
21. In the English medieval accounts, *cynopre* and *cynople* are used to describe a lake pigment; Erskine 1981: 144 and Salzman 1952: 168.
22. Such accounts may be the only record we now have of certain schemes, or may provide information about details which have subsequently been lost.
23. The first illustration of the paintings is an engraving by James Cole, published by Dart in 1726, while Keyser described the appalling condition of the paintings in 1878.

detailed records survive for polychrome sculpture, for example Stothard's 19th-century engravings which provided essential information regarding the alteration of verdigris on the de Ingham monument.[24]

More recent published sources

Tristram's seminal volumes on English wall painting of the 12th, 13th and 14th centuries each include a detailed overview of English medieval wall painting technology based on primary sources and on Tristram's own observations. No other overview of the techniques of English medieval wall painting exists. However, more recent studies based on the scientific examination of individual sites have been published, such as that for the Romanesque scheme in Kempley (Gloucestershire) (Rickerby 1990a), the 13th-century painting in the south transept of Winchester Cathedral (Curteis *et al.* 1994); the 14th-century paintings from St. Stephen's Chapel, Westminster Palace (van Geersdaele and Goldsworthy 1978), and the late 15th-century wall paintings in Eton College Chapel (Plummer 1986).

For the remainder of northern Europe, the chapter on western medieval wall painting technology in Mora *et al.* (1984) provides a useful outline of the original texts, and a brief discussion of techniques from preparatory drawing and setting out to the final paint layers. Demus' (1970) volume on Romanesque wall painting provides some useful technical details for the various schemes discussed. Aspects of wall painting technology in Germany are also covered in detail by Knoepfli *et al.* (1990), McGrath (1987) provides a study of the materials and techniques of six Irish medieval wall paintings,[25] and Nord and Tronner (2000) report on the analysis of seven medieval schemes of wall painting in Sweden. Site-specific studies are more common, and include those for the Carolingian and Romanesque schemes at Müstair (Switzerland) (Mairinger and Schreiner 1986; Emmenegger 1986), the paintings of *c*1130 at Idensen (Germany) (Matteini and Moles 1994), the Romanesque wall paintings from the chapter house of Santa Maria de Sigena (Spain) (Gasol Fargas 1999), the 13th-century scheme in the choir of Angers Cathedral (Demailly *et al.* 1998), and the paintings of *c*1332–49 on the inner face of the choir screens of Cologne Cathedral (Maul 1992).

Other invaluable references include the series *Artists' Pigments* (National Gallery of Art, Washington), which provides excellent studies of a number of important pigments employed by medieval artists and examples of works in which they have been detected.[26] The *Reclams Handbuch der künstlerischen Techniken* includes information regarding pigment identifications from wall paintings in Germany (Knoepfli and Emmenegger 1990), while studies of particular groups of pigments or alteration phenomena have also been published, for example Richter's (1988) study of salt greens in the Middle Ages, Rickerby's (1991) investigation of heat alterations to pigments painted in *fresco* technique, and studies of the alteration and re-conversion of lead white (Matteini and Moles 1981; Petushkova and Lyalikova 1986; Giovannoni *et al.* 1990; Koller *et al.* 1990).

24. See, for example, Stothard 1876.
25. A narrow palette of earth pigments, charcoal black and lime white was the sum of pigments detected at five of the sites, while a broad palette – discussed in detail in the present work in the context of the author's own findings – was found in the scheme of *c*1170 at Cashel. No analysis of original binding media was undertaken for any of the schemes.
26. For example, Fitzhugh 1986, Gettens *et al.* 1993a and Kühn 1993b.

In the context of polychrome sculpture, Brodrick's (1993) account of the painting techniques of early medieval sculpture provides a brief overview of the few northern European examples studied to date, and includes detailed analysis of the polychromy of the late Romanesque figures excavated from St. Mary's Abbey, York (*c*1180). Sinclair's comprehensive studies of the external polychromy of Exeter and Salisbury cathedrals are invaluable in providing an account of some of the most important *in situ* external medieval polychrome sculpture in England (Sinclair 1991, 1995). The traces of polychromy remaining on the Romanesque frieze on the west front of Lincoln have also been analysed (Larson 1992, 1993), as have samples from the polychromy of Wells west front (Sampson 1998), and the roof bosses of Exeter Cathedral (Hulbert 1991). Individual monuments have also been subjected to scientific examination, such as the 15th-century Fitzalan effigies at Arundel (Brodrick and Darrah 1986), and fragments of figure sculpture of *c*1475–90 from the Great Screen of Winchester Cathedral (Brodrick and Darrah 1989). Other recent studies of polychrome sculpture elsewhere in northern Europe include the south portal (*c*1216–20) of Lausanne Cathedral (Furlano *et al.* 1981), the 'Portail de la Mère Dieu' (*c*1220–60) of Amiens Cathedral (Weeks 1998), and the 13th-century tomb of Pfalzgraf Heinrich II, Maria Laach (Germany) (Oellermann 1993).

Similarly, studies of important early English panel paintings have been undertaken including the ceiling paterae (*c*1263–66) from the Painted Chamber of Westminster Palace (Liversidge *et al.* 1995; Lynn 1997), the Westminster Retable (*c*1270) (Binski and Freestone 1995), the Thornham Parva Retable (*c*1335) (Goldsmith and Bucklow 1998; Bucklow forthcoming), and the early 14th-century screen from Kingston Lacy (Dorset) (Tracy 1997; Plender and Marchant 2000). Other recent studies of early panel paintings elsewhere in northern Europe include the late 14th-century Wilton Diptych (English or French) (Roy 1997), the late 14th-century Estouteville Triptych (Germany) (Villers *et al.* 1997), and Plahter's studies of Norwegian altar frontals (Plahter 1995, 2002; Plahter and Plahter 1972).

The author's work and unpublished analytical sources

A wealth of unpublished analytical information is available in the form of recent reports, many produced by or archived in the Conservation of Wall Painting Department at the Courtauld Institute. These include important studies undertaken by the department on individual schemes, such as the wall paintings of *c*1100 at Hardham (West Sussex),[27] and studies of particular aspects of wall painting technology such as Welford's (1991) investigation of the phenomenon of dark flesh tones in English medieval wall painting, and Hluvko's (1991) study of red pigments. Extensive use has also been made of many conservation and analytical reports undertaken by individuals from outside the department, some of which have been amassed as part of Park's survey of English medieval wall painting. Examples of these include: Sinclair's unpublished technical examination of the polychromy of the Lady Chapel of Glastonbury Abbey;[28] University College London Painting Analysis report on paint samples from stone fragments from

27. Analysis by Howard in Sawdy *et al.* 1998.
28. Analysis by Sinclair in Sampson 1996.

Tewkesbury Abbey (UCL 1996); Allardyce's report on the wall painting in Durham Deanery (Allardyce *et al.* 1983); Darrah's (1992) analysis of the *Wheel of Fortune*, Rochester Cathedral; Curteis and Paine's (1995) report on the polychromy in the vaults of the Lady Chapel, Winchester Cathedral; and Grout's (1999) excellent study of the darkening of vermilion. Unpublished studies of wall paintings elsewhere in northern Europe are also included, such as Cadet's (1995) research on the Romanesque wall painting of Saint-Etienne, Auxerre (France), Ellwanger-Eckel's (1979) investigation of copper pigments in wall painting, and Kupfer's (1982) study of the Romanesque wall paintings at Nohant-Vicq (France). The analytical results from both published and unpublished sources have been tabulated (see Appendix 3).

Methods of scientific examination employed

Following a review of the primary and secondary sources, sites for examination were selected to provide a broad chronological, geographic and contextual spread, including schemes in cathedrals, parish churches and domestic contexts. Some sites were selected due to imminent conservation programmes which provided convenient access by scaffolding to otherwise inaccessible paintings.

The first stage in each investigation involved a thorough on-site inspection guided by an examination protocol (see Appendix 2 for an example of the proformas employed), and including graphic and photographic documentation. Visual inspection in normal, raking and ultraviolet (UV) light – and, as appropriate, at magnification with the aid of a video microscope and with infrared reflectography – provided information about preparatory techniques, present condition and previous interventions. This examination formed an essential preliminary to developing a sampling strategy, which aimed to maximise the information obtained from the painting with the minimum of invasive sampling. Following the detailed examination of the plastering and paint layers, a small number of minute samples (approximately $1 mm^3$) was taken to elucidate specific aspects of the original technique and of any subsequent redecoration: the stratigraphy of the paint layers and ground/s; and the type, application and present condition of the polychromy including the alteration of the pigments.

Paint samples were initially examined at 10–45× magnification with a binocular microscope, and their stratigraphy documented. A portion of the paint sample was then mounted as a polished cross-section (in polyester embedding resin), and as dispersions (in Meltmount, which has a refractive index of 1.662), and these were examined with a Leica DMRX optical microscope at 90–2500× magnification in incident, transmitted and UV light. The microscope was equipped with Leica 10× and 20× PL Fluotar objectives, a 50× PL Apo objective and a 1250× oil immersion objective. For UV examination, excitation filter BP 340–380 nm, dichroic mirror (RKP 400 nm) and suppression filter LP 425 nm was used. Photomicrographs were taken with a Leica DMRD camera at between 90× and 2500× magnification. Microchemical tests were undertaken to identify some metallic ions and functional groups, and histochemical tests were carried out on thin cross-sections to indicate the presence of oils and proteins.[29] To confirm

29. Many of the microchemical tests employed in the examination of paint samples are published by Plesters 1956. Additional microchemical tests included: 1,2 dihydroxycyclobutene-3,4-dione, which

identifications made with polarised light microscopy (PLM), additional instrumental analysis – including scanning electron microscopy (SEM/EDS) and Fourier transform infrared microspectroscopy (FTIR) was undertaken by the author, while gas chromatography-mass spectrometry (GC-MS), X-ray diffraction (XRD), fluorescence spectroscopy, direct temperature-resolved mass spectrometry (DTMS), high performance liquid chromatography (HPLC), lead isotope analysis, and Raman spectroscopy were kindly carried out by other researchers as described below.

A Leica Stereoscan 440i scanning electron microscope (SEM) with energy-dispersive X-ray (EDX) microanalysis, ISIS X-ray analysis programs and thin window germanium detector – providing analysis of elements down to oxygen – was used for electron micrographs, local microanalysis and to obtain elemental concentration maps. Samples for electron microprobe analysis were vacuum-coated with carbon to eliminate surface charging, while those for electron micrography were sputter-coated with gold.[30]

Identification of organic materials and additional confirmation of inorganic components were undertaken using FTIR. The analysis was performed on a Nicolet 510P FTIR spectrometer with a NicPlan IR microscope with a 32× Spectra-Tech Reflechromat Cassegrain objective and a tube factor of 10×. Samples of coatings, specific paint layers or components of particular paint layers to be analysed were manually separated and placed on sodium chloride or barium fluoride discs and reduced to a thin layer using a tiny stainless steel roller. The area of interest was delimited with an illumination masking device based on adjustable knife-edge apertures.[31] Each IR spectrum was collected in transmission mode at a resolution of $4cm^{-1}$ and the interferogram recorded as a sum of 120 scans using PCIR software. The recorded spectra were compared with standards prepared by the author and searched against the Getty Conservation Institute Standard Spectra Library.

GC-MS analysis of samples from some sites was kindly undertaken by Raymond White of the National Gallery,[32] and Dr M. P. Colombini of Pisa University. The method of analysis employed by Dr Colombini involved microwave-assisted acid hydrolysis of the sample, derivatisation of the freed amino acids with a silylating agent, and their quantitative determination by GC-MS.[33]

Analysis of red lake pigments was undertaken by J. Kirby and Dr Arie Wallert. Samples from the late 14th-century scheme in the chapter house of Westminster Abbey were analysed by J. Kirby using HPLC.[34] HPLC was undertaken using a Zorbax ODS (C18)

gives a yellow colour in the presence of copper; and lead nitrate and sodium rhodizonate, which produce a purple colour in the presence of sulphate ions (Matteini and Moles 1986). Preliminary staining for oil was undertaken using Sudan Black B which produces a blue colour when indicating a positive result. In UV light, proteins give a yellow fluorescence when stained with fluorescein isothiocyanate (FITC). Acid fuchsin was also used for preliminary identification of protein. For the analysis of paint media using histochemical tests, see Gay 1970 and Martin 1977.

30. For the principles and practice of scanning electron microscopy, see Goldstein *et al.* 1992. In particular, see pages 353–4 for the pathological overlaps encountered in energy-dispersive X-ray microanalysis. Especially relevant for this work are the interferences of: the lead Ma X-ray line with sulphur Ka and molybdenum La; barium La with titanium Ka; iron Kb and cobalt Ka; and manganese Kb and iron Ka.
31. See Pilc and White 1995 for a discussion of the application of FTIR microscopy in the analysis of paint binders.
32. See White and Pilc 1996 for the improved derivatisation developed for paint media at the National Gallery.
33. For the methodology of examination employed by Dr Colombini, see Colombini *et al.* 1998.
34. See Kirby and White 1996 for a discussion of the methodology used by J. Kirby at the National Gallery.

silyated silica reversed phase column and a HP1100 Diode array detector.[35] A sample from the Ante-Reliquary Chapel, Norwich Cathedral was analysed by Dr Wallert using fluorescence spectroscopy and confirmed the presence of madder lake.[36] DTMS was employed by Dr Wallert for the analysis of a sample from the early 14th-century paintings in the feretory of St. Albans Cathedral. DTMS analysis was performed on a JEOL SX-102 double focusing mass spectrometer, and data were acquired using a JEOL MP-7000 data system.

XRD analysis of lead-tin yellow samples was undertaken by Dr A. Roy of the National Gallery using a Debye-Scherrer camera with copper radiation (40 KeV and 35 milli-amps for 3½ hours), while XRD to confirm the presence of vivianite in a number of samples was carried out by Dr G. Cressy of the Mineralogy Department of the Natural History Museum, London, also using a Debye-Scherrer camera. The diffractograms were recorded on film, and measured against standards held by these institutions, and against standard Joint Committee for Power Diffraction Standards (JCPDS) files.

Lead isotope analysis of samples from the chapter house of Westminster Abbey was undertaken by Dr S. Stos-Gale of the Isotrace Laboratory, Oxford (see Stos-Gale 1999; Rohl 1996). Raman spectroscopy was employed to confirm the presence of cerussite in a small number of samples. This analysis was undertaken by Dr R. Withnall at University College, London, using an Olympus BH-2 microscope coupled to a Dilor-XY triple-grating spectrometer with an intensified photodiode array detector (see Best *et al.* 1995: 31–2).

Relevant results are presented in tables (Appendix 3) which indicate the methods of analysis employed in each instance and where, if at all, the results have been published.

35. For the sample preparation, eluents, method program and conditions, see Kirby 1999.
36. The principles of spectrofluometry and its application in the examination of paint samples are given by Wallert 1986b.

❦ THREE

Blue pigments

Ultramarine

Ultramarine was the characteristic brilliant blue mineral pigment almost universally employed in high-quality Romanesque wall painting in England. For centuries regarded as the most highly prized pigment, ultramarine, or lazurite – a complex sulphur-containing sodium aluminium silicate $[(Na,Ca)_8 (AlSiO_4)_6 (SO_4,S,Cl)_2]$ – is extracted from lapis lazuli. The derivation of its name from the Latin *lapis*, meaning stone, and *lazulus*, a Latinised form of the Persian for blue, *lajoard*, supports evidence that the source of lapis lazuli was the mines at Badākshān, now in north-east Afghanistan (Plesters 1993: 37).[1]

Cost

Ultramarine was as costly as gold, and in 13th- and 14th-century Italian panel painting was often supplied directly by the patron or made the subject of a separate payment (Bomford *et al.* 1990: 35, 202). High-quality ultramarine priced at four florins per ounce was specified for use by Filippino Lippi in the Strozzi Chapel, S. Maria Novella (Borsook 1980: 125). The cost of the ultramarine employed in English medieval wall painting is difficult to establish since no financial accounts relating to painting materials survive from the Romanesque period when it was most frequently used. Furthermore, the terminology for blue pigments in the Gothic period is often unclear, making it difficult to determine which blue pigment was actually purchased.[2] For example, although analysis has confirmed that ultramarine was employed in the 14th-century wall paintings from St. Stephen's Chapel, Westminster Palace, the relevant building accounts itemise only *azure* (Colvin 1963: 519).

1. The stone was imported into Egypt from pre-dynastic times and was known in antiquity, but was not then used as a pigment since Egyptian blue was readily available (Laurie 1913: 318).
2. In the Middle Ages the Latin word for blue was *lazurium* (which originally stemmed from the Persian *lajoard* that was also used for lapis lazuli); this became *azurium*, and finally *azure*. Thus, the general term for blue – *azure* – had its etymological derivation in the Persian for blue – referring to lapis lazuli – and yet also came to mean azurite.

The colour blue was indicative of purity, and therefore ultramarine was frequently specified for painting the Virgin's robe or other iconographically significant zones of painting. Lapis lazuli was regarded as the sapphire of the ancients and was said to symbolise heaven. The most highly prized examples of the mineral were those scattered with golden spots resulting from pyrite impurities, representing the blue of heaven and the host of stars (Kunz 1915: 282).

Cennini describes ultramarine as 'a colour illustrious, beautiful and most perfect, beyond all other colours', and advises that, on both wall and panel, metallic gold should be juxtaposed with the expensive mineral to attain both apparent and intrinsic richness (Cennini 1960: 36).

Optical, chemical and physical properties

Particles of extracted ultramarine are of irregular size and angular shape. In plane polarised transmitted light they are a translucent pale blue with a refractive index of 1.500. Particles of ultramarine are isotropic, but pigment samples often contain calcite as an impurity and these inclusions may display birefringence under crossed polars.

Its relatively low refractive index makes ultramarine a translucent glazing pigment when combined with oil, although the pigment retains its brilliant blue colour in aqueous media such as egg, glue or gum. The stability of the pigment in alkaline conditions also makes it exceptionally valuable in *buon fresco* painting. However, ultramarine is readily decomposed by acids, including weak organic acids such as acetic acid if contact is prolonged. 'Ultramarine sickness' – a greyish or yellowish mottled discoloration of the paint surface – may be attributable to acids from media or the environment, and is most often noted in samples of the synthetic pigment.[3]

Ultramarine is stable with regard to light and heat, and the latter property was sometimes used to test its purity in the medieval period. Since adulteration could occur during trade – and to test that natural azurite had not been substituted for the pigment – it was heated to redness. Under these conditions ultramarine remains stable, whereas azurite blackens.

Preparation

The separation of ultramarine from its mineral impurities requires a lengthy extraction process. One of the earliest surviving recipes for the purification of the pigment is in a 13th-century manuscript now at Gonville and Caius College, Cambridge (MS 181, *Azurium transmarinum hoc modo per pastillum afinatur*). There are two other 13th-century texts which describe making ultramarine from lapis lazuli *per pastillum*: the *Liber claritatis totius alkimicae* attributed to Geber (Bologna Cod. Lat. 164 (153)); and British Library, Ms. Sloane 342 (Thompson 1935a). By the 14th century, texts for its purification were more common, and the fullest account is provided by Cennini (1960: 37–8). He describes the process for grinding the mineral with wax, mastic and pine turpentine, washing it in lye and subsequently grading the purified extract. Generally, the highly prized largest particles of intense blue colour were extracted first. Unlike azurite, the colour of ultramarine is preserved in fine particle-size samples because the pigment has

3. For a comprehensive discussion of ultramarine and its synthetic equivalent, see Plesters 1993: 37–65.

such a high tinting strength: the colour intensity is dictated by the quantity of the polysulphide radical which produces the blue appearance. The blue-grey end product of the extraction, known as ultramarine ash, was however also valued in its own right as a pale blue-grey pigment (Plesters 1993: 39).[4]

Analytical evidence from this study confirms that ultramarine derived from lapis lazuli was used extensively in England long before any instructions for its extraction are found in European texts. This suggests that it was imported in the form of a prepared pigment, and the term *ultramarinus* – meaning oversea or from overseas – together with evidence from trade, supports this theory (Thompson 1956: 145). Ultramarine was probably imported via Venice and, as late as the 15th century, it was carried by a Venetian galley that docked in London (Miskimin 1975: 109–14).

Despite the most careful extraction, some mineral impurities – principally calcite – remain in the pigment. The presence of such impurities in panel painting may help to distinguish the natural mineral from the synthetic pigment. In wall painting, however, ultramarine was frequently applied in a lime matrix, and the presence of other mineral impurities may sometimes make it difficult to determine whether extracted ultramarine or a crushed sample of high-quality lapis lazuli was employed.

Use in English medieval wall painting

Analysis has confirmed that ultramarine was employed extensively for wall painting in England throughout the Romanesque period in both cathedrals and more modest parish church schemes. The earliest identified example is on excavated fragments from a scheme of *c*1090 at Lewes Priory (Howard and Park 1997).

Before this study, it seemed likely that its high cost may have made the pigment scarce in the Romanesque period, restricting its use to particularly important details. However, analytical evidence has confirmed that in a number of cathedral schemes, it was employed for painting extensive areas of background, as in the paintings of *c*1130–60 in the chapels of St. Gabriel and St. Anselm in Canterbury Cathedral (Plates 1–3).[5] The profligate use of so expensive a material underscores the importance of such decorative schemes. Extensive use of ultramarine continued into the late 12th century as in the scheme of *c*1170 at Cormac's Chapel, Cashel, Ireland (Plates 8 and 9) (Howard 1991);[6] in the paintings of *c*1175 in the Holy Sepulchre Chapel, Winchester Cathedral (Howard 1995a: 101); in the scheme of *c*1180 in the Galilee Chapel, Durham Cathedral, where it was employed for the backgrounds of the figures of the saints Cuthbert and Oswald (Howard 1988: 17); in the decoration of *c*1187 in the Lady Chapel at Glastonbury Abbey;[7] and in the 'Losinga' cycle of *c*1190 in Norwich Cathedral (Plate 12) (Howard and Gasol 1996).

Although ultramarine was less dominant in Gothic wall painting in England – when alternatives such as azurite and synthetic copper blue pigments take precedence – the

4. Ultramarine ash has been identified on the 13th-century Norwegian altar frontal Tingelstad III (Plahter 1995: 119).
5. The pigment has also been tentatively identified by C. Babington in the paintings of *c*1125–30 at Copford (Essex).
6. D. Park has recently redated this scheme from *c*1134 to *c*1170.
7. For the polychromy of the Lady Chapel at Glastonbury Abbey, see Sampson 1996; I am also grateful to Eddie Sinclair for providing details of her pigment analysis at this site.

continued use of the pigment is confirmed by its identification in the scheme of *c*1220 in the Holy Sepulchre Chapel, Winchester Cathedral (Howard 1995a: 102); the paintings of *c*1250 in the Ante-Reliquary Chapel, Norwich Cathedral (Park and Howard 1996); 13th-century decoration in the Guardian Angels' Chapel, Winchester Cathedral (where both ultramarine and azurite have been identified);[8] excavated 13th-century wall painting fragments from Clarendon Palace, Wiltshire (Ashurst and James 1988); on the painted coffin-lid of Archbishop Walter de Gray at York Minster of *c*1255 (Ramm *et al.* 1971); and in the splendid mid-14th-century wall paintings from St. Stephen's Chapel, Westminster (van Geersdaele and Goldsworthy 1978). Preliminary identification of ultramarine has also been made in the paintings of *c*1350 at St. Nicholas, Harbledown (Kent).[9]

Application

Directly on the plaster or over a lime white preparation

Ultramarine was generally applied directly over a lime white preparation in the paintings of *c*1090 at Lewes Priory. However, examination of some paint samples indicates that a few details in ultramarine were also applied over a subtle pink colour – vermilion combined with lime white – which may represent flesh tones (Howard and Park 1997). The use of linear detailing in natural ultramarine over a pink or yellow underpaint is paralleled in the paintings of *c*1125 at Ickleton (Cambridgeshire), the scheme of *c*1130 in St. Gabriel's Chapel, Canterbury Cathedral, and that of *c*1170 at Cormac's Chapel, Cashel. In all these examples the pigment is used for the representation of hair and beards.

At Ickleton, ultramarine was employed specifically for the hair and beard of Christ in each scene, and it seems likely that such use of the mineral pigment to pick out details of iconographically important figures is due to both symbolic and economic factors (Plate 6). Examination of a cross-section indicates that the pigment was combined with lime and applied in a single layer approximately 150μm thick directly over the underlying coarse plaster. The pigment here is of exceptionally high quality, consisting of extremely small particles, only 5 × 3μm (Plate 7). Since the colour of ultramarine is preserved even in fine particle-size samples such as this, it is clear that even for use in limited areas in parish church schemes extracted ultramarine was generally employed rather than material obtained from crushing the mineral lapis lazuli.

Ultramarine was also applied directly to the plaster substrate in the paintings of *c*1180 in the Galilee Chapel of Durham Cathedral. Here, a sample taken from the blue background of the St. Cuthbert figure indicated that high-quality ultramarine combined in a lime matrix was applied in a single layer 50μm thick directly over the lime plaster substrate (Howard 1988: 17). In the decoration of *c*1187 in the Lady Chapel at Glastonbury Abbey, the pigment was applied over a white ground (Sampson 1996). A white ground or plaster layer was also employed beneath ultramarine in the early Gothic

8. Unpublished analysis by H. Howard.
9. Here a blue sample from below an angel's wing was analysed by PLM and SEM/EDX, the results confirming the presence of ultramarine; unpublished analysis by H. Howard for Tom Organ, May 1992.

scheme in the Holy Sepulchre Chapel at Winchester Cathedral. Here a sample taken from the *Nativity* roundel on the south segment of the eastern vault confirmed the presence of natural ultramarine applied over a lime ground.[10] Ultramarine combined with lime white and applied over a white ground to produce a lighter and more luminous colour was employed in the soffit paintings in the 'Losinga' scheme at Norwich Cathedral (Howard and Gasol 1996: 23).

Examination of a paint sample from a blue flower on the scrollwork surrounding the roundels on the vault of the Guardian Angel's Chapel, Winchester Cathedral, also indicated that ultramarine combined with lime white was applied in a layer 90µm thick over a lime ground. Since the scrollwork was added to the existing scheme of roundels in *c*1280, it is clear that these paint layers date from this medieval redecoration of the vault.[11]

Directly over the lime plaster and used as a glaze

Ultramarine used for the background of the bishop's head of *c*1250 in the south-east corner of the Ante-Reliquary Chapel, Norwich Cathedral, was applied on a thin skim of fine plaster applied directly to the stone support (Plate 18). Here lime white was combined with the blue pigment to form a layer 75µm thick, which appears to have been subsequently glazed with high-quality ultramarine (Howard 1988: 17).

Over coloured underpaint

To enrich the blue effect by the reduction of light scatter, ultramarine was often applied over a grey or black underpaint.[12] Mixtures of charcoal black and white – which would themselves have had a blue tinge thus considerably enhancing the blue colour[13] – were employed for such local grounds in St. Gabriel's Chapel, Canterbury Cathedral; Cormac's Chapel, Cashel;[14] the Romanesque scheme in the Holy Sepulchre Chapel, Winchester Cathedral (Howard 1995a: 101); in the late 12th-century decoration in the Lady Chapel at Glastonbury Abbey (Sampson 1996); and in the 'Losinga' cycle of *c*1190 in Norwich Cathedral (Howard and Gasol 1996).

In order to achieve subtle coloristic effects, the colour of the underpaint varies from area to area in some schemes; for example, a dark red-purple (charcoal black combined with red iron oxide) was also used at Cormac's Chapel and in the Canterbury paintings (Plates 2, 3, 8 and 9). In both cases the intention was presumably to produce a purple tint. In addition, while a grey underpaint was employed for many parts of the splendid 'Losinga' cycle of wall paintings at Norwich Cathedral, a yellow underpaint was employed in other areas (Plates 12 and 13) (Howard and Gasol 1996).

Examination of a paint sample taken from the blue background of one of the angel roundels in the western segment of the vault of the Guardian Angels' Chapel, Winchester Cathedral, showed that, unusually, ultramarine was applied over azurite on a yellow

10. The lime ground was added during the detachment and replacement of sections of the scheme in the 1960s; Howard 1988: pl. 24.
11. Unpublished analysis by H. Howard.
12. This would have been necessary if the pigment was crudely extracted, with the dark underpaint serving to enhance the colour of the impure sample.
13. See the discussion of 'false blue' in chapter nine on black pigments.
14. For St. Gabriel's Chapel, see Cather and Howard 1994. Analysis of a sample taken from the Canterbury paintings by G. Hedley also indicated that in some areas ultramarine was applied directly on the lime white plaster; Hedley 1979. For Cashel, see Howard 1991.

underpaint.[15] Evidence from another sample taken from this area indicated that ultramarine was combined with azurite and applied on a secondary underpaint of azurite over yellow earth. In this instance it is likely that the azurite underpaint was employed for economic reasons, ensuring that only a fine glaze layer of the more expensive ultramarine need be applied. Since the early 13th-century decoration on the vault was redecorated in *c*1280 it is unclear to which phase of decoration these blue paint layers belong. However, as noted above, ultramarine was also identified on a flower from the scrollwork which belongs to the later phase of painting, while ultramarine on a lime ground was applied over the lowest red roundel of the 1230s phase of decoration on the south side of the east segment of the vault, confirming that ultramarine also formed part of the 1280s redecoration in this area.

Mixed with other pigments

To extend its coloristic range further, ultramarine was also combined with other pigments, as in the St. Gabriel's Chapel scheme of *c*1130 where it was combined with yellow iron oxide and lime white and applied over yellow and grey underpaints to produce various green tints. This is particularly evident on the soffits of the entrance arches to the apse, where the pale green appearance in the meander pattern is due to a combination of ultramarine and yellow earth particles in a lime matrix applied over the lime plaster. A similar combination, except in this case applied over a local ground of charcoal black and lime white, was used to achieve the green background of one of the soffit roundels (Cather and Howard 1994: 145).

In the late 12th-century paintings of the Holy Sepulchre Chapel, Winchester Cathedral, ultramarine was mixed with charcoal black and lime white to produce the deep purple shadow of Christ's drapery in the *Harrowing of Hell* (Howard 1995a: 101). It also seems likely that ultramarine and a red lake pigment were combined to produce a purple colour in the late Romanesque 'Losinga' paintings in Norwich Cathedral. Examination of the mineral blue employed here indicated the presence of some unusual columnar-shaped inclusions in the intense blue paint layer. Elemental analysis of these particles by SEM/EDX confirmed the presence of magnesium, potassium, aluminium and silica (suggesting the presence of muscovite or a similar species), and also equant-shaped particles of sodalite – an isomorph of lazurite. Since ground lapis lazuli contains not only the blue-coloured mineral lazurite, but may also contain silicate minerals such as diopside, forsterite, muscovite and wollastonite, it initially seemed possible that in this case ground lapis lazuli rather than extracted ultramarine may have been employed. However, further examination of the blue mineral in UV light indicated that the large (45 × 6μm) columnar-shaped inclusions – among some others in the paint layer – produce a brilliant orange fluorescence (Plate 14).[16] Since ground lapis lazuli – unlike extracted ultramarine – is known to have a spotty orange fluorescence in UV light of long wavelengths (400–315 nm) (Sinkankas 1972: 340), some fluorescence would not be surprising. However, the level of fluorescence in this sample is greater than might be expected, particularly when compared with other samples taken from Romanesque and

15. Unpublished analysis by H. Howard.
16. For examination of paint samples in UV light, the following conditions were employed: excitation filter BP 340–380 nm, dichroic mirror RKP 400 nm and suppression filter LP 425 nm.

'transitional' wall paintings.[17] Some red lake pigments, such as madder lake, produce an orange fluorescence in UV light, and close examination of one of the samples from the 'Losinga' cycle in normal light revealed the slightest trace of a pink hue at the base of the ultramarine layer. Given the intense blue of the mineral pigment employed here, it seems more likely that extracted ultramarine – combined with a lake pigment which has now almost completely faded – was employed here rather than crushed lapis lazuli.

Binding media

Analysis of the original binding media at St. Gabriel's Chapel, Canterbury Cathedral, indicated the presence of a small quantity of protein in both the ultramarine layer and the red and charcoal black underpainting in a sample taken from the blue robe of *Christ in Majesty* (Plate 3), while a small amount of linseed oil was also detected in a sample taken from the blue background (Cather and Howard 1994: 146). Although ultramarine is stable in alkaline conditions and so may be applied directly to the fresh plaster or combined with lime white or lime milk, the incorporation of additional organic binding media in these layers applied over tinted local grounds suggests that the artist did not wish to compromise the intensity of the colour by combining ultramarine with lime white. While the addition of an aqueous binding medium, such as a protein, would have maintained the intense blue colour, the relatively low refractive index of ultramarine makes it rather translucent when combined with oil alone (Plesters 1993: 44). Similarly, recent analysis of two 13th-century panels from the Painted Chamber of Westminster Palace indicates that ultramarine, employed for the glaze of the blue background, may have been ground in egg before being bound in linseed oil. Here, the egg component is also likely to have been included to intensify the pigment's bright blue colour.[18]

Comparison with its use in other media and in continental wall painting

The earliest identified occurrence of ground lapis lazuli used as a pigment is in late 6th-century paintings in the cave temples at Bāmiyān, approximately two hundred miles from the mines at Badākshān. Examination indicated that the pigment here was applied in a glue medium over an underpainting of charcoal black and gypsum, combined in varying proportions to provide a tonal underpainting (Gettens 1938a: 187–9). This relatively sophisticated technique probably enhanced the blue coloration of the specimen, which contained the natural colourless impurities of lapis lazuli. It seems likely that a high-quality sample of the mineral had simply been ground up for use as a pigment in this instance, since grinding and washing poorer quality specimens results in a dull blue-grey colour (Thompson 1956: 146).

The earliest identified example of purified lazurite – extracted from lapis lazuli – in manuscript illumination is in Queen Melisend's Psalter of *c*1131–44 (British Library MS Egerton 1139), produced in the Holy Land in a Crusader context (Laurie 1913: 334–5; Setton 1977: 125). The date of this manuscript corresponds quite closely to the earliest instances of the use of extracted ultramarine in Romanesque wall painting in England.

17. The ultramarine identified at the following sites was examined in UV light for comparison: Ickleton parish church *c*1125; St. Gabriel's Chapel, Canterbury Cathedral *c*1130; Cormac's Chapel, Cashel, *c*1170; the Holy Sepulchre Chapel, Winchester Cathedral, schemes of *c*1175 and *c*1220; and the Galilee Chapel, Durham Cathedral, *c*1180.

18. Analysis by R. White in Lynn 1997: 22.

The frequent use of the pigment in English Romanesque wall painting is paralleled elsewhere in northern Europe; for example, it was employed extensively in the scheme of *c*1130 at Idensen, Lower Saxony (Plates 4 and 5). Indeed the mixture of charcoal black and white employed as an underpaint for ultramarine at Idensen mirrors that used in the approximately contemporary paintings in St. Gabriel's Chapel at Canterbury and in the scheme of *c*1170 in Cormac's Chapel at Cashel.[19] Coloured underpaints have also been detected elsewhere, as in the scheme of *c*1165 at Pürgg (Austria) where ultramarine was applied over a dark red underpaint for a purple effect, and over grey where a dark blue colour was intended (Plates 10 and 11) (Howard 1995d).

The extensive use of ultramarine is also paralleled in Romanesque polychromed sculpture, as in the frieze from the west front of Lincoln Cathedral (*c*1140); on the limestone figure of Moses of *c*1180–85 from St. Mary's Abbey, York, where high-quality ultramarine was applied over a grey underpaint (Plate 20);[20] and on a 12th-century Virgin and Child from Auvergne, where a thin layer of lapis lazuli was applied over a ground of charcoal black combined with lead white for the Virgin's mantle (Kargère 2002: 509). Ultramarine was also extensively employed for the polychromy of the early Gothic south portal of Lausanne Cathedral of *c*1216–20 (Plate 21). Here it was applied, both pure and mixed with lead white, over a lead white ground. In other areas where a darker blue was required, ultramarine was applied over a grey underpaint of carbon black combined with lead white (Furlano *et al.* 1981: 16). Ultramarine has also been identified on the Adisham reredos, recently redated by C. Tracy on the basis of dendrochronological evidence to *c*1180 (Howard 1993c).

Although Cennini (1960: 33) recommends combining ultramarine with orpiment for use in panel painting, examples of the use of ultramarine to produce mixed greens is extremely unusual. The only identified example in English medieval wall painting is in the St. Gabriel's Chapel scheme of *c*1130, where ultramarine was combined with yellow iron oxide and lime white and applied over yellow and grey underpaints to produce various green tints.

Cennini also describes combining lapis lazuli with red lake – both kermes and brazil – to improve the colour of poor-quality ultramarine (Cennini 1960: 38–9). This may have been a long-used artist's or colourman's trick to improve the blue colour of the mineral pigment, and it is interesting to speculate whether the red lake which appears to have been incorporated in samples from the 'Losinga' scheme was employed to improve the colour of ultramarine, which retained considerable quantities of mineral impurity.

Discussion

As has been shown, ultramarine was the mineral blue normally employed in English Romanesque wall painting (Howard 1988) although azurite has been identified at Kempley, (Gloucestershire) (Rickerby 1990a: 255–6), and in a few continental wall paintings of the period.[21] Furthermore, despite the dominance of alternatives to ultrama-

19. For Idensen, see Matteini and Moles 1994; for St. Gabriel's Chapel, see Cather and Howard 1994; and for Cashel, see Howard 1991.
20. For the Lincoln frieze, see analysis by Roy and White in Larson 1992 and 1993: 34; for the York sculptures, see Brodrick 1993.
21. See the section on azurite (pp. 40–50) for details of occurrences of this pigment in both Romanesque and Gothic wall painting.

rine – such as azurite, indigo and synthetic copper blue – from the early 13th century, its continued use in English wall painting is confirmed by its identification in a number of important schemes including the paintings of *c*1220 in the Holy Sepulchre Chapel, Winchester Cathedral; paintings of *c*1250 in the Ante-Reliquary Chapel, Norwich Cathedral; and the mid-14th-century wall paintings from St. Stephen's Chapel, Westminster (Plate 19) (Howard 1995a: 102; Park and Howard 1996; van Geersdaele and Goldsworthy 1978). It is significant that all these examples are in important and costly paintings, suggesting that the price of ultramarine may have risen sharply in the 13th century making other options preferable for more modest schemes.

However, ultramarine was commonly employed in panel painting from the 13th century, notable examples being the ceiling paterae from the Painted Chamber at Westminster Palace of *c*1263, the Westminster Retable of *c*1270 (Plate 22), and the late 14th-century Wilton Diptych in which both natural ultramarine and azurite have been identified.[22] It is clear therefore that ultramarine was still available in England well after its use in wall painting began to decline.

It is significant that the majority of identified examples of the use of ultramarine in wall painting in England fall within the Romanesque period; moreover, with the exception of the azurite at Kempley, it is the only mineral blue identified in English wall painting of this period. Together with the evidence from identified examples on polychromed sculpture and the extent of the areas painted in this expensive mineral blue, this points to a source of highly purified ultramarine readily available in England in the Romanesque period, and not perhaps at quite such a premium as previously supposed.

Vivianite

The blue iron phosphate mineral vivianite ($Fe^{2+}_3 (PO_4)_2 8H_20$) occurs naturally in oxidised portions of some metalliferous ore deposits, as at Wheal Jane near Truro and mines around St. Agnes in Cornwall (Birley and Young 1993), where it generally appears as dark indigo-blue to blackish green crystals (Collins 1892; Rudler 1905; Embrey and Symes 1987). Vivianite is also found in an earthy form, frequently associated with bones, decaying wood and other organic, phosphate-rich environments (Palache *et al.* 1951). As early as the eighteenth century, Delaval (1777) noted the association of iron and peat in blue earthy deposits, which were presumably vivianite.

Cost

No records of the purchase of this material for use as a pigment appear to survive, but it is clear from identified examples that in the medieval period, vivianite was employed in paintings only where the mineral was found locally. Several sources of vivianite are located close to Winchester where the pigment was used for three schemes of wall painting in the medieval period. The source must therefore either have been known to the artists, or sold to them by someone relatively local.

22. For the ceiling panels from the Painted Chamber at Westminster, see Liversidge *et al.* 1995 and Lynn 1997. I am grateful to Dr Ashok Roy for information on the technique of the Westminster Retable; and for the Wilton Diptych, see Roy 1997.

Optical and physical properties

In its blue form the mineral displays a distinctive blue-yellow pleochroism (Faye *et al.* 1968). The refractive indices of the blue form of the mineral have been given as: α 1.5809, β 1.6038 and γ 1.6361; and α 1.5766, β 1.6050 and γ 1.6267 (Watson 1918: 160). The mineral is generally stable, although it may be colourless when initially brought to the surface (Watson 1918).

Preparation

No medieval instructions relating to the preparation of vivianite survive, however, it is likely that, as with other mineral pigments, the purest samples would have been selected and the crystals then ground to form particles of a size suitable for painting. Grinding may not have been necessary for the earthy form of vivianite, but some form of washing or purification may have been undertaken.

Use in English medieval wall painting

The use of vivianite at Winchester Cathedral – in both the Romanesque and early Gothic schemes in the Holy Sepulchre Chapel and also in the Guardian Angels' Chapel – is particularly surprising. Not only has there been no previous confirmation of its use in English medieval wall painting, but at Winchester it was clearly selected for its distinctive coloristic qualities rather than as a cheaper alternative to other mineral blues (Howard 1996a). In each case, initial identification of vivianite by polarised light microscopy – typified by blue-yellow pleochroism (Plates 30 and 31) – was confirmed by XRD.[23]

In the Holy Sepulchre Chapel, the characteristic deep 'indigo' blue of vivianite was used extensively in both schemes: for the blue areas of the panelled backgrounds, and for numerous details such as the central details on the vair lining of Nicodemus' cloak in the Romanesque *Deposition*. By contrast, rather more limited use was made of the pigment in the 13th-century scheme in the Guardian Angels' Chapel, where it has been identified, for example, in a sample from the wavy line outlining the largest roundel on the north side of the east segment of the vault.

In each of these schemes vivianite was employed in addition to natural ultramarine; and as well as azurite in the Guardian Angels' Chapel. In order to increase its coloristic potential, the pigment was applied in a variety of ways.

23. I am grateful to Dr G. Cressy of the Mineralogy Department of the Natural History Museum, London, for confirming the presence of vivianite using a Debye-Scherrer camera. The experimental film pattern matched that of a standard held at the Natural History Museum. Results published by the Joint Committee for Power Diffraction Standards (JCPDS) for synthetic vivianite (file 30-662) give strong d(Å) lines at 6.73, 3.21, 7.93, 4.90, 4.08, 2.99, 2.73 and 2.71 for vivianite. Analysis by Dr Cressey also suggested the presence of metavivianite in some samples, and this may relate to the presence of a yellow alteration product of vivianite. JCPDS file 26-1137 gives strong d(Å) lines at 6.71, 8.59, 4.86, 3.87, 2.77, 2.90, 4.27 and 3.07 for metavivianite.

Application

Over a white ground

In the 13th-century scheme in the Holy Sepulchre Chapel, vivianite was combined with lead white and applied over a lime white ground for details on the eastern of the two arches on the north side, and for details of a prophet's drapery on the soffit of the same arch. It also seems likely that a white ground was employed beneath the vivianite used for the background of Christ as *Pantocrator* on the vault (Plates 35 and 36), although the original plaster and ground were removed during the detachment of this area in the 1960s restoration.

Over coloured underpaint

Elsewhere in the early 13th-century scheme in the Holy Sepulchre Chapel a grey or black underpaint was employed, presumably to produce a darker blue effect by decreasing the light scatter from the lime plaster substrate, and thereby increasing the covering power of the vivianite (Plate 33). This is apparent in a sample taken from the background of the *Deposition*, where a charcoal black underpaint was detected. A grey underpaint of charcoal black combined with lime white was used for a blue linear detail on Joseph of Arimathea's leg in the same scene.

In the 12th-century scheme in the chapel, vivianite was applied over natural ultramarine for the blue sections of the panelled backgrounds and for numerous details, such as the central details on the vair lining of Nicodemus' cloak in the *Deposition*, where it was contrasted with the pale blue of natural ultramarine combined with lime white (Plates 28 and 29). The cross-section from a sample taken from this lining shows a layer of vivianite, which has altered and now appears green, over natural ultramarine. In another sample traces of the now yellow-green pigment are found within the ultramarine layer, confirming that the vivianite is not a later addition. Although this blue iron phosphate mineral now appears green, an examination of the approximately contemporary Morgan Leaf from the Winchester Bible shows that the vair lining motifs were represented in pale blue with a dark blue centre, and it is clear that this was also the intention in the Holy Sepulchre Chapel paintings.

Comparison with its use in other media and in continental wall painting

In other English medieval polychromy, only a preliminary identification of the mineral has been made on an Anglo-Saxon stone sculpture fragment from York.[24] In recent years, however, vivianite has been identified in medieval painting in Germany, as on the Romanesque lectern at Freudenstädter (*c*1150), where it was applied over a grey underpaint of lead white combined with carbon black (Richter 1988); on a seated Virgin Mary of *c*1250, where the pigment was employed for the lining of the Virgin's robe (Westhoff 2002: 37); on the 14th-century polychromed portals of Holy Cross Minster, Schwäbisch Gmünd (Hauff 1988); and on a late 14th-century wall painting at Urschalling, where the patchy blue-brown appearance of the decoration is thought to be due to an alteration of the mineral.[25] Analysis of medieval wall paintings in Sweden

24. J. Darrah, pers. comm., 1987.
25. Analysis by H. Kühn; M. Alof, pers. comm., 1988; see also Pursche 2000.

suggested that vivianite was employed for the 13th-century scheme at Mästerby, Gotland (Nord and Tronner 2000: 97, 99). Here yellowish-green iron phosphates were identified which are likely to be an alteration of vivianite. In panel painting, vivianite has recently been identified as an underpaint for azurite in the 13th-century Norwegian altar frontal Tingelstad III,[26] and on a 13th-century painting from Regensburg (Germany).[27]

Discussion

Vivianite was employed as a pigment only where a local source of the mineral was accessible in the medieval period, and it is interesting in this respect that its use in medieval polychromy has been confirmed, for example, in Germany and England where mineral deposits are well known. Numerous occurrences of the mineral in England have been reported, most recently by Birley and Young (1993) and Smith (1994). It is found as small crystals at Whale Chine on the Isle of Wight, and in its earthy form at Fordingbridge in Hampshire, both close to Winchester.[28] Either of these could be the local source used by the painters of the Holy Sepulchre Chapel and the Guardian Angels' Chapel in Winchester Cathedral. They are unlikely, however, to have been substantial enough to yield an abundant lasting supply of the pigment, and it is perhaps more likely that a local source of vivianite came to light fortuitously at around the time the painting was executed and that this material was found to be eminently suitable for the intended purpose.

Pigment alteration

The inclusion of vivianite in the palette of both the Romanesque and Gothic schemes in the Holy Sepulchre Chapel at Winchester is of considerable interest. Since vivianite is known to be generally stable in its blue form, it is significant that in the chapel paintings the particles have partially altered to a yellow colour, thus producing an overall green appearance to areas, such as the backgrounds, which would originally have been blue (Plates 28–34).

When the mineral is in the ground and unexposed it is often colourless. The change of the mineral on initial exposure from colourless to blue is due to increased ferric ion concentrations, and it has been established that mechanical grinding of the colourless crystals, heating in air, storage in a vacuum, or chemical treatment of samples can produce a more rapid conversion from ferrous to ferric ions and so to a blue colour (Hanzel *et al.* 1990). It is therefore interesting to speculate whether grinding of the blue mineral to produce a particle size suitable for use as a pigment, or a chemical treatment used during conservation, may have contributed to an additional increase in ferric ion concentration and ultimately to a further colour alteration from blue to yellow. It may well be significant that the pigment has retained its blue colour (Plate 36) only where it has been protected from the varied conservation treatments, which include waxing, solvent cleaning, facing for detachment with a proteinaceous adhesive and subsequent cleaning in the case of the 13th-century scheme, and consolidation with lime-water of both the 12th- and 13th-century schemes.

---

26. I am grateful to Unn Plahter for sharing her recent findings with me.
27. Burmester and Krekel 2001.
28. P. Tandy, pers. comm., 1994.

Examination of polished cross-sections at 1000× magnification indicates that it is the largest particles (19 × 7µm) that have remained blue in colour. Although yellow particles of similar dimensions (15 × 5µm) are present, there are clear divisions along cleavage planes within these, allowing for increased oxidation. At high magnification a yellow halo can be seen surrounding some of the large blue particles, while cleavage planes within the crystal are marked by a yellow line (Plates 33 and 34).

Identification of the yellow oxidation product of vivianite by XRD has proved extremely difficult since the minute samples taken from the wall painting contain a high proportion of calcite and a mixture of vivianite with the alteration product.

However, XRD analysis indicated the presence of metavivianite – a polymorph of vivianite – and it is possible that this relates to the yellow alteration product. The alteration of vivianite, which causes severe distortion of the colour balance in paintings where it has been employed, clearly requires further investigation. Once the mechanism for its alteration has been determined, it may be possible to take preventive measures to reduce the risk of further colour change from blue to yellow. As it is, it seems likely that this colour change will continue at Winchester.

Egyptian blue

The presence of Egyptian blue in the palette of the English medieval wall painter has yet to be confirmed. However, the identification of this pigment on an Anglo-Saxon capital – dated variously to the early 7th and to the 9th century – from St. Augustine's Abbey, Canterbury,[29] and on excavated fragments of sculpture dated to the mid 12th-century from St. Albans Abbey chapter house, makes a brief discussion of other instances of its use in the medieval period appropriate here (Biddle and Kjølbye-Biddle 1980).

Also known as blue frit or Pompeiian blue, Egyptian blue is the calcium-copper tetrasilicate $Ca\ CuSi_4O_{10}$.[30] Its first known use was in the Egyptian Old Kingdom (4th Dynasty), and it continued to be favoured as a mural pigment throughout antiquity. Until quite recently, it was assumed that the technology of its manufacture was lost at the end of the Roman period, and that isolated later instances where it was found to be employed represented a reuse of Roman material. However this theory has been re-evaluated in the light of a number of recent identifications of the pigment in post-Roman wall paintings. Lazzarini's (1982: 84) identification of Egyptian blue on the mandorla and in the background of the painting of the *Ascension* of *c*847–50 in the lower church of S. Clemente in Rome extended the ascertained use of the pigment by five centuries (Osborne 1984: 43). Egyptian blue has subsequently been identified in early 9th-century wall painting fragments from S. Vincenzo al Volturno, Molise;[31] the paintings at Castelseprio (Leveto

29. Initial identification of the pigment by PLM (H. Howard) was confirmed by SEM/EDX (A. Roy) and published by Brain and Welford 1991: 34–5. Welford also identified gold leaf, yellow and red earth pigments, and examination indicated that yellow earth pigment was used as a ground for gold leaf.
30. Details of the technology of the pigment are given by Tite *et al.* 1982; and its characteristics, use and criteria for identification are provided by Riederer 1997.
31. XRF analysis of the Egyptian blue from this site indicated that the same source of copper ore used in its manufacture may have been employed for the pigment found in the House of Livia, Rome; Howard 2001.

1985: 81);[32] paintings of the first half of the 9th century in the atrium of Sta Maria Antiqua, Rome;[33] and in the Carolingian paintings of *c*800 at Müstair, Switzerland (Arnold *et al.* 1986: 190; Mairinger and Schreiner 1986: 195). At Müstair, Egyptian blue was employed for backgrounds, haloes, mandorlas, in the folds of robes and – as at Castelseprio – as an underpainting for flesh tones (Emmenegger 1986: 197).

The identification of Egyptian blue on Romanesque sculptural polychromy from St. Albans is quite extraordinary – indeed it is the latest known instance of the use of this pigment. Identified by XRD, the pigment was detected on the drapery of a small figural sculptural, dated to 1151–66 (Kahn 1986: 86, pl. XXIXa).[34] Given the importance of Verulamium, it is of course possible that in this instance the Egyptian blue is reused Roman material. However, the identification of the pigment at diverse geographic locations in continental wall painting and in polychromed sculpture in England seems to point to a survival of the technology for the manufacture of Egyptian blue at least into the Carolingian and Anglo-Saxon periods. An increased awareness of the possibility of its use in the medieval period, together with the greater sophistication of analytical techniques now employed, make further identifications of this pigment increasingly likely.

Azurite

The blue mineral pigment typically used in high-quality English wall paintings from the middle of the 13th century is azurite. Azurite ($2CuCO_3 \cdot Cu[OH]_2$) is the natural blue pigment derived from a basic copper carbonate mineral of the same name. In antiquity the mineral was known as *Armenian stone*, when Armenia and Spain were the chief sources (Pliny 1984: 297).[35] The usual name for azurite during the medieval period was *azure*, or occasionally *citramarine azure* to distinguish it from *ultramarine azure* (Thompson 1956: 130). It was also known as *azurrum almaneum*, presumably owing to sources of the mineral pigment in Germany.[36] In the Middle Ages, important deposits of the mineral were exploited in Hungary and France and imported into England, although whether as a prepared pigment or in mineral form is not known (Miskimin 1975: 109–14). By the mid-15th century Laurie found the pigment to be of a brighter blue and finer quality than previously employed in manuscript illumination, so a new source may have been discovered at that time (Laurie 1913: 328).

Cost

The pigment was valued according to the intensity of its blue colour, and a receipt from the mid-15th century in Italy indicates that the finest azurite was then priced at three ducats per pound, while that of poorer quality cost only one (Dunkerton *et al.* 1987: 34).

32. These paintings are now reasonably firmly dated to the first half of the 9th century; see Leveto-Jabr 1987; Bertelli 1988; and Sibilia *et al.* 1988.
33. J. Osborne pers. comm.
34. XRD analysis was undertaken by Joyce Plesters of the National Gallery.
35. Azurite may have been employed as a pigment as early as the Fourth Dynasty in Egypt, though it was not widely used then or in antiquity owing to the availability of Egyptian blue; Gettens and Fitzhugh 1993a: 23–5.
36. Germany was the main source of the pigment in the 14th century; Bomford *et al.* 1990: 36–7.

Azure is frequently mentioned in the English documentary sources, but it is often not possible to be certain whether natural azurite or another mineral blue – or indeed a synthetic copper blue pigment – is indicated. For example, payments made for works at the court at Westminster in 1265 include one to William Page for ½lb of *azure* together with other colours, the total cost of which was 21s 10½d. Since the individual prices of the various pigments listed are not provided it is difficult to determine the relative value of the *azure*. In addition, *azure* is specified in the accounts for the mid-14th-century wall paintings in St. Stephen's Chapel, Westminster Palace, for which we know ultramarine was employed (Colvin 1963: 519; van Geersdaele and Goldsworthy 1978).

The Sacrist Rolls of *c*1345–46 for Ely Cathedral provide details of the purchase of two grades of *azure* – one at four times the price of the other – in adjacent entries in a list of expenses for painting:

in iij libr. de azure empt 7s 6d

in j libr. qu. de azure empt. 10s (Chapman 1907: II: 136)

Since in the same accounts 400 gold foils cost only 4s, it is clear that the *azure* at 10s per pound was indeed an extremely expensive commodity and was probably high-quality natural azurite. The cheaper variety may also have been azurite, but of much lower quality – perhaps of smaller particle size and therefore less intense colour – or it may have been another blue pigment such as a synthetic copper blue.

The Exeter Cathedral accounts for 1302–03 again only provide an overall cost (£26) for the materials purchased for painting '49 bosses, 8 corbels and other portions of the vault'. What is important here is that only gold, silver and *azure* are actually specified among other colours, suggesting that an expensive blue pigment – probably azurite – was intended (Erskine 1981: 24). This supposition was confirmed by recent analysis of the early 14th-century polychromy of the roof bosses, which indicated that both azurite and indigo were employed for their decoration (Hulbert 1991: 193–4). Again, in 1325–26, *azure* at 9s per lb is specified for what must have been an extremely elaborate and expensive decoration of the great cross for the cathedral, since the materials purchased for its polychromy also included 1000 gold foils at 36s 8d and 1000 silver foils at 5s (Erskine 1981: 172).

In one instance, the quality of the *azure* to be employed for the robes of the Virgin is specified. Fine *azure* – again probably azurite – is detailed for the Virgin's drapery in depictions of the *Coronation of the Virgin*, the *Annunciation* and other subjects to be painted in the oratory leading to the chapel of Vaudreuil Castle in Normandy. The contract, drawn up in 1356 between the king and Jehan Coste the painter, also indicates that everything is to be painted in fine oil colour (Evans 1948: 198).

Optical, chemical and physical properties

Unless well graded, particles of azurite are often very irregular in size – the largest commonly around 20 × 30μm – and of broken appearance with some conchoidal fractures. In plane polarised transmitted light the largest particles are deep blue in colour, while the smaller particles are much paler. Azurite has a monoclinic crystal structure and its refractive indices are α 1.730, β 1.758, and γ 1.838. In transmitted light with crossed polars, azurite displays strong birefringence and is highly refracting. It is common to see particles of malachite and cuprite present as impurities in a sample of the pigment since in nature these species are closely associated with azurite.

Azurite is stable in alkaline conditions unless heated, but is readily decomposed by even dilute acids, producing carbon dioxide gas. The mineral is stable to light but turns to black cupric oxide (CuO, tenorite) when heated to 200–500 °C.[37] The exact temperature of decomposition depends on both particle size and on the rate of heating, but recent experiments indicate that azurite completely blackens at temperatures above 300 °C (Rickerby 1991: 43).

Azurite is stable in normal atmospheres but will blacken in the presence of moisture and sulphur dioxide. In the presence of moisture it may also be partially converted to malachite. Other pigment alterations include conversion to green copper chlorides, to copper oxalate hydrate (Moffat *et al.* 1985: 237), and to brownish black copper oxide.[38] Discoloured binding media and varnishes may also be the cause of perceived colour change.

It has been suggested that the use of azurite in *buon fresco* painting may cause a colour change from blue to green, as observed in Cimabue's frescoes in the Upper Church at Assisi. However, experiments to test the stability of azurite under these conditions have shown no sign of alteration (Malaguzzi-Valerji 1973), and alteration of azurite has not been observed in samples where the pigment was applied combined with lime white or over a lime plaster support.

Preparation

The stone itself is a splendid blue colour, in nature often associated with cuprite and malachite (the green copper carbonate), and these and other impurities may be apparent in the pigment unless it is very carefully prepared. Indeed, the colour of azurite is dependent not only on the absence of impurities, but also on particle size: the coarser the particles, the deeper and more impressive the colour.[39] When very coarsely ground, it is too gritty to be used as a pigment, but when too finely ground the intensity of the blue colour is reduced and it appears a pale greenish-blue. Modelling of pictorial surfaces could therefore be achieved by careful grading of particle size, without the addition of white pigment to azurite.

In the medieval period, ingenious methods were developed for separating the mineral impurities from azurite particles of various grades, which usually involved suspending the powdered pigment in a liquid, stirring it and then allowing particles to gradually settle out according to their size. This process was observed carefully to determine the correct moment to decant a certain proportion of the liquid.[40] One of the earliest instructions for the washing and tempering of azurite is given in the Latin text of the late 12th-century manuscript British Library Cotton Titus D. XXIV, which provides technical

37. For a detailed description of the pigment's characteristics and use, see Gettens and Fitzhugh 1993a.
38. A partial alteration to dark copper oxide due to adverse cleaning techniques has been observed; see Bomford *et al.* 1986: 50, and Cheetham 1984: 56.
39. Detailed research of the particle sizes of azurite and ultramarine employed in panel painting. has been undertaken to determine possible differences between various workshops, though no firm conclusions were reached; van Asperen de Boer 1974.
40. The preparation of azurite and ultramarine is discussed by Wallert (1991a), who cites an illuminator's manual – Ambruogio di ser Pietro da Siena of *c*1462 – which details the separation technique.

details relating to manuscript illumination (Petzold 1995; Hunt 1995). Gum arabic solution and weak lye are suggested for washing and grading the pigment, which is subsequently tempered with gum arabic and a little size (Hunt 1995: 205, 209).

Use in English medieval wall painting

Despite the evidence from documentary sources, until quite recently it was assumed that although azurite was employed in continental examples, it was not used – or has not survived – in English medieval wall painting (Caiger-Smith 1963: 122). Indeed, before 1988 the only confirmed example of the use of azurite was in the early 16th-century wall painting in the church of St. Mary le Crypt, Gloucester, where the pigment was applied in a layer 200μm thick over a grey underpaint (Katkov and Oldenbourg 1982).

Research by the author in 1988 indicated that azurite was the most commonly employed mineral blue from the beginning of the 13th century, and confirmed the presence of the pigment in the paintings of *c*1250 at Horsham St. Faith Priory, Norfolk; in the scheme of *c*1300 in the Ante-Reliquary Chapel, Norwich Cathedral; and in the paintings of *c*1395 in the Byward Tower, Tower of London (Howard 1988: 19–35). This study has confirmed the presence of azurite in a number of other English Gothic wall paintings – including the late 14th-century *Apocalypse* cycle in the chapter house of Westminster Abbey[41] – while important work by Rickerby resulted in the identification of azurite in the paintings of *c*1130 at Kempley, Gloucestershire (Rickerby 1990a: 255–6).

Although unique in English Romanesque wall painting, the azurite at Kempley has early parallels elsewhere in Europe: in the late 10th-century paintings at St. Georg in Oberzell, Reichenau, Germany;[42] in the Romanesque scheme of paintings at Mårslet, Denmark (Graebe *et al.* 1986);[43] and in the Romanesque Revelation cycle at Saint Hilaire-le-Grand at Poitiers, where azurite of inferior quality has been found (Taralon 1986b: 288, n. 140).[44] These exceptional and geographically dispersed occurrences show that azurite was certainly available, although it is also clear that until the beginning of the 13th century ultramarine was generally selected in preference.

Application

Directly on the plaster or over a white preparation

The natural azurite employed for the elaborate vinescroll of *c*1300 on the vault of the Ante-Reliquary Chapel in Norwich Cathedral is a pure, unadulterated product of extremely high quality (Plates 42 and 43). Initial inspection of the vault revealed that the blue appearance varied considerably, suggesting that several different methods of application had been employed. However, the wide range of particle size was found to be responsible for the varied appearance. The largest intensely dark blue particles measure

41. Unpublished analysis by H. Howard 1998.
42. Here, the azurite has partially altered to form paratacamite; see analysis by Kühn in Knoepfli and Emmenegger 1990: 176.
43. The only blue pigment previously noted in Danish wall paintings of this period is ultramarine.
44. A rather tantalising reference to copper-based blues and greens is also made in Larson's article on the polychromy of the 12th-century frieze from the west front of Lincoln Cathedral, though ultramarine has also been detected; see Larson 1992: 188.

75 × 50µm,[45] indicating a pigment of quite exceptional quality which would have produced a beautiful surface texture – sparkling and scintillating as the crystals were caught by the light (Plate 44).

At Norwich, these large, coarsely ground particles of azurite were combined with others of widely ranging size – the smallest measuring only 10 × 5µm – and applied in a single paint layer over a ground of lead white combined with lime white. Of particular significance is the considerable variation in the thickness of the paint, which appears to be due to the partial, and in some places complete, loss of the uppermost portion of the layer. This damage may have occurred during the invasive conservation treatments to which the scheme has been subjected.[46] The present rather mottled appearance of the paintings is due both to the variation in the thickness of the paint layer and to the colour loss apparent around the perimeter of some of the particles, which could be due to the use of adverse cleaning materials.

In the late 14th-century scheme in the Byward Tower, a sample taken from the Virgin's robe indicated that azurite combined with lead white, and with inclusions of malachite and yellow iron oxide, was applied in a layer 65µm thick over lead white. This local white ground is itself painted over verdigris and copper resinate which form the background of the painting (Howard 1988: 19–35). Azurite in a lead white matrix and applied over a white ground was also employed for St. John's drapery in the approximately contemporary *Apocalypse* cycle in the chapter house of Westminster Abbey.[47]

Over coloured underpaint

The large particle size and consequent low covering power of azurite meant that several layers would be required to produce a strong, saturated blue if applied over a white ground. However, the use of a toned underpaint reduces the light scatter from the plaster or ground, enhances the colour, and results in considerable cost savings in the use of this expensive pigment.

Economies in the use of this pigment were sometimes achieved by other, less equitable, means. For example, the natural mineral applied over a local ground of verdigris in one of the narrative scenes of *c*1250 at Horsham St. Faith Priory (Norfolk) appears to have been adulterated with its cheaper, synthetic equivalent, presumably by the artist or supplier (Plates 39–41) (Howard 1988: 33). Synthetic copper blues have been identified in English wall painting from the 13th century onward,[48] and the use of this pigment to bulk out natural azurite has also been identified in a shell palette from Greyfriars, Norwich, dated to *c*1300, which may have been employed for large-scale painting (Plates 60–62).[49]

45. In their monograph on azurite and its synthetic equivalent, Gettens and Fitzhugh (1993a: 26) suggest that the largest particles of azurite were commonly around 20 × 30µm.
46. These interventions include the uncovering of the paintings from beneath limewash in the 19th century, when they were also presumably varnished; a wax treatment in the 1930s, and then removal of the coatings and consolidation with lime-water and casein in the 1960s.
47. North-west bay, arch 2, scene 5; Howard 2000 and forthcoming.
48. See section on synthetic copper blues (pp. 50–57) for details of the use of these pigments.
49. The azurite particles in the Greyfriars palette are of average size, with a few large inclusions; Howard and Park forthcoming b; see also a short note on the palette in the Annual Report of the Norfolk Archaeological Unit for 1996–7: 37 and 62. Azurite of larger-than-average particle size (the majority 15 × 30µm, with a few of approximately 30 × 50µm) has also been identified in an oyster shell palette dated to between the mid-14th and mid-16th century, excavated from the Carmelite Friary, Coventry. Azurite is the only blue pigment to have been identified in other shell palettes, such as those from Clarendon Palace and St. Augustine's Abbey, Canterbury; see Howard 1997a.

A green underpaint for azurite has also been identified in the paintings of *c*1220–30 on the entrance arch of St. George's Chapel, Ely Cathedral, where green copper silicate chrysocolla was apparently employed.[50] The inclusion of chrysocolla in the palette is particularly interesting, not only because this pigment has not previously been identified in English medieval wall painting, but also since it was clearly selected for its distinctive coloristic qualities.[51] The pigment was employed as an underpaint beneath azurite to produce the deep blue background, and was also combined with natural azurite to give a deep green colour.

Red, however, is the colour most typically used as an underpaint for azurite since it shifts the blue toward a richer, more purple hue. This was clearly the intention of the artist responsible for the painting of St. William of York (*c*1302–08) in the feretory of St. Albans Cathedral, where a complex series of local grounds was employed to achieve subtle coloristic effects for the saint's blue dalmatic (Plates 45 and 46). The drapery is composed of three layers over an initial white lime ground. First is an underpaint of red earth applied to enhance the saturation and alter the tone of the translucent blue toward a deeper, purplish colour. The blue was then applied in two layers: first a layer of synthetic copper blue mixed with varying amounts of lime white to produce the modelling of the drapery, and above this a layer of high-quality azurite which would have produced a sparkling and brilliant blue effect (Howard 1993a). A red underpaint of red lead combined with lead white was also employed beneath azurite for St. John's robe in the late 14th-century *Apocalypse* cycle in the chapter house of Westminster Abbey. Although the thick wax coating on the surface of the painting now produces a slightly green cast to areas of the blue pigment, a warm purple-blue was the colour originally intended here.[52]

Pigment mixtures

Purple hues were sometimes produced by combining azurite with red pigments, as in the chasuble of the corresponding archbishop on the south-east pier of the feretory of St. Albans Cathedral. The soft purple of the chasuble – contrasting with the strong orange of the maniple and amice – was achieved by mixing red lake and azurite and applying the mixture in a single layer over a lead white ground (Plates 47 and 48).[53] Similarly azurite and red lake were combined to produce the rich purple draperies of saints Peter and Andrew in the scheme of *c*1300 on the vault of the Ante-Reliquary Chapel, Norwich Cathedral, while for the more opaque, pale purple drapery of the central *Christ in Majesty*, lead white was added to the azurite-madder mixture (Park and Howard 1996; Howard 1995c: App. III). Azurite was also combined with a red lake pigment and lead white in the late 14th-century *Apocalypse* cycle in the chapter house of Westminster

50. These recently discovered paintings, on the soffit of the arch, show four seated figures in rectangular frames, each originally identified by an inscription. The lower figure on the north side has been identified by Miriam Gill as St. Lawrence.

51. The pigment was identified by SEM/EDX (which indicated the presence of silica, copper, calcium and oxygen), and by FTIR microspectroscopy (which provided no bands at 1400cm^{-1} which are indicative of carbonates [such as malachite], but rather, a broad band around 1100cm^{-1}, suggesting a silicate). For absolute confirmation of the presence of chrysocolla (copper silicate), XRD would be necessary. For a description of the pigment see Gettens and Stout 1966: 107.

52. South bay, arch 3, scene 68; Howard 2000 and forthcoming.

53. The blue fringe of the maniple is composed of a single layer of azurite lightened with lead white and applied over the lead white ground; see Howard 1993a.

Abbey, although here the purple layer was applied over tin leaf, presumably to produce an intense lustrous colour (Plates 49–52).[54]

Binding media

Despite its apparent stability in alkaline conditions, azurite was commonly applied *a secco* in a glue or egg medium, over an underpainting applied *buon fresco* or combined with lime white. Although the large particle size and consequent low covering power of the pigment is sufficient reason for the use of a toned or coloured underpainting – and thus for an additional binding medium in the azurite layer – other factors require careful consideration. For example, examination and analysis has determined that the exceptional survival of azurite on external wall paintings in Moldavia is due to the incorporation of a milk protein, probably casein, with lime in the paint layer. Since calcium caseinate is exceptionally weather-resistant, the blue and green paint layers with which the proteinaceous binding medium is most frequently associated have survived particularly well (Mora *et al.* 1984: 117).

Proteinaceous binding media have been found associated with blue and green copper-based pigments in a number of other instances. For example, analysis by Raymond White of wall paintings of *c*1270–80 at Angers Cathedral (France) demonstrated that within a single blue layer painted in oil, the individual azurite particles were coated in a proteinaceous medium only a few microns thick.[55] The use of a proteinaceous medium for azurite has also been established in the paintings of *c*1300 in the Ante-Reliquary Chapel, Norwich Cathedral (Howard 1995c); in addition, both protein and oil have been identified as constituents of the binding media of azurite-containing paint layers on the Cologne choir screen paintings of *c*1332–49 (Jägers and Schulze-Senger 1989: 187–8; Heritage 1990: 8).

This seemingly complicated method of application may be due to a number of factors. First, since the refractive indices of oil and azurite are similar, the pigment loses its intensity when applied in oil and is therefore normally applied in another medium. It is also possible that the degradation of copper-based pigments in the presence of drying oil was understood, and consequently the particles were isolated with glue or another form of protein before being bound in oil. Whether or not the property of amines (from protein) as inhibitors of copper corrosion was fully understood by the medieval artist is not clear, but this commonly found mode of application certainly suggests that, in practice, the properties of the pigments combined in various binders was well known.

Treatises provide much information concerning the use of azurite in wall painting; for example, Cennini is of the opinion that azurite is unsuitable for *buon fresco* painting (1960: 50), while in the early 12th century Theophilus recommends the use of veneda underpainting (1979: 25). Combining *azure* (possibly azurite) with egg for walls is suggested by Theophilus, and later by Cennini.[56] Of particular interest is Peter of St. Omer's description, dating from *c*1300, of repeatedly washing azure in egg white and

54. South-west bay, arch 2, scene 82; Howard 2000 and forthcoming.

55. Coating azurite particles in glue distemper was a technique commonly used in manuscript illumination. For the technique of the Angers paintings generally, see Subes-Picot 1992 and Demailly *et al.* 1998.

56. Theophilus (1979: 23) suggests the use of egg yolk combined with water as a blue glaze in wall painting, while Cennini (1960: 32, 36 and 54–5) indicates that azurite and gallorino should be tempered with egg for walls and panels.

then subsequently grinding the colour in oil for use on wood, or with egg and water (presumably egg tempera) for wall painting (Merrifield 1967: 1: 134).

Comparison with its use in other media and in continental wall painting

Azurite was also commonly employed for polychrome sculpture in England from the 14th century, as on the shrine of St. Alban of *c*1302–08 in St. Albans Cathedral; the de Bois tomb of *c*1380 at Ingham, Norfolk (Howard 1993d); early 15th-century polychromed stone fragments from Tewkesbury Abbey (UCL 1996); and in the 15th-century polychromy of the west front of Exeter Cathedral where it was applied in thick pure layers and combined with lead white and yellow to produce a blue-green colour (Sinclair 1991: 123, 1992: 10). For the Ingham tomb, an underpainting of vermilion was employed to provide a purple cast and to increase the covering power of the large particles (60 × 50μm) of natural azurite. Although red earth is commonly applied as an underpaint for azurite, this use of more costly vermilion is unusual; indeed no other examples have so far been identified in England although a similar stratigraphy was identified in the wall paintings of *c*1270–80 in Angers Cathedral (Stirnemann 1995: 141). It seems likely that at Ingham practical considerations prevailed over expense; since the upper portions of the crenellations were painted with vermilion, it was expedient to use the same pigment for underpainting azurite in adjacent areas. Perhaps with the intention of increasing its luminosity and brilliance, azurite was applied over silver leaf for certain details on the tomb. For example, the lozenge design on the cushion beneath the head of Lady Margaret de Bois was painted in *impasto* in azurite and vermilion over a layer of silver (Plates 53–56) (Howard 1993d).

A grey secondary underpaint of charcoal black combined with calcium sulphate was detected beneath azurite in several samples taken from the polychromed screens of *c*1370 enclosing the Chapel of Our Lady Undercroft in the crypt of Canterbury Cathedral. Again, the coloured underpaint was presumably employed to increase the covering power of the mineral blue pigment (Howard 1993b). Grey or black underpaint was also employed to intensify the blue colour of the azurite on the polychromed Great Screen of *c*1475–90 in Winchester Cathedral (Brodrick and Darrah 1989: 615), while in the 15th-century polychromy on the west front of Exeter Cathedral azurite was applied over indigo on the coats of arms (Sinclair 1991: 123, 1992: 10). Examination of the polychromy of the Winchester screen also indicated that azurite was combined with two red pigments – a lake and vermilion – to produce a rich purple colour (Brodrick and Darrah 1989: 615–16).

The use of a proteinaceous medium for azurite in wall painting is also paralleled in polychrome sculpture. FTIR analysis of an azurite sample from the de Bois tomb at Ingham suggested that the matrix for the azurite layer contained both a proteinaceous component and oil (Plate 56), which may indicate that either the binding medium was egg yolk (which contains both protein and oil), or that the mineral pigment was ground in a proteinaceous material such as size before being applied in oil (Howard 1993d).

In Gothic panel painting, azurite was sometimes employed for shadows in the flesh areas. Although relatively little original paint remains on the nave ceiling of *c*1220 at Peterborough Cathedral, of particular interest is the use of an underpaint of natural azurite combined with a little yellow iron oxide and lead white to produce the effects of modelling in the flesh tones (Plates 58 and 59) (Howard 1997b: 7). Despite the extensive and crude repainting which now disfigures the surface, traces of this green underpaint for the flesh are visible *in situ* with the aid of a videomicroscope, particularly in the face of St. Peter. Azurite was also used in the ceiling paterae of *c*1263–66 from the Painted Chamber of Westminster Palace, but here the pigment was combined with black and white and employed to outline the figures, producing a bluish undertone in areas such as the prophet's chin (Plate 57).[57] Azurite has also been detected as a component of flesh painting in the late 14th-century German 'Estouteville triptych', and appears to have been found frequently in flesh tones in other German and Flemish panel painting of this period (Villers *et al.* 1997: 674). However the technique of application in the 'Estouteville triptych' is quite different from the English panels, since the azurite is combined with lead white, vermilion, ochre and black in a single layer, with highlights and shadows applied wet-in-wet on top.

Azurite particles of varied size have been detected in the Thornham Parva Retable of *c*1335. In a sample taken from the blue chequered background of St. Paul, the largest azurite particles are found only in the upper portion of the paint layer, and it seems reasonable to assume that in this case a layer of paler coloured azurite – of smaller particle size – was applied initially as an underpainting for a layer containing larger, darker blue particles.[58] Large particles of high-quality azurite have also been found in the early 14th-century screen from Kingston Lacy (Dorset) (Tracy 1997: 28).

Discussion

The rather swift transition from ultramarine to azurite as the blue mineral pigment of choice for wall painting in England during the first half of the 13th century is remarkable. Whether this general shift was due to a preference in the Gothic period for the particular coloristic qualities and sparkling surface of azurite, or to a shortage of ultramarine – perhaps due to a change of, or blocking of trade routes – is unclear. Ultramarine was certainly still available in the 13th and 14th centuries, and it continued to be employed as in the paintings of *c*1352–63 in St. Stephen's Chapel, Westminster. Such examples are generally in important and presumably expensive commissions, which suggests that a sharp rise in cost in the early 13th century may have made ultramarine prohibitively expensive for most large-scale painting, thus leading to the use of more economic alternatives.

A continued high regard for ultramarine is supported by analytical evidence from the Norwegian altarpiece Tingelstad III (*c*1260–1300), where ultramarine was employed for drapery, while the cheaper azurite was used for larger areas of architecture and the mandorla, with vivianite as an underpainting.[59] Ultramarine and azurite also appear to have been employed concurrently in the Guardian Angels' Chapel at Winchester Cathedral. The vault scheme of the chapel depicts busts of angels in roundels which are datable stylistically to *c*1230, although foliate decoration was added over the large and medium-sized roundels when the vault was repainted in *c*1280 (Park and Welford 1993: 129). Scientific examina-

57. A more intense blue undertone of ultramarine combined with lead white was applied over a fiery red lead layer in some upper portions of the seraph, while a more complete pale blue underpaint is evident beneath the flesh tone of the prophet's hand; Liversidge *et al.* 1995: 499, and Lynn 1997: 22–3.
58. Goldsmith and Bucklow 1998, and Bucklow forthcoming.
59. Plahter 1995; and U. Plahter, pers. comm. Both ultramarine and azurite were also employed on the Heddal altarpiece of *c*1250–60.

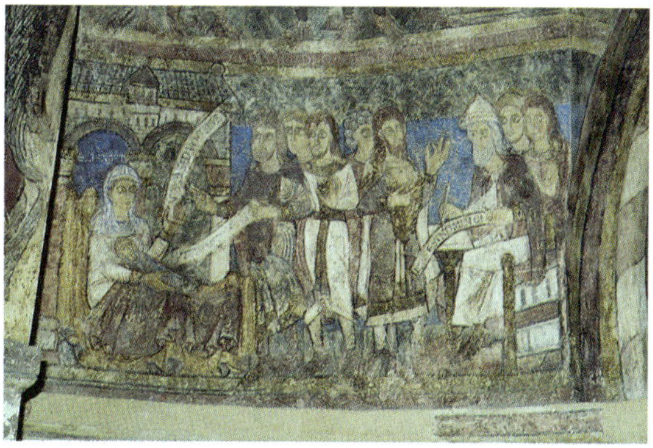

***Plate 1** St. Gabriel's Chapel, Canterbury Cathedral. Detail showing the* Naming of St. John *on the north side of the apse. Natural ultramarine was extensively employed for the background areas of this scheme of c1130. (© Crown copyright. NMR)*

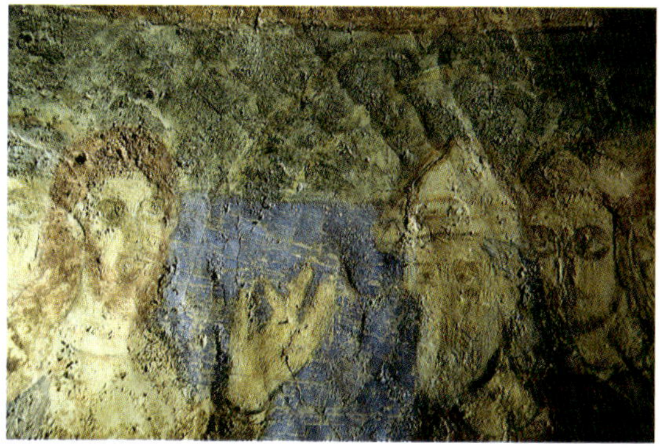

***Plate 2** Detail of Plate 1 photographed in raking light showing the undulating surface of the plaster and the rich blue of the ultramarine background. (© Courtauld Institute, Conservation of Wall Painting Department)*

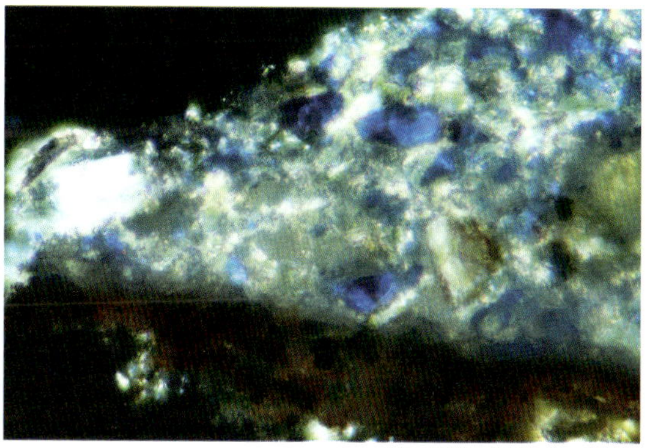

***Plate 3** St. Gabriel's Chapel, Canterbury Cathedral. Thin section of a sample taken from the blue robe of the* Christ in Majesty, *stained for protein with Amido black (Sol.1); photographed in transmitted light with crossed polars, original magnification 200×. Analysis of the original binding media confirmed the presence of protein in both the blue and red layers in this sample, while linseed oil was detected in a sample taken from the blue background of the scheme. (Photomicrograph: H. Howard; © Courtauld Institute, Conservation of Wall Painting Department)*

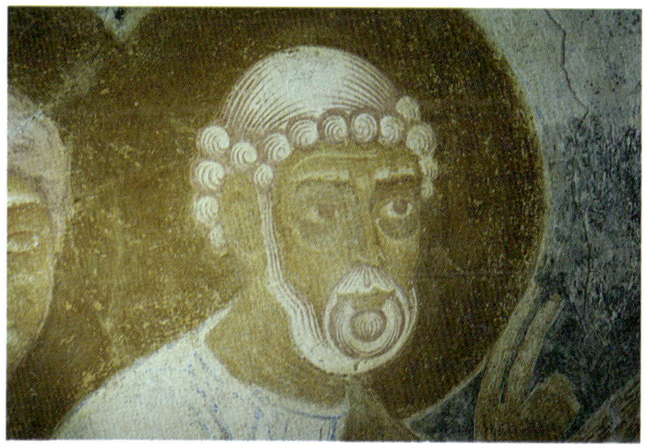

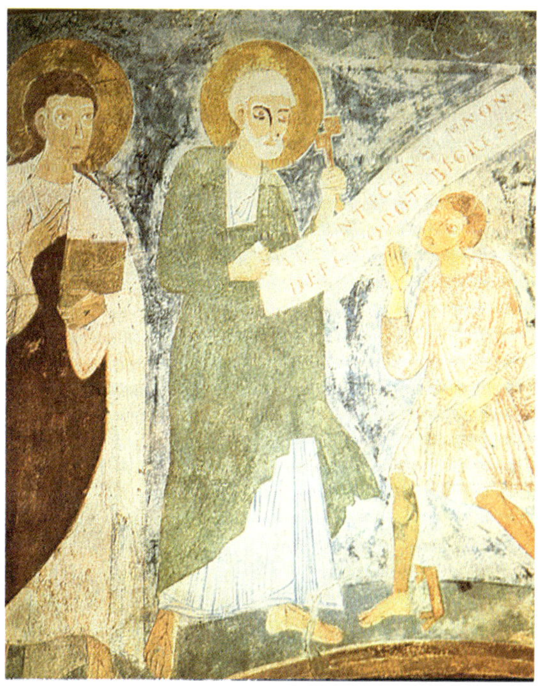

***Plates 4 and 5** Details of the scheme of c1130 at Idensen (Germany), where natural ultramarine is used extensively for the panelled backgrounds and other details. The use of charcoal black mixed with white as an underpaint for the ultramarine mirrors that in the approximately contemporary paintings in St. Gabriel's Chapel at Canterbury and in the scheme of c1170 at Cormac's Chapel, Cashel (Ireland). (Plate 4 © Courtauld Institute, Conservation of Wall Painting Department; Plate 5 © Dr Rolf-Jürgen Grote)*

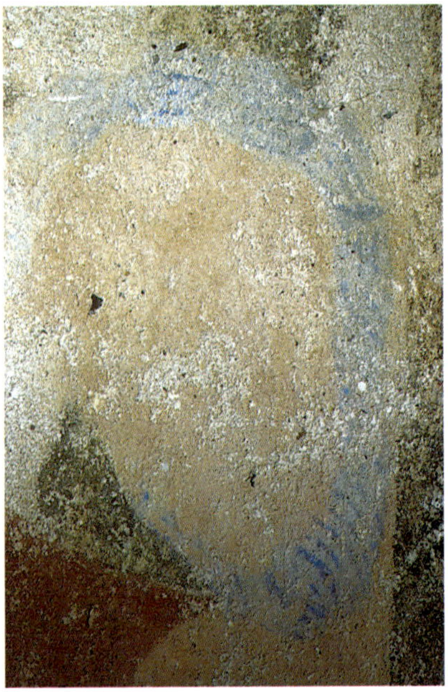

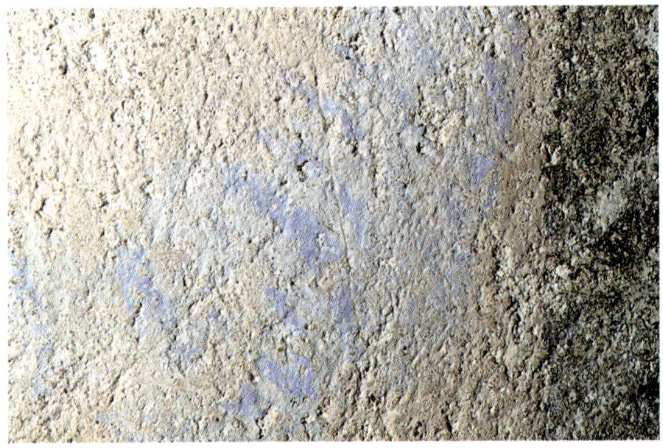

Plates 6 and 6a *Details of the* Crucifixion *above the north nave arcade at Ickleton (Cambridgeshire). Natural ultramarine was employed to pick out details of the hair and beard of Christ in each scene of this scheme of c1125. (Photographs: H. Howard; © Courtauld Institute, Conservation of Wall Painting Department)*

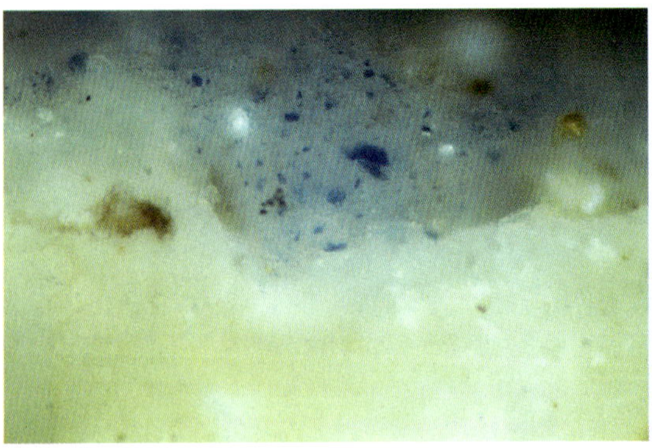

Plate 7 *Ickleton (Cambridgeshire). Cross-section of a sample taken from Christ's hair in the* Betrayal*, original magnification 200×. Ultramarine was combined with lime white and applied in a single layer approximately 150mm thick directly over the coarse underlying plaster. The blue pigment is of exceptionally high quality, consisting of extremely small particles only 5 × 3μm. (Photomicrograph: H. Howard; © Courtauld Institute, Conservation of Wall Painting Department)*

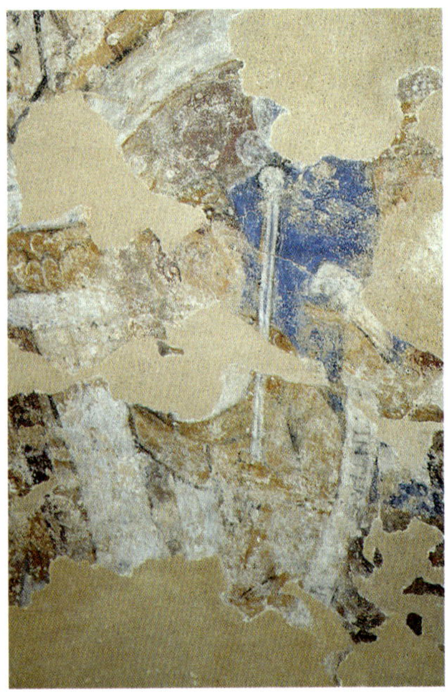

Plate 8 *Detail of the scheme of c1170 at Cormac's Chapel, Cashel (Ireland), where natural ultramarine was employed extensively for the backgrounds and other details. (Photograph: H. Howard; © Courtauld Institute, Conservation of Wall Painting Department)*

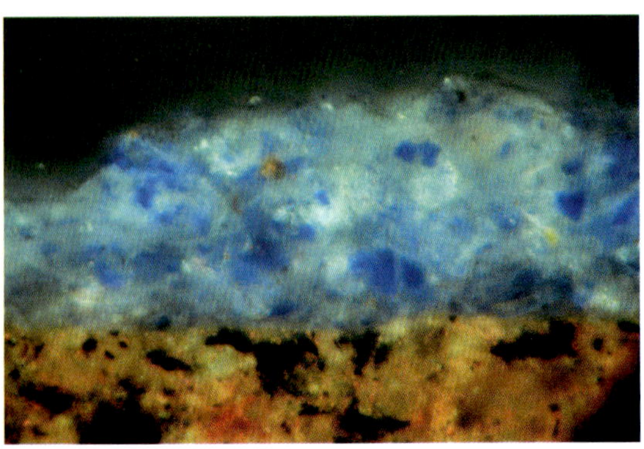

Plate 9 *Cormac's Chapel, Cashel (Ireland). Cross-section of a sample from the west side of the chancel arch, original magnification 400×. Natural ultramarine has been applied over a dark purple underpaint of charcoal black combined with red earth, presumably to produce a purple tint. (Photomicrograph: H. Howard; © Courtauld Institute, Conservation of Wall Painting Department)*

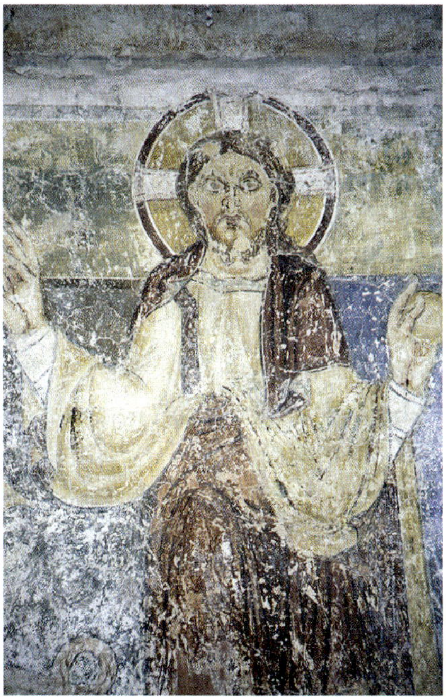

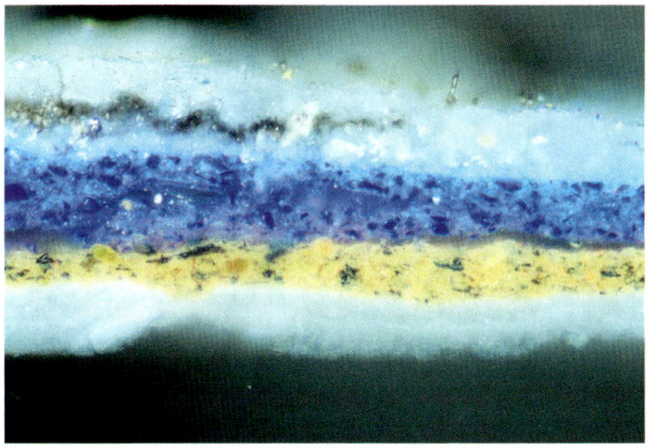

Plate 11 *Pürgg (Austria). Cross-section of a sample from Cain's drapery, original magnification 422×. Charcoal black and red earth were combined to produce the dark red colour of the drapery, while decorative elements were added in natural ultramarine. Traces of limewash are visible on the surface of the sample. (Photomicrograph: H. Howard; © Courtauld Institute, Conservation of Wall Painting Department)*

Plate 10 *Detail of the* Feeding of the Five Thousand *from the scheme of c1165 at Pürgg (Austria). Here ultramarine was also applied over coloured underpaint: over dark red where a purple effect was sought, and over grey where a dark blue colour was intended. (Photograph: H. Howard; © Courtauld Institute, Conservation of Wall Painting Department)*

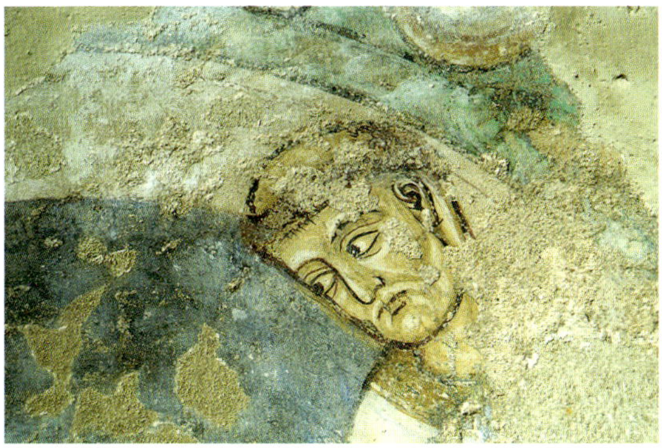

Plate 12 *Detail of the 'Losinga' cycle of c1190 at Norwich Cathedral showing Bishop Losinga and the intense ultramarine blue of the background. (Photograph: H. Howard; © Courtauld Institute, Conservation of Wall Painting Department)*

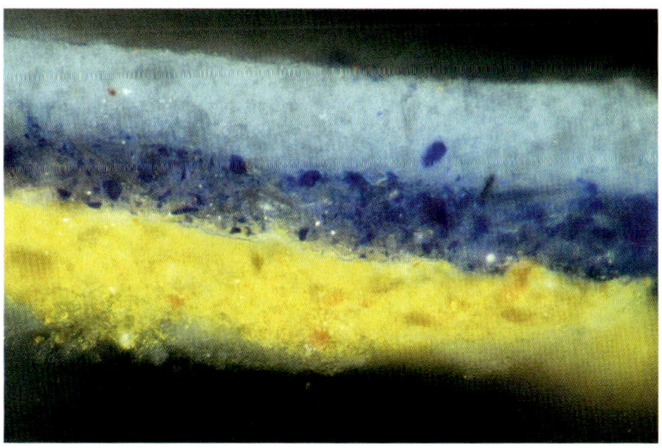

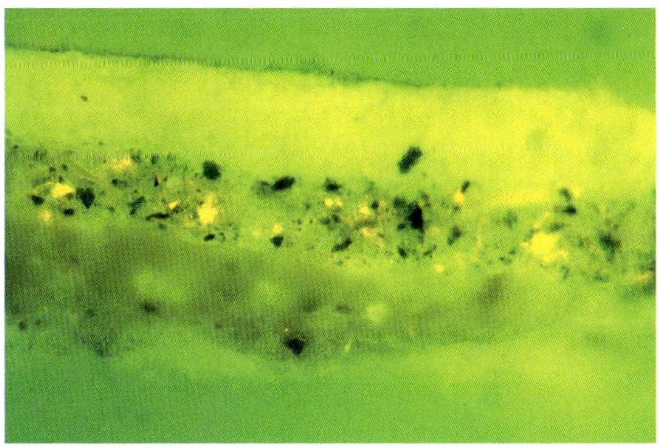

Plates 13 and 14 *'Losinga' cycle, Norwich Cathedral. Cross-section of a sample from the hem of the standing figure on the west segment of the vault, original magnification 340×. The drawing of the drapery was undertaken in yellow earth, and over this is a layer of ultramarine including other siliceous minerals and some unusual columnar-shaped inclusions. In UV light, these inclusions – among others in the paint layer – fluoresce a brilliant orange colour (Plate 14). Ground lapis lazuli is known to display a spotty fluorescence in UV of long wavelengths, but the level of fluorescence here is exceptional. In this instance, it is possible that a red lake pigment, which may also fluoresce in UV, was originally combined with the blue pigment, perhaps to produce a more purple tint. Although any red lake pigment has clearly faded substantially, close examination of one of the samples revealed the slightest trace of a pink hue at the base of the ultramarine layer. (Photomicrographs: H. Howard; © Courtauld Institute, Conservation of Wall Painting Department)*

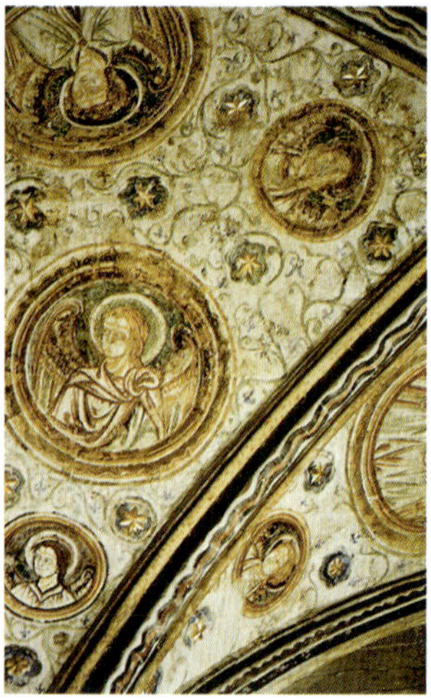

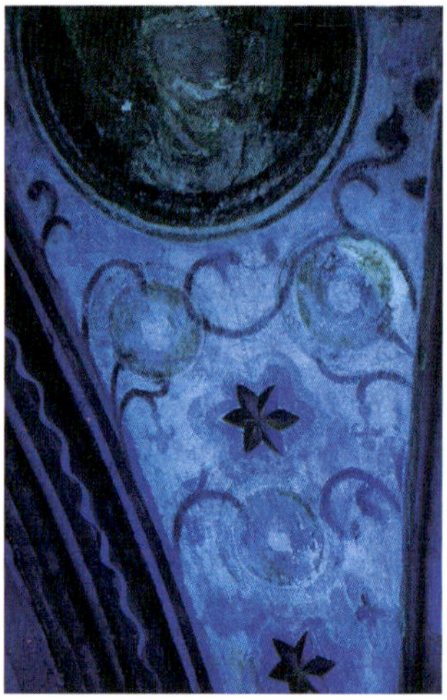

Plates 15 and 16 *Details of the scheme of c1230 on the vault of the Guardian Angels' Chapel, Winchester Cathedral. (©Crown copyright. NMR) In UV light (Plate 16), the scrollwork, which was added in c1280, is clearly visible. (© Courtauld Institute, Conservation of Wall Painting Department)*

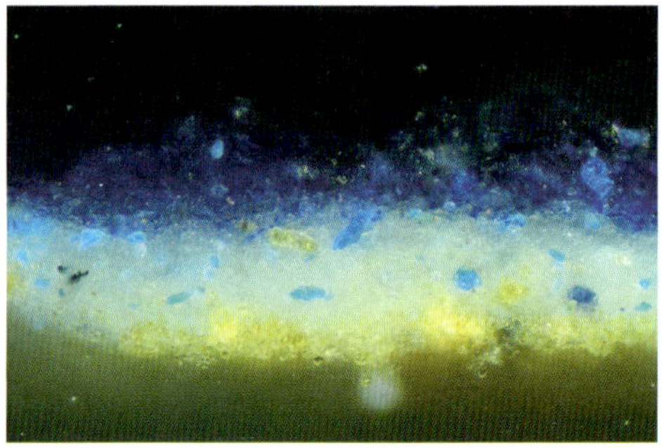

Plate 17 *Guardian Angels' Chapel, Winchester Cathedral. Cross-section of a sample from the lowest angel roundel on the west segment of the vault, original magnification 394×. Here, ultramarine combined with azurite was applied over azurite in a white matrix on a local yellow ground. Ultramarine was also identified in samples taken from a flower in the late 13th-century scrollwork. (Photomicrograph: H. Howard; © Courtauld Institute, Conservation of Wall Painting Department)*

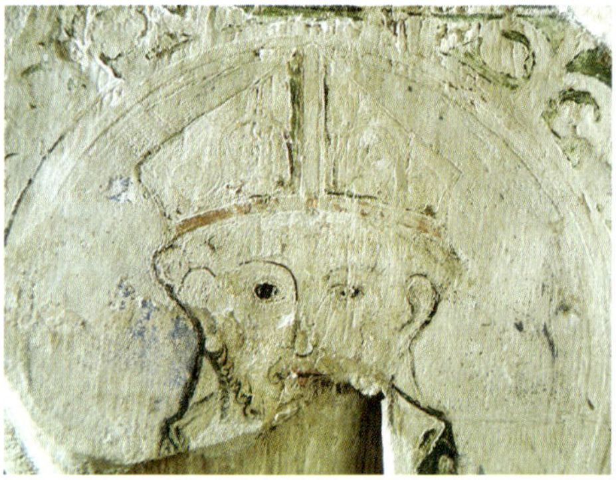

Plate 18 *Detail of the scheme of c1250 on the eastern arch of the Ante-Reliquary Chapel, Norwich Cathedral. Ultramarine, applied over a thin skim of plaster on the stone support, was used for the background of the bishop's head to the south of the arch. In raking light the incised details of the bishop's mitre and face are clearly visible. (Photograph: H. Howard; © Courtauld Institute, Conservation of Wall Painting Department)*

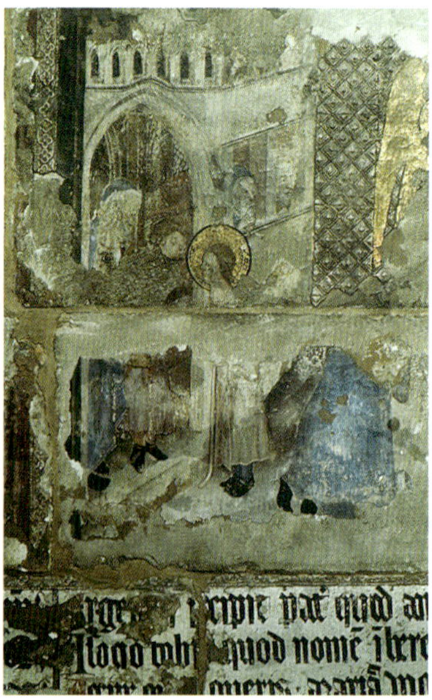

Plate 19 *Natural ultramarine continued to be employed in high-quality schemes of wall painting in the 14th century as in the important mid-14th-century paintings in St. Stephen's Chapel, Westminster Palace. (© Copyright The British Museum)*

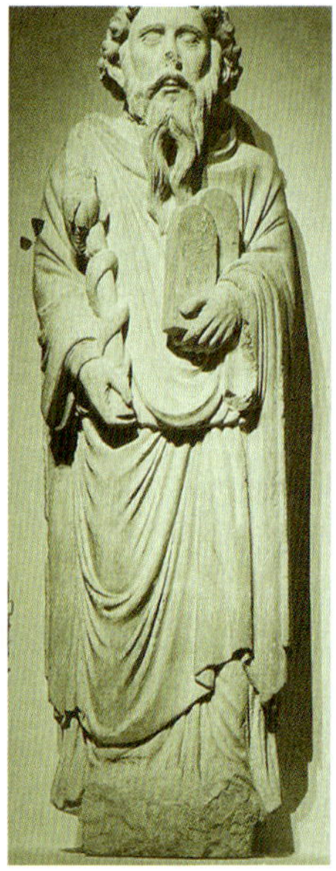

Plate 20 *The extensive use of natural ultramarine employed in Romanesque wall painting is paralleled in Romanesque sculpture as on the limestone figure of Moses (c1180–85) from St. Mary's Abbey, York. (© The Yorkshire Museum, York)*

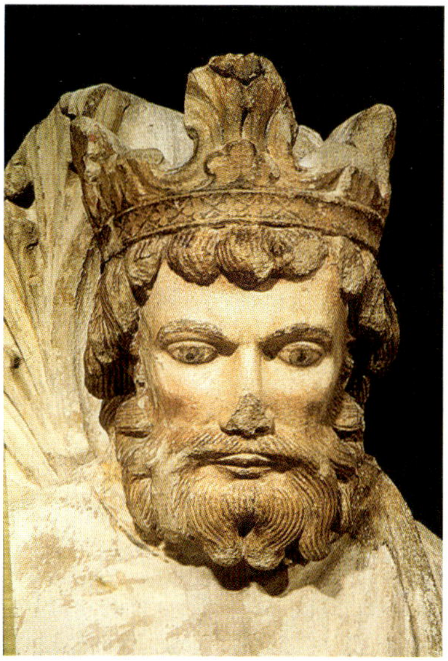

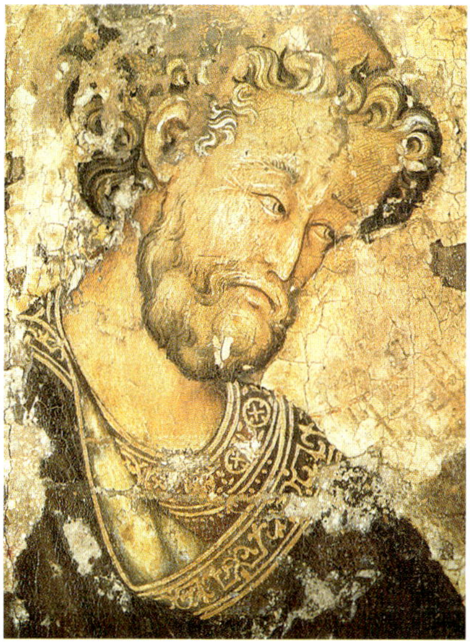

Plates 21 and 22
Ultramarine was also employed on the early Gothic south portal of Lausanne Cathedral in c1216–20. (©André Held). Ultramarine was commonly used in panel painting and notable examples include the Adisham reredos of c1180, the ceiling paterae from the Painted Chamber of Westminster Palace of c1263–66, and the Westminster Retable of c1270 (Plate 22). (© Dean and Chapter of Westminster)

Plate 23 *Detail of the* Traditio Legis *of c1100 on the east wall of the nave at Coombes (West Sussex). In this plate the 'false blue' fictive marbling at dado level is clearly visible. (© Courtauld Institute, Conservation of Wall Painting Department)*

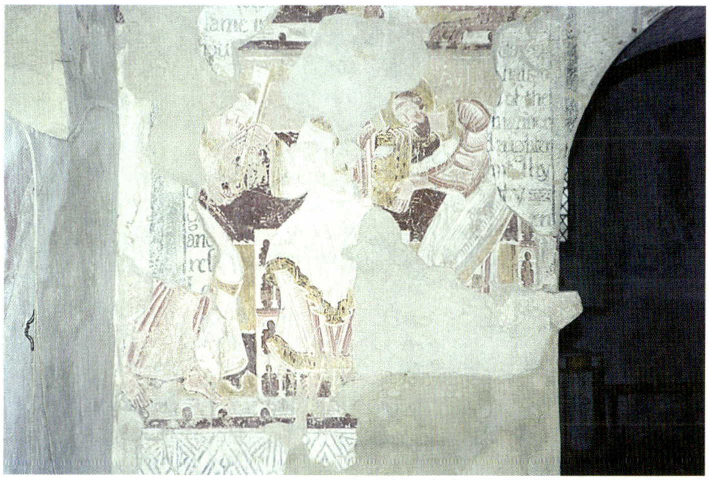

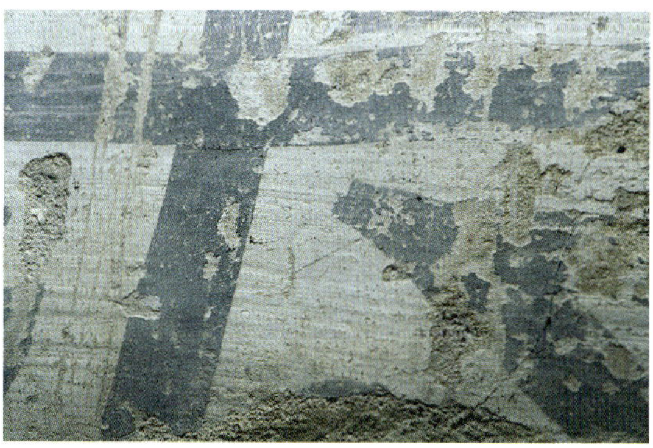

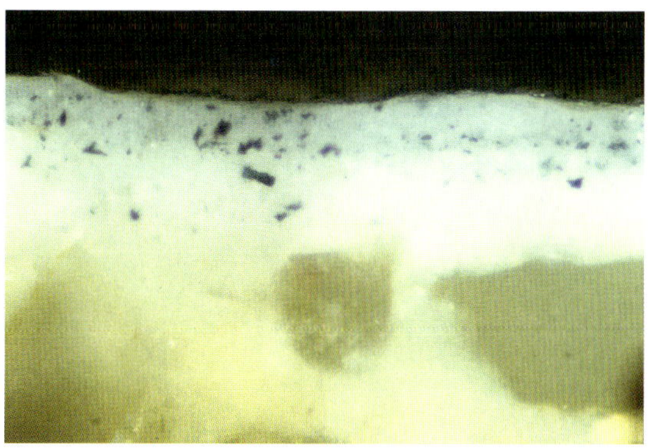

Plates 24 and 25 *Detail of the fictive marbling shown in Plate 23. Modelling effects have been created by varying the amounts of lime white combined with charcoal black, and a cross-section taken from a dark shadow clearly shows a higher proportion of charcoal black in the upper layer (Plate 25), original magnification 200×. The use of this pigment mixture is particularly prevalent in Romanesque wall painting and a much cheaper alternative to the costly mineral blues then available. (Photograph and photomicrograph: H. Howard; © Courtauld Institute, Conservation of Wall Painting Department)*

Plate 26 *'False blue' continued to be employed in the Gothic period, and a blue-black mixture made from charcoal black combined with lime white was also employed for the archbishop's shoes on the painted coffin lid of Archbishop Walter de Gray at York Minster of c1255. (© Crown copyright. NMR)*

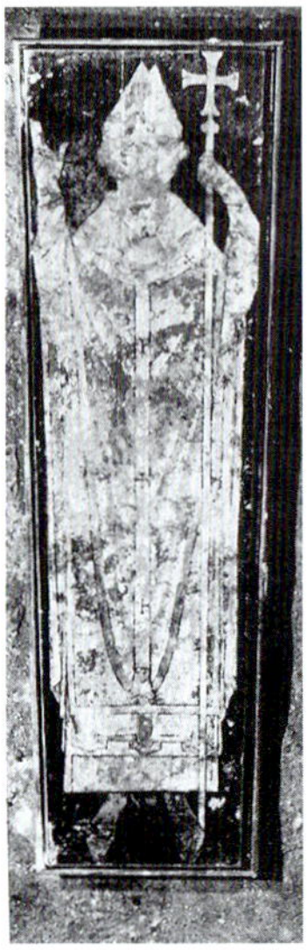

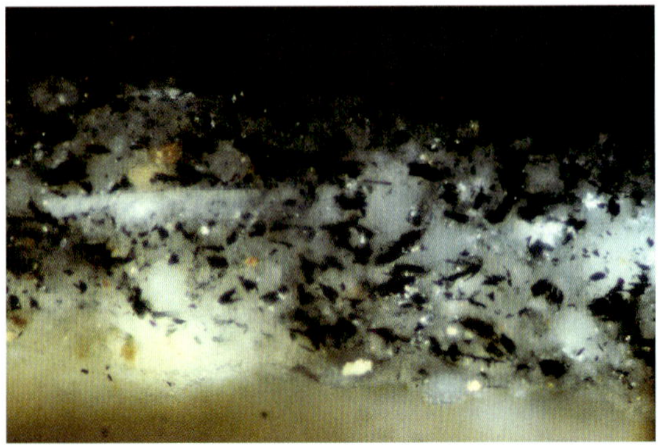

Plate 27 *'False blue' was also used to produce a blue-grey colour on a 13th-century polychromed stone column in Lichfield Cathedral. The cross-section shows particles of charcoal black combined with lime white and a little red earth, original magnification 400×. (Photomicrograph: H. Howard; © Courtauld Institute, Conservation of Wall Painting Department)*

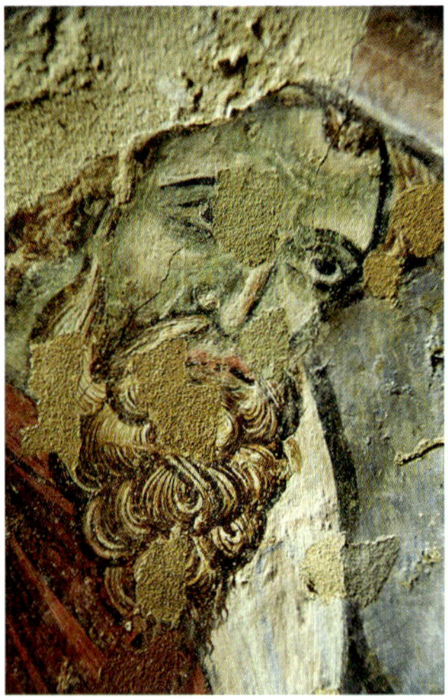

Plate 28 *Winchester Cathedral, Holy Sepulchre Chapel: detail of the head of Joseph of Arimathea in the* Deposition *on the east wall of the chapel (c1175). Vivianite was extensively employed for the blue portion of the panelled backgrounds and for numerous details such as the vair lining of Nicodemus' cloak. (© Courtauld Institute, Conservation of Wall Painting Department)*

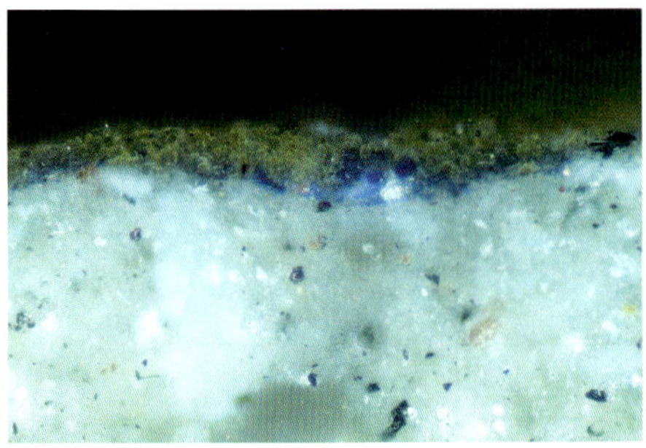

Plate 29 *Winchester Cathedral, Holy Sepulchre Chapel. Cross-section of a sample from the background of the* Deposition, *now an olive-green colour, original magnification 500×. A thin layer of natural ultramarine is visible over the grey plaster substrate, while the present green appearance is due to a layer of vivianite on the surface of the sample. Originally blue in colour, some of the particles of this vivianite have altered and now appear yellow, producing an overall green effect. (Photomicrograph: H. Howard; © Courtauld Institute, Conservation of Wall Painting Department)*

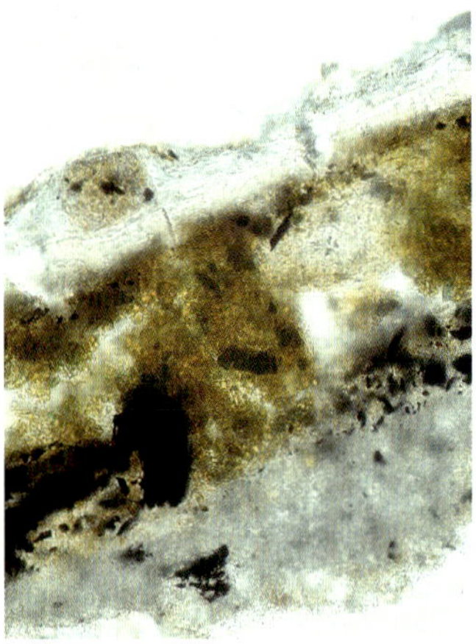

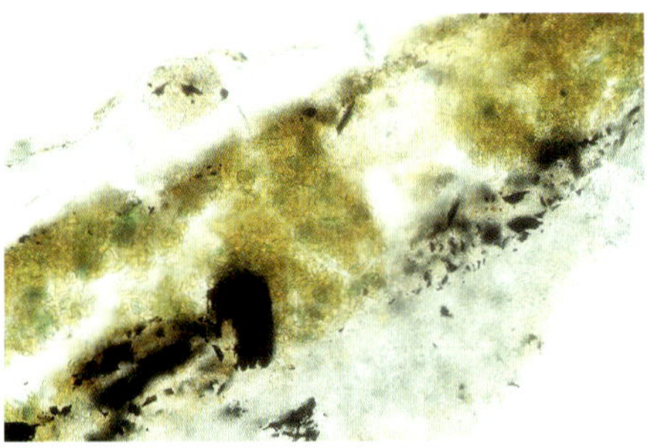

Plates 30 and 31 *Initial identification of vivianite at Winchester was made by polarised light microscopy, since the particles display a characteristic blue-yellow pleochroism. The thin section shown in these plates has been photographed in plane-polarised transmitted light and is shown in varying alignments as the microscope stage was rotated, original magnification 450×. The blue-yellow pleochroism is particularly visible in the large elongated particle which appears blue at the centre of Plate 30, while it is yellow in Plate 31. XRD provided further confirmation of the presence of vivianite. (Photomicrographs: H. Howard; © Courtauld Institute, Conservation of Wall Painting Department)*

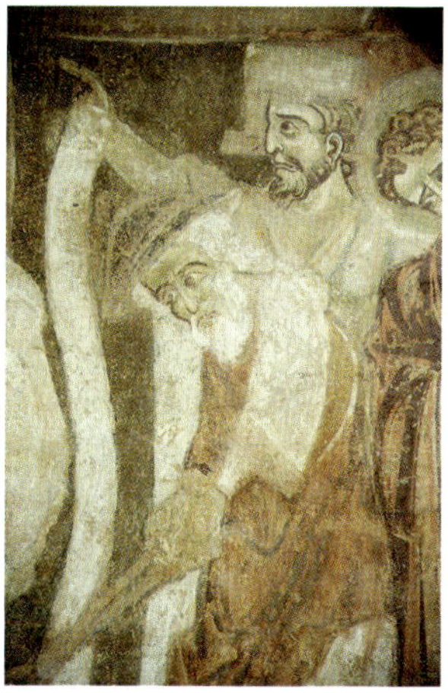

Plate 32 *Holy Sepulchre Chapel, Winchester Cathedral: detail of the* Deposition *of c1220 now transferred to the west end of the chapel. Vivianite was employed for the blue panelled backgrounds of this later scheme in the chapel and these areas now appear green. (© Courtauld Institute, Conservation of Wall Painting Department)*

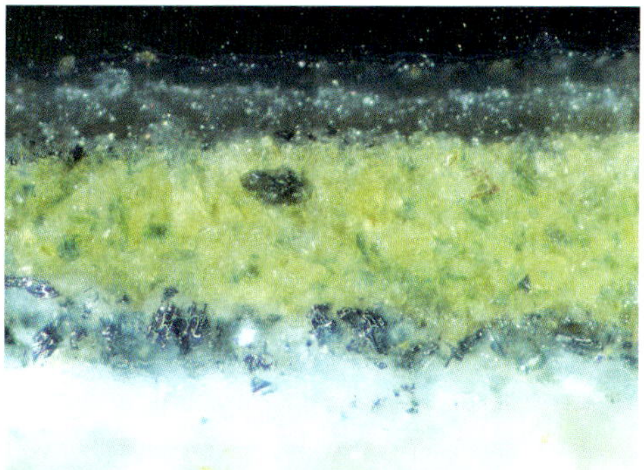

Plate 33 *Cross-section of a sample from the 'green' background shown in Plate 32, original magnification 470×. Vivianite was applied over grey underpaint of charcoal black combined with lime white which increased its covering power. The majority of the blue particles have now altered to yellow although the blue centres of the larger particles are still clearly visible. (Photomicrograph: H. Howard; © Courtauld Institute, Conservation of Wall Painting Department)*

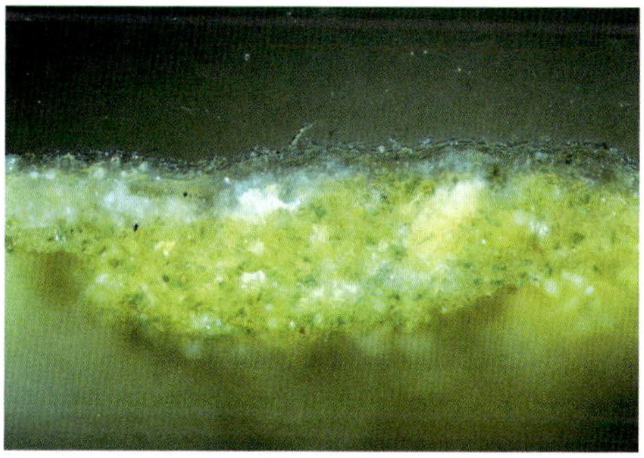

Plate 34 *Examination of polished cross-sections of the partially altered vivianite at high magnification indicates that it is the largest particles (19 × 7μm) which have remained blue in colour. Although yellow particles of similar dimensions are present, they show clear divisions along cleavage planes allowing for increased oxidation and therefore colour change. (Photomicrograph: H. Howard; © Courtauld Institute, Conservation of Wall Painting Department)*

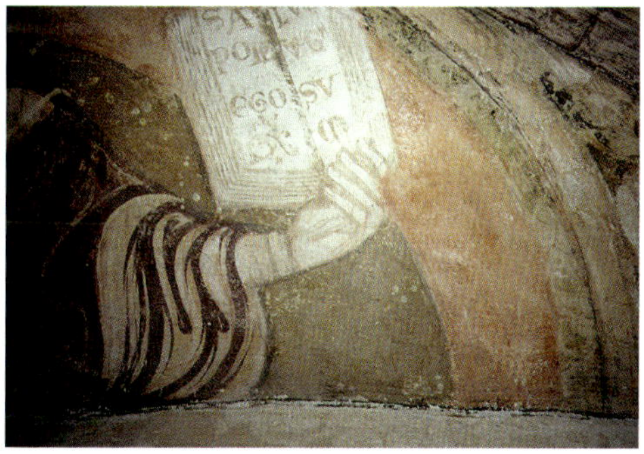

Plate 35 *Detail of the eastern vault of the Holy Sepulchre Chapel, Winchester Cathedral. Vivianite has retained its blue colour only where is has been protected from subsequent conservation treatments; for example, in the paint layer beneath the white dots on the background to the* Pantocrator. *(© Courtauld Institute, Conservation of Wall Painting Department)*

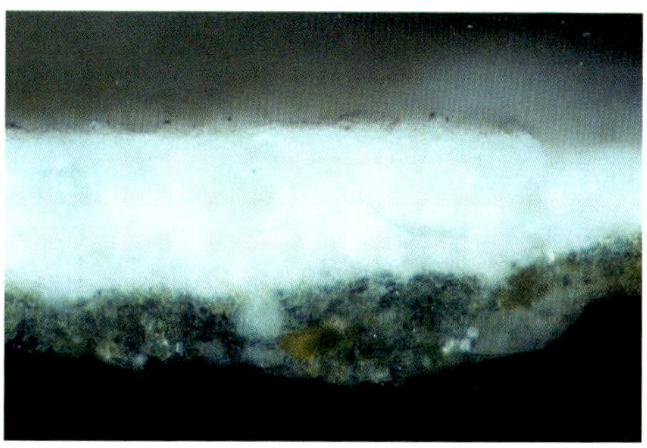

Plate 36 *A cross-section from one of the dots in Plate 35 shows the dark 'indigo' blue of vivianite beneath the white impasto layer, original magnification 500×. (Photomicrograph: H. Howard; © Courtauld Institute, Conservation of Wall Painting Department)*

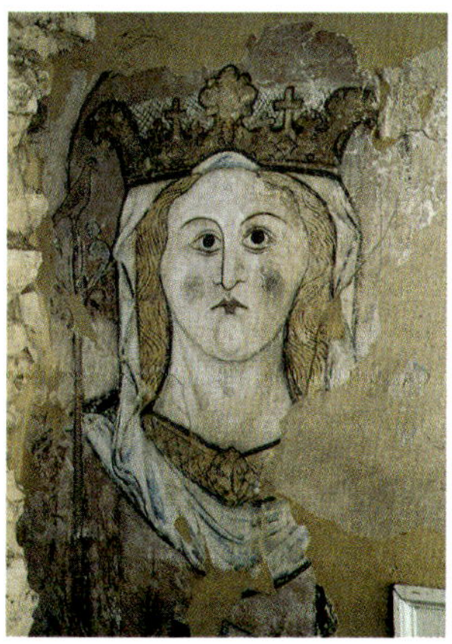

Plate 37 *Horsham St. Faith Priory (Norfolk): detail of the monumental figure of St. Faith on the east wall of the refectory (c1250). Although azurite has been identified in English wall painting as early as c1130, it was not commonly employed until the 13th century, as here at Horsham St. Faith. (© Crown copyright. NMR)*

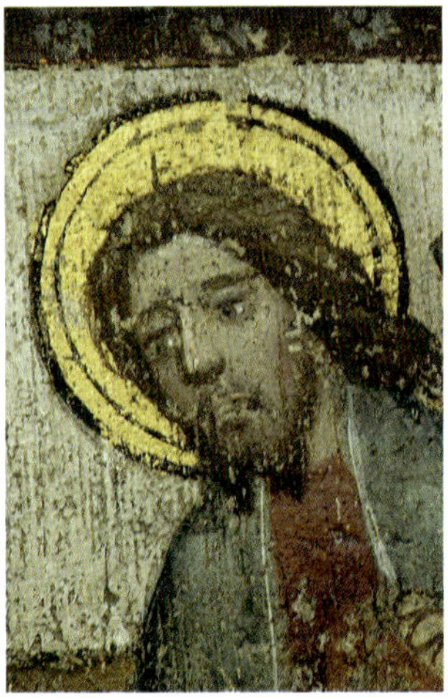

Plate 38 Westminster Abbey, chapter house, Apocalypse *cycle of c1375–1404: detail of* Banishment to Patmos. *Azurite combined with lead white was employed for St. John's blue drapery. (Photograph: R. Gowing; © Courtauld Institute, Conservation of Wall Painting Department)*

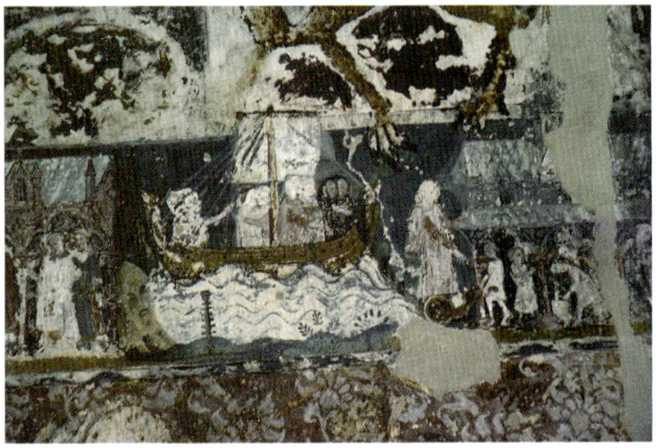

Plate 39 *Horsham St. Faith Priory (Norfolk): detail of the* Foundation Narrative *on the east wall of the refectory. (© Crown copyright. NMR)*

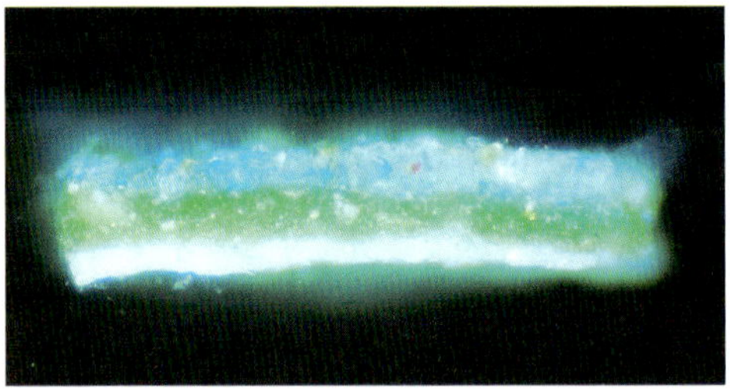

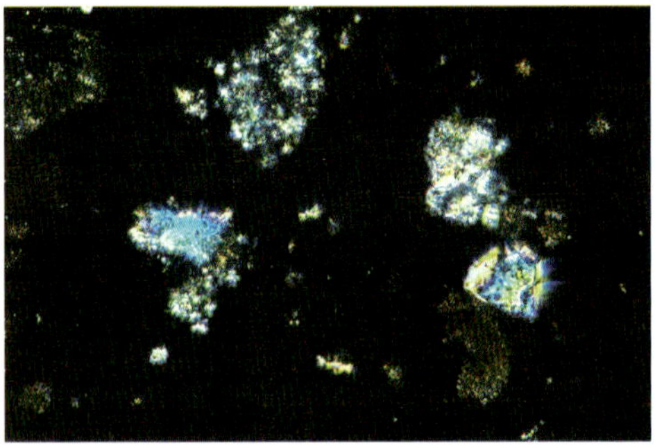

Plates 40 and 41 *Horsham St. Faith Priory (Norfolk). Cross-section of a sample taken from the first scene in the* Foundation Narrative, *original magnification 500×. Natural azurite, adulterated with its synthetic equivalent, has been applied over a layer of verdigris on a lead white ground. A dispersion of the blue pigment photographed in transmitted light (Plate 41) with crossed polars shows the angular particles of natural azurite; these display striations indicating their mineral origin. The synthetic copper carbonate has a combined aggregate structure seen at the top centre of the illustration, original magnification 400×. (Photomicrographs: H. Howard; © Courtauld Institute, Conservation of Wall Painting Department)*

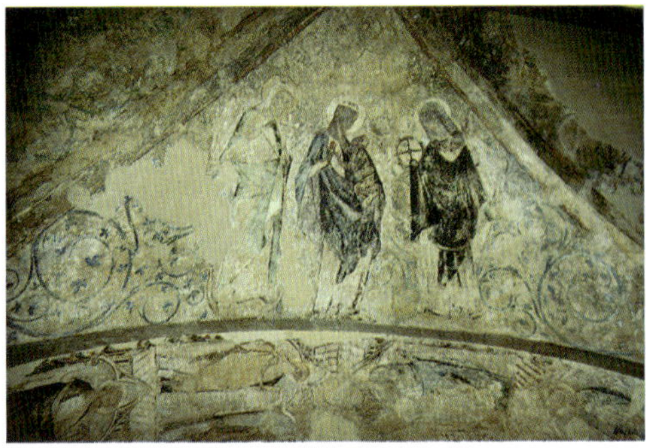

Plate 42 *Ante-Reliquary Chapel, Norwich Cathedral: detail of the scheme of c1300 on the west segment of the vault. (© Crown copyright. NMR)*

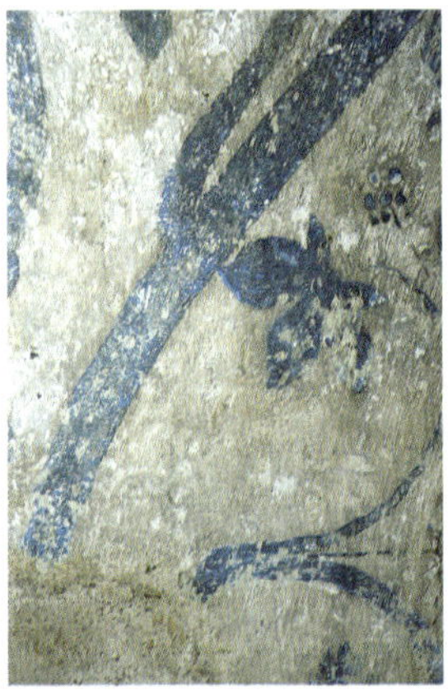

Plate 43 *Ante-Reliquary Chapel, Norwich Cathedral. Detail of the elaborate blue vinescroll surrounding the figures on the vault. (Photograph: H. Howard; © Courtauld Institute, Conservation of Wall Painting Department)*

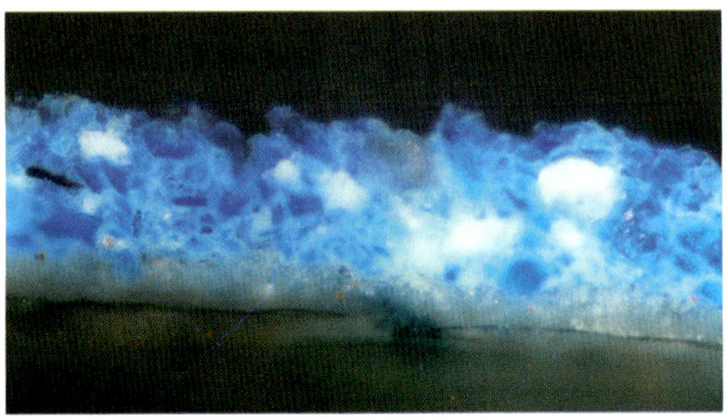

Plate 44 *Ante-Reliquary Chapel, Norwich Cathedral. Cross-section of a sample from the vinescroll showing the use of pure high-quality natural azurite. The largest particles, of an intense blue, measure some 75 × 50μm. Overall the pigment would have produced a beautiful surface texture, sparkling as the crystals were caught by the light. (Photomicrograph: H. Howard; © Courtauld Institute, Conservation of Wall Painting Department)*

tion and analysis of the scheme suggests that azurite in a white matrix was employed as an underpaint for ultramarine, or for ultramarine combined with azurite (Plates 15–17).[60]

The transition from ultramarine to azurite in 13th-century wall painting is exemplified by the two successive schemes in the Ante-Reliquary Chapel, Norwich Cathedral. Ultramarine was employed extensively both on the vault and on the east transverse arch of the chapel for the decoration of *c*1250, but only fifty years later a completely new scheme incorporating large quantities of azurite replaced the earlier painting.[61]

Pigment alterations

Examination of paint samples by PLM has provided some evidence for the transformation of azurite to malachite in two instances: in the scheme of *c*1130 at Kempley, Gloucestershire (Rickerby 1990a: 256), and on the late 14th-century polychromed screens surrounding the Chapel of Our Lady Undercroft in the crypt of Canterbury Cathedral (Howard 1993b). However, since malachite is frequently found as an impurity in samples of the blue pigment it is often difficult to determine whether a blue-green alteration has actually occurred.

Nevertheless, the alteration of azurite in the presence of humidity and chloride ions to form a green chlorine-rich alteration product has been confirmed in a number of instances. For example, this type of alteration has been detected in the scheme of *c*1300 in the Ante-Reliquary Chapel at Norwich, in samples taken from the shrine of St. Alban of *c*1302–08 in St. Albans Cathedral, and in the 14th-century *Apocalypse* cycle in the chapter house of Westminster Abbey.[62] This phenomenon is by no means confined to English wall painting; copper chlorides which are probably due to pigment alteration have also been identified in a number of Austrian wall paintings dating from the 13th to the 15th centuries (Kerber *et al.* 1972). For example, the transformation of azurite to form paratacamite has been detected in the paintings of *c*1263 on the vault of the west gallery of Gurk Cathedral.[63]

Examination of cross-sections from the shrine of St. Alban clearly demonstrate the blue-green alteration signalling the transformation, which in this case progresses further to a dark material visible in the centre of particles (Plates 63–66). Analysis of these samples by SEM/EDX showed that the green material consists of a copper chloride, while the dark centre of the particle is principally a copper oxide (Plates 67–72). It is curious that oxidation appears to have occurred at the centre of the particle.

The oxidation of azurite particles to form a dark alteration product (presumed to be tenorite) has also been identified in the late 14th-century wall paintings in the chapter house of Westminster Abbey. While a transformation of azurite to black cupric oxide (tenorite) has been detected in excavated medieval polychromed sculpture, in such cases the dark particles of tenorite have, as might be expected, developed on the surface of the

60. H. Howard, unpublished analytical investigation.

61. No accounts seem to refer directly to the two decorative schemes in the Ante-Reliquary Chapel, the Sacrist's rolls of the cathedral priory reveal that blue ('*azur*') and vermilion were purchased in 1276–77 for work that seems at least partly to have concerned polychromed sculpture; see Park and Howard 1996: 401, and Howard 1995c.

62. For the shrine of St. Alban, H. Howard, unpublished analysis. For the Ante-Reliquary Chapel, see Howard 1995c: 6, and for the *Apocalypse* cycle, H. Howard, unpublished analysis 1998.

63. For the analysis by F. Mairinger, see Ellwanger-Eckel 1979: 5.

blue pigment. In the Romanesque cycle of wall paintings in the chapter house of Santa Maria de Sigena (Spain), the transformation of azurite to a black alteration product, thought to be tenorite, is due to heat damage (Gasol Fargas 1999: 470). The transformation of azurite to tenorite in excavated polychrome sculpture has been shown to be due to the presence of calcium oxide and moisture which provided humid, alkaline conditions during burial (Gutscher *et al.* 1989).

Synthetic copper blues

The high cost of mineral blue pigments in the medieval period was a catalyst for artists to seek alternatives. 'False blue,' typically made from a combination of charcoal black and lime white, was widely employed in the Romanesque period, particularly in modest schemes in parish churches.[64] However, from the beginning of the Gothic period, a range of alternatives – including indigo and various synthetic copper blue pigments – seem to have suited both the wall painter's palette and the patron's purse.

Extensive use was made of the large variety of synthetic copper blue pigments, which fall into various categories depending on the nature and chemical form of the constituent materials. These include copper, vinegar, verdigris, sal ammoniac (ammonium chloride), lime, egg shells, oil of tartar (a saturated solution of potassium carbonate) and water.

The nomenclature of these synthetic blue pigments is almost as problematic as the determination of their precise composition, with the names verditer, silver blue and blue bice often used synonymously. However, verditer is generally considered to be artificial blue copper carbonate (synthetic azurite).[65] Although the other synthetic copper blues were also intended as replacements for azurite and natural ultramarine in the palette, they are generally copper acetates, or were manufactured from copper acetate. The name 'silver blue' is misleading, since the pigment formed in these recipes is principally neutral verdigris (copper(II) acetate monohydrate). This copper blue pigment is formed since traces of copper were present even in relatively pure forms of silver in the medieval period (Orna *et al.* 1985; Bomford *et al.* 1990: 24). During the 14th century the term 'bice' (bis, byce, bys) was an adjective meaning dark, thus 'azure bys' indicated a dark blue (Thompson 1956: 152).[66] From the 15th century the word bice was a noun meaning blue and so a wide variety of blue pigments may have been called bice (Thompson 1956: 152).

Preparation

The earliest manuscripts which provide recipes for synthetic copper blues are dated to the late 8th or early 9th centuries. One is part of the *Liber Pontificalis* in the Codex Lucensis 490 at Lucca, while the other is in the *Mappae Clavicula*. The Lucca manuscript outlines the process for the preparation of synthetic copper blues as follows: 'take pure

64. For a discussion of 'false blue', see chapter nine on black pigments.

65. 'Verditer' is frequently mentioned in Scottish accounts, particularly from the early 16th century; for example, in 1506 colours supplied by George Edwardson for Hathowy include 3½lb 'verditer' at 6s per lb; and in 1507 colours supplied by John Wright of Leith for Piers include 3lb 'viriditer' at 3s per lb; see Apted and Hannabus 1978: 148.

66. Gee (1984: 13) indicates that the term bis/byce suggested a dark grey or neutral blue.

copper plates and hang them over strong vinegar. Place them in the sun without touching them. After two weeks . . . collect the efflorescence'. The *Mappae Clavicula* gives two recipes. For the best blue, 'take a new pot and set in it sheets of the purest silver, cover and seal the pot. Set it in must discarded from the wine press and then cover it with must and keep it well for fifteen days. Then shake off the efflorescence which surrounds the silver'. For another blue, which was not considered to be of such high quality although perfectly serviceable for painting wood or plaster walls, 'take a flask of the purest copper and put lime into it halfway up, and then fill it with strong vinegar. Cover and seal it, then place it in the earth or some other warm place and leave it there for one month'.[67]

Without exception, these recipes would have produced basic copper acetates, copper(II) hydroxides or copper calcium acetates which are blue or blue-green in colour (Kühn 1993a: 132–3), and the procedures reappear in 12th-century texts, although usually as little more than copies of those in the earlier treatises.[68]

From the 14th century numerous recipes are found,[69] which are substantial variations and embellishments of the basic instructions from the Lucca manuscript and the *Mappae Clavicula*. However, the new recipes may be loosely classified into two basic types: those that replace metallic copper with verdigris as the starting material, and those that also include lime or egg shells and sal ammoniac and/or oil of tartar.

The substitution of verdigris for metallic copper was a substantial improvement, since the use of copper may produce a number of additional side reactions and therefore a very impure final product. The incorporation of sal ammoniac (ammonium chloride) in the basic recipe is first mentioned in a 14th-century manuscript (Cambridge, Trinity College MS 1451, p. 7) for making 'azure byzs' (Orna *et al.* 1980: 58). The author maintains that the blue produced by this method is superior to that made from other recipes. Although the processing time is halved, it seems unlikely that this in itself was seen as a substantial improvement, but rather that the addition of sal ammoniac produced a more stable dark blue product. The dark blue colour is due to the property of copper and ammonia to form deep blue cuprammonium compounds. In conjunction with lime it is possible to fix the cuprammonium salt in a relatively stable form, and some recipes recommend treatment with long steady heat which may increase the stability of this product.[70] Nevertheless, the main disadvantage of these copper ammonia–lime compounds was their tendency to change to green-blue through loss of ammonia content.[71]

Properties

The microscopic appearance of synthetic copper carbonates is of tiny particles of rather uniform size which may be slightly rounded and are pale blue in plane polarised

67. For these recipes and a full discussion of the primary sources, see Orna *et al.* 1980 and Scott *et al.* 2001: 82–9.
68. It is particularly interesting in this respect that although Theophilus provides no recipe for a synthetic copper blue, he describes a very similar process for the preparation of 'Spanish Green', by coating copper plates with pure hot vinegar and laying them in a wooden chest for a couple of weeks (Theophilus 1979: 41).
69. Listed by Thompson 1935c.
70. For example, see folio 48r of the 15th-century MS 1243 in the Biblioteca Riccardiana, Florence; Orna *et al.* 1980: 59.
71. Recent experiments have successfully produced an intense blue synthetic copper carbonate pigment using a recipe from the primary sources (Lombardi *et al.* 2000).

transmitted light. Under crossed polars they display birefringence (Gettens and Fitzhugh 1993a: 31). On the other hand, the structure and appearance of the other types of synthetic copper blue, which are generally copper acetates or were manufactured from copper acetates, are likely to be as varied as the recipes themselves. In many instances in which synthetic copper blues have been identified in English medieval wall paintings the pigment has an amorphous structure, appearing rather like a coloured stain on a lime white matrix, and often provides no clear XRD pattern when subjected to analysis. However, at high magnification (1000×) tiny blue particles with slightly rounded edges have been detected in a few samples (Plates 74 and 75).

Where an amorphous structure and stained appearance of a lime white matrix is evident it seems likely that the appearance may relate to the method of manufacture, since the process described in several recipes involves the grinding together of the reactants, in some cases to form a paste.[72] It is a matter of conjecture whether these blue materials were produced by simply grinding the reactants together, or by pounding together a synthetic blue pigment with a lime white matrix just before it was applied to the wall; in either case the result must have been much the same, except that the latter would probably have been lighter in tone.

Synthetic copper carbonates are stable in alkaline conditions unless heated, but are readily decomposed by even dilute acids, producing carbon dioxide gas. The physical and chemical properties of other synthetic copper blues are not reported, but since they are usually intrinsically bound with lime they would certainly be unstable in acidic conditions. The dark blue synthetic copper pigment formed by combining verdigris, lime and ammonium chloride was, however, known to lose its colour through loss of ammonia, and probably formed a pale blue-green alteration product.

Cost

The appearance of recipes for the more stable cuprammonium blues in early 14th-century treatises corresponds closely with the appearance of synthetic copper blues in other documentary sources. For example, the Westminster Palace accounts provide evidence for the use of synthetic copper blues in England in 1290, although the type of painting for which they were used is not specified. Both azurite and dark blue are itemised, '1½ oz pure azure – 2s 6d, and ¼ lb. azure biz – 7½ d', with the pure 'azure' – presumably azurite – approximately ten times the cost of the synthetic dark blue (Salzman 1952: 169). This accords well with the occurrence of synthetic copper blues in wall painting in England, for although a few late 13th-century examples are known, the accumulated analytical evidence suggests that the pigment did not become common until the early 14th century.

The royal accounts for 1365–66 (E.493, 2) list 12lb of verdigris followed by other pigments including 7lb *asure de bys* at 3s per lb purchased for the wall paintings which decorate St. John's Tower at Windsor Castle (Plate 107). Preliminary analysis of this scheme indicated the presence of several of the pigments listed in the accounts – including verdigris

72. For example, the method of production described in the Trinity College MS involves grinding together the reactants, vinegar, lime putty, sal ammoniac and copper, and then leaving them in an elevated temperature until the dark blue pigment is formed. Wunderlich (1998) discusses the recipes given in a 15th-century artist's booklet from Nuremberg which describe simply grinding together lime, verdigris and ammonium chloride with acetic acid to form a pale blue paste of copper calcium acetate.

and red lead – although no blue pigment was found (Brindle and Kerr 1997: 45). However, it is hoped that the *asure de bys* would be detected during any future examination of the wall paintings.

Use in English medieval wall painting

Although Caiger-Smith observed in the 1960s that artificially prepared blues – probably made from copper salts and ammonia fixed with lime – are occasionally found in English medieval wall paintings, there were no confirmed examples of their use before a study undertaken in 1988 by the author (Caiger-Smith 1963: 122; Howard 1988). At that time synthetic copper blue pigments were identified in the wall paintings at Horsham St. Faith Priory, Norfolk, of *c*1250; in the scheme of *c*1300 in the Ante-Reliquary Chapel, Norwich Cathedral;[73] in the scheme dated to the first half of the 14th century at Little Wenham, Suffolk (Park and Howard 1996: 396); in the 14th-century schemes on the chancel east wall at Brent Eleigh, Suffolk; in the 14th-century Doom over the chancel arch at Ickleton, Cambridgeshire; and in the 14th-century paintings at Great Hockham, Norfolk (Howard 1988: 33–4).

The blue pigment identified at Horsham St. Faith Priory remains, to date, the only confirmed example of a synthetic copper carbonate in English medieval wall painting. The pigment, combined with natural azurite, was identified in a sample taken from the first scene of the 'Foundation Narrative' on the east wall of the refectory (Plates 39 and 40). This combination would have the effect of bulking out the more expensive pigment, and may be a case of deliberately adulterated merchandise. It is interesting to speculate whether the 13th-century artist unknowingly purchased the adulterated pigment or made the addition himself. At moderate magnification, the combined aggregate structure of the synthetic product is easily distinguishable from the angular particles of the natural mineral (Plate 41). Both the natural and synthetic particles range in size from 5μm to a few at 20μm.

Since 1988 a synthetic copper blue pigment has been identified in the scheme of *c*1302–08 in the feretory of St. Albans Cathedral (Howard 1993a: 36). Analysis of paint samples from the 14th-century *Christ in Majesty* at Farringdon, Hampshire, also suggested that a synthetic copper blue pigment was employed for Christ's drapery and other details, but no details of the pigment's constituents or application are available.[74]

Recent research, and further examination of all the firmly identified examples, indicates that the synthetic blue-green material falls into two categories: (1) where the turquoise-blue pigment appears to have been a component of the original palette; and (2) instances where the material was found to be an alteration product of verdigris, usually following contact with limewash, or an alteration product of azurite.

Application

Directly over a lime ground

Synthetic copper blue appears to be a component of the original palette of the 14th-century wall paintings at Little Wenham parish church, Suffolk. Here the pigment was

73. In this case, subsequent analysis indicates that the 'synthetic' copper blue material here is in fact an alteration product of natural azurite.

74. The turquoise blue colour is reported to be similar to that found at Little Wenham; for the analysis undertaken by J. Darrah of the Victoria and Albert Museum, see Ballantyne 1988.

applied in a layer approximately 55µm thick over a buff-tinted lime ground for the drapery of the figures on the chancel east wall. In some parts of the samples the synthetic copper blue has no distinct particulate structure, and appears almost as a stain on the white lime matrix. However, at high magnification (500–1000×), tiny blue particles with slightly rounded edges are apparent in a few places (Plates 74 and 75). A similar rather stained blue-green paint layer was found in the 14th-century paintings in the north aisle of Great Hockham parish church, Norfolk. Here a blue-green layer 75µm thick was applied over a white lime ground, and details applied to the surface in charcoal black and red ochre. In both cases the blue-green colour is enhanced by the juxtaposition of warm orange-beige colours. The characteristic turquoise blue colour which is such a feature of both of these schemes may, or may not, relate closely to the appearance of the synthetic blue pigment when freshly applied, since it is possible that some alteration has occurred. For example, if the original pigment was a copper ammonia–lime compound and therefore originally dark blue in colour, it may have changed to green-blue through loss of ammonia content.

Binding media

In English medieval wall painting, synthetic copper blue pigments have generally been found combined with lime white. Through the carbonation of free calcium hydroxide, the lime white would have functioned as a binding medium. Unfortunately, invasive conservation treatments have compromised analysis of any original organic component of the binding medium in the samples examined from Little Wenham. However, it is possible that a proteinaceous component may have been included since amines (from protein) act as inhibitors of the corrosion of copper, and proteinaceous binding media have frequently been found to be associated with blue and green copper pigments.

Comparison with their use in other media

No examples of synthetic copper blue employed as part of the original palette have so far been detected in English medieval panel painting or polychrome sculpture.[75] Although a synthetic copper-based blue material has been detected in a few instances on Gothic polychromed sculpture – as on the late 14th-century polychromed screens surrounding the Chapel of Our Lady Undercroft in the crypt of Canterbury Cathedral (Howard 1993b: 32), and in the polychromy on the tomb of Sir Oliver de Ingham of *c*1344, at Ingham, Norfolk (Howard 1993d: 12–13) – in these examples it is clear that the blue colour represents an alteration of verdigris.

Pigment alteration

Significantly, examination and analysis of the blue alteration product on the Ingham tomb of *c*1344 indicates that the transformation appears to be associated with contact between the verdigris and overlying limewash layers: alteration appears to occur initially at the interface between the two layers, or where cracks in the underlying green paint

75. Although it seems likely that synthetic copper blues were employed for polychrome sculpture there is, as yet, no firm evidence to confirm this. Even the accounts for Exeter and Ely cathedrals, which frequently refer to *azure* or azura for painting areas such as the High Altar or the great cross, provide no details concerning the precise type of blue pigment employed. See, for example, Erskine 1981: 25, 134, 144, 172; and Chapman 1907: 136.

have provided a passage for further penetration of the calcium hydroxide (limewash).[76] A similar alteration has occurred on the polychromed stone screens surrounding the Chapel of Our Lady Undercroft, and in a sample taken from the mid-14th-century paintings over the chancel arch of Ickleton parish church. In the latter, a turquoise-blue layer 15μm thick has formed over the original paint layer of verdigris combined with lead white. It is perhaps significant that the turquoise-coloured alteration product is not apparent where an original dark green glaze over the verdigris layer remains intact. Where the glaze was lost, however, limewash layers applied subsequently have come into direct contact with the verdigris layer and effected the transformation.

Of particular interest is the similarity in appearance between the original pigment at Little Wenham, and the alteration product of verdigris in the presence of limewash as at Ingham. Perhaps this should not be surprising given the similarity in constituents of some of the recipes for the manufacture of the copper blues, and the materials which have come into contact on a wall painting or polychrome sculpture where such transformations have occurred. Thus, recipes for the manufacture of synthetic copper blues often call for a combination of verdigris, lime and ammonium chloride. The dark blue pigment formed was known to lose its colour through loss of ammonia, and although the resultant colour is not described, it seems likely that it would be similar in appearance to the alteration product formed from verdigris in contact with fresh limewash – a combination of materials similar to those for the preparation of the blue pigment with the exception of ammonium chloride (sal ammoniac).

It is, of course, possible that the synthetic copper blue employed at Little Wenham was made from verdigris and lime alone, and therefore would originally have been a pale blue-green colour. However, instrumental analysis undertaken on the samples from both Little Wenham and Great Hockham confirmed the presence of copper and calcium together with a high chloride content. This evidence suggests that the turquoise blue colour which is now apparent at these sites was not the intended colour, but rather that the present appearance is the result of the loss of ammonia content from the original blue pigment. Indeed, it is possible that the pale turquoise blue – which seems to complement so perfectly the Gothic mannerism of the paintings – may be an alteration of a dark blue copper ammonia lime compound.

The presence of chloride ions is not, however, conclusive proof that ammonium chloride was employed in the original manufacture of the now turquoise blue pigment. Chloride-containing salts were sometimes incorporated in the manufacture of a complex copper acetate-based green pigment,[77] and it is possible that a verdigris-like pigment manufactured in this way was combined with lime to produce a blue-green colour of the type now apparent at Little Wenham.

Following detailed examination and analysis, the 'synthetic' copper blue material detected in samples from the scheme of *c*1300 in the Ante-Reliquary Chapel in Norwich Cathedral was identified as a copper chloride-based alteration product of original azurite layers. Samples taken from areas of the vault and north side of the west transverse arch show various stages of the alteration process (Howard 1988: pl. 36; Howard 1995c: 6 and pl. 27).

---

76. For a full discussion of this pigment alteration, see the section on verdigris and copper resinate in chapter four.
77. See, for example, Theophilus 1979: 41.

At other sites there is insufficient analytical evidence to determine whether the turquoise blue pigment present represents part of the original palette or is an alteration of verdigris. This is the case at Farringdon, Hampshire, where analysis of paint samples from the 14th-century *Christ in Majesty* indicated the presence of a synthetic copper blue pigment employed for Christ's robe and for the background to the cross in his nimbus. The turquoise blue colour is described as similar in appearance to that found at Little Wenham, but no further details of its constituents or appearance are currently available (Ballantyne 1988). However, since iconographically Christ's robe is more likely to be blue than green, is seems that a synthetic copper blue was probably employed here, although as with the scheme at Little Wenham it is difficult to determine the precise hue of the blue colour originally applied.

At Brent Eleigh, near Little Wenham in Suffolk, the situation is still more complex. The paintings here were discovered beneath panelling, pigment and limewash layers in the 1960s, and the 14th-century paintings now visible on the east wall comprise two different schemes (Plates 76 and 79). The earlier, to the south of the altar, depicts the Harrowing of Hell with a tonsured donor figure. Partially overlapping this scheme and covering the area above the altar is a painted reredos in imitation of a panel painting. To the north of the altar, and forming part of the same scheme as the reredos, are two censing angels once attendant on a carved figure, set against a now dark background peppered with pale blue stars.[78]

Most areas of the background of the reredos appear bright turquoise blue, with tiny rosettes in pale blue painted on the surface. Analysis of a sample taken from the blue background indicated the presence of a synthetic copper blue pigment with a stained blue appearance, similar to that at Little Wenham (Plate 77). By contrast, examination of samples taken from one of the blue-green stars on a dark background to the north of the altar suggested the presence of verdigris, in what appears to be an oil medium (Howard 1988: App. III). The limewash applied over the surface of the painting has initiated a reaction resulting in the formation of the turquoise blue alteration product of verdigris in some areas.

In a few places, particularly at either side of the reredos and in a few small zones under the stars, a grey colour is visible. Analysis of these areas suggested the presence of dark copper oxide, which represents a further alteration by oxidation of both the synthetic copper blue pigment which forms the background of the reredos, and of the turquoise alteration product of verdigris.

Discussion

The stained appearance noted in most samples of the synthetic copper blue pigment identified is a phenomenon which is not fully understood. It may occur during the preparation of the pigment just before application, as it is possible that copper ions migrate into the white matrix during grinding, thus staining it blue. Alternatively, the process of the manufacture of the pigment – which in some cases is described as grinding slaked lime, water and verdigris into a paste – may result in blue copper calcium acetates of amorphous structure.

78. For a detailed discussion of these paintings, see Norton *et al.* 1987: 73–4.

Particularly with respect to the improvements in the method of manufacture of synthetic copper blue pigments in the 14th century discussed above, it is significant that all the identified examples date from the Gothic period, and it seems likely that it was very common at that time. The pigment occurs at sites ranging from a cathedral chapel to small parish churches. This, together with the extensive areas over which it was applied, even in parish church contexts, points to a low cost, as borne out by the Westminster accounts cited above.

Indigo

The principal source of indigo is the plant *Indigofera tinctoria* L., native to India. The other main source, although inferior in dye content, is European woad *Isatis tinctoria* L.[79]

Documentary evidence relating to native and imported woad

Archaeological evidence confirms that woad was grown in England from the Roman period.[80] Anglo-Saxon and medieval examples of woad deposits are known, for example those of the 9th–10th century at York (Tomlinson 1985). Extensive cultivation of woad in England in the medieval period left a trail of agricultural destruction because the plant rapidly exhausts the soil of nutrients. This led to the introduction of legislation to govern its cultivation in an attempt to reduce its pernicious economic effect (Thompson 1956: 136–8). However, this intensive cultivation was not undertaken for the production of a blue pigment, since the dye was mostly employed for dyeing cloth in the medieval period (Gettens and Stout 1966: 176).

Although archaeological and documentary evidence confirms that woad was cultivated in England in the Middle Ages, the earliest specialist woad-producing areas were in Picardy and later in Brabant and Lombardy. A record of woad merchants from Amiens trading in Bristol in 1294 suggests that woad was imported into England from these areas as early as the 13th century (Wilson 1937: 13). By the 15th century, Erfurt in Germany and the Toulouse region of France had become important centres of production, the latter with outlets at Bordeaux and elsewhere,[81] and documentary evidence confirms that quantities of woad were imported into Southampton and Bristol.[82]

The debt Court Rolls for Suffolk in 1328 detail the case of Thomas Gasconye, a merchant summoned for a debt of £11 owing on an order of woad. Even using calculations

79. *Indigofera* plants contain the colourless plant indican, indoxyl-D-glucoside. On hydrolysis this liberates indoxyl, which in turn oxidises to the dye indigo. In the case of woad, the indican has been shown to be indoxyl-5-ketogluconate; see Stowe *et al.* 1967: 547. Closely related to indigo is the expensive dyestuff Tyrian purple, obtained from Murex shells. In this case the dye is dibromoindigo; see Mills and White 1994: 150. For a detailed description of indigo and its preparation, see Schweppe 1997.
80. Indeed, Caesar noted that Britons painted themselves blue with woad (*vitrum*) which produces a terrifying appearance in battle; van der Veen *et al.* 1993: 367, 370.
81. For the woad industry at Erfurt, see Cardon 1990: 133–51. Woad from both the Toulousain and Albigeois was exported from Bordeaux; Vale 1990: 148.
82. Woad was included in the cargo brought by two Spanish ships docking at Southampton in 1352; Childs 1978: 29, 107–8.

based on the cost of indigo from Baghdad in the early 14th-century (18d per lb) (Erskine 1981: 134), this sum would have purchased approximately 145lb of woad (Thornton 1928: 174). The considerable amounts of woad exchanging hands were probably employed for dyeing cloth. This notion is supported by documents that detail items such as alum and wood ash – essential commodities for this process – which were imported with the woad.[83]

As early as the 12th century, indigo was imported into Europe from Baghdad. This important trading centre for Indian and Persian indigo gave rise to the medieval name for the pigment, *inde baudas*. With the increase in East Indian trade in the post-medieval period, and despite strict rulings against its importation, the indigenous species was threatened by increasing quantities of indigo arriving from India. By the 17th century trade in woad had fallen sharply in favour of indigo, the species with a higher dye content (Schweppe 1997: 83).

Documentary evidence relating to imported indigo

Documentary evidence confirms that indigo was imported into Europe from at least the middle of the 12th century, for example at Genoa in 1140. In 1228 indigo from Baghdad was recorded with other imports for customs at Marseilles, and again in 1276 at London (Salzman 1952: 169; Vetterli 1950). It was almost certainly the superior quality of the renowned Gulf or Baghdad indigo which justified the cost of its importation, and which made it many times the price of woad.

The superior Baghdad indigo is generally specified in documentary sources relating to painting in the medieval period. For example: in the Exeter Cathedral accounts of *c*1320–21, *ynde baudas* (Erskine 1981: 134); in the 14th-century accounts of the Countess of Artois, *d'inde de Baudas* (Richard 1887: 336); and for a 15th-century painting at York Minster, *blew ynde* (Gee 1984: 13). However, if a blue organic colorant such as woad was locally available, it is almost inconceivable that it would not have been pressed into service. In addition, it is not always possible to be certain which of a range of organic colorants was actually employed since materials were often named according to function, and so any blue dyestuff which served the same purpose may have been known as indigo (Thompson 1935b: 135). Unfortunately is it not possible to determine the source of the blue colorant – whether from *Indigofera tinctoria* (indigo) or *Isatis tinctoria* L. (woad) – once it has been prepared from the plant source.[84] Consequently, references in this section to the identification of indigo denote the identification of a blue organic colorant of the indigo–woad type.

The use of a range of blue colorants is confirmed by Theophilus who refers to indigo from the Orient, and also dyes produced from folium, mensc or elderberry – the last two described as alternatives to indigo, and both folium and mensc specified for use on a wall (Theophilus 1979: 15, 21, 25, 39).[85]

---

83. For example, the cargo of Genoese carracks docking at Bristol in 1381 included woad, wood ashes and alum (PRO Treaty Roll, 66), Wilson 1937: 39.
84. All analytical results by the author refer to indigo, but it is not possible to be sure that woad was not the origin of the blue dye.
85. Folium is derived from the plant turnsole – identified by Merrifield (1967: 1: cxc) as *Croton tinctoria,* though more likely to be *Helianthus annuus* L. (Cardon 1990: 182) – which appears blue when combined with lime, purple with weaker alkalis, and red in acidic conditions (pH 2–5). The lichen *Ochrolechia tartarea* (L.) is also mentioned in relation to making '*tournesol*'; see Cardon 1990: 305–6, 326–7. For the tempering of folium, see Thompson 1926: 301. Mensc is also derived from a plant, though of undetermined species.

Cost

Clearly much of the indigo imported into England would have been used in the dyeing industry, but evidence from building accounts confirms that the organic colorant was also employed for polychromy. There are numerous references to indigo in the English royal accounts of the second half of the 13th century, although these do not refer specifically to wall painting.[86] The Exeter Cathedral accounts for 1320–21 include an entry listing among other materials, 'one lb. of indigo of Baghdad (*inde baudas*) at 18d'. This record also indicates that indigo was six times the cost of the synthetic pigment lead white, although it was considerably cheaper than the '*azur*' also listed, which was bought from London at a cost of 3s 6d per lb (Erskine 1981: 134).

Some consistency in the price of indigo in the 14th century is suggested by a document recording the same price for indigo purchased to paint streamers for the king's fleet in 1350: '7lb of indigo (*indebandache*) at 18d per lb' (Staniland 1986: 241). Again, another blue purchased in addition to the indigo was much more expensive: at 6s 8d per lb, the *azure* purchased for the streamers was almost twice the price of that employed at Exeter. In the later medieval period *blew ynde* was listed with other pigments used for painting the central tower vault of York Minster in 1472,[87] and reference to indigo for the painting of ceilings is made in the Scottish treasury accounts of 1502: '8oz *inde* at 1s 2d per oz' (Apted and Hannabuss 1978: 140).

There is no documentary evidence to confirm that woad was employed for wall painting in England. However, an entry in the 14th-century accounts of the Countess of Artois provides what may be interpreted as a record of woad used for wall painting in France. Among the materials purchased for decorating rooms in the manor of the Count of Toussaint is an entry for indigo from Baghdad (*d'inde de Bandas*) and another for *inde*. That both were employed in the same decorative scheme suggests a variation in quality which is certainly borne out in the cost – the Baghdad indigo was seven times the price of the *inde* – and so it seems reasonable to assume that *inde* refers to the native species, woad (Richard 1887: 336).

Optical, chemical and physical properties

The optical and chemical properties of the pigment indigo depend largely on the substrate – lead white, gypsum or chalk – with which the dyestuff is associated.

The organic colorant itself is insoluble in common solvents such as water, alcohol and ether, but is soluble in chloroform, hot analine and phenols. The dyestuff is unaffected by dilute acids and alkalis, and is relatively stable in combination with oil, egg and glue.

Despite its permanence with respect to alkali, and although, as might be expected, indigo has frequently been identified in Indian wall painting, traditional Indian texts advise against its use on walls (Agrawal 1969–70: 111; Coomaraswamy 1934: 64). This is probably due to the photosensitivity of the dyestuff. Although indigo is the most lightfast of the natural dyes, it does fade at a slow constant rate (Crews 1987: 69–70).

86. A document of 1283 which refers to painting saddles indicates that both indigo and brazil lake were considered inferior (bad) colours, presumably due to their photosensitivity; see Tristram 1950: 399 and 409–10.
87. York Fabric Rolls, 77; Gee 1984: 13.

Preparation

Indigo and woad were primarily used as dyes, the pigment being the dried scum or foam collected from the surface of the dye-vat. Indigo is itself a very dark purple colour, and consequently white materials such as powdered marble, gypsum, egg shells or lead white were stained with it to produce a blue pigment – a practice described in classical antiquity.[88] The earliest medieval source which provides detailed instructions for this staining process is a 12th-century manuscript – Cambridge, St. John's College, MS 79 – cited by Thompson (1935a: 464), and from the 13th century such recipes were common.

Use in English medieval wall painting

The impermanence of indigo with respect to light led writers such as Tristram to assume that even if indigo had been employed for wall painting in the medieval period, it would by now have entirely faded away (Tristram 1950: 410). However, at the Reformation and subsequently, many English wall paintings were covered with limewash and other materials which would have protected the blue dye from light, and it is in paintings which have been revealed from behind such protective coverings that indigo has been found.

Indigo was first identified in a surviving English medieval wall painting as a result of the examination of the mid-13th-century *Wheel of Fortune* in Rochester Cathedral in 1987 (Darrah 1992: 2). A number of examples have subsequently been confirmed. The earliest in date, and the only instance so far in which indigo has been identified in English Romanesque wall painting, is in the decoration revealed during recent excavations under a property in Guildford High Street. This unusual building, dated stylistically to *c*1180, may represent the remains of the only extant 12th-century synagogue in England.[89]

The pigment appears to be quite common from the 13th century onward, and has been identified in, for example, the wall paintings of *c*1220 in the Eastbridge Hospital, Canterbury; the important late 14th-century paintings in the chapter house of Westminster Abbey; in the late 14th-century scheme in the Chapel of Our Lady Undercroft, Canterbury Cathedral; the scheme of *c*1470 in the Deanery at Durham, showing scenes of the Virgin; in the paintings of *c*1510 at St. Mark's, Bristol; and in the late medieval repainting of the figure of St. William of York in the feretory of St. Albans Cathedral.[90]

88. Vitruvius VII, 14, in Granger, Vol. 2, 1962: 129. Pliny (1984: 295) gives two sources of the pigment, namely indigo from India and skimmings from the dyer's vat of woad, and describes a test to confirm the presence of genuine indigo by observing the purple flame and the aroma of the sea produced on contact with hot coals.
89. Although the original function of the building is still uncertain, the room was first interpreted as a synagogue by David Keys. A detailed discussion of medieval synagogues in Europe (but excluding England and France) is provided by Krautheimer (1927). For the analytical investigation, see Howard 1996c.
90. For the Eastbridge paintings, see Howard 1994; and for both Westminster and Durham, see Howard 1988. For the Chapel of Our Lady Undercroft, see Howard 1993b: 32; for the St. Mark's scheme, see Gill and Howard 1997; and for the paintings in the feretory of St. Albans Cathedral, see Howard 1993a.

Application

Combined with white and applied over a white ground

Examination of a paint sample taken from the north wall of the 'synagogue' of *c*1180 at Guildford indicated that indigo was combined with lime white and applied in a layer 45µm thick over a lime white ground. A black linear detail in charcoal black and lime white was added to the surface (Howard 1996c: 4–5). For the blue-grey of the king's stockings in the mid-13th-century *Wheel of Fortune* in Rochester Cathedral, indigo was combined with chalk and applied over the white ground (Darrah 1992: 2).

Examination of a sample from the outer border of the nimbus of the St. Matthew symbol in the Eastbridge Hospital paintings confirmed that indigo combined with lead white was applied in a layer 75µm thick over a preliminary drawing in red earth on a calcium carbonate ground. The present green appearance of the paint layer is due to a resinous coating on the surface in which yellow particles from the centre of the halo have been incorporated (Howard 1994: 4 and pl. 16).

In the chapter house at Westminster, a sample from the blue robe of a figure in the south-east bay indicates that indigo was combined with lead white and applied over a lead white ground in a single layer 40µm thick. Close examination suggested that a few particles of bone black were incorporated in the blue paint layer, and although this is likely to be incidental it is also possible that they were added to produce a slightly darker blue effect (Howard 1988: 42). Indigo combined with lead white and applied over a white ground was also employed for St. John's drapery in the scene of the *Lamb on Mount Sion* from the *Apocalypse* cycle.[91]

Over coloured underpaint

Details in indigo combined with lead white were applied over a dark red layer of red earth and carbon black for the stencilled borders of the late 14th-century *Apocalypse* cycle in the chapter house of Westminster Abbey. Here the blue layer was applied in considerable *impasto*, and the red background colour would have had little or no impact on the colour of the stencilled detail. The blue organic colorant has also been identified in the *Majesty* in the eastern bay of the chapter house, where indigo was applied over a copper green layer for the lining of the drapery held by the angel in the third arch. Indigo combined with lead white was also applied over red earth beneath the nimbus of the angel at the extreme right of the first arch (Plates 82 and 83).[92]

In the paintings of *c*1510 at St. Mark's, Bristol, the use of coloured underpaints to react optically with subsequent paint layers has been employed to full effect. For the Virgin's robe in the *Nativity*, a thick black underpainting for the blue layer increased the intensity of the indigo by reducing the light scatter from the lead white ground, and thus made the dye less susceptible to fading (Plate 87) (Howard 1988: 42). Similarly a grey underpaint was applied beneath the pale purple (indigo combined with vermilion and lead white) of the veil of the female figure behind Christ in the *Noli me Tangere* (Plate 86) (Gill and Howard 1997: 100–101).

By contrast, in the late 15th-century scheme in the Deanery at Durham, a warm reddish underpaint was employed, presumably to provide a slightly purple cast to the Virgin's robe in the *Nativity*. Here the indigo was combined with lead white and applied

91. Howard 2000 and forthcoming.
92. Howard 2000 and forthcoming.

in two layers over the red underpaint (Howard 1988: 42–3). For the armour of a soldier in the *Resurrection*, indigo was combined with lead white and applied in a layer 60μm thick over a warm orange-yellow layer, which in turn was applied over a fine layer of bone black. The resultant blue-grey colour of this layer has an almost metallic sheen (Plates 84 and 85) (Howard 1988: 42).

Layered with yellow pigments to produce a green colour

Unusually, indigo was overlaid with a orpiment to produce the green colour at the base of the wheel in the mid-13th-century *Wheel of Fortune* in Rochester Cathedral.

Applied over natural azurite

Technical examination of the figure of St. William of York in the feretory of St. Albans Cathedral indicated that areas of his blue dalmatic were overpainted with indigo later in the medieval period. Its use here is particularly surprising since not only is indigo subject to fading but its small particle size and inherent dark blue colour produces a colour far less brilliant than the original azurite layer it covers (Plate 46) (Howard 1993a: 37).

Binding media

Indigo – often specifically 'Baghdad indigo' – is recommended as a pigment suitable for both wall and panel painting by Cennini. He suggests indigo as a substitute for azurite and in various pigment mixtures – with orpiment or with lead white – tempered with size. Although Cennini suggests the use of an additional binding medium for indigo, the organic colorant is stable in alkaline conditions, and he also describes imitating ultramarine with indigo combined with lime white applied in a fresco technique (Cennini 1960: 32, 36, 52). Since the majority of identified examples in English wall painting date from the Gothic period, the pigment was not generally applied in a *fresco* technique. However in the Guildford scheme of *c*1180 the pigment was combined with lime white and applied over a lime ground and was therefore adhered by the carbonation of calcium hydroxide.

The binding media associated specifically with indigo have not been identified in other instances where the pigment has been detected in English wall painting. However, the presence of oil was confirmed by instrumental analysis in layers containing lead white and verdigris in the Eastbridge Hospital scheme, linseed oil by GC-MS in the St. William of York at St. Albans, and oil – used both as a sealant and as a binding medium – by FTIR analysis in the paintings at St. Mark's, Bristol. It seems likely, therefore, particularly given the stability of indigo in a range of organic media, that these were commonly employed to bind the pigment in Gothic wall paintings.

Comparison with its use in other media and in continental wall painting

Although indigo has only been detected in one example of English Romanesque wall painting, it has also been identified on the carved frieze on the west front of Lincoln Minster (*c*1140), and on the column figures of *c*1180–85 from St. Mary's Abbey, York.[93] A dark red underpaint was employed beneath the indigo on the Lincoln frieze, and here

93. For the York figures, see Brodrick 1993: 23–7.

the indigo was combined with verdigris and copper resinate, presumably to form a very dark green paint layer.[94]

As with wall painting, there are more numerous examples of indigo in polychrome sculpture of the Gothic period. It has been identified, for example, on the 14th-century roof bosses of Exeter Cathedral, where it was employed in addition to natural azurite (Hulbert 1991: 195). It was also extensively employed for the polychromy on the west front of the same cathedral. In the 14th-century phase of decoration, the pigment was used for details such as beards and the wings of a small angel in the south porch, while in the 15th-century polychromy, indigo was employed for the lining of a head-dress, and as an underpaint for natural azurite (Sinclair 1991: 123, 1992: 10). In the latter case, the organic colorant would have increased the covering power of the more expensive mineral blue by reducing light scatter. The azurite would have required an additional organic binding medium in this type of stratigraphy, and would therefore have been more susceptible to loss than the underlying indigo layer. Although pure conjecture, it is therefore interesting to speculate whether indigo was employed as an underpaint for azurite elsewhere, and that the mineral blue was lost leaving the organic colorant exposed.

Indigo on a red underpaint has also recently been identified in the lining of a cloak of one of the figures from the 15th-century Great Screen in Winchester Cathedral (Brodrick and Darrah 1989: 616–17), and, mixed with lead white, on the roughly contemporary polychromed wooden figure of Christ from Cartmel Fell.[95] The pigment has also been detected on a series of wooden beams, with heraldic decoration dated to the 15th century, found reused in the roof of Ducklington Rectory, Oxfordshire (Steane 1985: 203).

The presence of indigo has also been confirmed in a number of Gothic panel paintings, as on altar frontals of *c*1300 from Trondheim and Nedstryn in Norway. In the latter, azurite was employed for the garments of figures while indigo was used only for the wavy border inside the frame, while on the Trondheim frontal indigo was combined with azurite and lead white and employed as an underpaint for natural azurite (Plahter 1995). Use of indigo together with both natural ultramarine and azurite has also been confirmed on the late 14th-century Wilton Diptych (Roy 1997: 133–4), and indigo has also been detected on the 15th-century Herlin altarpiece from St. James, Rothenburg (Broekman-Bolstijn 1970).

Discussion

It is significant that in all the examples in which indigo has been identified in wall painting, they have been protected from light for a considerable period. Residues of a calcium carbonate layer are present over the polychromy in many samples from the Guildford painting, representing either the remains of a ground for later repainting or a limewash layer used to cover the decoration at some time before the room was sealed off, perhaps shortly after the Jews were expelled from Guildford in 1275 (Boas *et al.* 1996: 14). The Eastbridge Hospital paintings were covered over following the Reformation, and the royal arms of James I (1603–1625) and the arms of Archbishop George Abbott (1611–

94. Information on the identification of indigo on the Lincoln frieze was provided in a lecture (3 Dec. 1987) by Josephine Darrah of the Victoria and Albert Museum.
95. Analysis by Plesters and Roy, in Behrens 1982: 133.

1633) painted over them on a limewash ground.[96] Both the Westminster and Durham schemes were panelled over, while that at Bristol was covered during structural alterations to the building in the post-Reformation period, and not subsequently revealed again until 1824 (Gill and Howard 1997: 97–8).

It is possible that indigo may have been used very widely in wall paintings. In many cases the organic colorant would have faded completely before any limewash or other covering provided protection from light; in others, it may have faded since being revealed. No doubt as more scientific examination of medieval wall painting is undertaken, further cases of its use will be confirmed. It is interesting to speculate, for example, whether the trace of blue observed on the paintings of *c*1479–87 in Eton College Chapel may be indigo (Plate 88). Blue was only detected on the sword of the figure furthest west on the south wall, an area largely protected from direct light (Plummer 1986: 36). Routine analysis of the pigments employed in wall painting – particularly those which have been recently uncovered – is therefore of primary importance in order that those with materials susceptible to photodeterioration may be protected from light.

In both wall painting and polychrome sculpture, the number of identified examples increases substantially in the Gothic period, and indigo was clearly widely available at this time. This increase may relate to greater use of the pigment at this time, or to a slightly greater survival rate of the photosensitive organic colorant from polychromy of a later date.

Pigment alteration

Documentary evidence shows that in the case of the Westminster chapter house scheme the blue has faded substantially since the paintings were discovered in 1847. In the early 20th century, W. R. Lethaby described the principal colours of the recently revealed scheme as blue, green, brown and red (Noppen 1931: 151). The blue dye was still apparent almost a hundred years later when Tristram waxed the paintings and observed the deep blue diapered drapery of the angels (Tristram 1955: 201). Indeed a watercolour copy of the paintings shows swathes of blue employed, for example, for the seraphs' wings and for the drapery behind Christ in the Majesty cycle in the eastern bay. However, on initial inspection in 1987, the overall colour bias appeared red-green, with little blue apparent, although a more detailed examination at magnification revealed that a faded and dirty blue remained on Christ's robe in the *Apocalypse* cycle and in scenes in the south-east arcade. Subsequent analysis confirmed the presence of indigo, although the pigment was notably faded towards the upper surface of the paint layer.[97]

As with other organic colorants, loss of colour is directly proportional to exposure to UV radiation and, as has been shown for red lake pigments, fading is generally more evident where the dye was combined with white and applied over a white ground, as in the Eastbridge Hospital and Westminster examples (Howard 1988: 42, 1994: 4).

96. Later in the 17th century a fireplace and chimney were constructed against the wall; see Medhurst 1993.
97. For a recent study showing the cumulative effect of incident daylight illumination on these important paintings, see Gowing 1997: 33–42.

FOUR

Green pigments

Green earth

The pigment green earth consists principally of the siliceous, iron-rich clay minerals glauconite and celadonite. Although the chemical composition of these two minerals varies slightly, both may be represented by the general formula $K[(Al, Fe^{III}), (Fe^{II}, Mg)](AlSi_3, Si_4)\, O_{10}(OH)_2$. Although similar in composition, glauconite and celadonite generally do not occur together since they are formed under very different geological conditions: glauconite in marine sediments, and celadonite as a secondary mineral in certain basalts. Other minerals with similar structures – such as chlorite and montmorillonite – have also on occasion been employed as green earth pigments.[1]

Although sources of green earth minerals are common throughout the world, their quality varies considerably. Among those most renowned, presumably for their purity and richness of colour, were green earth from Smyrna (Asia Minor) in Antiquity;[2] Creta Cirina, known also as Theodote (Eraclius 1996: 125, n. 17); and Verona.[3] From the 19th century, numerous sources are known including those in Bohemia, Cyprus, Saxony, France, and England (from the Mendips and Cornwall). Analysis suggests that the geological source in France may consist principally of glauconite, while in Cornwall both septechlorite and cronstedtite have been identified (Grissom 1986: 149).

Cost

Recent research indicates that green earths were imported from north Italy and Cyprus for use as pigments in Gallo-Roman wall paintings in southern France (Delamare *et al.* 1990; Odin 1988: 75–6). In medieval England it is uncertain whether local or imported green earths were employed, or how much they may have cost.

However, the green earth pigment recently identified in wall painting fragments of *c*1190 from Glasgow Cathedral was found to be celadonite (Howard 1998a; Park

1. For a full discussion of the characteristics and use of this pigment, see Grissom 1986.
2. Vitruvius Book VII, Chapter VII, in Granger 1962: 113.
3. This source was documented from 1574.

1998; Park and Howard forthcoming). In this instance a few particles of celadonite were also detected in the lime plaster substrate, where it was employed as aggregate with other materials such as quartz, feldspar, and red and yellow earths. Aggregate for plaster would typically have been obtained locally available material, and one nearby is Ailsa Craig, situated at the entrance of the Firth of Clyde. This distinctive blue-grey tertiary microgranite boss is rich in glauconite-celadonite and other accessory minerals. During the late medieval period the large sea-bird population of this island was harvested to provide food for the abbey of Crossraguel on the mainland, and it is unlikely that the green mineral was passed unnoticed. In addition, the longshore drift carries pebbles of the microgranite from the island to the mainland north-east of Ailsa Craig (Harrison *et al.* 1987: 5, 5, 15; Craig 1991: 479–80). It is possible therefore that in this instance the green mineral may have been collected locally, in which case the green pigment employed at Glasgow was probably low in cost.

Clearly, analysis of the geological source of these and other mineral pigments would be an interesting avenue for further research.

Optical, chemical and physical properties

Celadonite and glauconite are not distinguishable from each other by PLM. In plane polarised transmitted light both minerals appear as rather rounded translucent particles of varying shades of green, sometimes with yellow or brownish particles. The refractive indices of the monoclinic crystals increase with ferric iron content and are given by Grissom (1986: 150) as $\alpha = 1.592–1.610$, and $\beta = \gamma = 1.614–1.641$.[4] The particles often have a grainy texture, flat crystal faces and display green-yellow (α) and blue-green-olive green (β and γ) pleochroism. In crossed polarised light the extinction is undulose and the particles display low to moderate birefringence (Grissom 1986: 151–2).

Green earth pigments are stable in the alkaline conditions associated with the *buon fresco* technique, although they are partially soluble in acids and more concentrated alkalis. They are compatible with all other types of binding media and pigments. Green earth is stable with respect to light, although on heating to around 500 °C burnt green earth is produced which is a warm, dark brown colour (Rickerby 1991: 40). Due to the layer silicate structure of these clay minerals they are highly absorbent. Glauconite commonly weathers to form mixtures of limonite and geothite, and will eventually form montmorillonite with the loss of potassium ions (Gribble and Hall 1992: 111).

Preparation

The pigment green earth is prepared by washing and grinding iron-rich clay minerals, principally glauconite and celadonite. Cennini recommends grinding the pigment with clear water (Cennini 1960: 30). To eliminate impurities, acid treatment and levigation may also on occasion be employed (Grissom 1986: 150).

4. Gribble and Hall (1992: 111) give the refractive indices for glauconite as $\alpha = 1.56–1.61$ and $\beta = 1.61–1.65$.

Use in English medieval wall painting

Confirmed examples of the use of green earth in English wall painting occur principally within the Romanesque period, although there are isolated examples later in the medieval period. It appears that the slightly soft green colour of the earth pigment accorded well with the Romanesque painter's palette, while the more intense chroma of verdigris ensured the more frequent selection of that pigment by the Gothic wall painter.

Although green earth has occasionally been identified in Romano-British wall painting,[5] the earliest identified use of the pigment in English medieval wall painting is in the scheme of *c*1090 in the Cluniac Priory at Lewes, Sussex. Analysis of excavated fragments from this scheme indicated not only that green earth was included in the palette, but that it was employed to lay out the design together with lime white, red and yellow earth pigments, and lime white combined with earth colours, in layers typically 20–30μm thick. Other identifications in Romanesque wall painting in England include the scheme of *c*1130 in St. Gabriel's Chapel, Canterbury Cathedral; the late 12th-century paintings in the Holy Sepulchre Chapel at Winchester; and in the 'Losinga' cycle of *c*1190 in Norwich Cathedral. On the wall painting fragments of *c*1190 recently excavated from the crypt of Glasgow Cathedral, celadonite was identified by FTIR.[6]

Green earth has been identified in a few later medieval wall paintings, including a scheme of *c*1220 in the crypt of Worcester Cathedral.[7] The pigment has also been detected on fragments of 13th-century wall painting excavated from the Hospital of St. Giles at Brompton Bridge (North Yorkshire) (Stewart 1995), and tentatively identified in the 14th-century paintings in Prior Crauden's Chapel, Ely Cathedral.[8]

Application

Over white ground or directly over the plaster support

In the scheme of *c*1175 in the Holy Sepulchre Chapel at Winchester, green earth combined with lime white and applied directly over the lime plaster was employed for areas such as the Magdalene's drapery. The dark green shadows of the drapery were modelled in green earth combined with charcoal black and lime white, and finally decorative details in red lead were applied in *impasto* on the surface.[9] In the scheme of *c*1220 in the crypt of Worcester Cathedral, the pigment was combined with lime white and applied over a white ground to provide a background colour for the figures of bishops.

Over coloured underpaint

In the scheme of *c*1130 in St. Gabriel's Chapel, Canterbury Cathedral, evidence suggests that green earth may have been employed for the panelled backgrounds of the scenes

5. Analysis by FTIR suggested the presence of celadonite on Roman fragments excavated from the Bedern, York (analysis by Howard, in Wong forthcoming). In addition, green earth was identified in the remains of a bowl found in a Roman dump at Wroxeter (Hughes and Heyworth 1990), and on 1st–4th century wall paintings in the baths of the Roman villa at Bignor, Sussex (Grissom 1986: 162).
6. The FTIR spectrum of celadonite [$K(Fe^{3-}, Al^{3+}) (Mg^{2+}, Fe^{2+}) Si_4O_{10} (OH)_2$] has characteristic absorptions at 3601cm^{-1}, 3578cm^{-1}, 3556cm^{-1}, 3533cm^{-1}, 838cm^{-1}, and 798cm^{-1}. For the technique of the Glasgow fragments, see Howard 1998a, Park 1998, and Park and Howard forthcoming.
7. Analysis by Howard and Kakoulli in Park 1997.
8. Unpublished analysis by C. Babington, 1987.
9. Analysis by J. Darrah in Hluvko 1991: 27–8.

(Rees Jones 1936: 6–7; Cather and Howard 1994). Although little of the pigment now survives, a striking feature of the scheme – particularly when compared with the mid-12th-century *St. Paul and the Viper* in the Chapel of St. Anselm above – is the almost complete absence of greens.[10] This is likely to be a relatively recent phenomenon, since in the 19th century Scott Robertson described the background as follows:

> the broad band of neutral blue, speckled with white, which runs across the panel, may represent sky . . . the narrower borders of green may represent a floor or ceiling (Scott Robertson 1880: 69).

Earlier this century, a paint sample taken from the then green border of the *Naming of St. John* indicated the presence of green earth applied over grey underpaint (Rees Jones 1936: 6–7). These narrow borders now appear grey, and recent samples confirm that any original green pigment – presumably applied *a secco* – has been lost.[11] Since green earth is stable in alkaline conditions, the reason for its application *a secco* may have been that the grey underpaint was necessary to increase its covering power, or that it was originally applied combined with an alkali-sensitive pigment.[12] In the late 12th-century wall painting fragments from Glasgow Cathedral green earth also appears to have been applied over grey underpaint (Howard 1998a: pl. 18).

Examination of samples from the late 11th-century fragments from Lewes Priory indicated a complex stratigraphy with well-defined linear details in black applied over the green earth, which was in turn applied over yellow or red-pink underpaint. This use of green over red or pink underpaint is surprising, since it would reduce the intensity of the green colour. When thinly applied, however, green earth has a low covering power and is rather translucent, and perhaps a subtle interaction between the pink and green was therefore intended here to represent the shadows of flesh tones.

Used as an underpaint for flesh tones

The use of green earth to produce shadows in areas of flesh painting has been confirmed in only one instance in English Romanesque wall painting: in the scheme of *c*1175 in the Holy Sepulchre Chapel at Winchester (Plate 89). Here, it is particularly clearly visible in the semi-naked figure of Christ in the *Entombment*,[13] and is among a number of features in these paintings suggesting strong Byzantine influence.[14] The pigment was also

10. Preliminary examination of pigments in the *St. Paul and the Viper* suggested the presence of malachite, and a green made from combining charcoal black and yellow pigments; Gärtner and Organ 1989: 11. Similarly, a mixture of charcoal black, yellow earth and lime white was used for the green detail of Christ's foot in the *Feeding of the Five Thousand* (*c*1165) at Pürgg, Austria; Howard 1995d: 4.
11. By contrast, a 'green' colour has survived on the soffits of the apsidal arches, where its pale green appearance is due to a combination of ultramarine and yellow earth in a lime matrix.
12. Cennini (1960: 30, 33–4) recommends green earth combined with water or lime white for wall painting.
13. Green, produced by the optical mixing of yellow ochre and black, was employed to build up the flesh tones in the *St. Paul and the Viper* in St. Anselm's Chapel, Canterbury Cathedral; Gärtner and Organ 1989: 11. Green underdrawing also produces effects of modelling in the 12th-century scheme in the north transept of Chichester Cathedral; Gärtner 1988: 2.
14. The Byzantinising style of the Winchester paintings and their relationship with artists of the Winchester Bible has been discussed in detail by Park 1983.

employed to produce the shadows of flesh tints in one other instance elsewhere in Britain in this period, in the late 12th century fragments of wall painting from Glasgow Cathedral (Howard 1998a; Park 1998).

For areas of flesh painting in the Holy Sepulchre Chapel, multiple layers were applied to produce complex effects of modelling. For example, the mid-dark flesh tone was achieved by applying five layers of earth pigment – green, red and yellow earth employed singly or in combination with lime white and charcoal black. Generally in this painting, it appears that flesh tones were initially established in varying yellow hues, consisting of yellow earth pigments combined with lime white and charcoal black. Following this, the shadows were painted in green earth applied in layers approximately 45µm thick.[15] Finally, for the mid-dark flesh tone of Christ's foot, a layer consisting of red, yellow and green earths combined with charcoal black was employed, while for the drop of blood on his hand, a layer of vermilion was applied over a trace of yellow iron oxide (Howard 1995a: 101).

In the 'Losinga' cycle of paintings of *c*1190 in Norwich Cathedral, the subtle yellow-green of green earth was juxtaposed with the more intense chroma of 'salt green'. The interplay of the two greens is clearly visible in a sample taken from the floral motif between the two lower roundels on the soffit of the western arch, where green earth combined with lime white was applied in a layer approximately 30µm thick over atacamite, also combined with lime white. A watercolour copy of the scheme made by G. E. Fox at the beginning of the 20th century, when the paintings were in better condition, indicates that the careful layering of greens of varying hue was then visible in other parts of the scheme. For example, in the *Simony* roundel a warm yellow-green was applied over a paler cooler green layer to model Ranulf Flambard's drapery.[16] The two pigments employed to achieve this effect were also presumably green earth and atacamite.

Used as a local ground beneath other pigments

Tentative identification of green earth, applied as a local ground beneath successive layers of yellow ochre and carbon black, was made in the 14th-century paintings in Prior Crauden's Chapel, Ely Cathedral.[17]

Binding media

In the Romanesque period, green earth was frequently applied either directly over the lime plaster or in combination with lime white. In the latter case the lime white pigment would have provided binding capacity within the paint layer, as well as contributing to the overall tone of the pigment mixture. Although green earth is stable in alkaline conditions, it was frequently applied *a secco* over coloured underpaint. The rationale for this particular method of application is probably associated with efforts to increase the covering power of the pigment, or to increase the intensity of particularly pale green examples. Green earth is compatible with all types of organic binding media but no instrumental analysis of the medium employed in cases where it was applied *a secco* has been undertaken.

---

15. The rationale for the use of green for shading flesh tones is likely to be the more lively neutral colour achieved by combining or layering green and pink, by contrast with the dull appearance which results from darkening pink by the addition of black or grey.
16. Fox's watercolour is in the Library of the Society of Antiquaries of London. For the technique of this scheme, see Howard and Gasol 1996.
17. Unpublished analysis by C. Babington, 1987.

Comparison with its use in other media and in continental wall painting

The presence of green earth has been confirmed in various Romanesque wall paintings elsewhere in northern Europe, as in the scheme of *c*1089 in the west choir of Lambach, Austria (Plate 90) (Hammer 1987: 45); in the late Romanesque scheme at Müstair, Switzerland (Emmenegger 1986: 198); in the wall paintings at St. Bavo's Abbey, Ghent (Grissom 1986: 163); in the 12th-century schemes at Berzé-la-Ville (Grissom 1986: 163); and in the crypt of the Abbey of St. Philibert, Tournus, France (Rollier-Hanselmann 1997: 66). In the 12th-century scheme in the crypt of Auxerre Cathedral (France), a very pale green colour was achieved by applying glauconite *a secco* over yellow underpaint (Cadet 1995: 37). In Romanesque polychromed sculpture, green earth was employed for the green belly of the lion-capital from Vossestrand, Norway (Plahter 1990: 278).

Notable occurrences of green earth in panel painting are cited by Grissom (1986: 162–3), although one recent identification worthy of note is in the late 14th-century Wilton Diptych, where a copper green likely to be copper resinate was also employed (Roy 1997: 133).

Discussion

The use of green underpainting for flesh tones is recorded as early as the 11th century as in a manuscript from Mount Athos (Schäfer 1855; Grissom 1986: 143). Theophilus, writing in the early 12th century, describes a mixture called *prasinus* – a combination of green and black which may or may not have included green earth – for painting the shadows of flesh tones, although he provides no clear indication of the method or stratigraphy of its application.[18]

Green underpaint for flesh tones was employed in a number of instances in French Romanesque wall painting, for example in the schemes at Berzé-la-Ville, Tournus, and the cathedral at Nevers.[19]

The method employed in the Romanesque scheme in the Holy Sepulchre Chapel at Winchester, whereby the mid tones were initially established, followed by green shadows and finally highlights, diverges from the Byzantine tradition described in the treatise by Dionysius. Although Dionysius recommends a mixture of green earth, dark [yellow] ochre, lime white and black for underpainting flesh colour on a wall, he indicates that this shading should be applied first, and over this dark yellow-green layer, a medium flesh colour was to be thinly applied, merging with the green underpaint around the edges. Successively lighter flesh colours were then applied, finishing with highlights in white (Dionysius 1974: 8, 14).

Cennini, writing some two hundred years after the completion of the Romanesque paintings in the Holy Sepulchre Chapel, provides a precise methodology for the use of green earth for the underpainting of flesh:

18. Since the preparation of *prasinus* involved straining the water-saturated material carefully through a cloth, an organic green pigment is suggested rather than the mineral green earth. This organic green may relate to the leek-green gem *prasitis* of Theophratus, or the *prasius* of Pliny, see Theophilus 1979: 16. Theophilus recommends *prasinus* for painting on a new wall, which suggests that the green pigment is not sensitive to alkaline conditions.
19. The term *pseudo-verdaccio* is used by Rollier-Hanselmann (1997: 61) to describe the green underpainting of portions of faces, hands and feet in these schemes.

> then take a little terre-verte in another dish, well thinned out . . . start shading under the chin, and mostly on the side where the face is to be darkest; and go on by shaping up the under side of the mouth . . . and in this way you pick out the whole of the face and the hands, wherever flesh colour is to come.

Over this shading, lime white thinned with water is systematically employed to suggest highlights, and pink applied for cheeks and lips. Finally, a wash of flesh colour and reinforcement of white for the highlights is applied. Cennini goes on to say that the method by which flesh colour is applied, followed by shading in *verdaccio*[20] and flesh colour, then followed by a few highlights – as at Winchester – was used only by those who knew little of the artist's profession (Cennini 1960: 45–6).

The particular hue of green earth varies considerably from source to source, and the natural variety was further extended by using the pigment as a substrate for dyes, usually to produce a more intense green colour (Grissom 1986: 147). Although no organic green or yellow colorants have as yet been identified in English medieval wall painting,[21] evidence from documentary sources suggests that they were certainly known from antiquity and employed in manuscript illumination in the medieval period.[22] For example, a Latin text in the late 12th-century manuscript B.L. Cotton Titus D. XXIV provides instructions for the tempering of moss green, saffron green (croceum viride) and green earth. This manuscript, from the Cistercian abbey of Rufford (Nottinghamshire), contains a version of *De Coloribus et mixtionibus,* and is particularly significant since it is the earliest technical treatise with instructions in the vernacular. The Anglo-Norman instructions – which are contemporary with the Latin – describe the preparation of moss green in detail.[23] Recent analysis of the Book of Kells (*c*800) has confirmed the presence of a sap green pigment.[24] Theophilus (1979: 38) recommends the use of sap of iris, leek or cabbage for manuscript illumination.

For wall painting, Eraclius describes green earth mixed with mallow distempered with vinegar or good wine, and recommends this mixture for painting on fresh plaster (Eraclius 1967: 236–7). It is not clear whether Eraclius intended to intensify the colour of the green earth pigment by dying it with an organic colorant from the mallow plant or, since the name for malachite is derived from the Greek malach meaning mallow, that he was suggesting a mixture of green earth and the mineral copper green pigment. Precisely this mixture of green earth and malachite was identified in the scheme of *c*1089 in the west choir at Lambach (Austria), although here it was applied *a secco* (Plate 90) (Hammer 1987: 45). Both green earth and a copper green have also been detected in the Romanesque scheme at Müstair, Switzerland, although no details have been published of how they were applied (Emmeneggger 1986: 198).

20. Thompson suggests that *verdaccio* is a dark greenish or brownish tone; Cennini 1960: xv.
21. Recent research suggests the presence of an organic yellow colorant in late and post-medieval domestic wall painting; see Shekede 1998: table 6.14. A yellow lake pigment has also been identified in the 13th-century polychromy of the west front of Wells Cathedral; Sampson 1998: 119.
22. Pliny (1984: 296–7) describes the inexpensive pigment *Appian*, which is made from green earth enhanced to appear brighter green in colour, rather like malachite. It has been suggested that the name *Appian* stems from *apianum* or *apiacum*, indicating that it may have been made from parsley.
23. For a full discussion of the important early versions of *De Coloribus*, particularly those dated to the 11th and 12th centuries from Benedictine houses in Canterbury, see Petzold 1995. For Cotton Titus D. XXIV, see also Hunt 1995.
24. See Fuchs and Oltrogge 1994: 134, 136–7.

In view of these findings it is important to consider a note by Professor Rees Jones concerning his analysis of the 12th-century scheme in St. Gabriel's Chapel at Canterbury, in which he indicated that the colour of the green pigment suggested malachite (Rees Jones 1936: 6–7). Since this observation was not confirmed chemically, and only green earth was identified in the single sample taken from the then green border of the *Naming of St. John*, it remains purely speculative to suggest that a mixture of green earth and malachite may have been employed and applied *a secco* as in the Romanesque scheme at Lambach, Austria (Hammer 1987: 45).

Overall it is significant that, as in English wall painting, the use of green earth in other northern European wall painting declined sharply at the end of the Romanesque period. Although the pigment continued to be employed in Gothic wall painting, it was used much less frequently than malachite or verdigris, and it is likely that the greater intensity of these and other copper greens proved more compatible with the Gothic palette than the rather soft warm green of green earth. Moreover, since the quantity of additional organic binding media increased in wall painting from the mid-13th century, it was no longer necessary to select a pigment stable in alkaline conditions.

Pigment alterations

Conversion of ferrous iron (Fe II) to ferric iron (Fe III) is probably the basis for the most commonly documented heat alteration of green earth, that is to form brown or 'burnt green earth'.[25] In his study of heat alteration of pigments painted in *buon fresco*, Rickerby (1991: 40, 42) documented alteration to light brown at 500 °C.[26] His work also indicated that green earth is one of the most heat-sensitive pigments.

Salt green

Copper chloride or 'salt green' is the name frequently used to describe the pigment which contains one or more of the four copper(II) hydroxychloride isomorphs – atacamite, paratacamite, botallackite and clinoatacamite – with the general formula [$Cu_2Cl(OH)_3$] (Pollard *et al.* 1989: 557–63; Scott 2000: 47–50).[27] XRD analysis undertaken by Riederer has indicated that copper chloride pigments most frequently occur either as atacamite or as a mixture of atacamite and paratacamite (Riederer 1974: 107).[28]

25. There is a higher ferrous to ferric ion ratio in celadonite than glauconite; Grissom 1986: 144.
26. Kauffman and Dillings (1950: 233) documented endothermic reactions occurring between 125 and 200 °C, at 350 °C, between 500 and 600 °C and at 775 °C.
27. The fourth copper(II) hydroxychloride isomorph clinoatacamite has recently been identified by Dr David Scott of the Getty Conservation Institute. As a consequence of his research he thinks it is likely that some identifications of paratacamite may be incorrect, and on analysis by XRD will prove to be clinoatacamite.
28. His work indicates that atacamite produces strong interferences of equal intensity at d = 5.40 Å and d = 5.00 Å, while for mixtures of atacamite and paratacamite, the d 5.00 Å line is weaker than that at d = 5.40 Å.

Atacamite occurs naturally in England, as at Land's End in Cornwall, where its presence may be due to the ingress of sea water into copper deposits.[29] However, given the homogeneous nature of the granular particles in all the samples examined, it seems likely that the pigment used in English medieval wall painting was produced synthetically.

Cost

There are no indications of the cost of this pigment in English accounts, nor is it possible to determine from the documentary sources that this particular copper pigment was employed for wall painting. However, the process of its manufacture was certainly known from an early date, and given the synthetic nature of the green pigment it is likely to have been relatively inexpensive.

Optical, chemical and physical properties

Optically, synthetic copper chlorides – which are a mixture of cupric basic acetate (verdigris) and copper hydroxy-chlorides (atacamite and paratacamite) – somewhat resemble verdigris. The mineral atacamite itself has an orthorhombic crystal system and is characterised by its green colour and granular crystalline aggregates. The refractive indices for atacamite are given as $\alpha = 1.831$, $\beta = 1.861$, $\gamma = 1.880$ (Klein and Hurlbut 1985: 325). The other three isomers are pale green or pale bluish-green in colour. Paratacamite has a rhombohedral crystal system while clinoatacamite and botallackite, the least stable, are monoclinic (Scott 2000: 40). For the characteristics of verdigris, see the section below on that pigment.

Verdigris is said to be incompatible with orpiment and lead-containing pigments (Thompson 1956: 168).[30] However, mixtures of verdigris and lead white have frequently been identified in English medieval wall painting and display no adverse effects. Copper chloride greens have, however, shown signs of oxidation to form brown-black copper oxides. The conditions necessary for this alteration are not fully understood, but alkalinity and the presence of an oxidising agent are thought to be important factors.[31]

Preparation

Pliny describes an artificial method of producing his 'chrysocolla, or gold-solder' on the walls of copper mines, by the infiltration of [sea] water in winter to produce what is likely to have been a copper chloride green.[32] Another, and doubtless more common, method of producing the pigment (at least in the Middle Ages) is that outlined by Eraclius and

29. Botallackite, an isomorph of atacamite and paratacamite, is found at Botallack in Cornwall. I am grateful to Brian Young of the British Geological Survey for this information.
30. Thompson also mentions another variant of these synthetic copper greens based on verdigris: Rouen green or *viride rotomagense*, prepared by coating copper with soap and exposing it to the acidic vapours of vinegar. The resultant green pigment would probably have been a complex mixture of copper salts of fatty acids.
31. I am indebted to Dr David Scott of the Getty Conservation Institute for this information.
32. Although gold-solder is translated by Rackham as malachite (basic copper carbonate), the shafts in which Pliny proposes that the process takes place are clearly near the sea; it therefore seems likely that it is sea water which would infiltrate the copper mines thus encouraging the formation of copper chlorides rather than copper carbonate (Pliny 1984: 4 n. a, 61, 67). This suggestion that copper

Theophilus, who describe the preparation of salt green *viride salsum* by placing sheets of copper – which have been smeared with honey and coated with common salt (sodium chloride) – together with hot vinegar or urine in a well-sealed vessel.[33] It is clear from this recipe that cupric chlorides and acetates may be formed, and recent spectrometric analysis of the copper chloride green in the 12th-century wall paintings at Idensen (Germany) confirmed the presence of copper hydroxy-chlorides (atacamite and paratacamite) and cupric basic acetate (verdigris), suggesting that a similar method of production may have been employed in this case (Matteini and Moles 1994: 84).

Use in English medieval wall painting

Although the presence of copper chloride greens in English medieval wall painting has been confirmed in a number of instances in recent years, it is likely that in many cases the copper chloride is an alteration product of azurite, malachite or a synthetic copper carbonate pigment in the presence of humidity and chloride ions. Thus, malachite – now substantially altered to paratacamite – has been identified on excavated fragments of *c*1180 from the chapter house of Sherborne Abbey (Howard and Park forthcoming a). Similarly, a copper chloride has been identified in the polychromy of *c*1302–08 on the Shrine of St. Alban, and in the wall paintings of *c*1300 in the Ante-Reliquary Chapel of Norwich Cathedral, but in both cases appears to be an alteration product of azurite.[34]

However, in the scheme of *c*1100 at Hardham (West Sussex) and the 'Losinga' cycle of paintings of *c*1190 in Norwich Cathedral, a green copper chloride or 'salt green' was clearly incorporated as part of the original palette. These two instances represent the first certain identifications of atacamite employed as part of the original palette in English Romanesque wall painting. The use of salt greens continued later in the medieval period in England, although the only identified example to date is in the wall paintings of *c*1510 at St. Mark's, Bristol (Gill and Howard 1997).

At Hardham, all the samples examined show clear evidence of the alteration of the matrix of the green pigment from calcium carbonate to form calcium sulphate, but since the chloride ions are associated specifically with the pigment particles alone – and not with the matrix – it is clear that an original 'salt green' was employed. FTIR analysis of the green pigment at Hardham suggested the presence of atacamite.[35] Similarly at Norwich, SEM/EDX analysis indicated that the chloride ions are associated only with the green particles (and not the surrounding matrix), while examination of the pigment by PLM indicated the presence of granular particles of chromatic homogeneity combined with an excess of calcite. Additional analysis by FTIR microspectroscopy confirmed the presence of atacamite ($Cu_2Cl[OH]_3$, confirmatory peaks at 3440, 3340, 982, 943, 913

---

chlorides are formed is supported by Pliny's description of the artificial pigment as hygroscopic: 'it is a substance that absorbs water' (Pliny 1984 :67–9). He goes on to describe the dyeing of *gold-solder* with a yellow organic colorant to enhance the colour of the pigment which is thereafter called *orobitis*, and recommends its application over a coloured ground to enhance the green colour (1984: 71).

33. Eraclius' recipe for *viridis color cum sale* corresponds closely with Theophilus' *viride salsum*; see Eraclius 1967: 236, and Theophilus 1979: 41.
34. For the Shrine of St. Alban, H. Howard, unpublished analysis; for the Ante-Reliquary Chapel, see Howard 1995c: 6, and the section on azurite in chapter three of this book.
35. Some paratacamite may also be present.

and 892cm^{-1}) (Tennent and Antonio 1981: 18/23/3) combined with calcium carbonate, and also some calcium sulphate and copper oxalate (Howard and Gasol 1996: 21–2).

In two other instances where copper chloride greens have been identified it is unclear whether they were applied as original pigments or whether they represent an alteration product. FTIR analysis of samples from the 13th-century paintings on the vault of the Guardian Angels' Chapel, Winchester Cathedral, suggested the presence of paratacamite, although it is not yet certain whether a pigment alteration has occurred.[36] Recent FTIR analysis of a sample from the early Gothic scheme in the Holy Sepulchre Chapel, also in Winchester Cathedral, has indicated that one of the green pigments may be a mixture of atacamite and paratacamite.[37] In this case, the copper chloride green pigment was identified in the trefoil frame surrounding the *Pantocrator* on the vault where it was applied in a layer 35µm thick directly over a white ground, while on the south wall of the chapel the pigment was applied in two layers 50µm thick, the upper containing a higher concentration of green particles. Although it is not yet certain whether the copper chloride green in the chapel is a conversion product or part of the original palette, its use for the bright green border above the *Washing of Feet* scene on the south wall may suggest a considered application of specific pigment mixtures rather than a pigment alteration.

Application

Over white ground or directly on the plaster support

In the 'Losinga' scheme the pigment was generally applied over a lime white ground. For example, atacamite was combined with lime white and applied in a layer 150µm thick over a lime ground for the green mantle of the standing figure (left as viewed) on the north side of the west segment of the vault (Plate 99). A similar pigment mixture, 425µm thick, was employed for the lower drapery of the right-hand figure, and in both cases the yellow underdrawing used to set out the figures is evident. In this scheme the pigment was clearly employed for its particular coloristic qualities in conjunction with green earth, as indicated by the subtle juxtaposition of details in the latter pigment over a layer of copper chloride green combined with lime white in a sample taken from the foliage on the soffit of the western arch. This application method supports analytical evidence which suggests that the copper chloride green was specifically applied as a pigment and that it is not an alteration product.[38]

Over coloured underpaint

Although the pigment was generally applied over a white ground in the 'Losinga' cycle, in one instance a grey underpaint of charcoal black combined with lime white was employed. For the lowest roundel on the soffit of the western arch, atacamite combined in a lime white matrix was applied in a layer 160µm thick over a dark grey underpaint to produce the brilliant green of the cathedral's roof (Plate 100).

---

36. Examination of the fingerprint region of the FTIR spectrum indicated the presence of strong bands at 987cm^{-1}, 925cm^{-1}, 861cm^{-1}, and 829cm^{-1}, suggesting the presence of paratacamite.
37. H. Howard, unpublished analysis.
38. Examination of a watercolour copy of the paintings made by G.E. Fox in 1906, when they were in better condition, indicates that this combination of the warm yellow-green of green earth over the paler atacamite layer was also used to model Flambard's drapery in the upper roundel on the western arch; Howard and Gasol 1996: 21–2.

At Hardham, the green pigment was employed only for a few important details – such as the nimbi of the *Agnus Dei* and censing angels, and Virgin and St. Gabriel in the *Annunciation* and *Visitation* on the east wall of the nave – and a variety of coloured underpaints was employed to achieve subtle coloristic effects (Plates 91 and 92). For the Virgin's drapery in the *Flight into Egypt* on the nave north wall, the copper chloride green was combined with lime white, and applied over grey underpaint which would have improved its covering power and thereby enhanced the intensity of the green (Plate 93).

Unusually, red was used as an underpainting for the nimbus of the *Agnus Dei* at Hardham (Plates 96 and 97), while a pale yellow layer is present beneath the green pigment in the nimbus of the southern censing angel (Plate 98), and white beneath Gabriel's nimbus in the Annunciation (Plates 91 and 92).[39] The variety of coloured underpaints employed beneath the green pigment suggests that particular coloristic effects were sought in each instance.

In the early 16th-century paintings at St. Mark's, Bristol, the copper green pigment was combined with lead white and applied in a layer 60μm thick over a grey underpaint of carbon black combined with lead white to produce the rich green background of Christ in the *Trinity* scene. The exact composition of the copper chloride green here remains to be established, but its appearance – similar to that of verdigris – and the elemental analysis provided by SEM/EDX analysis, which confirmed the presence of copper, chlorine and lead (from the lead white matrix), strongly suggest the presence of salt green (Howard 1995b: 6, pls. 1 and 2; Gill and Howard 1997: 101).

Binding media

In both the Hardham and 'Losinga' schemes the copper chloride pigment was combined with lime white, which not only affected the overall hue but would also have acted as a binder. Instrumental analysis undertaken to determine the presence of any original binding medium in the 'Losinga' scheme provided no evidence for an organic component (White 1998). By contrast, in the St. Mark's scheme, although the binding medium incorporated with the copper chloride green was not specifically examined, oil was identified both as a sealant and as a binding medium in all samples analysed (Gill and Howard 1997: 101).

Comparison with its use in other media and in continental wall painting

Copper chloride greens have frequently been identified in manuscript illumination,[40] and in panel painting,[41] and there are also confirmed instances of the use of the pigment in continental wall painting. In the scheme of *c*1130 at Idensen (Germany), atacamite combined with copper(II) hydroxy-acetate has been identified (Matteini and Moles 1994: 83–4); copper chloride green associated with green earth has been identified in the 12th-century wall paintings at Berzé-la-Ville (France) (Rollier-Hanselmann 1997: 67); and the presence of salt greens has also been confirmed in Danish wall paintings of

39. Analysis by Howard in Sawdy *et al.* 1998: 35.

40. See, van't Hul-Ehrnreich and Hallebeek (1972: 5) who have also identified paratacamite and calumetite (a basic copper chloride) in 14th–15th-century wall paintings; and Naumova and Pisareva 1994: 278.

41. Copper chloride-containing greens have been identified in the Norwegian altar frontals of Odda and Dale II, both dated to after *c*1300; Plahter 1995: 119.

*c*1200–25 (Richter 1988: 174).[42] Atacamite has also been identified in a number of medieval wall paintings in Sweden: in the 13th-century schemes at Garde, Källunge and Mästerby in Gotland; and also in paintings of *c*1500 at Härkeberga, Härnevi and Täby (Nord and Tronner 2000: 99). However, malachite was also identified in all the schemes with the exception of Garde and Källunge. Where both malachite and atacamite are present it seems likely that the copper chloride green represents an alteration of malachite. Only at Garde and Källunge is atacamite likely to have been incorporated in the original palette.

At Idensen, distinct crystals of the copper chloride green pigment were found combined with lime white, calcium sulphate and a little charcoal black to produce a pale green paint layer. This was applied over *veneda* to improve its covering power, thus increasing the intensity of the green colour (Matteini and Moles 1994: 83–4). The application of the pigment at Hardham – combined with lime white and applied over grey underpaint – almost exactly parallels that found in these slightly later paintings at Idensen.

In Romanesque polychrome sculpture, copper chloride greens have been identified on the 11th-century Hørning Plank from Hørning stave church, Denmark; on the Aby crucifix of *c*1050–1100, also in Denmark (Richter 1988: 174); on the monumental wooden doors from St. Maria im Kapitol, Cologne of *c*1050 (Jägers 1991: 214); and on the Romanesque lectern at Freudenstädter of *c*1150 (Richter 1988: 174).

Use of the pigment continues later in the medieval period in England, for example, it appears to have been employed on the polychromed image screen on the west front of Exeter Cathedral.[43]

Discussion

The presence of original copper chloride greens has now been confirmed in English medieval wall painting. Considered in conjunction with the evidence from elsewhere in northern Europe, it seems likely that the pigment was more common in the Romanesque than the Gothic period, although more confirmed examples are needed to support this supposition.

Copper chlorides which result from the alteration of azurite and malachite have also been identified in English medieval wall painting and elsewhere in northern Europe. For example, copper chlorides which are due to pigment alteration have been identified in a number of Austrian Gothic wall paintings.[44] It is therefore important, not only that instrumental analysis is undertaken to identify the presence of copper chlorides, but also that sufficient evidence is available to determine whether the material forms part of the original palette or represents an alteration product of other copper-based pigments.

42. Richter also cites an identification of copper chloride greens in wall paintings of unspecified date at Lübeck, Germany.
43. The phase of the polychromy in which the copper chloride was detected is not specified. SEM/EDX analysis of the Exeter polychromy by A. Roy indicated a high chloride content in the green pigment, which may be the result of a pigment alteration; see Sinclair 1992: 10.
44. Kerber *et al.* 1972. The transformation of azurite to form paratacamite was detected in the paintings of *c*1263 on the vault of the Bishop's Chapel, Gurk Cathedral; for the analysis by F. Mairinger, see Ellwanger-Eckel 1979: 5.

Pigment alterations

The transformation of a copper chloride green to form a dark alteration product, presumed to be tenorite, has recently been detected in the early Romanesque wall paintings at Hardham, West Sussex (Plates 92, 95, 96 and 98).[45] A similar dark oxidation product of copper chloride has been identified in the polychromy of *c*1302–08 on the Shrine of St. Alban, although in this example the copper chloride is itself an alteration product of azurite.[46] Since the formation of tenorite requires an oxidising and alkaline environment – pH 8–12[47] – it is likely that lime-water, which was probably used to consolidate the Hardham paintings during the 1960s, would have provided the necessary alkaline environment while microbiological factors may have initiated oxidation reactions.[48]

Malachite

Malachite, the green mineral basic copper carbonate ($CuCO_3.Cu(OH)_2$), frequently occurs in nature in large, relatively pure masses which may simply be coarsely ground and levigated for use as a pigment. The most frequently occurring mineral impurities associated with malachite are azurite (blue basic copper carbonate, $2CuCCO_3.Cu(OH)_2$), chrysocolla (copper silicate) and copper chlorides. Geologically, azurite is the parent species, while malachite – containing a higher proportion of chemically combined water – usually represents an alteration (Thompson 1956: 161; Klein and Hurlbut 1985: 341). The derivation of malachite was clearly understood by Cennini who called the mineral blue-green (*verde azurro*) and was unable to consider the green pigment completely natural since it was 'formed out of' azurite (Cennini 1960: 31). The name malachite is derived from the Greek *malach* meaning mallow, which was probably descriptive of its colour (Klein and Hurlbut 1985: 341).[49]

Malachite is found in the oxidised upper zones of copper ore deposits, usually those associated with limestone. Although the medieval deposits of the mineral are not known, they are presumably similar to those for azurite which are described in detail in the documentary sources for that pigment.[50] Azurite deposits in east central Europe, at Chessy-

45. Analysis by Howard in Sawdy *et al.* 1998.
46. Unpublished analysis by H. Howard.
47. Dr David Scott, pers. comm.
48. A major campaign of conservation was undertaken by E. Baker between 1961 and 1963 involving uncovering, removal of wax and varnish, and mortar repairs. Although lime-watering of the paintings is not specified, this treatment usually formed an integral part of the conservation procedures employed by her team. For a summary of the conservation history of these paintings, see Sawdy *et al.* 1998: 14–17. As part of this current examination, analysis and remedial treatment of these paintings being undertaken by the Courtauld Institute, a bioremediation study is also being carried out, for which see Kyi and Drew 1998.
49. Since the green mineral was used for soldering in antiquity, Pliny uses the term *gold-solder* or *chrysocolla* for malachite, although it remains unclear whether he was referring to copper chlorides in some instances. See Pliny 1984: 4 n. a, 61, 67–9, and the discussion of this terminology in copper chloride greens above.
50. Armenia was an important source of both azurite and malachite in antiquity. In addition, it is likely that the blue and green pigments were exported from Cyprus since, in antiquity, it had the largest

les-Mines near Lyons, and in Hungary were known in the Middle Ages, and in addition, the nomenclature for azurite, *azurrum almaneum* suggests that other sources in Germany were also utilised (Thompson 1956: 130; Gettens and Fitzhugh 1993b: 184).

There are major copper deposits in Devon and Cornwall, the Lake District, and in the Welsh mountains, although there are fewer coloured copper minerals (oxidation products) in the latter two areas. One particularly well-known deposit is at Alderley near Macclesfield, which is known to have been worked at an early date – certainly since the 16th century – when it was used as a source for the metal ore. The azurite and malachite at this site are in a soft earthy form and would therefore have been ideally suited for use as pigments.[51]

Cost

There are no references in the medieval accounts which can be directly related to malachite. However, as with azurite, its colour is directly related to particle size, and so the pigment would presumably have been graded and priced according to intensity of colour.

Optical, chemical and physical properties

Monoclinic crystals of malachite – which may be pseudomorphs after azurite – are often slender and prismatic although seldom distinct, and may appear as granular or earthy forms. Since the pigment is prepared by crushing or grinding the mineral species, the particles have a fractured appearance and may retain evidence of the characteristic pale green to almost black banding which is a feature of the mineral. In plane polarised transmitted light the particles appear rather pale green with high relief. Larger particles display green to almost colourless pleochroism and the refractive indices are $\alpha = 1.655$, $\beta = 1.875$, and $\gamma = 1.909$.[52] With crossed polars, particles of malachite display birefringence and clear extinction.

Malachite is decomposed by dilute acids with the evolution of carbon dioxide gas. The pigment is stable in contact with dilute alkali when cool, but will darken if heated with the reagent. The green pigment is stable to light but decomposes thermally with loss of water and carbon dioxide. Recent experiments have indicated that malachite alters to a brown colour at 300 °C and that complete transformation to black copper oxide (tenorite) occurs from 400 °C (Rickerby 1991: 43).

A synthetic form of malachite, prepared by precipitation, has been reported in a number of cases.[53] In these instances the particles are globular in form, although the FTIR spectrum and XRD patterns are identical to those of natural malachite.

copper mining industry in the world; Kakoulli 1995: 5. Pliny also intimates that a number of cheap imitations for the mineral malachite existed; in discussing a recently discovered green pigment he exclaims, 'just as if there were too few spurious varieties of it already', see Pliny 1984: 297, 217.

51. I am grateful to B. Young of the British Geological Survey for information on copper deposits in the British Isles.

52. For full details of the pigment and its characteristics, see Gettens and Fitzhugh 1993b.

53. Identifications of the synthetic form of malachite have previously been made in 15th-century Italian panel paintings; see Gettens and Fitzhugh 1993b: 197. Synthetic malachite has also recently been detected by the present author in a 14th-century wall painting at Judenburg, Austria; see Binder 1997: plates 61–2.

Preparation

The pigment is prepared by crushing and coarsely grinding the mineral, followed by washing and levigation to remove impurities. Coarse grinding is essential since malachite, like azurite, loses its colour when finely ground. Cennini describes the pigment as 'looking like sand', and indicates that it should be ground very little with a light touch and washed in water several times (Cennini 1960: 31–2).

Use in English medieval wall painting

Copper green pigments have frequently been identified in English polychromy of the medieval period, although in few instances has instrumental analysis been undertaken to confirm the particular species present.[54] In the past it was a common assumption that any green copper-containing mineral was malachite, and similarly that any copper blue mineral was azurite. In nature this is not the case, since copper sulphate and copper chloride minerals are also common and may well be intimately mixed with malachite and azurite.

The presence of malachite has been frequently suggested in English medieval wall painting. Preliminary identification of its use in the Romanesque period has been made, for example, in the scheme of *c*1130 at Kempley, Gloucestershire; in the mid-12th-century painting in St. Anselm's Chapel's at Canterbury; in the painting of *c*1140 in the north transept of Chichester Cathedral; and on excavated fragments of *c*1180 from the chapter house of Sherborne Abbey, Dorset. However, a brief review of the methods of examination undertaken to determine the presence of malachite in these instances reveals that in most cases insufficiently detailed analysis has been undertaken to allow full characterisation of the species of copper green present.

Microscopic examination of the pigments and a positive microchemical test for copper suggested the presence of both malachite and azurite in the Kempley paintings (Rickerby 1990a: 255–6). Here, the blue and green copper pigments were detected in details such as haloes and the pupils of the eyes of some of the apostles, and in addition, slight traces were also found on the draperies. Examination of the paint samples also suggested that, in some zones, the azurite had partially converted to form malachite. Indeed, it was suggested that the presence of malachite in some of the samples from Kempley was due to alteration from azurite, and firm identification of malachite as part of the original palette at this site remains to be confirmed (Rickerby 1990a: 255–6).

Similarly, PLM was employed initially to detect the presence of malachite in a sample taken from the chamfer of the capital – of *c*1130 although possibly repainted later in the medieval period – on the south side of the entrance to the apse of St. Gabriel's Chapel, Canterbury Cathedral. Here the green copper-based pigment is combined with lead white and applied over a local ground of red earth, lead white and charcoal black (Cather and Howard 1994: 145). However, in this instance, subsequent analysis by FTIR confirmed the presence of verdigris rather than malachite.

Microscopic examination alone was considered to indicate the presence of malachite in the *St. Paul and the Viper* in St. Anslem's Chapel, also at Canterbury. In this painting

54. Identification has sometimes been made by a microchemical test for copper and a green appearance of the pigment under the microscope. It is clear that instrumental analysis should be undertaken to clarify the identification of copper green pigments in such cases.

Plate 45 St. Albans Cathedral: detail of St. William of York on the north-east pier of the feretory (c1302–08). (Photograph: H. Howard; © Courtauld Institute, Conservation of Wall Painting Department)

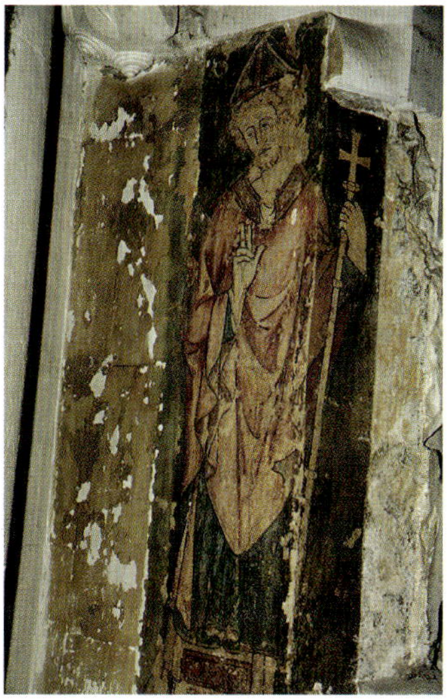

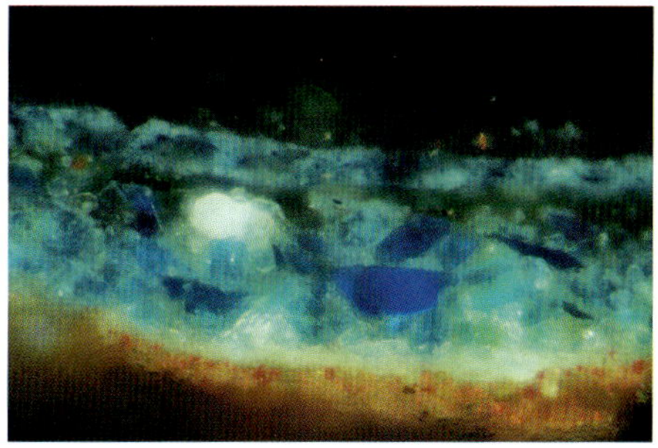

Plate 46 St. Albans Cathedral. Cross-section of a sample from St. William's blue dalmatic, original magnification 400×. Natural azurite has been applied over red underpaint, presumably to provide a richer, more purple hue. Later in the medieval period, the sparkling blue of this high-quality azurite was covered with a duller, darker layer of indigo. (Photomicrograph: H. Howard; © Courtauld Institute, Conservation of Wall Painting Department)

Plate 47 St. Albans Cathedral: detail of an unidentified archbishop on the south-east pier of the feretory. (Photograph: H. Howard; © Courtauld Institute, Conservation of Wall Painting Department)

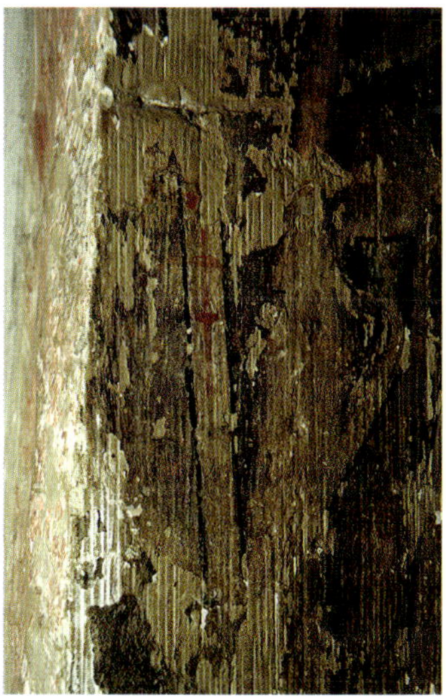

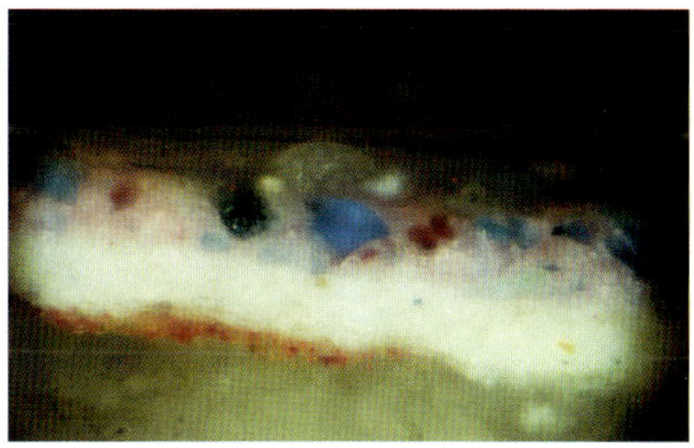

Plate 48 Cross-section of a sample taken from the purple chasuble of the archbishop shown in Plate 47, original magnification 400×. The soft purple colour was achieved by combining azurite with a red lake pigment and a little lead white, and applying the mixture in a single layer over a lead white ground. (Photomicrograph: H. Howard; © Courtauld Institute, Conservation of Wall Painting Department)

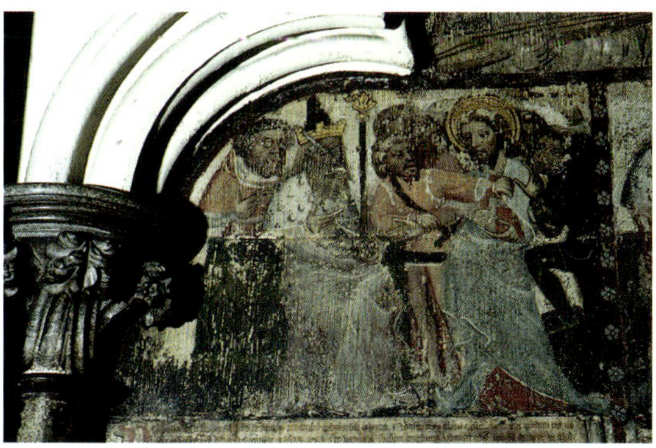

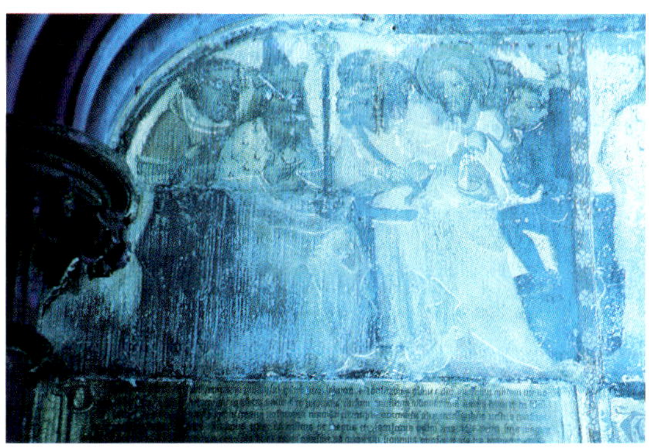

Plates 49 and 50 Westminster Abbey, chapter house, Apocalypse *cycle of c1375–1404: detail of* St. John the Evangelist before Domitian. *Both azurite and indigo have been identified in this cycle of wall paintings, with azurite employed extensively for areas such as St. John's blue drapery. In UV light (Plate 50), the modern wax coating – which is particularly thick in hollows in the paint layer and areas such as the eyes which have been gouged – appears a milky white colour. (Photographs: Howard/Kakoulli; © Courtauld Institute, Conservation of Wall Painting Department)*

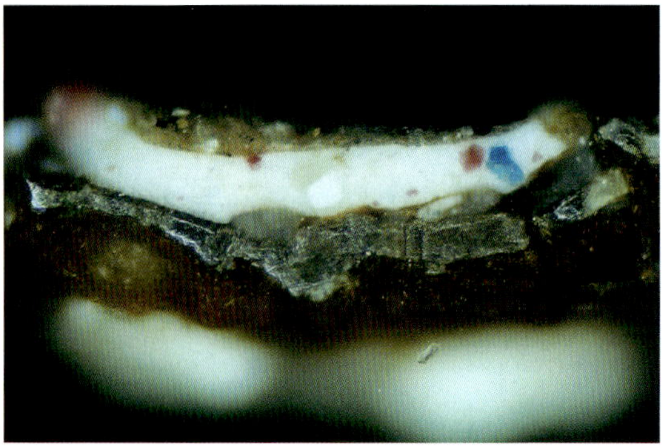

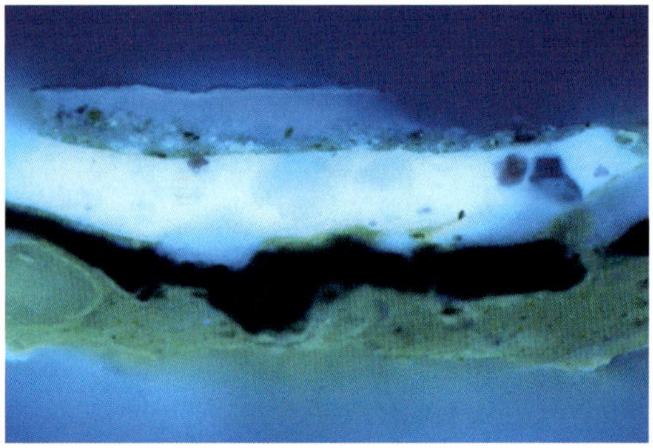

Plates 51 and 52 *Westminster Abbey, chapter house. Cross-section of a sample taken from another scene in the* Apocalypse *cycle, where azurite has been combined with red lake and lead white to produce a purple colour. This mixture was applied over a layer of tin leaf, presumably to provide additional luminosity to the purple colour. In UV light (Plate 52), the organic layer beneath the tin leaf (which appears as a dark zone in this photomicrograph) fluoresces an orange/yellow colour. The waxy surface coating fluoresces a milky blue colour, original magnification 382×. (Photomicrographs: H. Howard; © Courtauld Institute, Conservation of Wall Painting Department)*

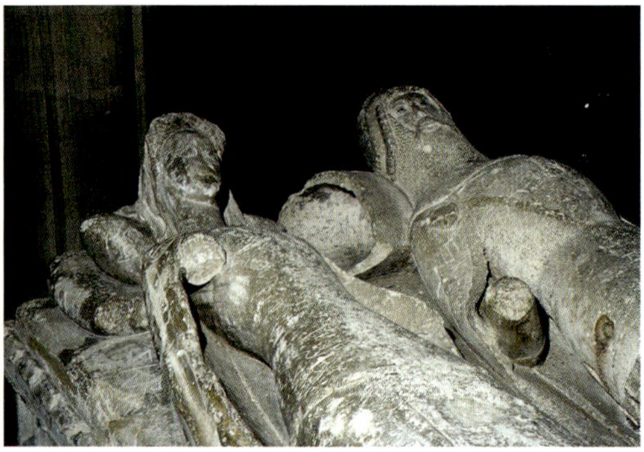

Plate 53 *Ingham (Norfolk): tomb of Sir Roger and Lady Margaret de Bois (c1380). (Photograph: H. Howard; © Courtauld Institute, Conservation of Wall Painting Department)*

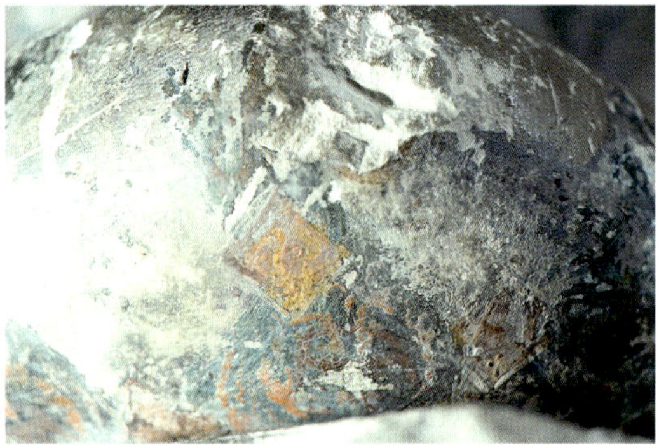

Plate 54 *Ingham (Norfolk), de Bois tomb. Despite severe damage to the polychromy, traces remain in the carved detail as on the cushion below Lady Margaret's head. (© John Burbidge)*

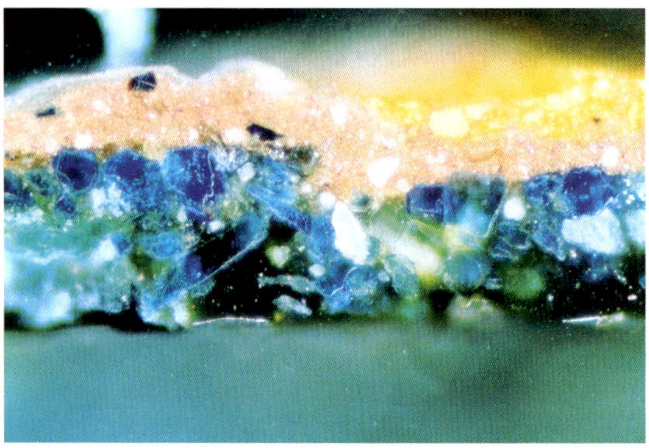

Plate 55 *Cross-section of a sample from the red lozenge pattern shown in Plate 54, original magnification 500×. Here, perhaps with the intention of increasing its luminosity and brilliance, azurite was applied over silver leaf. Red impasto linear details were then added in vermilion and red lead. (Photomicrograph: H. Howard; © Courtauld Institute, Conservation of Wall Painting Department)*

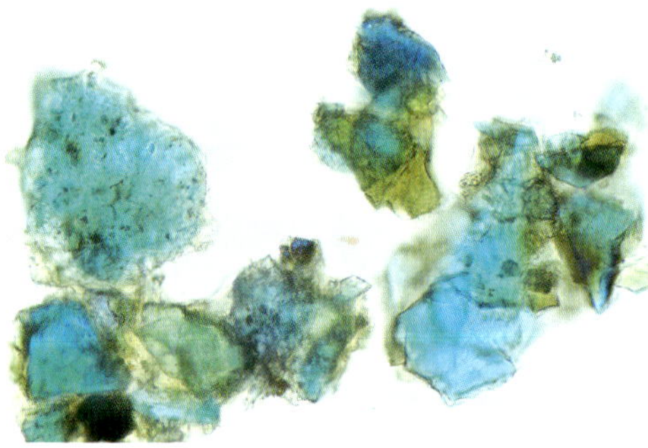

Plate 56 *Dispersion of azurite particles from the cross-section shown in Plate 55, photographed in plane-polarised transmitted light. The large particles, typically 60 × 50μm, attest to the high quality of the mineral blue pigment. The now-yellow binding medium is particularly apparent at the right side of the plate. Analysis by FTIR established that the original binding medium contained both oil and a proteinaceous component. This may indicate that the medium is egg yolk, or that the blue particles were ground in protein before being bound in oil (as found in the wall paintings of c1270–80 in the choir of Angers Cathedral). (Photomicrograph: H. Howard; © Courtauld Institute, Conservation of Wall Painting Department)*

Plate 57 Azurite was also employed in Gothic panel painting. In the ceiling paterae of c1263–66 from the Painted Chamber, Westminster Palace, azurite was combined with black and white and used to outline the figures. (© Copyright The British Museum)

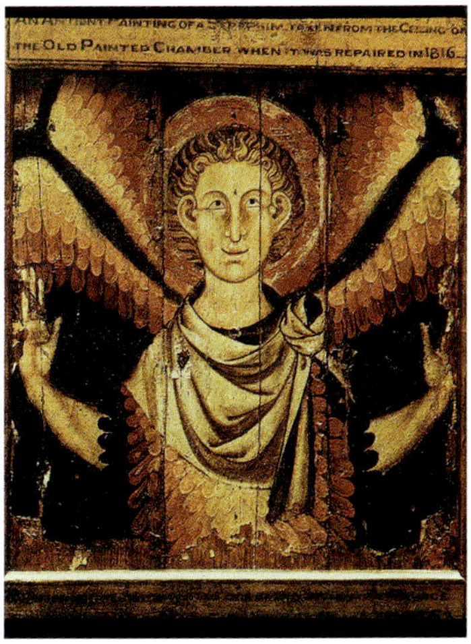

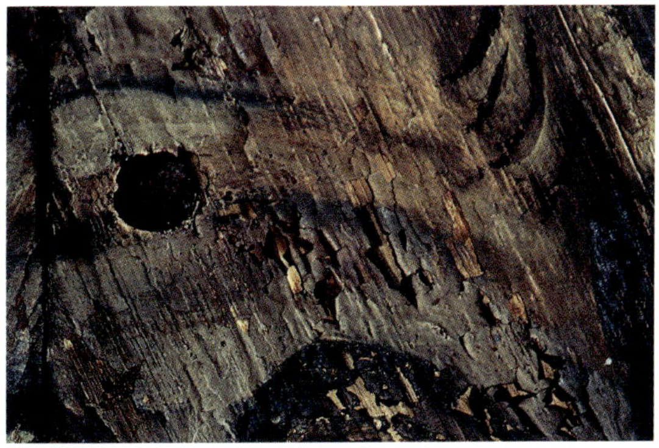

Plate 58 Peterborough Cathedral, nave ceiling (c1220): detail of St. Peter's head, showing the crude repainting and severe cupping and lifting of the paint layers before the recent phase of conservation by the Perry Lithgow Partnership. Although very few original paint layers remain on the ceiling, natural azurite combined with a little yellow earth and lead white were combined to produce effects of modelling in the flesh tones. (Photograph: Howard/Heritage; © Courtauld Institute, Conservation of Wall Painting Department)

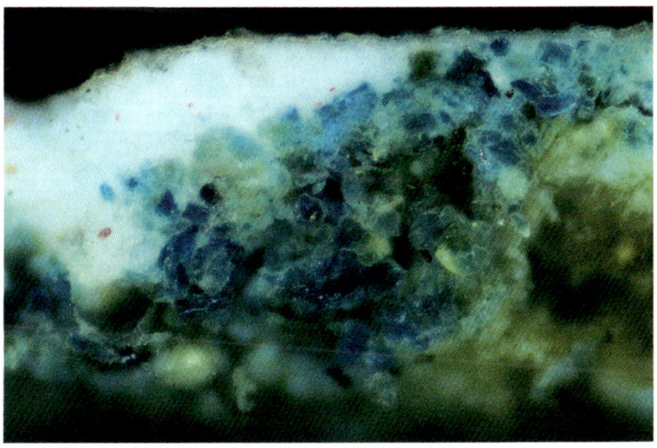

Plate 59 Peterborough Cathedral, nave ceiling. Cross-section of a sample from St. Peter's left temple, original magnification 500×. A layer of dark green underpainting, produced by mixing natural azurite, yellow earth and lead white, was employed to produce the tonal modelling for the flesh tones of vermilion combined with lead white. (Photomicrograph: H. Howard; ©Courtauld Institute, Conservation of Wall Painting Department)

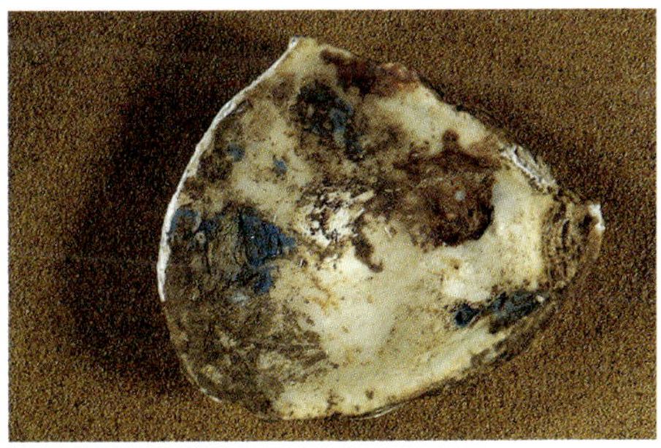

Plate 60 Oyster-shell palette of c1300 excavated from the Greyfriars, Norwich, which may have been used for large-scale painting. (Photograph: H. Howard; © Courtauld Institute, Conservation of Wall Painting Department)

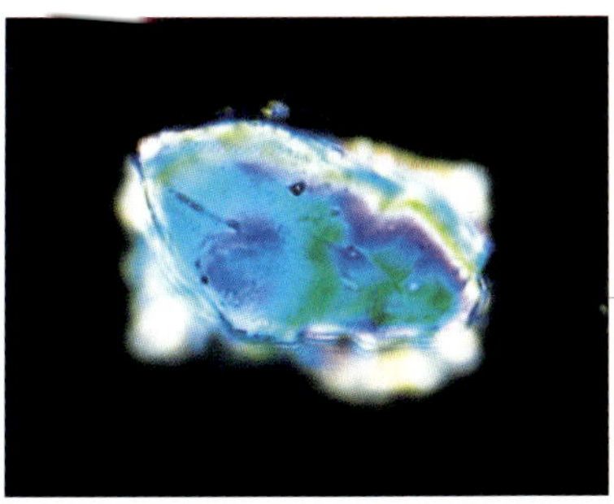

Plate 61 Oyster-shell palette from Greyfriars, Norwich. The azurite particles in the Norwich palette are of average size with a few larger inclusions. This natural azurite particle has been photographed in transmitted light with crossed polars, original magnification 1500×. Examination suggests that the mineral blue pigment here was bulked out with its synthetic equivalent. This practice has also been found in the azurite employed for the mid-13th-century scheme at Horsham St. Faith Priory, Norfolk (Plates 37, 39, 40 and 41). (Photomicrograph: H. Howard; © Courtauld Institute, Conservation of Wall Painting Department)

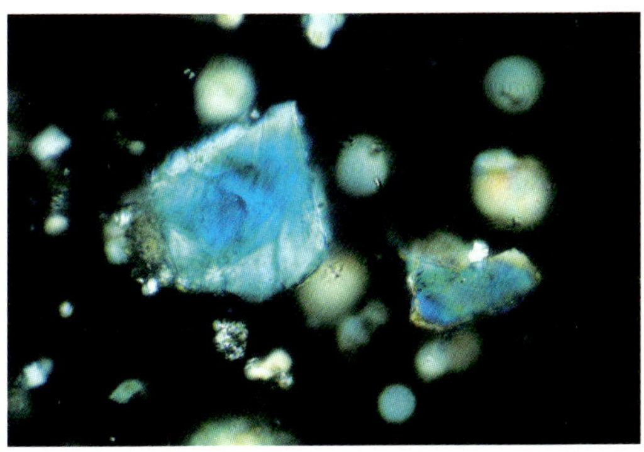

Plate 62 Natural azurite also occurs in an oyster-shell palette (of between the mid-14th and mid-16th century), excavated from the Carmelite Friary, Coventry. Here the particles are slightly larger than average size (the majority are 30 × 15μm, with others of approximately 50 × 30μm). These particles have been photographed in transmitted light with crossed polars, original magnification 1380×. (Photomicrograph: H. Howard; © Courtauld Institute, Conservation of Wall Painting Department)

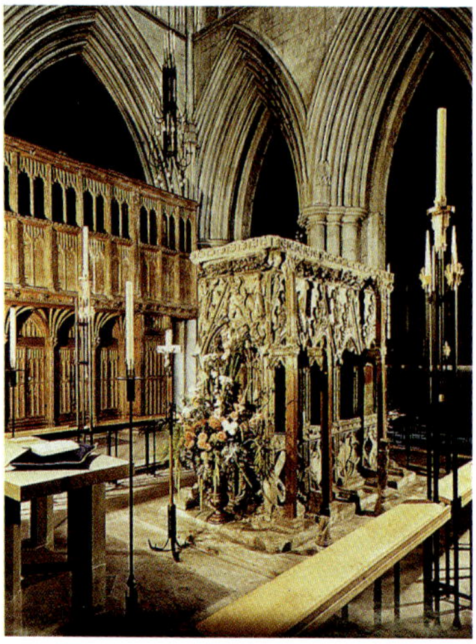

Plate 63 St. Albans Cathedral, feretory: Shrine of St. Alban (c1302–08). (© The Cathedral and Abbey Church of St. Alban)

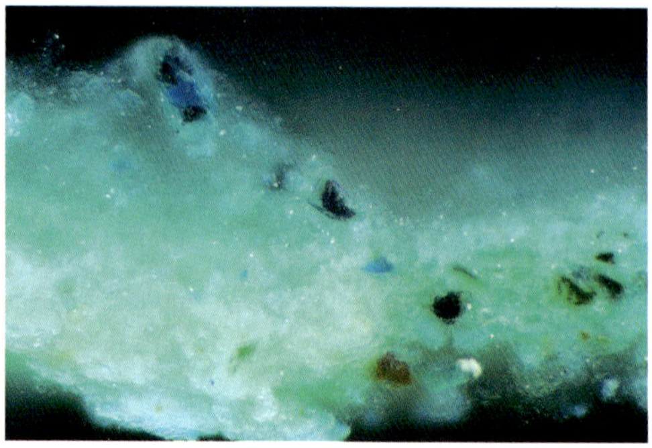

Plate 64 St. Albans Cathedral, feretory: Shrine of St. Alban. The alteration of azurite to form a copper-chloride alteration product has been identified in a number of instances, notably in samples from the shrine. Cross-sections clearly demonstrate the blue-green alteration signalling the transformation, which in this case progresses further to a dark material visible at the centre of the particles. A single blue particle with a green halo is visible at the centre of the cross-section, while green, copper chloride-rich particles with dark centres are also visible, original magnification 400×. (Photomicrograph: H. Howard; © Courtauld Institute, Conservation of Wall Painting Department)

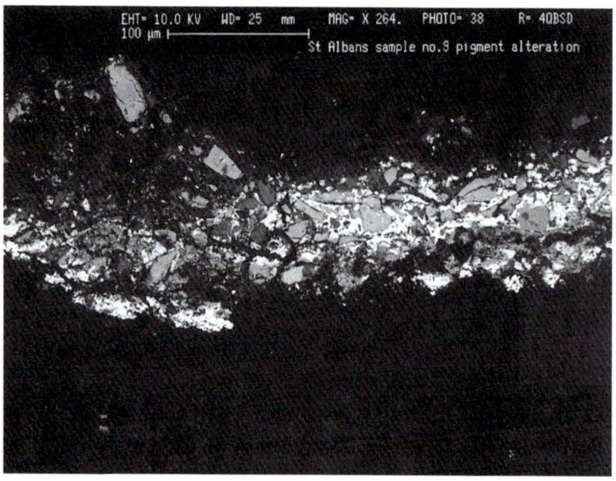

Plate 65 SEM micrograph showing a backscatter image of the sample in Plate 64. Details of the particles at top left of this image are shown in the subsequent plates. (SEM micrograph: H. Howard; © Courtauld Institute, Conservation of Wall Painting Department)

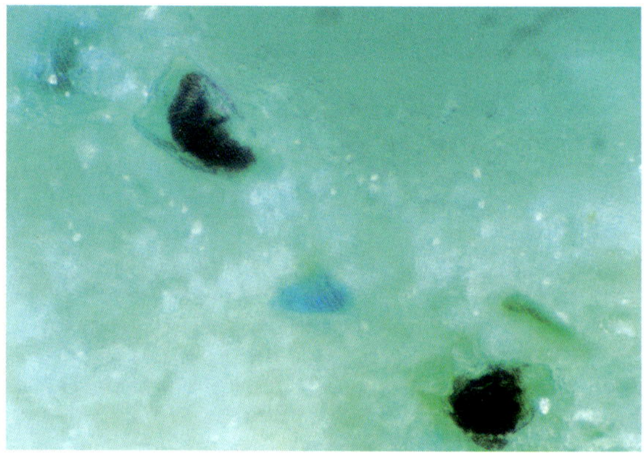

Plate 66 Detail of the cross-section in Plate 64, original magnification 1000×. The subsequent four plates relate to the particle at the top left of this image. (Photomicrograph: H. Howard; © Courtauld Institute, Conservation of Wall Painting Department)

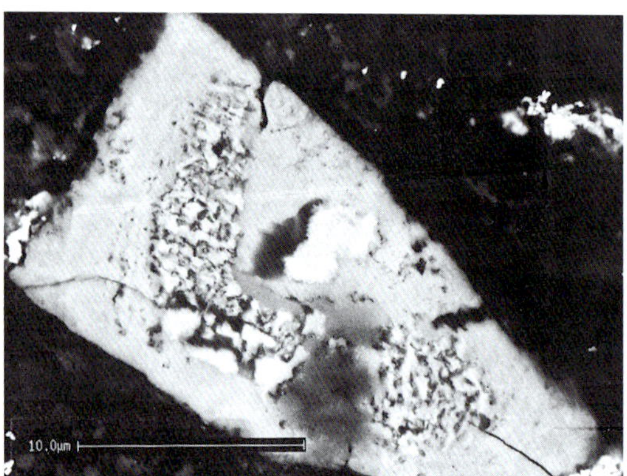

Plate 67 SEM micrograph showing a backscatter image of the green particle with a dark centre at the top left of Plate 66 The disrupted structure of the centre of the particle is clearly visible. (SEM micrograph: H. Howard; © Courtauld Institute, Conservation of Wall Painting Department)

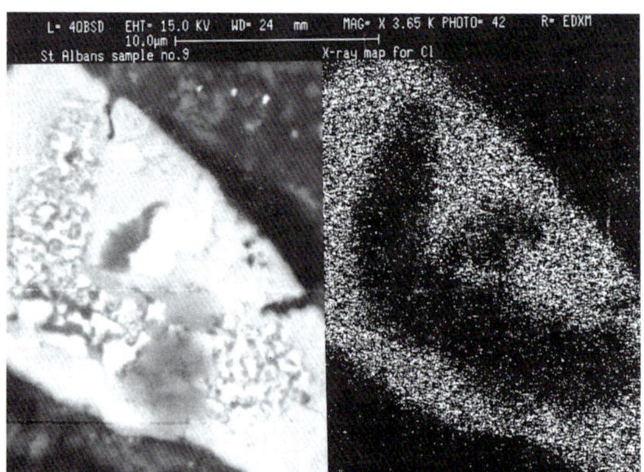

Plate 68 X-ray map showing the distribution of chlorine which is clearly present around the green perimeter of the particle, although not in its dark centre. (SEM micrograph: H. Howard; © Courtauld Institute, Conservation of Wall Painting Department)

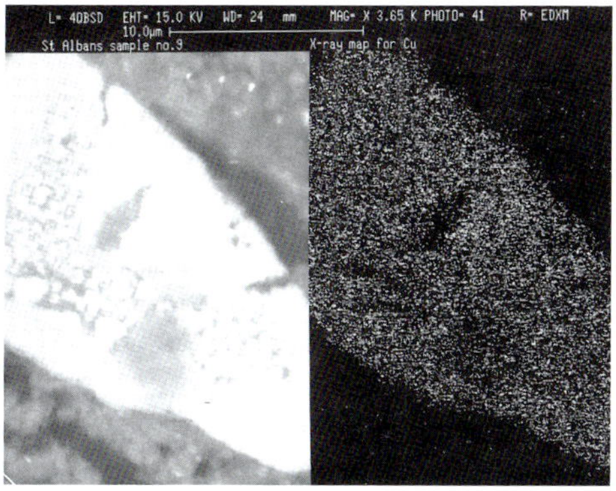

Plate 69 SEM micrograph with X-ray map for copper showing that its distribution is fairly consistent throughout the green and dark brown areas of the particle. (SEM micrograph: H. Howard; © Courtauld Institute, Conservation of Wall Painting Department)

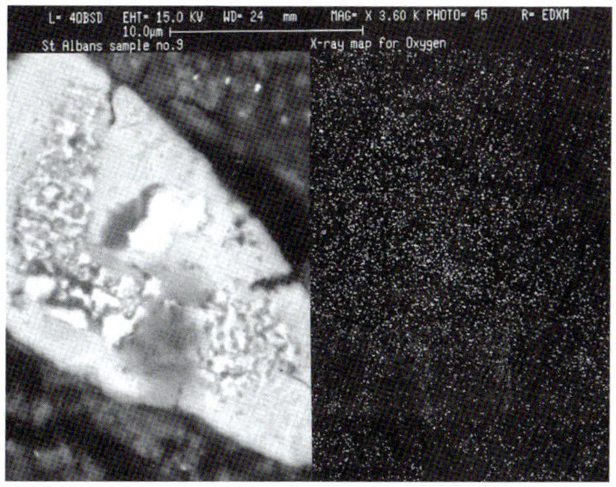

Plate 70 SEM micrograph with X-ray map for oxygen showing a slightly increased concentration in the centre of the particle. (SEM micrograph: H. Howard; © Courtauld Institute, Conservation of Wall Painting Department)

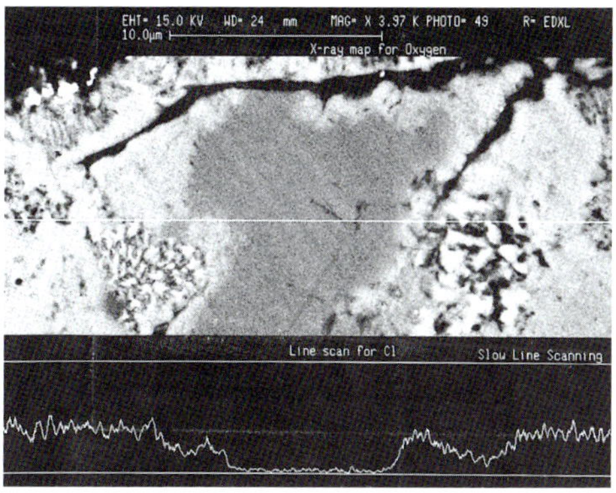

Plate 71 SEM backscatter image of the particle at top left of Plate 65, with an X-ray line scan for chlorine. The line scan confirms the increased concentration of chlorine around the green perimeter of the particle. (SEM micrograph: H. Howard; © Courtauld Institute, Conservation of Wall Painting Department)

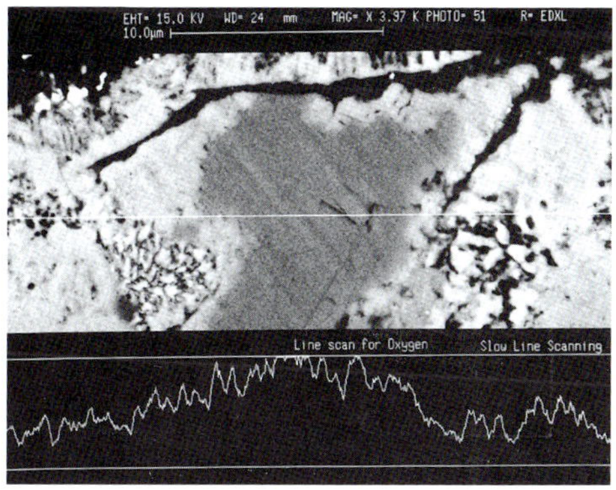

Plate 72 SEM backscatter image of the same particle as in Plate 71, but here with an X-ray line scan for oxygen. The line scan confirms the increased concentration of oxygen in the dark brown centre of the particle. The analysis suggests that while the green alteration product of azurite particles on the Shrine of St. Alban consists of a copper chloride-based material, the dark centre of the particle is principally copper oxide. (SEM micrograph: H. Howard; © Courtauld Institute, Conservation of Wall Painting Department)

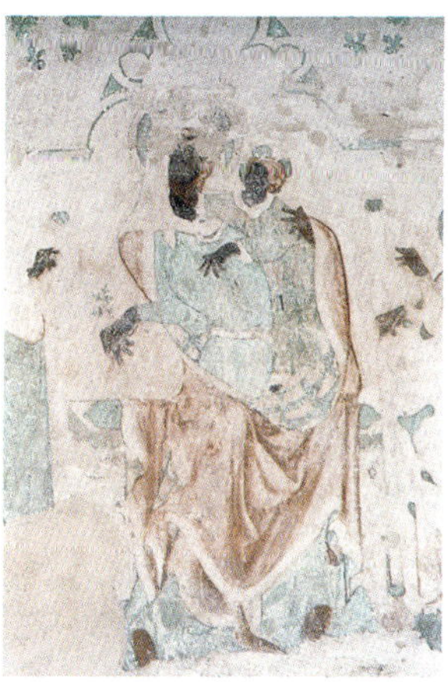

Plate 73 Little Wenham (Suffolk): detail of the early 14th-century wall paintings on the chancel east wall, showing the turquoise blue drapery and darkened flesh tones which are such features of the scheme. (© Courtauld Institute, Conservation of Wall Painting Department)

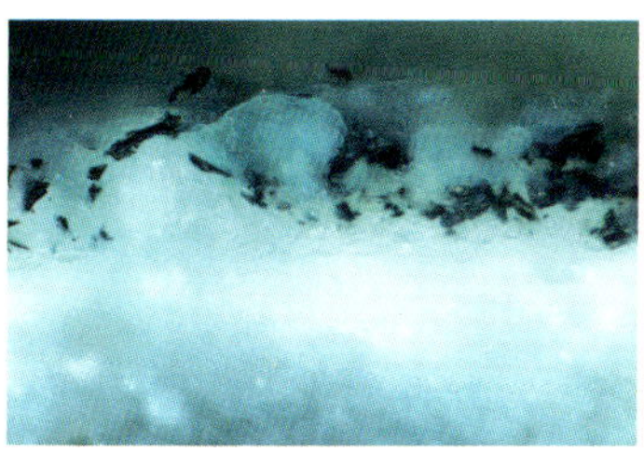

Plate 74 Little Wenham (Suffolk). Cross-section of a sample from the drapery of one of the female saints on the east wall of the chancel, original magnification 1000×. At low magnification the synthetic copper blue pigment appears to have no particulate structure, appearing almost as a stain on the white lime matrix. However, at high magnification, as here, tiny blue particles with slightly rounded edges are apparent in a few places. The blue paint layer has been applied directly over a lime ground and details in charcoal black applied on the surface. (Photomicrograph: H. Howard; © Courtauld Institute, Conservation of Wall Painting Department)

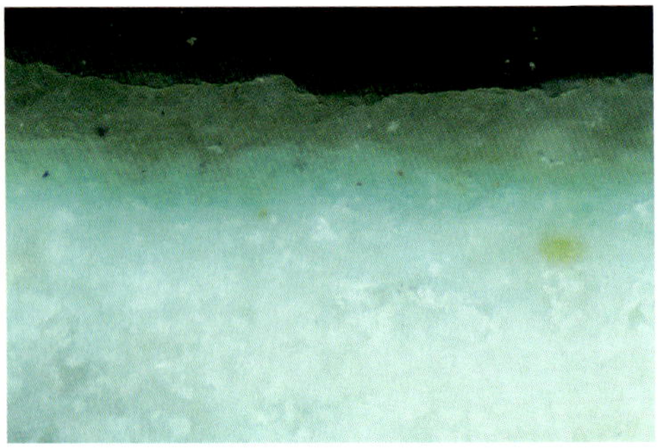

Plate 75 *Little Wenham (Suffolk). A sample from the blue border at the south end of the chancel east wall indicates the presence of the same synthetic copper blue pigment. A few tiny rounded particles are just visible, original magnification 500×. (Photomicrograph: H. Howard; © Courtauld Institute, Conservation of Wall Painting Department)*

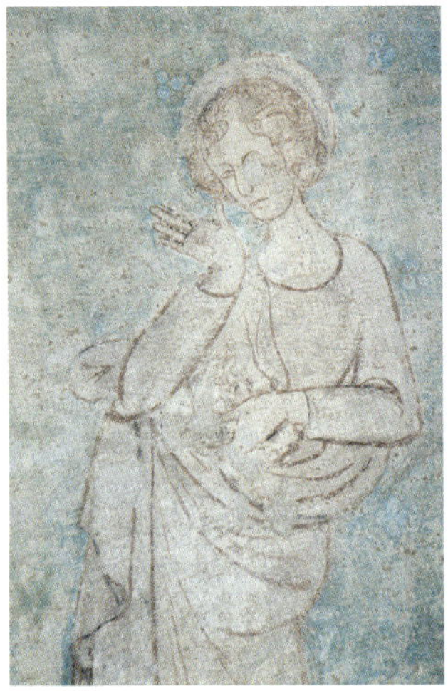

Plate 76 *Brent Eleigh (Suffolk): detail of the fictive reredos of c1335 on the east wall of the chancel. The background appears bright turquoise blue with tiny rosettes painted in pale blue on the surface. In a few places, particularly at either side of the reredos and beneath some of the rosettes, a grey colour is visible. (© Courtauld Institute, Conservation of Wall Painting Department)*

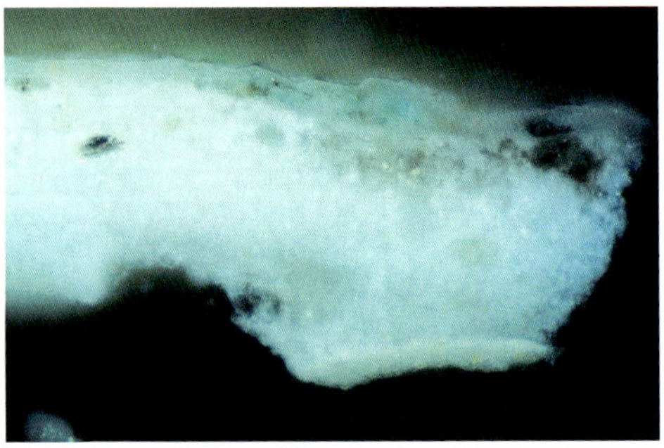

Plate 77 *Brent Eleigh (Suffolk). Examination of a cross-section from the blue background indicated the presence of a synthetic copper blue pigment, similar to that found at neighbouring Little Wenham. The sample also includes grey-brown particles, original magnification 700×. Analysis of these areas suggested the presence of copper oxide, indicating an alteration of the synthetic blue pigment. (Photomicrograph: H. Howard; © Courtauld Institute, Conservation of Wall Painting Department)*

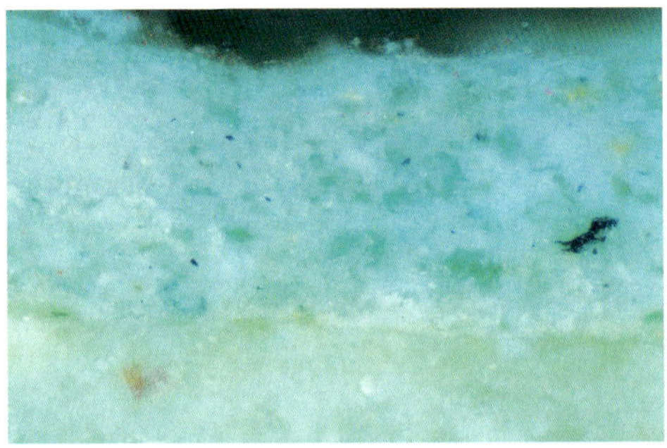

Plate 78 *A similar 'stained blue' pigment occurs in the 14th-century paintings in the north aisle of Great Hockham Church (Norfolk). The cross-section shows that the blue-green paint layer (75µm thick) was applied over a lime white ground. The dark particles are inclusions of charcoal black. (Photomicrograph: H. Howard; ©Courtauld Institute, Conservation of Wall Painting Department)*

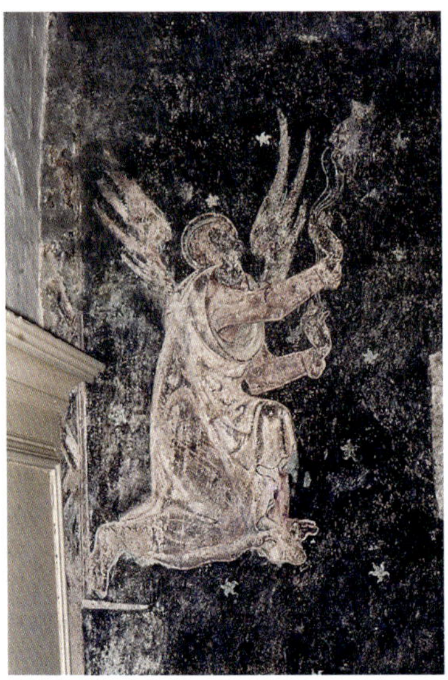

Plate 79 *Brent Eleigh (Suffolk), chancel east wall: detail of a censing angel set against a now dark background scattered with stars. This painting forms part of the same scheme as the fictive reredos illustrated in Plate 76. (© Crown copyright. NMR)*

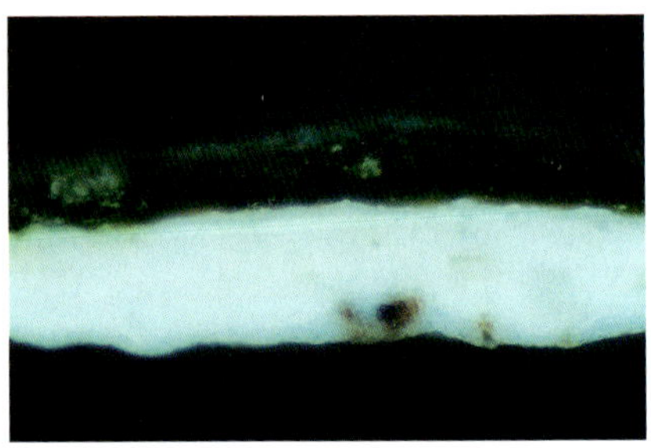

Plate 80 *Brent Eleigh (Suffolk), chancel east wall. Cross-section of a sample from a pale blue area on the dark background. Here verdigris has been applied over the white ground and the pale blue layer apparent at the surface appears to be an alteration product of this. In this instance, limewash applied over the surface of the painting has probably initiated a reaction resulting in the formation of the turquoise blue alteration product of verdigris in some areas. (Photomicrograph: H. Howard; © Courtauld Institute, Conservation of Wall Painting Department)*

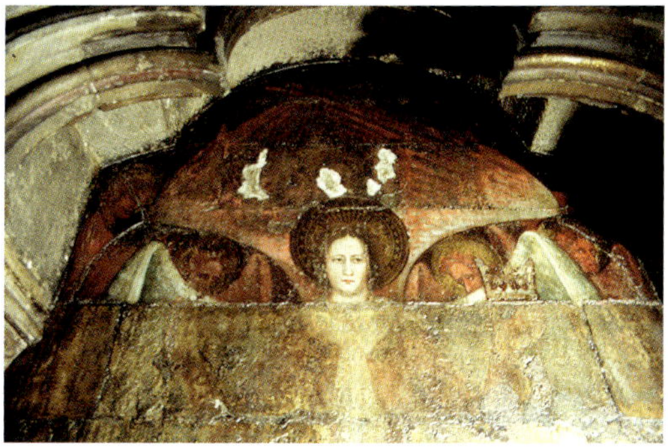

Plate 81 *Westminster Abbey, chapter house, east bay: detail of* Seraph *(c1375–1404). (Photograph: Howard/Kakoulli; © Courtauld Institute, Conservation of Wall Painting Department)*

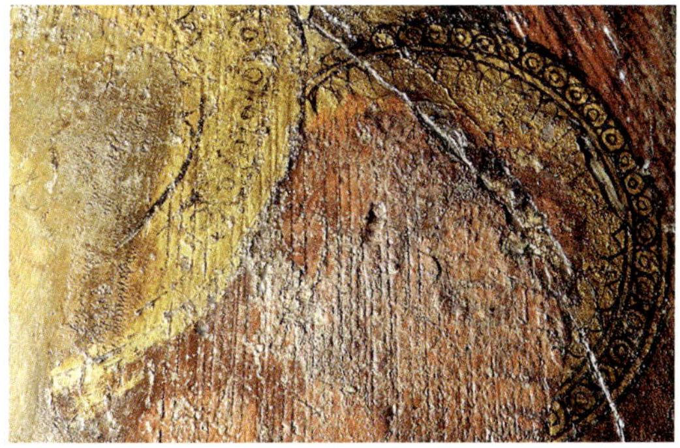

Plate 82 *Westminster Abbey, chapter house. Detail of the angels' nimbi in the first arch of the east bay. (Photograph: Howard/Kakoulli; © Courtauld Institute, Conservation of Wall Painting Department)*

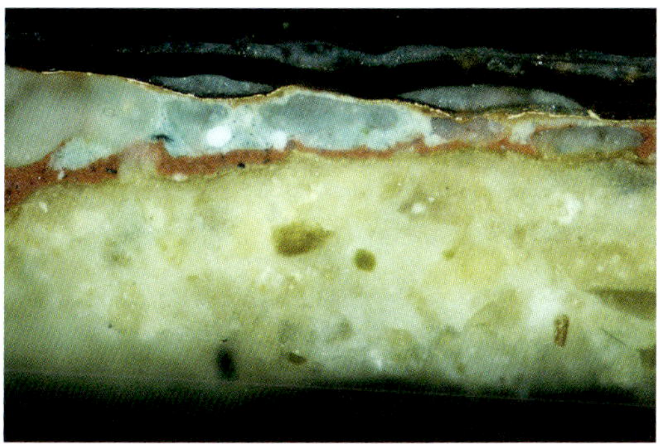

Plate 83 *Cross-section of a sample from the nimbus on the right side of Plate 82, original magnification 180×. Red earth was applied over the plaster support, and over this a layer of indigo combined with lead white. A thin layer of red 'bole'– comprising red earth combined with red lead – has then been applied in preparation for the gilding. Thick layers of wax and other surface accretions are evident on the surface of the sample. Indigo was employed quite extensively throughout the* Majesty *and the* Apocalypse *cycle in the chapter house. (Photomicrograph: H. Howard; © Courtauld Institute, Conservation of Wall Painting Department)*

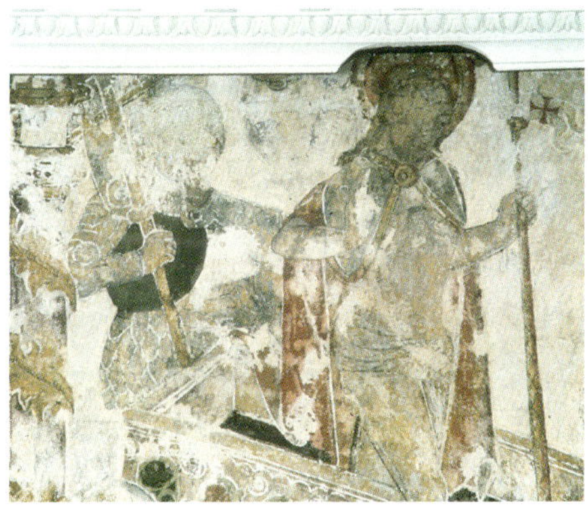

Plate 84 *Durham Deanery, detail of the paintings of c1470 featuring the Virgin: the* Resurrection. *(© Crown copyright. NMR)*

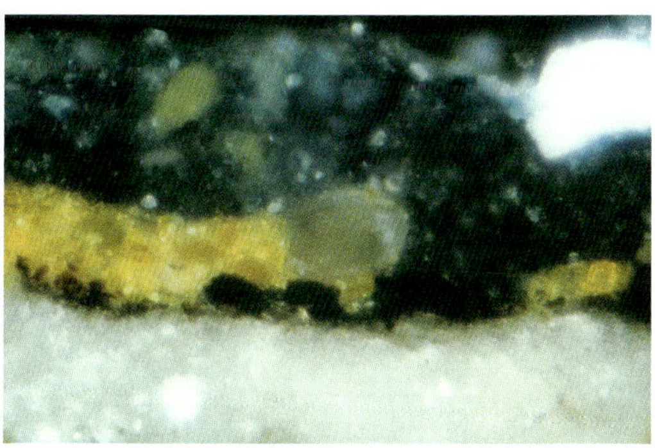

Plate 85 *Durham Deanery. For the armour of one of the soldiers in the* Resurrection, *indigo was combined with lead white and applied in a layer 60µm thick over a warm orange-yellow layer, which in turn was applied over a fine layer of bone black. The resultant blue-grey colour of this layer has an almost metallic sheen, original magnification 400×. (Photomicrograph: H. Howard; © Courtauld Institute, Conservation of Wall Painting Department)*

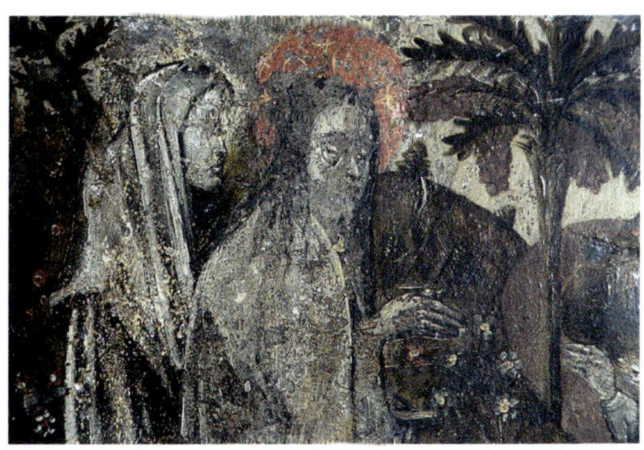

Plate 86 *The Lord Mayor's Chapel (St. Mark's), Bristol: detail of the paintings of c1510, showing the* Noli me Tangere. *(Photograph: H. Howard; © Courtauld Institute, Conservation of Wall Painting Department)*

the use of the green pigment is extensive, for example in the panelled background, although the stratigraphy of its application is not specified (Gärtner and Organ 1989: 11). In the same way, the presence of malachite was suggested in the paintings of *c*1140 in the north transept of Chichester Cathedral. Here the green diagonal lines of the diaper pattern which forms the background to the seated bishop were considered to have been painted in malachite (Gärtner 1988: 2).

In only one case has instrumental analysis confirmed the presence of malachite in English Romanesque wall painting: on fragments of *c*1180 excavated from the chapter house of Sherborne Abbey. Examination of traces of green pigment at 500–1000× magnification in transmitted light – both plane polarised and with crossed polars – indicated the presence of some particles retaining the characteristics of malachite, while others appeared more granular with less distinct birefringence. FTIR analysis confirmed the presence of malachite, while additional investigation of the granular particles confirmed the presence of paratacamite which, given the excavated nature of the fragments, is likely to be an alteration produce of the malachite (Howard and Park forthcoming a).

Malachite apparently continued to be employed in the Gothic period. In England, the pigment has been provisionally identified in only a few instances, and in only one case has instrumental analysis been employed to confirm identifications made by PLM. Analysis by SEM/EDX was employed to examine the green pigment used for the angel's wing in the 14th-century *Annunciation* at St. Nicholas' Hospital Church, Harbledown. Here the identification of malachite is particularly surprising since ultramarine was the blue mineral pigment selected for this important scheme.[55] Other instances where malachite has been tentatively identified by PLM only include the background of the early 14th-century figure of St. William of York, in the feretory of St. Albans Cathedral; the 14th-century scheme at Willingham, Cambridgeshire; and the 15th-century painting *St. Michael Weighing Souls* at Cowlinge, Suffolk.[56]

Application

Over white ground or directly over the plaster support

A trace of a copper green, likely to be malachite, applied over a lime white ground was detected in a sample taken from the green background at the feet of the Virgin in the 15th-century scheme at Cowlinge.[57]

Over coloured underpaint

The traces of paratacamite and malachite which remain on the late 12th-century fragments of wall painting from Sherborne Abbey were applied over a grey underpaint of charcoal black combined with lime white. This use of a coloured underpaint would have increased the covering power and enhanced the colour of the green pigment (Howard and Park forthcoming a).

The original background of the St. William figure at St. Albans consists of what appears to be malachite combined with lead white applied over a secondary ground of

55. Analysis by Howard in Organ 1992.
56. For the technique of St. William of York, see Howard 1993a; for the Willingham scheme, see Wehlte 1954 (here the method of pigment identification is not specified, but is presumed to be by PLM); and for the Cowlinge paintings, see analysis by Howard in Organ and Sawdy 1991.
57. Analysis by Howard in Organ and Sawdy 1991.

yellow earth, in some areas combined with a little red. This yellow underpaint would have provided warmth to the naturally cool malachite green of the background. The malachite layer was originally overlaid with a copper green glaze, but this was over-painted later in the medieval period with verdigris combined with lead white (Plate 101) (Howard 1993a: 35–6).

Mixed with other pigments

Azurite apparently combined with malachite and lime white and applied over a dark grey underpaint was detected by PLM in a second sample taken from the background at the feet of the Virgin in the 15th-century *St. Michael Weighing Souls* at Cowlinge.

Binding media

Malachite was sometimes applied over coloured underpaint, a method of application advised by for example, Cennini, who suggests the application of a range of greens made from malachite tempered with egg yolk over black underpaint to represent foliage and trees (Cennini 1960: 31, 56). Since malachite is stable in alkaline conditions and may therefore be applied directly to fresh plaster or combined with lime, this application *a secco* is likely to be due to its large particle size and consequent low covering power rather than to any presumed incompatibility with lime. Indeed, examination of a number of samples indicated that malachite was combined with lime white. For example, the FTIR spectrum of the green pigment from Sherborne Abbey confirms the presence of malachite and paratacamite in a calcium carbonate matrix.

By contrast, GC-MS analysis of a sample taken from the green background of the St. William of York figure at St. Albans confirmed the presence of linseed oil, which had been prepolymerised to promote drying with the addition of a very small amount of conifer resin.[58] Although malachite has an average hiding power in aqueous media it is rather translucent when combined with oil. This characteristic could be exploited to achieve particular effects, such as in the representation of water; Cennini advises that a glaze of malachite or green earth in oil should be applied over an underpainting depicting details such as fish, and that highlights in white to suggest movement in the surface of water should be applied on top (Cennini 1960: 96).

Comparison with its use in other media and in continental wall painting

Unfortunately, the analytical techniques employed are not always specified where malachite has been identified in continental wall paintings. Malachite has apparently been found in the paintings of *c*1025–28 in the Cathedral of Verden, Germany, and in the scheme of *c*1089 in the west choir of Lambach, Austria, where the pigment was mixed with green earth and applied *a secco*. It may also have been identified in early 12th-century paintings at Kappel, Germany; in the paintings of *c*1160–70 at Perschen, Germany; and in the scheme of *c*1165 in the Johanneskapelle at Pürgg, Austria (Plate 102).[59] Examples from the Gothic period

58. I am grateful to Raymond White for carrying out this analysis for me. GC-MS was undertaken following benzene-ether extraction; see Howard 1993a: 37.

59. For the examples in southern Germany – identified by H. Kühn – at Verden, Kappel, and Perschen Oberpfalz, see Gettens and Fitzhugh 1993b: 198. The Lambach paintings are published by Hammer 1987: 45. Pigment sampling at Pürgg was undertaken by Ivo Hammer in 1977. For the technique of the scheme at Pürgg generally, see Howard 1995d.

include the splendid 13th-century scheme in the choir at Angers Cathedral, where a copper green has been tentatively identified as malachite (Subes-Picot 1992; Stirnemann 1995);[60] 13th-century paintings at Mästerby, Sweden (Nord and Tronner 2000: 99); 14th-century paintings at Aeschi, Switzerland; the scheme of *c*1440 at Linden, Germany; and a wall painting of *c*1484–90 at Sachsen Cathedral, Freiburg;[61] an early 15th-century painting at Risinge, Sweden; and schemes of *c*1500 at Härkeberga, Härnevi and Täby, Sweden (Nord and Tronner 2000: 99). An interesting and unusual identification of spherical malachite has been made recently in the 14th-century *Crucifixion* on the north wall of the choir at Judenburg, Austria. Here a layer of the synthetic green pigment some 500μm thick was applied directly to the plaster support for the robe of the woman to the left of the figure of Mary Magdalene.[62]

In polychrome sculpture, malachite has been identified on the Portail Royal of Chartres Cathedral of *c*1140–50 (Nonfarmale and Manaresi 1987: 268, 274), and on the 13th-century tomb of Heinrich II at Maria Laach (Oellermann 1993: 234). Since details of application are not provided, it is difficult to draw any conclusions regarding the use of the pigment in these instances.

SEM/EDX analysis was employed to confirm the elemental constituents of the green pigment used on the painted stone screens of *c*1370 surrounding the Chapel of Our Lady Undercroft in Canterbury Cathedral. Here both azurite and malachite were identified, but close examination of paint cross-sections suggested that the malachite may represent an alteration of azurite, and was not therefore a component of the original palette.[63]

Discussion

Several authors have indicated that malachite was widely employed in the Middle Ages. For example Laurie (1913: 326) suggests that it was a common and universal green and that it was used in manuscript illumination in England and on the continent, while Thompson (1956: 161) highlights the frequency of its use throughout the Middle Ages.

In addition, the presence of malachite has been frequently suggested following analysis by PLM and microchemical tests, and published data resulting from this indicate that the pigment was equally common in Romanesque and Gothic wall paintings (see Appendix 3, Table 2). However, it is increasingly clear that these analytical methods are insufficient to confirm the presence of malachite, and that where instrumental analysis has been undertaken, other copper greens have often been found. The apparent frequency of its identification is surprising for a number of reasons. Not only are no mineral sources for the pigment specified in the medieval period, but also, until Cennini, there are no references to the pigment in medieval treatises. The apparent frequency of its occurrence in Romanesque wall painting in England is also remarkable, since it is clear that instances of the use of its sister mineral, azurite, are extremely unusual.

It is possible, however, to account for some of these difficulties. Since in nature azurite and malachite are so frequently associated, it may have been deemed unnecessary to list mineral sources of the green pigment separately since it was well understood that it would be found with azurite. Regarding its absence in the documentary sources, a green

60. See also Demailly *et al.* 1998 who list only a copper green.
61. Analysis at these sites was undertaken by H. Kühn, and published by Gettens and Fitzhugh 1993b.
62. See PLM and FTIR analysis by H. Howard in Binder 1997: appendix /SMJ2029 and plate 62.
63. Analysis by SEM/EDX was undertaken by Dr A Roy; see Howard 1993b: 32.

pigment known as *viride terrestre* is mentioned frequently in the texts. Thompson has suggested that this material may be distinct from green earth *(terre verte)*, and that *viride terrestre* may refer to malachite (Thompson 1956: 162). The apparently common use of malachite and the paucity of azurite in English Romanesque wall painting is certainly strange. However, it is possible that since the blue mineral ultramarine was supremely suited to the Romanesque palette – and may have been more widely available and perhaps not at such a premium as previously supposed – azurite may have been used less.

Alternatively, it is quite likely that in some instances malachite has been erroneously identified. Thompson (1956: 161) was the first to suggest this possibility, and recent analysis suggests that in a number of instances this may indeed be the case. An analytical study of green copper-based pigments in easel painting has indicated that green pigments are frequently very heterogeneous in composition. This heterogeneity may result from natural variety in the mineral species or the impurity of raw materials and the methods of manufacture employed for synthetic pigments in the medieval period (Martin and Eveno 1992: 785).

In recent years, the employment of instrumental techniques in the examination and analysis of original painting materials has produced results of increased specificity. Whereas in the past it was relatively simple to detect the presence of a copper green by means such as microchemical tests, the identification of a chloride or other anions in green pigments often eluded the analyst. A large number of copper chloride greens have now been detected in wall paintings, suggesting that some of the previously unspecified copper greens may prove to be copper chlorides or other such compounds rather than malachite. Copper compounds which have been recently identified include posnjakite (basic copper sulphate, $Cu_4SO_4(OH)_6$) and natural pseudo malachite (a basic copper phosphate, $Cu_5(PO_4)_2(OH)_4$) in early 16th-century Russian wall paintings;[64] posnjakite in medieval wall painting in Germany (Richter 1988: 174); and antlerite ($CuSO_4 \cdot Cu(OH)_2$) in unspecified wall paintings of 14th–15th century date (van't Hul-Ehrnreich and Hallebeek 1972). Whether a mineral or synthetic pigment is present, it is increasingly clear that detailed instrumental analysis should be undertaken on a number of grains of pigment from each sample to determine the often complex nature of the materials present.

Pigment alterations

The alteration of both malachite and azurite to form a copper chloride-rich material in the presence of humidity and chloride ions has been noted in a number of instances. For example, malachite – now substantially altered to paratacamite – has been identified on the excavated fragments of *c*1180 from the chapter house of Sherborne Abbey (Howard and Park forthcoming a). This phenomenon is by no means confined to English wall painting. Copper chlorides which are probably due to a similar pigment alteration have been identified in a number of Austrian wall paintings from the 13th to the 15th centuries (Kerber *et al.* 1972).

64. Naumova *et al.* 1990 suggest that the presence of these unusual compounds may be indicative of the impure materials employed for the synthesis of pigments.

Verdigris and copper resinate

Verdigris is the generic term for a range of copper acetates (copper ethanoates) of varying composition and colour. A recipe for the preparation of verdigris, *viride grecum*, is given in the *Mappae Clavicula*, and again by Eraclius, although under the name *viride cupri vel aeris*.[65] Theophilus provides a detailed description of the method of manufacture of *viride hispanicum*: by coating thin, clean sheets of copper with hot vinegar and then sealing them in a wooden chest one third full of warm vinegar, and keeping the whole apparatus warm in a pile of dung for about two weeks (Theophilus 1979: 41).

These basic early recipes are repeated in various forms in later medieval texts. The impurity of the final product is hinted at by Jehan le Begue who suggests that either copper or brass may be hung over vinegar. Alternatively, he suggests putting a little urine in the bottom of a brass vase and leaving it in the sun until it dries out, and then scraping off the verdigris (Merrifield 1967: I: 48, 66). Since neither the copper itself nor its alloys would have been particularly pure, it is clear that a green or blue-green product of mixed composition would have resulted from these instructions.

The green pigment may be classified into two types: basic and neutral verdigris. Basic verdigris is formed by the type of process described by Theophilus, that is by exposing copper to the vapour of acetic acid, air and water. The resultant pigment has been shown by Gauthier (1958: 17–19; 27–43) to contain one or more of four slightly different copper acetates, three of which are blue and one of which is green (d):

(a) $[Cu(CH_3COO)_2]_2.Cu(OH)_2.5H_2O$
(b) $Cu(CH_3COO)_2.Cu(OH)_2.5H_2O$
(c) $Cu(CH_3COO)_2.[Cu(OH)_2]_2$
(d) $Cu(CH_3COO)_2.[Cu(OH)_2]_3.2H_2O$

According to Gauthier, manufactured verdigris could be any one of these compounds, or a mixture of a and b, or of b and d.

More recent analysis has shown, however, that it is only possible to synthesise three of the above four basic salts – (b) was not reproducible (Rahn-Koltermann *et al.* 1991). These investigations also determined the presence of a new basic salt $Cu(CH_3COO)_2.[Cu(OH)_2]_4.3H_20$, and suggest that there may be a continuum based on the formula $[Cu(CH_3COO)_2]_x[Cu(OH)_2]_y.zH_2O)$ in which typical values are $x = 1$, $y = 3$ and $z = 2$.[66]

Neutral or distilled verdigris is the neutral copper(II) acetate monohydrate ($Cu(CH_3COO)_2.H_2O$), formed when unrefined basic copper acetates are dissolved in acetic acid and recrystallised by slow evaporation on wooden sticks. Cennini describes grinding basic verdigris in vinegar, which would certainly have produced the neutral and more stable form of the pigment (Cennini 1960: 33).

65. See *Mappae Clavicula* (1974: 42), which suggests hanging copper leaf over sharp vinegar and leaving it in the sun for two weeks; Eraclius 1967: 238–9 indicates that strips of copper placed in a basin with white wine vinegar and left for two to three months would produce an excellent green colour.
66. For a full discussion of the composition, properties and manufacture of verdigris, see Kühn 1993a; Scott *et al.* 2001; and van Eikema Hommes 2001: 164–70.

Cost

Verdigris appears to have been relatively cheap, its cost approximately the same as that of other synthetic pigments such as vermilion. This is clearly indicated in the accounts for Exeter Cathedral for *c*1320–21 which itemise verdigris at 7d per lb and vermilion at 8d per lb, probably for use in painting sculpture (Erskine 1981: 134; Hulbert 1998: 36–7). Verdigris is also recorded in the accounts for the mid-14th-century wall paintings of St. Stephen's Chapel, Westminster. Here three slightly different prices are given for the pigment, purchased by three different painters; the cost ranges from 10d per lb to 1s 2d per lb, which may reflect differences in quality (Smith 1807: 220–21).

Optical, chemical and physical properties

Scott *et al.* (2001: 74–7) have characterised the optical properties of neutral verdigris and the basic copper acetates (a–d) (see also Kühn 1993a: 132–5):

> Neutral verdigris is deep blue-green, and when mounted in Meltmount (RI 1.662) and examined in plane polarised transmitted light is characterised by light-blue tabular crystalline fragments, which may exhibit conchoidal fracture, clear relief and a refractive index less than 1.662. With crossed polars, the particles are birefringent, exhibiting brown, yellow and dark blue colours and clear extinction.
>
> Basic copper acetate (a) is pale blue-green or blue in colour. When examined as a dispersion in Meltmount in plane polarised transmitted light the particles appear as turquoise fibrous crystals. With crossed polars the brush-like crystal aggregates display blue, yellow and green birefringence.
>
> Basic copper acetate (b) is a deep blue-green colour. Examination by polarised light microscopy indicated the presence of at least two different crystal types. A few irregular crystalline fragments of pale blue-green colour were present, though the majority of particles appear to be aggregate bundles of fine fibers, some of which are curved. Examination in plane polarised transmitted light indicated that the refractive indices are 1.548 and 1.53. With crossed polars many of the particles display intense birefringence and undulose extinction.
>
> Basic copper acetate (c) is pale turquoise. In plane polarised transmitted light the particles appear as a mass of light-green acicular crystals. The majority of the crystals are long prismatic needles with refractive indices of approximately 1.56 and 1.58 and a positive sign of elongation. They exhibit parallel extinction when examined with crossed polars.
>
> Basic copper acetate (d) is a light green-blue colour. Examination by polarised light microscopy indicated that the particles are transparent to light-green crystalline fragments with refractive indices close to 1.662. With crossed polars the particles appear intense light-blue and pale straw coloured, and microfissures twinkle on rotation with no clear extinction.

Copper acetates are stable to light, but on heating they darken with the formation of black cupric oxide. Verdigris has frequently been found in mixtures with lead white and lead-tin yellow with no sign of alteration. However, combinations of verdigris and sulphur-rich pigments such as ultramarine in aqueous binding media may alter, producing a darkened effect due to the presence of copper sulphide.

Verdigris is prone to colour change from blue to green, particularly in the first month or so after manufacture. This type of change is more pronounced for basic verdigris, and where a water-based medium is employed. Cennini's explicit direction for the production of the neutral form, and for combining it with an egg tempera medium, suggests that this change was well understood (Cennini 1960: 33). The alteration of verdigris in association with various binding media is complex (see discussion in Scott *et al.* 2001: 76–8).

Copper acetates are soluble in dilute acids, and are unstable in alkaline conditions forming blue copper hydroxides. This incompatibility with lime prompted Cennini among others to warn against using verdigris in *buon fresco* painting (Cennini 1960: 50).

Verdigris reacts with oil, proteins and resins, forming oleates, proteinates and resinates of copper. It has generally been assumed that the majority of green copper-based glazes consist of copper resinate of the type typically made by dissolving verdigris in hot varnish and/or pine resin. However, mixtures of oil and verdigris alone have poor covering power, and this combination has also been used to full effect as a glaze.[67] Indeed, an important new study (van Eikema Hommes 2001) examining historical recipes for glazes with verdigris used in European easel painting indicates that for glazes with distilled verdigris, the pigment was most commonly combined cold with heat-bodied linseed oil. Although a little resin was often added to the mixture to improve its glazing properties, the quantities were small, certainly not sufficient to produce a copper-resinate glaze. Van Eikema Hommes (2001) found that, though the composition of verdigris-based glazes varied considerable, there is little evidence to suggest that glazes incorporating verdigris were of the copper resinate type; verdigris in oil was clearly used in preference.

When verdigris in oil is combined with lead white, lead-tin yellow or other pigments of high refractive index, the resultant mixture produces paint layers of high opacity.[68] In addition, the siccative properties of verdigris when combined with an oil medium to form copper oleates made the pigment particularly useful, especially when combined with mixtures of black pigment and oil which never dry on their own. All these properties were fully exploited in English medieval wall painting.

Use in English medieval wall painting

Perhaps the instability of verdigris in the presence of fresh lime explains its extremely limited use in Romanesque wall painting in England. Indeed, in the only instance where verdigris has been identified in a sample which may date from this period – from the capital on the south side of the entrance arch to the apse of St. Gabriel's Chapel, Canterbury Cathedral – it is possible that this area was repainted later in the medieval period since it stood partially outside the wall which blocked the apse at the end of the 12th century (Cather and Howard 1994: 145). By contrast, in the Gothic period when organic binding media became increasingly prevalent, the intense green of verdigris became the most popular of the green pigments.

The pigment was cheap, readily available, and widely employed from the first quarter of the 13th century. Examples occur in the wall paintings of *c*1220 in the refectory of the

67. The discrete particles of verdigris in a verdigris-oil glaze are often difficult to distinguish under the microscope, particularly in an aged paint film where a reaction between the copper pigment and the oil may have occurred producing transparent copper salts of fatty acids.

68. Verdigris is commonly found in mixtures with lead white, contrary to Cennino's (1960: 33) warning that the two pigments are incompatible.

Eastbridge Hospital, Canterbury (Howard 1994: 4); on mid-13th-century fragments of excavated wall plaster from Clarendon Palace (Ashurst and James 1988: 253); in the scheme of *c*1250 at Horsham St. Faith Priory, Norfolk (Babington and James 1988: 15–16); in the wall paintings of *c*1300 in the Ante-Reliquary Chapel, Norwich Cathedral (Howard 1995c; Park and Howard 1996); in the mid-14th-century scheme over the chancel arch at Ickleton, Cambridgeshire (Howard 1988: 33); in the approximately contemporary wall painting in the tomb recess of St. William of Perth in Rochester Cathedral (Rickerby 1988: 11); in the royal wall paintings of *c*1352–63 from St. Stephen's Chapel, Westminster (Smith 1807: 223–5; van Geersdaele and Goldsworthy 1978); in both the 14th- and 15th-century schemes at Willingham, Cambridgeshire (Curteis 1995a: 9); in the wall paintings of *c*1400 at Bradfield Combust, Suffolk (Rickerby 1990b: 5); and in the early 15th-century scheme in the chapel of Farleigh Hungerford Castle, Somerset (Howard *et al.* 1998).

Application

Directly over the plaster substrate, or over a white ground

In the early 13th-century scheme in the Eastbridge Hospital, Canterbury, verdigris was applied directly over a lime white ground for the border of Christ's mandorla and for the green background of the inscription above the symbol of St. Matthew (Howard 1994: App. 1). A smooth white lime plaster was detected beneath a layer of verdigris combined with lead white for St. Faith's drapery on the east wall of the refectory at Horsham St. Faith. In this instance, a green glaze layer was applied to the surface to provide translucency and enrich the green appearance. Verdigris was also applied over a lead white ground for St. Christopher's robe in the mid-14th-century scheme at Willingham, and for the angel's wings in the *Assumption of the Virgin* dated to the 15th century in the same church (Curteis 1995a: 9).

Over coloured underpaint

In St. Gabriel's Chapel, Canterbury Cathedral, the green copper-based pigment is combined with lead white and applied over an underpaint of red earth, lead white and charcoal black (Cather and Howard 1994: 145). In the 14th-century scheme over the chancel arch at Ickleton, verdigris combined with lead white was applied over red underpaint consisting of red earth combined with a little red lead (Howard 1988: pl. 52). A local ground of yellow earth has been detected beneath verdigris in a sample taken from the St. Christopher of *c*1400 at Bradfield Combust (Rickerby 1990b: 5).

Mixed with other pigments

Verdigris combined with lead white has been frequently detected in English medieval wall painting. The purpose was presumably to produce an opaque green layer of mid tone that could be glazed with a layer of verdigris in oil if greater translucency or depth of colour was required. In the early 15th-century scheme at Farleigh Hungerford Castle, verdigris was combined with lead white and charcoal black in the oil-rich but opaque paint layer used for the lining of St. George's cloak (Plates 108 and 109). In the same painting, verdigris was also combined with lead-tin yellow to produce an opaque brilliant lime green colour in the zone directly above the saint (Howard *et al.* 1998: 60–62).

Use of verdigris in opaque paint layers and as a glaze

In most examples in English medieval wall painting, verdigris was used both as a component of an opaque paint layer and as a glaze. Even in the later medieval repainting of the green background of the early 14th-century 'St. William of York' in the feretory of St. Albans Cathedral, where a glaze layer does not appear to be present, the basic colour was manipulated by varying the proportions of medium and of the added lead white pigment to produce a brilliant green appearance. Here the synthetic green pigment was mixed with lead white and applied in two layers in a manner analogous to the original paint layers beneath, with the upper layer containing less white and more medium and therefore producing a more translucent effect (Howard 1993a: 37).

Where a green glaze was applied over a layer of verdigris combined with lead white, a translucent deep green colour was produced. Despite previous conservation treatments which impeded analysis of the binding medium employed in the scheme of *c*1300 in the Ante-Reliquary Chapel, Norwich Cathedral, it was possible with FTIR analysis to determine the presence of verdigris in an oil medium in one area: the green drapery of one of the apostles on the south side of the western arch. Here, over the reflective lead white ground, verdigris combined with lead white in oil was applied to produce an opaque green paint layer. This was subsequently glazed with verdigris combined with charcoal black in an oil medium, resulting in a richly translucent dark green effect (Howard 1995c: 5, fig. 2 and App. II). Similarly, verdigris combined with lead white in oil and glazed with what appears to be verdigris in oil occurs in the early 15th-century kneeling knight on the south wall of the chapel at Farleigh Hungerford Castle (Plate 110) (Howard *et al.* 1998: 61). A green glaze, possibly of verdigris, was also employed for details on St. George's helm on the east wall of the same chapel. Here, a gold effect was achieved by applying a yellow glaze over silver leaf, and further embellishment provided by raised curvilinear details made from a second layer of silver leaf highlighted with a green glaze (Plates 183–185).

In other instances, verdigris was employed purely as a glaze layer; for example, for raised decorative elements on the canopy of the niche of the reredos of *c*1370 in the Chapel of Our Lady Undercroft at Canterbury Cathedral, a luminous effect was produced by glazing silver leaf with verdigris in oil (Plates 105 and 106) (Howard 1993b: 32).

Binding media

In all cases where instrumental analysis has been undertaken of verdigris in English medieval wall painting, an oil medium has been identified. For example, FTIR analysis determined that verdigris was combined with a drying oil in the early 13th-century paintings in the Eastbridge Hospital, Canterbury (Howard 1994). GC-MS analysis undertaken on samples from the mid-13th-century scheme at Horsham St. Faith Priory confirmed the presence of linseed oil in the drapery of the monumental figure of St. Faith on the east wall, while prepolymerised linseed oil was detected in the green pigment employed for the 15th-century repainting in the area above the boat in the *Return to England* scene of the *Foundation Narrative* (Plate 103) (White 1989). As discussed above, the presence of verdigris in oil was also detected by FTIR in the scheme of *c*1300 in the Ante-Reliquary Chapel, Norwich Cathedral (Howard 1995c: 5 and fig. 2). This finding supports Tristram's supposition (1955: 65) that the original binding medium was oil or an oil emulsion.

For the vine leaves painted in the recess of the tomb of St. William of Perth (*c*1340) at Rochester Cathedral, verdigris was bound with walnut oil and subsequently glazed with copper resinate (Plate 104).[69] The identification of walnut oil in this instance is exceptional since in all other instances linseed oil is the only type of oil identified by instrumental analysis in English wall painting throughout the medieval period.[70] Analysis by FTIR also detected the presence of a drying oil combined with verdigris in original paint layers in the St. George at Farleigh Hungerford Castle.

Comparison with its use in other media and in continental wall painting

By contrast with analytical evidence from England, verdigris has been detected in several Romanesque wall paintings on the continent; for example at Saint-Aignan, Brinay-sur-Cher and Notre-Dame-la-Grande, Poitiers in France.[71] The pigment has also been detected in the Gothic period, as in the splendid 14th-century paintings on the choir screens of Cologne Cathedral, although no details of its application are provided (Jägers and Schulze-Senger 1989: 187–8).

Use of verdigris in polychrome sculpture

Verdigris was employed extensively in Gothic sculptural polychromy, as on the de Ingham tomb of *c*1344 at Ingham, Norfolk, where it was applied in an opaque paint layer combined with lead white for the naturalistic vinescroll, and over silver leaf for a more luminous effect. For the de Bois tomb of *c*1380, also at Ingham, verdigris was applied over a ground of lead white combined with yellow earth for the niches on the tomb chest (Howard 1993d: App. I, II). The pigment has also been detected in 14th- and 15th-century polychromy on the west front of Exeter Cathedral (Sinclair 1991: 117, 121–3); on fragments of sculpture of *c*1475–90 from the Great Screen at Winchester Cathedral (Brodrick and Darrah 1989: 616); and on the 15th-century Fitzalan effigies at Arundel (Brodrick and Darrah 1986: 73).

Continental examples include the 13th-century tomb of Count Palatine Henry II at Maria Laach (Germany) (Oellermann 1993), and the 14th-century portal sculpture of the Holy Cross Minster in Schwäbisch Gmünd (Hauff 1988: 715).

Use in panel painting

Verdigris has been identified in the reredos at Adisham, Kent – recently redated to *c*1180 – where it was combined with lead white and applied over a lead white ground (Howard

69. Analysis by K. Groen of the Hamilton Kerr Institute, published by Rickerby 1988: 11. Given the exceptional nature of this result, it would be important to reassess the analytical data and reconfirm the identification.
70. In Italian easel painting, there appears to have been a preference for walnut oil until the 16th century when a change was made to the better drying but more strongly yellowing linseed oil (White 1995: 135). In northern medieval panels, there are few confirmed instances of its use, although it has been identified in the German 'Estouteville Triptych' of *c*1400 (Villers *et al.* 1997). However, the archival records of Royal French wall paintings in the 14th century record materials for wall paintings which include walnut and linseed oil. A. H. van Buren (1986: 101, n. 40) suggests that her forthcoming research on wall paintings from Hesdin and Broederlam will indicate that many 14th-century examples were painted in oil on a gesso ground on wood. She refers in particular to a set of hanging panels, painted with scenes of St. John the Evangelist, which constituted the wall decoration in the Chapel of the Leugemeete in Ghent.
71. Analysis by the LRMH; see Rollier-Hanselmann 1997: 67.

1993c: App. II).[72] The use of verdigris has also been confirmed in English 13th-century panel painting. It was employed in the original decoration of the nave ceiling at Peterborough Cathedral of *c*1220, where it has been detected in a sample taken from the Psaltery Player; here, it was applied over a local ground of red earth combined with lead white and carbon black (Howard 1997b: table 1). It has also been identified in the ceiling paterae from the Painted Chamber, Westminster Palace, of *c*1263 (Lynn 1997: 20). Later examples include the early 14th-century painted screen from Kingston Lacy, Dorset (Tracy 1997: 28), and the German late 14th-century Estouteville Triptych (Villers *et al.* 1997).

Discussion

In medieval wall painting in England and elsewhere in northern Europe copper greens are commonly reported, although with fewer confirmed examples of verdigris. In England, the rapid increase in the use of the pigment is mirrored by an increase in the use of organic binding media from the early 13th century. Indeed, given the instability of verdigris in alkaline conditions it would seem an inappropriate choice in Romanesque wall painting when the carbonation of lime was generally the principal binding mechanism despite the limited use of additional binding media. Verdigris was however well suited to Gothic painting technology, and in English painting of that period was often selected for works in various media. Where instrumental analysis has been undertaken, it shows that oil was employed as the binding medium for verdigris.

Pigment alterations

An amorphous pale blue alteration product of verdigris has sometimes been detected in both wall painting and polychrome sculpture. This alteration seems to be associated with contact between the verdigris and overlying layers of fresh limewash or lime plaster. Since verdigris is known to be unstable in alkaline conditions – forming copper hydroxides – this is not surprising. Alteration appears to occur initially at the interface between the layers, or where cracks in the underlying green paint have provided a passage for further penetration of the calcium hydroxide from the fresh limewash or plaster. The phenomenon has been noted in several Gothic wall paintings in England: at Brent Eleigh in Suffolk (Howard 1988: App. II); in the Ante Reliquary Chapel, Norwich Cathedral (Howard 1995c; Park and Howard 1996); at Ickleton, Cambridgeshire (Howard 1988: 33); and at Farleigh Hungerford Castle (Howard *et al.* 1998: 62–3).

In the mid-14th-century paintings over the chancel arch at Ickleton, the turquoise-blue alteration product has formed in a layer 15μm thick over the original paint layer of verdigris combined with lead white. It is significant that the transformation product is not apparent where an original dark green glaze remains intact over the verdigris layer. Where the glaze was lost, however, subsequent limewash layers have come into direct contact with the verdigris layer and effected the transformation (Plate 117).

In polychrome sculpture, the phenomenon has been detected on the tomb of Sir Oliver de Ingham (*c*1344) at Ingham, Norfolk, where the extent to which this alteration

72. The reredos has recently been redated by C. Tracy to *c*1180 on the basis of dendrochronological evidence.

has occurred varies enormously. In one case a trace of the pale blue alteration product can be seen at the interface of the verdigris and the overlying limewash (Plate 113). Where the alteration is in a more advanced stage (Plate 114), the once-green layer directly beneath the limewash has almost completely changed to a pale blue colour; in addition, amorphous blue areas can be seen within a second underlying layer of verdigris. In a third sample, taken from the recess behind the effigy, almost complete transformation has occurred, and only a trace of green coloration at the base of the blue layer indicates the original appearance of the paint layer (Plate 115) (Howard 1993d).

It is significant that in Stothard's engraving of the tomb – first published in 1812 and made before the application of the limewash – all the present blue areas are recorded as green. However, by the middle of the 20th century, partial alteration of the verdigris had clearly already occurred, since Tristram records the scroll motif on the tomb chest as blue (Tristram 1955: 186). Comparison of the FTIR spectra from the green pigment and the blue alteration product in one of the samples from the tomb suggests that the oil medium for the green material may have been leached from the now-blue sample.

The band centered at around 1560–1610cm^{-1} – caused by stretching vibration of the acetate group – is indicative of verdigris. This band is considerably reduced in the spectrum of the alteration product. In addition, bands at 1740, 2856 and 2930cm^{-1}, which suggest that oil is the binding medium of the green pigment, are almost completely absent from the spectra of the alteration product, indicating that little of the original medium is now present. Leaching of the binding medium results in a low pigment-to-binder ratio, and the consequent open porous structure is likely to encourage chemical reactions to occur. A similar alteration had been detected on the late 14th-century polychromed stone screens surrounding the Chapel of Our Lady Undercroft in the crypt of Canterbury Cathedral (Howard 1993b: 32).

Further analysis of the pale blue alteration product from the de Ingham tomb and also from other sites indicates that it includes a copper chloride component. For example, examination of a sample from the early 15th-century scheme at Farleigh Hungerford Castle indicated the transformation of verdigris to a pale blue-green copper chloride-containing material (the copper chloride component was identified by FTIR as atacamite, Plates 109–112). In such cases it is important to try to determine the origin of the chloride ions. It is possible that salt was added to the limewash layers subsequently applied to the surface of the painting. However, chloride-containing salts were sometimes incorporated in the manufacture of a complex copper acetate-based green pigment,[74] and it is possible that a verdigris-like pigment produced in this way was originally employed at Farleigh Hungerford.

Verdigris with a high level of chlorine has also been identified in the polychromy of the west front of Exeter Cathedral.[75] The author is currently attempting to characterise the chlorine-rich alteration product of verdigris, and to investigate the mechanism for the alteration, which at Ingham and Farleigh Hungerford is so clearly associated with the application of alkaline material (limewash or lime plaster) to the surface.

73. The strong peaks at 1740, 2856 and 2930cm^{-1} indicate that oil binds the green pigment particles.
74. See, for example, Theophilus 1979: 41.
75. The precise date of this polychromy is not specified; Sinclair 1992: 10.

Copper resinate

Oils, resins and proteins react with verdigris to form copper oleates, copper resinates and copper protein compounds, and these were employed to form translucent green glazes, especially from the 15th century onward. Until recently it has been supposed that, of these verdigris-based glazes, copper resinates were the most frequently employed.

Copper resinate is a transparent, amorphous green which consists principally of copper salts of resin acids, but its precise composition depends on the method employed for its manufacture (Kühn 1993a: 148–55). A typical recipe for the preparation of copper resinate is provided by Birelli (1601: 370–71).

> Take one pound of fine white turpentine resin (pine resin), three ounces of mastic . . . and half an ounce of new wax. Put everything together in a small newly glazed pot, and boil . . . mixing it all the time. This done . . . strain . . . and then add to these things an ounce of verdigris . . . (translation by Joyce Plesters).

Woudhuysen-Keller divides the various recipes for the preparation of copper resinate into two principal types: first, those in which the verdigris is dissolved in pine resin to form a glassy material which may then be ground in oil or varnish, and secondly, those in which verdigris is dissolved in hot oil and pine resin. She also discusses the type of additives which were employed to modify the strong colour of the pigment, including yellow lakes and juice of the herb rue (*ruta graveolens* L.).[76]

Verdigris in an oil medium can itself be employed as a glaze. Since the verdigris does not dissolve in the oil, discrete particles of the pigment are usually present, though they may be difficult to distinguish under the microscope, particularly in an aged paint film where a reaction between the copper pigment and the oil may have occurred producing transparent copper salts of fatty acids. It is therefore often difficult to distinguish between copper resinate and other verdigris-based glazes by microscopy.

Since definite identification of a copper resinate glaze depends on instrumental analysis (usually GC-MS or FTIR) determining the presence of copper salts of resin acids and a high percentage of resin in the binding medium, there are very few instances in which the presence of copper resinate has been reliably confirmed in wall painting.

FTIR analysis undertaken on samples from the scheme of *c*1220 in the Eastbridge Hospital, Canterbury, did however suggest the presence of a resinous component and therefore the possible presence of a copper resinate glaze. In a sample from the border of Christ's mandorla, a layer of verdigris combined with lead white was found to be glazed with verdigris in oil with an additional resinous component, probably copper resinate. Elsewhere in the scheme, verdigris combined with lead white in oil was employed for the rich green background of the Evangelist Symbols (Howard 1994: 4–6, pls. 17–19).

In addition, instrumental analysis of organic components from the mid-14th-century wall painting in the recess of the tomb of St. William of Perth at Rochester Cathedral showed that the naturalistic vine leaves and tendrils were painted in verdigris combined with oil and then covered with a copper resinate glaze (Rickerby 1988: 11).

The presence of copper resinate has been suggested in a number of other instances, but without instrumental analysis to confirm the presence of copper salts of resin acids in the

76. For a discussion of various methods for preparing this pigment, see Woudhuysen-Keller and Woudhuysen 1998.

glaze layer. For instance, in the mid-14th-century scheme over the chancel arch at Ickleton, Cambridgeshire, a layer of verdigris combined with lead white appears to have been glazed with a copper resinate or copper oleate layer (Howard 1988: 33 and pls. 50–52). In this instance, it is the characteristic brown discoloration at the surface of the translucent green layer which may signal the presence of copper resinate.[77] Similarly, in the St. Christopher of *c*1400 at Bradfield Combust, Suffolk, a zone of brown discoloration at the surface of a layer of verdigris also suggests that a copper resinate glaze, now altered, may be present. In this instance the verdigris was applied over yellow underpaint (Rickerby 1990a: 5).

A very translucent copper-rich glaze, possibly containing copper resinate, has also been detected on the early 15th-century painting in a tomb recess in the south choir aisle of Lichfield Cathedral.[78] Other northern European examples include the important paintings of *c*1260 in the choir of Angers Cathedral, where a copper green and a copper glaze, probably copper resinate, have been detected (Subes-Picot 1992: 89; Demailly *et al.* 1998: 11).

However, an important new study of recipes for glazes with verdigris used in easel painting indicates that the majority of green glazes consist not of copper resinate as is often presupposed. The composition of the glazes employed varies, but they were generally made by combining verdigris, without heating, in an oil medium, principally linseed oil which was often pre-polymerised (heat-bodied). Where resin was detected, the component is only a small additional amount incorporated to improve the glazing properties of the paint (van Eikema Hommes 2001: 164, 190). It is likely therefore that similar findings will prevail also for wall painting once sufficient instrumental analysis has been undertaken on these glaze layers.

Regarding the discoloration of 'copper resinate' glazes, experiments by van den Berg *et al.* (2000: 20) have shown that alkaline conditions make copper glazes containing dissolved verdigris more vulnerable to discoloration. Chalk-containing layers on the surface of these green glazes were found to be the most discoloured. The presence of chalk in these uppermost layers is attributed to the presence of yellow lake pigments – incorporated to influence the strong green of the verdigris-based glaze –which have faded. These results have profound implications for wall paintings. First, since verdigris-based glazes are frequently in an overall alkaline environment in wall paintings and would therefore be particularly susceptible to discoloration, and secondly, since yellow lake pigments have yet to be identified in medieval wall painting in England, their presence would be unexpected in a verdigris-based glaze layer. Thus the presence of chalk in surface layers over a green glaze would not generally be attributed to a faded lake pigment.

Use of copper resinate in polychrome sculpture

Copper resinate has been tentatively identified on a number of medieval sculptures. For example in the 13th-century polychromy on the west front of Wells Cathedral, the brilliant green background of the *Coronation of the Virgin* is painted in a glazing pigment,

77. The brown discoloration often detected at the surface of copper resinate glazes is thought to be due to a photocatalytic reaction, see Kühn 1993a: 151 and also van den Berg *et al.* 2000.
78. S. Cather, 'Throne of mercy: an early 15th-century wall painting in Lichfield Cathedral', unpublished lecture.

probably copper resinate, over a lead white ground (Roy and White 1982; Sampson 1998: 118–19). Copper resinate has also been detected on 14th- and 15th-century polychromy on the west front of Exeter Cathedral, although here the pigment has altered to a slightly brown colour (Sinclair 1991: 122–3). The presence of verdigris in a dissolved form – either copper resinate or copper oleate – has also been suggested in samples from a group of early 15th-century stone fragments from Tewkesbury Abbey which may originate from a late medieval high-altarpiece.[79] For example, for the surcoat of one figure, verdigris with the possible addition of copper resinate or oleate was applied directly over a lead white ground (UCL 1996). Similarly, analysis suggested that a copper resinate-type green glaze was used to saturate an intense green colour on the hilt of the sword of a 14th-century effigy at Erwarton, Suffolk (Roy 1984). Preliminary identification of copper resinate combined with verdigris and lead white, and overlaid with a copper resinate glaze, has also been made on polychromed fragments of *c*1475–90 from the Great Screen of Winchester Cathedral (Brodrick and Darrah 1989: 616–17).

79. This group of limestone sculptures is now stored in the room above the north porch of the Abbey. Catalogued by Dr Richard Morris, the most important among the group are twenty fragments of a figure series of the lords of the manor of Tewkesbury, two heads of prophets and the head of an angel.

❦ FIVE

Red pigments

Vermilion, red lakes and red lead are discussed in detail in this chapter. Earth pigments, however, are examined in the next chapter due to the wide range of hues available – from purple to red, orange, yellow and brown.

Vermilion

In the medieval period, two forms of vermilion (mercuric sulphide, a – HgS) were available: the mineral cinnabar, which was simply washed and ground for use as a pigment, and its synthetic equivalent, dry-process vermilion.[1] Nomenclature of the pigment in the Middle Ages is often confusing, with the terms *vermiculus*,[2] *cinnabaris*,[3] *cenobrium* and *minium* employed by various authors.[4] In antiquity *cinnabaris* and *minium* were used synonymously, while red lead was properly called *minium secundarium* (Pliny 1984: 84–5).[5] Eraclius uses both *cinobrium* and *vermiculus*, though it seems unlikely that he uses the different terms to distinguish the ground mineral from the synthetic pigment, since

1. It is difficult to distinguish cinnabar from dry process vermilion by PLM except perhaps by inert mineral impurities which may be present in cinnabar. During the late 17th century the 'wet-process' manufacture of vermilion was discovered: this was easier and cheaper than 'dry process' or sublimation and swiftly became the favoured method. For a description of the pigment and its characteristics, see Gettens *et al.* 1993c: 163–5.
2. *Vermiculus* is derived from the Latin *vemes*, a name originally used for the insect kermes from which a brilliant red dye is made; Gettens *et al.* 1993c: 159.
3. The word cinnabar may be of Indian origin and has been used to describe a bright red resin known as dragon's blood; Gettens *et al.* 1993c: 159.
4. Theophilus (1979: 40) used the name *cinobrium*; Peter of St. Omer uses the term *vermiculum* (Merrifield 1967: 138–9).
5. Although *minium* is now taken to mean red lead, some confusion in the terminology for the two red pigments persisted throughout the Middle Ages (Gettens *et al.* 1993c: 159). In addition, the red lake pigment kermes was known in the medieval period as vermilion and vermeil (English) or vermeil (French), synonyms which can be traced to the ancient *vermiculum* (little worm).

his *cinobrium* is incorporated in a mordant (Eraclius 1967: 238–9, 250–53; 1996: 118, n. 4).

In antiquity the best quality, highly crystalline mineral cinnabar was exported to Rome from the famous mines at Almaden, Spain.[6] Vermilion from Spain is mentioned again in a 14th-century treatise, suggesting that it may have continued to be an important source of the mineral throughout the medieval period (Thompson 1926: 281, 297). Although sources of natural vermilion – cinnabar – are now known in England, it seems unlikely that these would have been employed in the medieval period.[7] Any identified examples of the pigment in English wall painting are therefore likely to be from imported high-quality cinnabar or to have been synthesised by sublimation.

Cost

Examination of the documentary evidence from the Gothic period suggests that vermilion may not have been so expensive as previously supposed. The Exeter Cathedral accounts of 1320–21, for example, document the cost of painting materials – probably for polychrome sculpture – including 4lb vermilion (*vermilioun*) at 2s 8d (8d per lb) and red lake (*cinople*), which at 4s 9d for ¾lb is approximately six and a half times more expensive. The vermilion is also cheaper than indigo at 18d per lb, although it is approximately the same cost as verdigris at 2s 4d for 4lb (Erskine 1981: 134). This price range supports the notion that the vermilion was generally a synthetic material. However, the same accounts show that the other readily available synthetic red pigment, red lead, was even cheaper, and at only 5s 2½d for 25lb was approximately one-third of the price of vermilion (Erskine 1981: 131).

Evidence for the price range of red pigments is provided by the mid-14th-century accounts relating to the wall paintings in St. Stephen's Chapel, Westminster. Red lead and ochre are extremely cheap, at 4d and 2d per lb respectively, while 1lb *vermelon* cost 1s 8d. A higher-quality pigment probably accounts for the higher price of vermilion at Westminster than at Exeter, although it should be noted that it is still only 1/20th the cost of the red lake pigment purchased for these royal paintings (Smith 1807: 218; Salzman 1952: 168). It is also possible that a variation in colour or in particle size, and therefore presumably quality, may have accounted for the difference in price of vermilion in this case.

The relative cost of various red pigments, purchased together with oil and varnish, is also outlined in the Sacrist's Rolls of Ely Cathedral for 1341–42:[8]

j qr cinopro [red lake]	4 s
iij lbr vermilion prec. lbr xd	2s 6d
vj lbr rubei plumbi [red lead]	2s

6. So highly prized was this material that it was not refined *in situ*, but up to 2000lb per annum was exported under seal for purification in Rome. Thus vermilion was extremely costly in antiquity; a price of 70 sesterces per lb was set by law; Pliny 1984: 85–9.
7. The mineral, which has been found in the Pennine orefields and the Lake District, is usually poor in quality, exists in small quantities and appears a reddish brown colour. I am grateful to Brian Young of the British Archaeological Survey for discussing sources of the pigment.
8. Unfortunately the type of painting for which these materials were purchased is not specified; Chapman 1907, vol. 2: 121.

Red lead is the cheapest at 4d per lb, while vermilion cost 10d for the same quantity. What is striking is that the red lake pigment cost 16s per lb, making it nineteen times more expensive than vermilion and forty-eight times more expensive than red lead. Thus, it is clear that although vermilion was by no means the cheapest red pigment available in the Gothic period, it was far less expensive than red lake.

Optical, chemical and physical properties

Unless carefully graded, particles of both dry-process vermilion and natural cinnabar range widely in size and have a fractured appearance. In plane polarised transmitted light the particles appear bright cherry red with high relief. Vermilion (α – HgS) has a hexagonal crystal structure and high refractive indices ($\alpha = 2.819$ and $\gamma = 3.146$). In transmitted light with crossed polars, particles of vermilion are highly birefringent and display distinct extinction.

Vermilion is stable in acidic conditions but it will decompose to form mercuric chloride if warmed with aqua regia (a mixture of nitric and hydrochloric acids). It is also stable in alkaline conditions and may therefore be applied directly to fresh plaster over a lime ground or in combination with lime white (Gettens *et al.* 1993c).

Since vermilion is a sulphide of mercury it is potentially reactive when combined in pigment mixtures. In practice, however, it appears to be fairly stable even when combined with lead-based pigments with little evidence of the formation of black lead sulphide (Gettens and Sterner 1941). Indeed, recent research indicates that combining red lead with vermilion may inhibit alteration of the latter to its dark form (Grout 1999: 71).

The dark form of vermilion was thought by Eibner to result from a physical change of the hexagonal crystal structure of cinnabar (red α-HgS) to the black cubic form of mercury sulphide, metacinnabar (black α'-HgS). However, this has not been convincingly confirmed by X-ray diffraction analysis, leading Daniels (1987: 281–2) to conclude that the darkened material may be an amorphous form of black metacinnabar. XRD analysis may also be unsuccessful because the darkened material is often present as an extremely thin film over the red particles (Grout and Burnstock 2000: 15). Recent analysis of the alteration product on panel paintings indicates that the darkened crust on the surface of the red pigment contains both greyish-white and black particles. Analysis by Raman microscopy identified the greyish-white particles as the mercury chloride compound calomel (Hg_2Cl_2) (Spring and Grout 2002: 54).

Where the darkening of vermilion does occur, various factors have been implicated, including: exposure to light wavelengths of 400–570 nm; the presence of impurities in vermilion (principally arsenic for dry-process vermilion, alkali metal halides for wet process vermilion, and copper-based impurities and chlorine for cinnabar); exposure to halides; high RH; and trituration (Spring and Grout 2002; Grout and Burnstock 2000; McCormack 2000; Grout 1999; Daniels 1987; Feller 1967; Eibner 1914, 1917a–f, 1918).

Preparation

The recipe for artificial dry-process vermilion first appears in the West in an 8th-century manuscript, Lucca MS 490, which describes combining two parts of mercury with one of sulphur and heating the black mixture slowly in a flask.[9] A detailed description of the sign

9. The process for the manufacture of vermilion may have originated in China; Gettens *et al.* 1993c: 161.

of adequate roasting – namely the appearance of vaporised mercury sulphide which then recondenses in its red form at the top of the flask – is provided, together with a specification of the type of flask required and a warning against over-firing. There are numerous subsequent texts, both of pure alchemy and providing technical information for painters, which contain recipes following the general pattern of that in the Lucca MS, although some incorporate additional detail or reverse the ratio of the constituent materials.[10] For example, Theophilus (1979: 40) provides a recipe following the Lucca ratio of two parts mercury to one of sulphur, while Peter of St. Omer gives two recipes reversing the balance of the mixture and providing further information concerning vapours emitted during firing.[11] By the 14th century, the manufacture of vermilion was sufficiently well established for details of the process to be considered 'tedious', and Cennini recommends purchase of the pigment from the local apothecary (Cennini 1960: 24).[12]

A number of authors warn against the adulteration of vermilion. Pliny, for example, not only indicates that it was adulterated in various ways – including adding goat's blood, crushed service berries or red lead – but adds that painters may also pilfer vermilion by collecting the dense mineral as it settled out of the water with which they cleaned their brushes (Pliny 1984: 89–91). Cennini reiterates the warning, indicating that vermilion may be fraudulently combined with red lead or brick dust, and for this reason suggests that it is purchased intact rather than ground (Cennini 1960: 24).

Use in English medieval wall painting

The strong blue of ultramarine and the rich copper greens which played such an important role in the Romanesque wall painters' palette were well balanced by the intense red of vermilion. Although red lead became more common in the Gothic period, vermilion continued to be employed extensively.

Vermilion has been identified in a number of English Romanesque wall paintings. In several cathedral schemes it was used extensively, as in the Holy Sepulchre Chapel of Winchester Cathedral, the 'Losinga' scheme in Norwich Cathedral, and in St. Gabriel's Chapel, Canterbury Cathedral. In the last of these the pigment was applied both alone and combined with red iron oxide. Only once has vermilion been identified in a parish church scheme of this period – at Copford in Essex – although there are examples of its use in monastic settings as at Lewes Priory and in the chapter house of Sherborne Abbey, where excavated fragments painted in vermilion and carbon black on a white ground may represent fragments of drapery from large-scale figure subjects.[13] The pigment was also extensively employed in the late Romanesque wall painting fragments from Glasgow Cathedral (Park 1998; Howard 1998a).

---

10. For a discussion of relevant texts, see Thompson 1933.
11. For Peter of St. Omer, see Merrifield 1967: I: 138–40. Thompson (1933: 66–7) indicates that a ratio of two of sulphur to one of mercury is most frequently specified, and that, although in practice an excess of sulphur was incorporated, the theoretical ratio for combining mercury with sulphur is 6¼:1.
12. Wet-process vermilion – introduced in 1687 – was made by heating black amorphous mercuric sulphide in a solution of ammonium or potassium sulphide to precipitate the red form; Gettens *et al.* 1993c: 163.
13. Tentative identification of vermilion at Copford was made by Babington; Organ *et al.* 1991. For the Lewes Priory scheme, see Howard and Park 1997; and for Sherborne Abbey, see Howard and Park forthcoming a.

Scientific examination has also shown that this medium-priced and readily available pigment was widely used in wall painting throughout the Gothic period in parish churches, cathedrals and in domestic schemes. For example, it has been identified in the mid-13th-century *Wheel of Fortune* at Rochester Cathedral (Darrah 1992); in the roughly contemporary first scheme of the Ante-Reliquary Chapel, Norwich Cathedral (Park and Howard 1996; Howard 1995c); the paintings of *c*1302–08 in the feretory of St. Albans Cathedral (Howard 1993a); in the scheme of *c*1310 in the refectory of Bushmead Priory, Bedfordshire (Park 1986: 72); in the domestic wall paintings of *c*1330 at Longthorpe Tower, Cambridgeshire;[14] in the polychromy of *c*1370 on the vault of the Chapel of Our Lady Undercroft, Canterbury Cathedral (Howard 1993b); in the paintings of *c*1395 in the Byward Tower, Tower of London (Hluvko 1991: 40); in the 14th-century decoration at Fiddleford Manor, Dorset (Hluvko 1991: 39);[15] in the late 14th-century wall paintings in the chapter house of Westminster Abbey (Hluvko 1991: 40); the scheme of *c*1400 at Bradfield Combust, Suffolk (Rickerby 1990b); in the *Last Judgement* of *c*1435–60 at Holy Trinity, Coventry (Howard 1996b: 5); and in the paintings of *c*1510 at St. Mark's, Bristol (Gill and Howard 1997).

Application

Vermilion applied over a white ground and mixed with lime

Unlike certain other mineral pigments such as azurite, vermilion has excellent covering power owing to its high refractive index. Consequently, even when finely ground it retains a high chroma when applied over a white ground.[16] Despite Cennini's assertion that vermilion is better applied on panels than on walls (1960: 24), the pigment is stable in alkaline conditions and so may be applied in *buon fresco* painting or over a fresh lime ground.

The application of vermilion over a lime white ground or directly on the plaster substrate has been confirmed in a number of instances, as in the scheme of *c*1130 in the apse of St. Gabriel's Chapel at Canterbury. Here, vermilion over lime white was employed for the background to the seraph on the southern buttress, while it was also employed in a lime white matrix and applied directly over the lime plaster substrate to produced the salmon pink colour of the prophet's drapery on the west wall of the apse (Cather and Howard 1994: 145).

In the scheme of *c*1170 at Cormac's Chapel, Cashel (Ireland), vermilion combined with lime white was applied over lime white to produce the brilliant red background of the west segment of the vault. In the late 12th-century paintings in the Holy Sepulchre Chapel, Winchester Cathedral, the pigment was applied directly to the lime plaster for St. John's drapery in the *Deposition* on the east wall (Hluvko 1991: 41), while in the 'Losinga' cycle of *c*1190 at Norwich, vermilion was combined with lime white and applied both over a lime white ground and the plaster substrate for the panelled background of the vault.[17] Examination of the late Romanesque fragments from Glasgow Cathedral also indicated that vermilion was combined with lime white and applied both over a lime white ground and directly to the plaster substrate (Howard 1998a).

---

14. Unpublished analysis by H. Howard.
15. C. Babington, pers. comm.
16. Cennini (1960: 24) indicates that even if the pigment were ground in clear water daily for twenty years its appearance would be improved.
17. For Cashel, see McGrath 1987 and Howard 1991; and for the 'Losinga' cycle, see Howard and Gasol 1996.

The application of vermilion over the plaster substrate or a white ground is also found throughout the Gothic period; for instance, it was applied directly over the plaster in the mid-13th-century scheme in the Ante-Reliquary Chapel, Norwich Cathedral (Howard 1995c: 4, pl. 6). Similarly, vermilion is applied directly on the plaster skim for the angel in the 14th-century *Annunciation* at Fiddleford Manor, Dorset (Hluvko 1991: 39). For the red background to the left of St. Faith's face in the mid-13th-century *Wheel of Fortune* in the choir of Rochester Cathedral, vermilion was applied over the white plaster support (Darrah 1992). The pigment was applied over a lead white ground to produce the brilliant red of the angel's wing in the *Nativity* of *c*1510 in the Lord Mayor's Chapel, Bristol (Plate 134) (Gill and Howard 1997: 100–101).

Over coloured underpaint

Despite its high covering power, vermilion was also applied over coloured underpaint, presumably to produce a slight variation in its hue. In Cormac's Chapel, Cashel, both grey and yellow underpaints were employed which may have produced slightly cooler or more orange effects respectively (Howard 1991), while grey underpaint was also detected in one of the Romanesque fragments from Glasgow. Vermilion was also applied over a yellow earth pigment to produce the brilliant red colour of Joseph of Arimathea's robe in the late 12th-century *Deposition* in the Holy Sepulchre Chapel at Winchester Cathedral (Howard 1995a: 101).

In some instances the presence of local coloured grounds beneath vermilion may be incidental – that is, they were colours employed in surrounding areas over which vermilion was applied for details. This is likely for the drapery of the mid-13th-century angel on the south side of the eastern arch of the Ante-Reliquary Chapel in Norwich Cathedral, where linear details in vermilion were applied over yellow earth combined with lead white (Howard 1995c: App. 1 and pl. 10). Similarly, in the St. Christopher of *c*1400 at Bradfield Combust, Suffolk, vermilion was applied over yellow ochre for a detail in the water at the saint's feet (Rickerby 1990b: 4–5). Unusually, details in vermilion appear to have been applied over a layer of carbon black on tin leaf for the nimbus of a *Musician Angel* in the north-west bay of the chapter house of Westminster Abbey. Since the tin leaf has now oxidised and there is no evidence of the glaze or varnish layer which must once have covered its surface, it is difficult to determine the original stratigraphy of this area, and what the function might have been of the dark layer beneath the vermilion.

Applied over red underpaint

A complex tonal underpainting in red earth was employed for St. William's chasuble in the paintings of *c*1302–08 in the feretory of St. Alban's Cathedral. Here, only the very darkest shades of the drapery were added in vermilion over this red layer (Howard 1993a: 35–7). In other instances where red underpaint was employed, it may have extended the more expensive red pigment. This intention is particularly clear where vermilion was both combined with other red pigments and applied over red underpaint, as in the early 15th-century tomb recess in the south choir aisle in Lichfield Cathedral. Here, red lead combined with red ochre and a little vermilion was applied over red ochre, while the thinnest trace of pure vermilion was scumbled across the surface.[18]

18. S. Cather, 'Throne of mercy: an early 15th-century wall painting in Lichfield Cathedral', unpublished lecture.

Examination of early 13th-century wall painting fragments excavated from the crypt of Worcester Cathedral indicated that vermilion combined with red lead was applied in a layer 20μm thick over red lead.[19] Despite the high covering power of vermilion, this use of a red lead preparation for vermilion may have provided a slightly orange cast to the latter, in addition to extending the more expensive red pigment. At Rochester Cathedral, a thin layer of vermilion over red lead was detected on the hem of the king's mantle below the wheel in the mid-13th-century *Wheel of Fortune* (Darrah 1992). Similarly, a rather thin layer of vermilion was applied over a substantial layer of red lead for the late 14th-century decoration of the vault of the Chapel of Our Lady Undercroft in Canterbury Cathedral (Plates 124–127).[20] Vermilion of varying particle size – some extremely large – was applied over a preparation of red lead combined with lead white in the St. Eustace wall painting of *c*1480, also in Canterbury Cathedral (Hluvko 1991: 41).

Vermilion combined in other pigment mixtures

Despite its potential reactivity when combined with pigments such as lead white and red lead, vermilion appears to be fairly stable, particularly when an oil-based medium is employed (Gettens and Sterner 1941). In a number of instances vermilion was combined with red lead, as in the St. Gabriel's Chapel scheme at Canterbury Cathedral, where this mixture was applied over a secondary ground of lead white for areas of red drapery in the *Birth of the Baptist* (Cather and Howard 1994: 144). For the intense orange-red highlight on Nicodemus' stocking in the late 12th-century Holy Sepulchre Chapel paintings at Winchester, vermilion was combined with red lead and applied over red lead underpaint (Howard 1995a: 101).

Where vermilion was combined with red lead, the overall effect would certainly have been to increase the orange cast of the vermilion, while also extending the more expensive red pigment. This mixture was employed to good effect for the inner robe and wings of St. Michael, and for highlights of the fleurs-de-lys in the paintings of *c*1395 in the Byward Tower, Tower of London (Hluvko 1991: 42–3, pls. 57–58). Here, however, the overall opulence of the scheme makes it likely that these pigments were selected on the basis of colour rather than cost.

Other examples of vermilion combined with red lead include the mid-13th-century scheme at the Hospital of St. Giles, Brompton Bridge (North Yorks), where lead white was also combined with the red pigments (Stewart 1995: 214); the cross-staff of the early 14th-century archbishop on the south side of the feretory of St. Albans Cathedral, where the mixture was applied over a lead white ground (Howard 1993a: 38); and the 14th-century paintings at No. 9 Silver Street, Ely, where the red pigments were combined for the stems of the vinescroll ornament (Courtauld Institute of Art 1990: 6–15). Analysis of samples from the late 14th-century *Majesty* and *Apocalypse* paintings in the chapter house of Westminster Abbey indicated that vermilion combined with red lead was glazed with red lake. In this instance the red lake layer would have served both to saturate and heighten the rich red colour (Hluvko 1991: 43).

Other pigment mixtures, where colour variation was clearly sought, have also been identified. For example, vermilion combined with a small quantity of yellow ochre, presumably to produce a warm orange-red colour, was tentatively identified in a sample taken from

19. Analysis by Howard and Kakoulli in Park 1997: no. 4, fig. 4.24.
20. Although red seems an unusual colour with which to represent the sky, it was also employed in medieval woodwork set with gilt stars at Durham Cathedral; Howard 1993b: 32–3.

draperies in the scheme of *c*1125–35 in the apse of Copford, Essex (Organ *et al.* 1991). In the 12th-century paintings in Cormac's Chapel, Cashel, vermilion was combined with charcoal black and red iron oxide to produce a dark purple colour above the blind arcading on the south wall of the chancel (Howard 1991). A similar pigment mixture was employed for the dark purple-black background to a wing on one of the late 12th-century fragments from Glasgow Cathedral, while charcoal black and vermilion with the addition of a considerable quantity of lime white was used to achieve a pale purple colour on another of these fragments (Howard 1998a). In 13th-century fragments excavated from Worcester Cathedral, and in the St. George of *c*1400 at Bradfield Combust, Suffolk, vermilion was mixed with a carbon black pigment to produce a purple colour (Park 1997; Rickerby 1990b: 4–5). In the paintings of *c*1510 at St. Mark's, Bristol, a combination of vermilion, indigo and lead white was applied over grey underpaint of carbon black and lead white for the pale purple veil of a female figure, possibly the Virgin, behind Christ in the *Noli me Tangere* (Gill and Howard 1997: 100–101). Preliminary examination of samples from the paintings of *c*1330 in Prior Crauden's Chapel, Ely Cathedral, suggested that azurite combined with lead white applied over a layer of vermilion may have been intended to produce a subtle purple effect (Babington 1987; Binski and Park 1986).

Most unusually, vermilion combined with red iron oxide has been identified in the *sinopia* of the Holy Sepulchre Chapel paintings of *c*1175 (Howard 1995a: 101). However, in the late 12th-century paintings from Glasgow Cathedral, traces of a red line are evident at the interface of two plaster layers on one fragment, while extensive areas of red pigment are apparent on the lower plaster layer of another. Samples taken from the red paint on the latter confirmed the presence of both vermilion and red earth. Although the traces of red paint on the penultimate plaster layer at Glasgow may represent a *sinopia*, it is also possible that a very simple scheme in red was applied shortly before the more complex overall decoration – including figurative elements – was executed. However, since very few painted fragments have survived at Glasgow, it is difficult to draw firm conclusions regarding the possible presence of more than one decorative scheme (Howard 1998a; Park 1998). The high cost of vermilion by comparison with red iron oxides would be sufficient reason for the rarity of its use in zones within the stratigraphy – such as *sinopie* – which had no impact on the final appearance.

Vermilion was combined with varying quantities of lead white in a number of layers to produce the modelling of Christ's red drapery in the late 14th-century *Majesty* in the chapter house of Westminster Abbey (Plates 128 and 129).

Vermilion for flesh tones

In the Romanesque period, vermilion was employed in a variety of ways to produce different flesh tones. For example, in the scheme of *c*1090 at Lewes Priory, subtle pinks were achieved by combining vermilion with lime white, and for some details pure vermilion or ultramarine was applied over this pink colour. A similar mixture, sometimes with the addition of a yellow earth pigment, was employed for the flesh tones in the Romanesque paintings of Cormac's Chapel, Cashel.[21]

21. This mixture accords well with Theophilus' recommendation (1979: 23) that a combination of yellow ochre, cinnabar and lime white should be employed for flesh tones in wall painting. Particles of ultramarine were also found within a paint sample from Cormac's Chapel, which may have come from the adjacent blue drapery, or may have been an intentional addition to the mixture; McGrath 1987: 101.

A complex mixture of vermilion combined with red and yellow earth pigments and carbon black was also employed for Nicodemus' flesh tone in the 12th-century *Deposition* in the Holy Sepulchre Chapel, Winchester Cathedral (Hluvko 1991: 40–41). While Cennini advised against the incorporation of vermilion in flesh tones on the wall, he suggests that it should be used for blood and wounds.[22] Vermilion used for precisely this purpose occurs in this Winchester scheme, applied in a layer 20µm thick over the flesh tone for the blood on Christ's hand (Howard 1995a: 101). By contrast, in the earlier 12th-century paintings in the apse of St. Gabriel's Chapel, Canterbury Cathedral, the shadow on an angel's foot was achieved by applying vermilion and a little charcoal black *over* the flesh tone which consists of two layers of yellow earth combined with lime white, the upper containing a greater proportion of yellow earth (Cather and Howard 1994: 144).

Vermilion was combined with lead white in various quantities to produce a range of flesh tones throughout the Gothic period. For example, in the scheme of *c*1300 in the Ante-Reliquary Chapel, Norwich Cathedral, this pigment mixture was applied over a ground of lead white combined with calcium carbonate for St. Paul's flesh colour (Howard 1995c: App. III and pl. 22). Vermilion combined with lead white was also employed for Christ's hand in the *Majesty* of *c*1372–1404 in the chapter house of Westminster Abbey, while a more complex stratigraphy was used for one of the cherubim. Here, vermilion in lead white was applied over a thick bluish layer of charcoal black combined with chalk over a local ground of lead white with a little vermilion (Hluvko 1991: 41–2).[23] The same pigment mixture continued to be employed later in the Gothic period; for example, in the early 16th-century painted niches in the Lady Chapel of Winchester Cathedral, where vermilion was combined with lead white for the flesh tints (Paine 1996: 4). For the rendering of fresh blood in the *Last Judgement* of *c*1435–60 at Holy Trinity, Coventry, a combination of vermilion with red lead and red lake was applied over a thin layer of carbon black on the flesh tone (Plates 132 and 133).

Binding media

The stability of vermilion in alkaline conditions makes this intense red pigment entirely suitable for use in *buon fresco* painting and in combinations with lime white. Vermilion in a lime white matrix has frequently been detected in wall painting in the British Isles, particularly in the Romanesque period, as in the fragments of *c*1190 from Glasgow Cathedral where this combination was applied directly over the lime plaster substrate (Howard 1998a: table 1). Instrumental analysis of a sample of vermilion combined with lime white from the approximately contemporary 'Losinga' cycle in Norwich Cathedral provided no evidence of any original organic binding medium.[24]

However, analysis has indicated that in other cases additional organic binding media were employed for the pigment in 12th-century wall painting. A ninhydrin test suggested the presence of glue distemper or casein in a sample of vermilion combined with red lead

22. Cennini (1960: 23–4,45–6, 95) suggests that the fine red earth pigment *cinabrese* be used in its place.
23. A similar stratigraphy has been identified in the pale flesh tones of the angel in the first arch of the east bay, except that here a grey-green preparation of charcoal black combined with yellow in a white matrix was applied beneath vermilion combined with lead white; Howard 2000 and forthcoming.
24. The sample was subjected to GC-MS analysis and a ninhydrin test. Traces of wax from a previous conservation intervention were detected; White 1998: 1.

taken from a highlight on Nicodemus' right leg in the late 12th-century scheme in the Holy Sepulchre Chapel, Winchester Cathedral (White 1998: 1). Contemporary treatises also recommend an additional binding medium for vermilion; for example, Theophilus (1979: 32–3) recommends that red lead, lead white and vermilion should be ground with egg white.[25]

It is clear that, as with other pigments, the quantities of organic binding material used with vermilion increased steadily from the beginning of the 13th century. GC-MS analysis of a sample from the scheme of *c*1230 on the vault of the Guardian Angels' Chapel at Winchester suggested the presence of a little drying oil combined with vermilion which has partially altered to form metacinnabar (Plates 120 and 121).[26] By the 14th century complex layering of pigments and combinations of media became widely used to produce rich effects. For example, a proteinaceous material appears to have been used to bind specific pigments or for particular layers such as the vermilion layer in a sample from the de Ingham tomb (Plates 172 and 173). In this photomicrograph, the proteinaceous component in the vermilion layer has been stained to appear yellow in UV light, while the oil-bound red lake glaze over the vermilion appears dark.[27]

Comparison with its use in other media and in continental wall painting

The frequent use of vermilion in English wall painting throughout the medieval period is mirrored elsewhere in northern Europe. For example, the pigment has been identified in the late 12th-century scheme at Müstair (Switzerland), where it was applied over red iron oxide (Plate 118) (Mairinger and Schreiner 1986: 198). The same stratigraphy was employed for the brilliant red of Christ's drapery in the *Majestas Domini* of *c*1130 at Idensen (Germany), but here the red iron oxide-based underpaint was combined with lime white (Matteini and Moles 1994: 83). This application closely follows the recommendation in Theophilus' contemporary treatise for red mixed with lime as an underpaint beneath vermilion on walls (Theophilus 1979: 25).

The pigment has also been identified in the Romanesque scheme at Berzé-la-Ville (France) (Rollier-Hanselmann 1997), while from the Gothic period it is used in the important mid-13th-century paintings in the choir of Angers Cathedral – both alone and combined with red lead (Demailly *et al.* 1998: 11) – and in the choirstall scheme of *c*1332–49 in Cologne Cathedral (Jägers and Schulze-Senger 1989: 187–8; Maul 1992).

The presence of vermilion has also been confirmed in polychromed sculpture throughout the medieval period and in panel painting from the 13th century. A thick lead white underpaint was used for a layer of vermilion applied in an egg tempera medium for the polychromy of *c*1225–40 of figure 193 on the west front of Wells Cathedral (Plate 135) (Sampson 1998: 118–19; Roy and White 1982), while examination of a late 13th-century pendant from a canopy in the chapter house of York Minster indicated that

25. Although referring to manuscript illumination, MS Cotton Titus D.XXIV – a compilation made at the Cistercian abbey of Rufford (Nottinghamshire) at the end of the 12th century – suggests that vermilion should be combined with egg; Petzold 1995: 52.
26. This scheme was partially repainted in *c*1280, but the sample (from the lowest rosette on the south side of the western segment of the vault) appeared to be from the original scheme. For results of the GC-MS analysis, see White 1998: 1.
27. For the analysis of this tomb, see Howard 1993d. Instrumental analysis of the red lead layer below the vermilion confirmed the presence of glue as the binding medium.

vermilion was applied over a white ground, the lower layer consisting of calcium sulphate and the upper of lead white (Heywood and Oosterwijk 1993: 54–5). Vermilion applied over a white ground was identified on the crucifix at Hemse (Sweden) of *c*1170–90 (Tängeberg 1984: 24–44).

Vermilion combined with red lead has been identified on the polychromed sculpture of *c*1180–85 excavated from St. Mary's Abbey, York. For example, vermilion combined with red lead and a little lead white was employed for the pallium of the figure of St. John. Details of the cuff of this figure were applied in gold leaf on a pale pink layer of vermilion combined with red earth over a lead white ground (Brodrick 1993: 23). Both vermilion and red lead have also been identified on the 12th-century frieze on the west front of Lincoln Cathedral (Larson 1992: 188, 1993: 34), while vermilion combined with a little red lead over a lead white ground was employed for the Martyrdom Doorway of *c*1236 in the Great Cloister of Canterbury Cathedral (Plate 136) (Hluvko 1991: 41–2). Vermilion combined with red lead was also used for the polychromy of *c*1321–49 in the Lady Chapel of Ely Cathedral (Hluvko 1991: 42), for a 14th-century statue from the Edgar Tower at Worcester,[28] and for the late 15th-century bosses in the vault of the Lady Chapel at Winchester (Curteis and Paine 1995: 10). Vermilion has also been detected in the polychromy of *c*1220–60 on the '*Portail de la Mère Dieu*' of Amiens Cathedral, where it was employed for the cheek patches and lips, particularly of the jamb figures (Weeks 1998: 102). A mixture of vermilion and red lead was used on the facade portals of Ferrara Cathedral (Rossi Manaresi 1987: 174–84). Where extremely large surface areas required painting, as on external facades, red lead would probably have been added to extend the more expensive pigment.

Vermilion and lead white were combined to produce flesh tones on polychromed sculpture as on the Great Screen of *c*1475–90 at Winchester Cathedral. Here, for areas such as the lips of the Christ Child, a little red lead was incorporated into the mixture which was then applied over an orange underpaint of red lead (Brodrick and Darrah 1989: 615–17).

In the 14th-century polychromy of the de Ingham tomb at Ingham, Norfolk, vermilion was combined with charcoal black, red lake and lead white and applied over lead white, red lake and gold leaf on raised decorative elements (Howard 1993d: App. 6). With such complex pigment mixtures and stratigraphies it is hard to determine the overall intended effect.

Examination of the 14th- and 15th-century polychromy on the west front of Exeter Cathedral indicated that while rather finely ground vermilion was employed for the flesh tints, the pigment was coarsely ground for the drapery of the central figure in the uppermost tier (C17) (Sinclair 1991: 121–2). By contrast, on a 14th-century effigy at Erwarton (Suffolk), both coarse and fine vermilion particles are present within the same paint layer.[29] These variations in particle size – particularly the clear distinction between the use of the coarsely ground pigment for draperies at Exeter, while only fine particles were introduced into flesh tints – are likely to be representative of the varying grades of vermilion.

28. Analysis by Howard and Kakoulli in Park 1997: cat. 30, fig. 30.2.
29. A few scattered particles of vermilion were also detected in the ground layers; Roy 1984.

Panels

Vermilion was applied over a thick lead white ground on a capital of the Adisham reredos, recently redated to *c*1180.[30] In addition, examination of the small number of surviving early medieval English panel paintings suggests that some aspects of the application methodology for vermilion are comparable with those of contemporary wall painting. For example vermilion was combined with lead white and a little red iron oxide to produce flesh tones on the Westminster Retable of *c*1270.[31] The use of a bluish-grey underpainting for flesh tones in vermilion – as detected in the late 14th-century chapter house wall paintings at Westminster – is paralleled in panel painting, for example on the nave ceiling of *c*1220 at Peterborough Cathedral and on the 13th-century ceiling paterae from the Painted Chamber at Westminster.[32] Unusually vermilion was employed for underdrawing in the Thornham Parva Retable (*c*1335). Evidence suggests that an initial preparatory drawing in vermilion was executed on the chalk ground below a lead white imprimatura.[33]

Discussion

Vermilion was extensively employed for wall painting and polychrome sculpture in England throughout the medieval period, although there is little evidence for its use in parish church schemes in the Romanesque period. In the Gothic period the pigment was employed in cathedrals, parish churches and domestic settings, and documentary evidence suggests that at this time the pigment was by no means the most expensive red available. It is tempting to suggest that the paucity of the use of vermilion in parish church schemes in the Romanesque period was due to a much inflated price, perhaps owing to the cost of importing mineral cinnabar rather than using the synthetic equivalent, dry-process vermilion. Approximate equivalence with the cost of other synthetic pigments in the 14th-century accounts clearly indicates that dry-process vermilion was generally employed at this time, although a 14th-century treatise does mention good quality vermilion from Spain (Thompson 1926: 281, 297). Unfortunately no documentary evidence survives from the Romanesque period to confirm whether the mineral species was imported but it is clear that recipes for the manufacture of vermilion were available from the 8th century.[34]

Although vermilion is stable in alkaline conditions and has been identified combined with lime white and applied over a lime plaster, additional organic binding media for vermilion have also been identified as early as the 12th century. These may have been incorporated to assist in the wetting-out of the pigment, or simply because it was applied *a secco*. Throughout the Gothic period there is a steady increase in the quantity of organic binding media employed with the pigment.

30. The red pigment was also combined with lead white to form a paler red with details in charcoal black on the surface; see Howard 1993c: App. II. The reredos has been redated by C. Tracy on the basis of dendrochronological evidence.
31. Dr A. Roy, pers. comm.
32. For the Peterborough ceiling, see Howard 1997b; and for the Westminster panels, see Lynn 1997.
33. Examination by infrared reflectography also indicated the presence of a carbon black underdrawing in the figures to the left of Christ and partially for the figure of Christ himself; Goldsmith and Bucklow 1998: 40–41; and Bucklow forthcoming.
34. As in the Lucca MS 490; see Gettens *et al.* 1993c: 160–61.

Pigment alterations

An investigation into the causes of the darkening of vermilion by Eibner early last century concluded that the pigment produced by the wet process is more prone to darkening than its dry-process equivalent, and that darkening is probably light-induced. In addition, he detected greater darkening with an oil medium than with watercolour (gum) due to increased light penetration. Significantly, Eibner also found that wet-process vermilion prepared in the presence of thiosulphate is considerably more prone to darkening (Eibner 1914, 1917 a–f, 1918). This early work was reviewed in the 1960s by Feller who confirmed the direct relationship between light and alteration for wet-process vermilion, and that the intensity of light of wavelengths 400–570 nm was the determining factor in darkening. Feller (1967) detected greater darkening in linseed oil than in acrylic emulsion and concluded that with a low pigment volume concentration, darkening was greater since more light penetrates the paint film. He also confirmed that the traditional practice of glazing vermilion with lake pigments may protect the underlying layer from darkening since the intensity of transmitted light is reduced. Daniels (1987) found that alkali metal halide impurities increase the rate of darkening for wet-process vermilion and also indicated that the rate of darkening increases with high RH. He suggests that salts from ground water and high RH may accelerate the alteration of vermilion in wall paintings (Daniels 1987: 282).

McCormack (2000) found that some mineral specimens of cinnabar contain 0.05–1.0% chlorine and are photosensitive, darkening in the presence of light. Significantly, he also confirmed that cinnabar from the historical sources in Italy and Spain which contain no chlorine become photosensitive and darkened following exposure to halides.

Research by Grout and Burnstock (2000) has shown that although wet process vermilion is more prone to darkening than mineral cinnabar or the pigment prepared by sublimation, the latter also shows considerable alteration following a relatively short ageing period.[35] Their study shows that although light is a contributory factor in the process, darkening is largely dependent on an interaction between vermilion and impurities present (Grout and Burnstock 2000: 20; Grout 1999: 67–70).[36] The work confirms a number of contributory factors indicated by earlier researchers but,[37] since their tests included all three types of vermilion bound in both oil and egg tempera, the results are particularly significant in the context of medieval polychromy.

Further research by Spring and Grout (2002) on samples of darkened vermilion from panel paintings dating from the 14th to the 17th centuries indicates that the darkened crust on the surface of the red pigment contains both greyish-white and black particles. Analysis by Raman microscopy identified the greyish-white particles as the mercury chloride compound calomel (Hg_2Cl_2). Since no chlorine was found in the unchanged red pigment below the surface crust, Spring and Grout suggest that alteration is due to a reaction with chloride, most likely from sodium chloride originating from human skin debris, which is present in atmospheric dirt. Furthermore they propose a mechanism for alteration by which deposition of hygroscopic dirt particles containing an acidic solution

35. Samples examined by Grout and Burnstock were artificially aged for the equivalent of 31–45 years under museum conditions.
36. This evidence for a multifactorial cause of darkening is supported by the results of recent research by Saunders and Kirby (1996: 89) who detected colour change in vermilion not exposed to light.
37. Namely that more darkening occurs in an oil medium than in tempera, that there is greater darkening in medium rich samples, and accelerated alteration with increased RH.

of chloride ions on the surface of a painting promotes a reaction with vermilion and forms the photosensitive mercury sulphochloride compound corderoite ($Hg_3S_2Cl_2$). This in turn alters by a process of photodegradation to form products including calomel (Hg_2Cl_2) and a black material, perhaps mercury sulphide.[38]

The irregular darkening of vermilion often produces a rather patchy effect in paintings. Grout and Burnstock's study provides important information that may help to explain this phenomenon. They found preferential darkening for large particles of dry-process vermilion and cinnabar and, significantly, that a dark line was produced at the interface of arsenic-rich seams and mercury sulphide in the dry-process pigment. Darkening focused around such seams of impurity. Thus in dry-process vermilion and mineral cinnabar, where copper-containing impurities were found, the patchy presence of such impurities would lead to uneven darkening, whereas with the pigment prepared by the wet process alteration would be more uniform since the impurities are present throughout (Grout 1999: 35–7; Grout and Burnstock 2000: 18–19).[39] Variable binding between pigment and medium also appeared to cause significant patchy alteration in oil-rich films (Grout 1999: 33; Grout and Burnstock 2000: 17–18).

The patchy alteration of vermilion is apparent in various English medieval wall paintings. In some instances the presence of a patchy red-brown material, in areas which otherwise appear brilliant red, may signal the pigment alteration. This is evident in the scheme of *c*1300 on the vault of the Ante-Reliquary Chapel of Norwich Cathedral, where dark zones are apparent in the drapery of St. Paul (Plates 122 and 123), and in St. Peter's staff where examination of paint cross-sections indicated that darkening had occurred in the upper portion of the layer (Howard 1995c: 6, pls. 25 and 26).

Many of the five-petalled rosettes decorating the masonry pattern of *c*1310 in the refectory of Bushmead Priory now appear a grey-black colour, although originally painted in vermilion (Park 1986: 72). This phenomenon is also evident in the splendid cycle of domestic wall paintings of *c*1330 at Longthorpe Tower. For example, vermilion has darkened to a depth of 80µm in a sample taken from the shield with lions rampant towards the east end of the south wall.[40]

Until the conservation programme of 1986–91, the background of the attached suns and stars dated to the 1370s on the vault of the Chapel of Our Lady Undercroft was blue-black in appearance. Although prior to this intervention it was generally considered that the background must have originally been painted blue or black, Winfield recorded that the colour was, 'a brilliant red, which is probably vermilion, now badly blackened.[41] Recent analysis confirmed that vermilion was indeed employed, and that it was applied in a thin layer over a red lead ground to produce a rich orange-red effect. However, the dark appearance of the surface was not due to the darkening of vermilion, but resulted from a layer of carbon black – probably lamp black from candle burning – combined with wax in a sulphated crust on the surface (Plates 124–127) (Howard 1993b: 32). It is particularly significant that the presence of metacinnabar was not found to be the cause

38. Spring and Grout (2002) were not able to firmly identify the black product of alteration by Raman microscopy.
39. The dry-process vermilion employed for the study was obtained from China. Microscopic and SEM/EDX examination of the particles revealed the presence of yellow, arsenic-rich seams – perhaps consisting of orpiment – which appeared to contribute to the alteration process.
40. Unpublished analysis by H. Howard.
41. Unpublished report on the paintings at Canterbury Cathedral by D. Winfield, 1977: 6.

of the dark appearance of the vault of this chapel since it has been suggested that red lead may inhibit the darkening of vermilion (Grout 1999: 71). Certainly vermilion in this particular pigment combination appears to be unaltered in a number of instances including: the scheme of *c*1130 in the apse of St. Gabriel's Chapel, Canterbury Cathedral; the Romanesque paintings in the Holy Sepulchre Chapel, Winchester Cathedral; the paintings of *c*1395 in the Byward Tower, Tower of London; and in the *Last Judgement* of *c*1435–60 at Holy Trinity, Coventry (Plates 132 and 133). In addition, the presence of a red lake glaze over vermilion and red lead in the late 14th-century paintings in the chapter house of Westminster Abbey would have provided additional protection against the darkening of the vermilion.

Red lake pigments

Lake pigments are prepared from red organic dyestuffs – extracted from both plant and insect sources – adsorbed onto, or co-precipitated with, a colourless inorganic substrate. The relative translucency or opacity of a lake pigment is partly dependent on the substrate selected: hydrated alumina, calcium carbonate, calcium sulphate, lead white or kaolin.[42] Among the most important red lakes in the medieval period were kermes – derived from a scale insect – and madder and brazil, both extracted from plants.

Madder lake was prepared from the roots of the plant *Rubia tinctorum* L., native to Europe. The dyestuff is found principally under the bark of its root and of related species such as wild madder, *Rubia peregrina* L.[43] The organic colorant madder contains at least nineteen anthraquinones including alizarin and purpurin. Historical terminology for the dye includes: erythrodanon (Greek); erythrodanum (Latin); warentia in one of Charlemagne's decrees, the *Capitulare de vilis*; warencia; garancia and rubea radix (Eraclius); rubia tinctura in the medieval period; and rubia in the *Mappae Clavicula*.[44]

Documentary evidence indicates that madder was produced in France in the medieval period, particularly in the areas of Marseilles and Avignon, and it is mentioned in the statutes of the city of Marseilles as early as 1287 (Merrifield 1967: 1: clxxx). The Netherlands, especially around Zeeland, were celebrated for the quality of the madder produced, and from the 14th century the industry there was both substantial and valued.[45] That madder was cultivated in England in the early medieval period is confirmed by remains of the plant excavated from waterlogged deposits dated to the 9th–10th century at York (Tomlinson 1985). It is also clear that Norwich Cathedral Priory was involved in the production of madder throughout the 14th century, since the Communarius Rolls for 1339 record profits received from the growth and sale of madder, and in 1387 the gardener in the monastic garden sold 8s 2d of madder (McLean 1981: 38). Contemporary Court Rolls for Suffolk also provide details of a debt of 27s for 'mader', and in 1386 John Webbe is named as a 'maderer' (Thornton 1928: 174). In

42. Peter of Omer also suggests wheat flour as a support for red lake; Merrifield 1967: 1: 144.
43. J. Kirby, pers. comm.
44. For the nomenclature and pigment characteristics, see Schweppe and Winter 1997.
45. The Reymerswaal regulations, for example, indicate that the grades of madder and the quantity of dirt that the dyestuff was permitted to contain had been determined by 1480; Kirby and White 1996: 65–66.

addition, a madder market – situated near the church of St. John Maddermarket – existed in Norwich from at least the 14th century (Spence and Kent 1970).

Brazilwood (*Caesalpinia sappan* L.), from which the red colorant brasilein was extracted, was mainly exported from Ceylon via Alexandria into Europe in the medieval period. The name brazil may be derived from the Arabic *braza* for fire or glow, from the Italian *brage* for red colour, or from the Spanish *brasas* also meaning red colour. Brazil appears in a tax list for Genoa for the first time in 1140, and in medieval documents it was known variously as *brexilium*, *bresilium*, *berxinum* and *versinum*.[46]

Kermes was dyestuff extracted from the insect *Kermes vermilio* Planchon which is parasitic on the kermes oak (*Quercus coccifera* L.), native to the Mediterranean and Middle East. The principal constituent of this dyestuff is kermesic acid. The name kermes is thought to have originated from the Sanskrit *kirmidja* (meaning engendered by a worm). Kermes was known in the medieval period as vermilion and vermeil (in England) or vermeil (in France), synonyms which can be traced to the ancient *vermiculum* (little worm). The term *grain* was also used to denote kermes in the medieval period, and Wallert (1991b: 74, 81) carefully distinguishes *kermes grain* from an organic lake pigment '*lacca di cimatura di grana*' prepared from *cimatura de grana*.[47] *Grana* (grain) is described in Italian illuminators' treatises as being extremely costly and employed for dyeing the most luxurious textiles such as silk.[48]

Other types of red organic colorant included the resin-like material secreted by the scale insect *Kerria lacca* Kerr which inhabits host trees in India and Southeast Asia, and Polish cochineal from the insect *Porphyrophora polonica* L. which lives on the roots of the perennial knawel (*Scleranthus perennis* L.), common in eastern Europe.[49] The resin-like lac lake from India – better known today by the refined product shellac – was obtained from the scale insect *Kerria lacca* Kerr and a number of other related insects. It was imported into Spain and Provence for dyeing as early as 1220 (Merrifield 1967: 1: clxxviii).[50] The red dyestuff extracted from this resin-like substance contains laccaic acid A., which has a chemical structure related to kermesic and carminic acids.[51] Polish cochineal was also known as Polish Kermes, German Kermes, root Kermes, cremesino, and St. John's blood. The organic colorant extracted from the insect contains both kermesic and carminic acid (Kirby 1977: 36). While Polish cochineal was known in the Middle Ages, cochineal from the insect *Dactylopius coccus* Costa was only introduced into Europe from the New World in about 1523, becoming widely used in the 17th century.[52]

---

46. For brazil lake, see Wallert 1986.
47. *Cimatura di grana* – the shearings of silk dyed with *grana* – were employed for preparing lake pigment. Although *grana* was probably principally kermes, some Polish cochineal may also have been present; J. Kirby, pers. comm.
48. For the nomenclature and characteristics of kermes, see Schweppe and Roosen-Runge 1986, and Wallert 1991b.
49. Another Old World cochineal, known as Armenian cochineal (*Porphyrophora hamelii* Brandt), may also have been employed for dyeing, and silks dyed with it may have been imported into England. I am grateful to J. Kirby of the National Gallery for this information.
50. J. Kirby suggests that it may have been known much earlier than this.
51. The most obvious and important components of the red resin-like material are laccaic acids A and B, though B, C, D and E are also present (Mills and White 1994: 144). See also Schweppe and Roosen-Runge 1986: 258.
52. The dyestuff was probably only known in Spain at this date; J. Kirby, pers. comm. For cochineal lakes, see Schweppe and Roosen-Runge 1986.

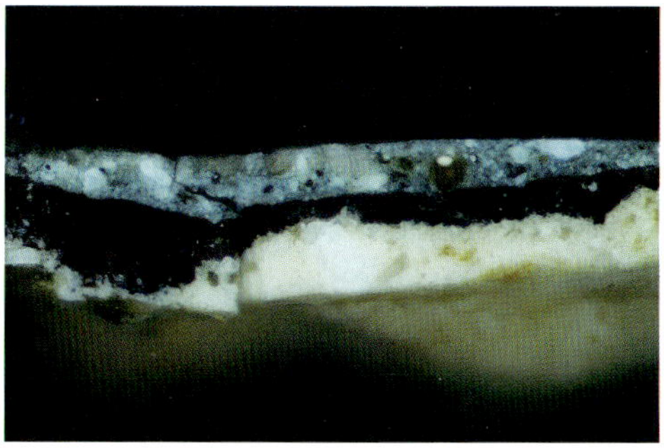

Plate 87 The Lord Mayor's Chapel (St. Mark's), Bristol. Cross-section of a sample from the Virgin's drapery in the Nativity, where thick black underpaint was applied beneath a layer of indigo combined with lead white. This increased the intensity of the blue colour by reducing the light scatter from the lead white ground below, and also made the dye less susceptible to fading. Similarly, grey underpaint was employed beneath the pale purple (indigo combined with vermilion and lead white) of the veil of the female figure behind Christ in the Noli me Tangere. (Photomicrograph: H. Howard; © Courtauld Institute, Conservation of Wall Painting Department)

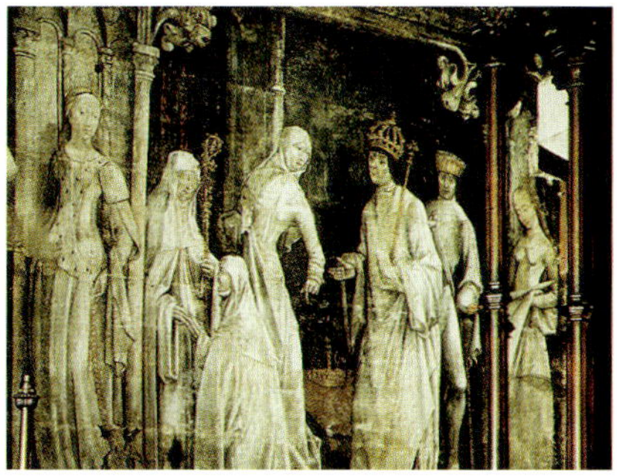

Plate 88 Eton College Chapel: detail of the scheme of c. 1479–87. Although the presence of indigo has yet to be confirmed in these paintings, it is interesting to speculate whether the trace of blue – apparently much faded – that has been observed in the scheme may indeed be indigo. The blue colour was only detected on the sword of the figure farthest west on the south wall, an area largely protected from direct light. (© Crown copyright. NMR)

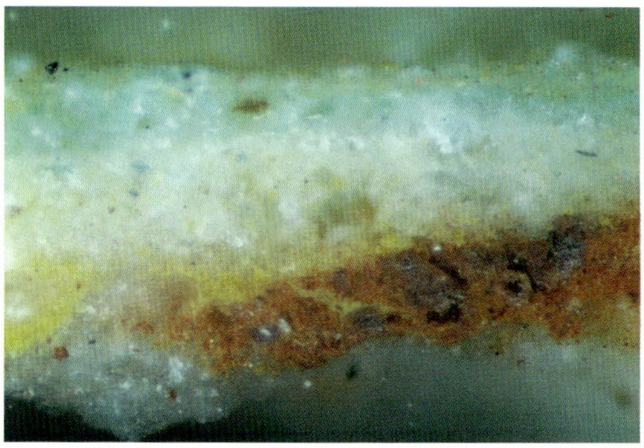

Plate 89 Confirmed examples of the use of green earth in English medieval wall painting fall principally in the Romanesque period. The pigment was employed, for example, in the scheme of c1175 in the Holy Sepulchre Chapel of Winchester Cathedral, for flesh tones (Plate 28) and other areas such as drapery. This cross-section of a sample taken from a shadow in Christ's flesh in the Entombment shows a layer of green earth applied over lime white and red and yellow earth pigments, original magnification 410×. (Photomicrograph: H. Howard; © Courtauld Institute, Conservation of Wall Painting Department)

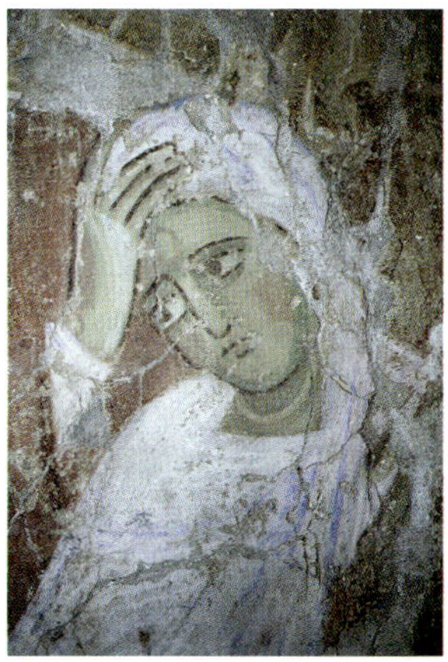

Plate 90 Lambach (Austria): scheme of c1089 in the west choir, where a mixture of green earth and malachite has been identified. The malachite may have been added to intensify the colour of the green earth pigment. (Photograph: A. Heritage; © Courtauld Institute, Conservation of Wall Painting Department)

Plate 91 Hardham (West Sussex): detail of the Annunciation on the east wall of the nave (c1100), showing the use of a copper chloride green in Gabriel's halo and for details on his drapery. (Photograph: Sawdy/Heritage; © Courtauld Institute, Conservation of Wall Painting Department)

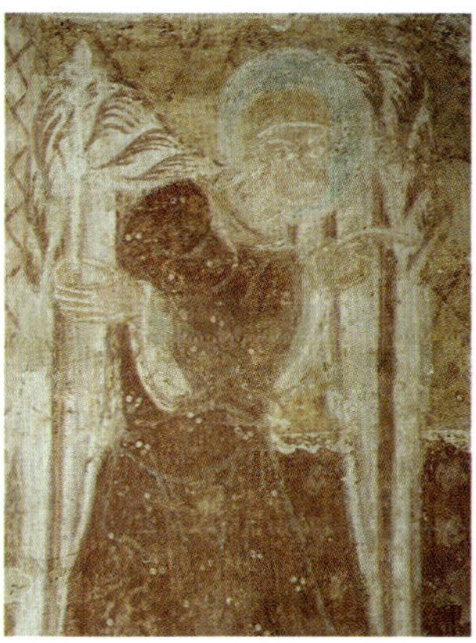

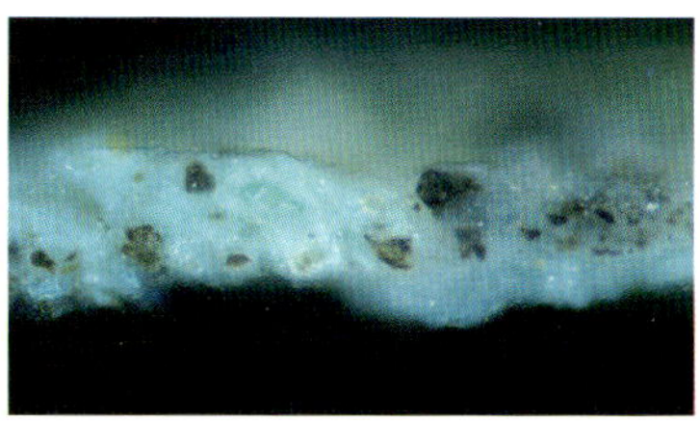

Plate 92 Hardham (West Sussex). Cross-section of a sample from Gabriel's halo, showing the copper chloride green in a lime white matrix, original magnification 500×. The copper chloride green has undergone partial alteration to form what appears to be dark copper oxide. (Photomicrograph: H. Howard; © Courtauld Institute, Conservation of Wall Painting Department)

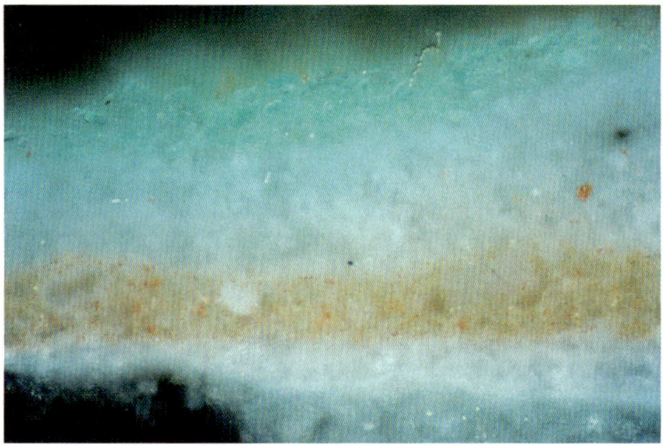

Plate 93 Hardham (West Sussex). Cross-section of a sample from the drapery of the Virgin *in the* Flight into Egypt *on the nave north wall. Here, copper chloride green combined with lime white was applied over very pale grey underpaint of charcoal black combined with lime white on a pink preparation. The grey underpaint would have improved the covering power of the green pigment and thereby enhanced its appearance. (Photomicrograph: H. Howard; © Courtauld Institute, Conservation of Wall Painting Department)*

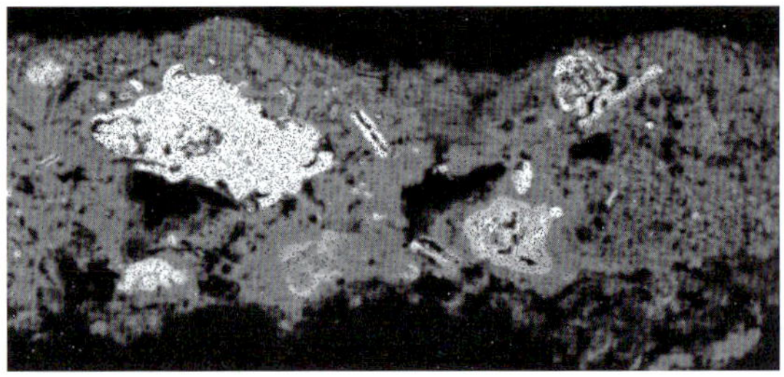

Plate 94 Backscatter SEM micrograph of the sample shown in Plate 92 with X-ray map for copper. The copper is clearly associated with the green and brown (partially altered) particles. (SEM micrograph: H. Howard; © Courtauld Institute, Conservation of Wall Painting Department)

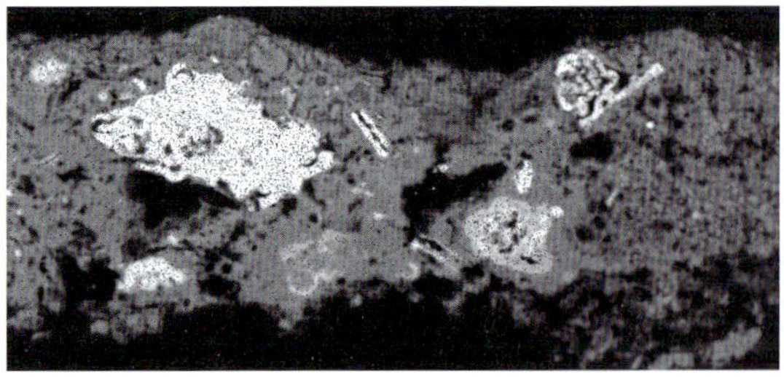

Plate 95 Backscatter SEM micrograph of the same sample with X-ray map for chlorine. The chlorine is associated with the copper-rich particles. (SEM micrograph: H. Howard; © Courtauld Institute, Conservation of Wall Painting Department)

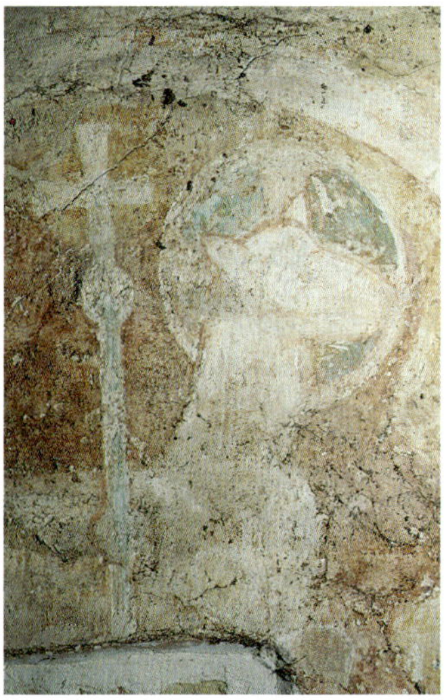

Plate 96 Hardham (West Sussex): detail of the Agnus Dei *on the east wall of the nave (c1100), showing the use of a copper chloride green for the nimbus and cross-staff. (Photograph: Sawdy/ Heritage; © Courtauld Institute, Conservation of Wall Painting Department)*

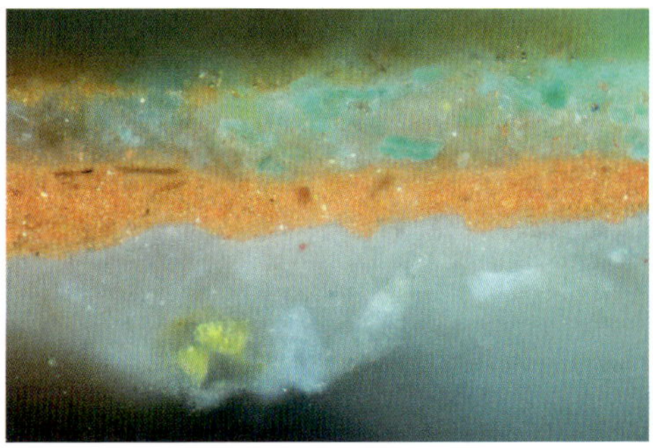

Plate 97 Hardham (West Sussex). Cross-section of a sample from the nimbus of the Agnus Dei, *showing a copper chloride green in a lime white matrix applied over a local red earth ground, original magnification 426×. (Photomicrograph: H. Howard; © Courtauld Institute, Conservation of Wall Painting Department)*

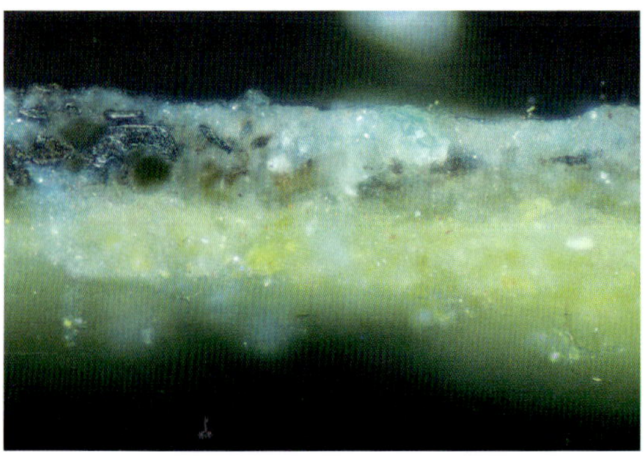

Plate 98 Hardham (West Sussex). Cross-section of a sample from the nimbus of the southern angel censing the Agnus Dei, *where yellow earth was employed beneath the green pigment, original magnification 466×. The variety of coloured underpaints used beneath the copper chloride green suggests that particular coloristic effects were sought in each instance. In this sample the green pigment has undergone partial transformation to form an alteration product presumed to be tenorite. (Photomicrograph: H. Howard; © Courtauld Institute, Conservation of Wall Painting Department)*

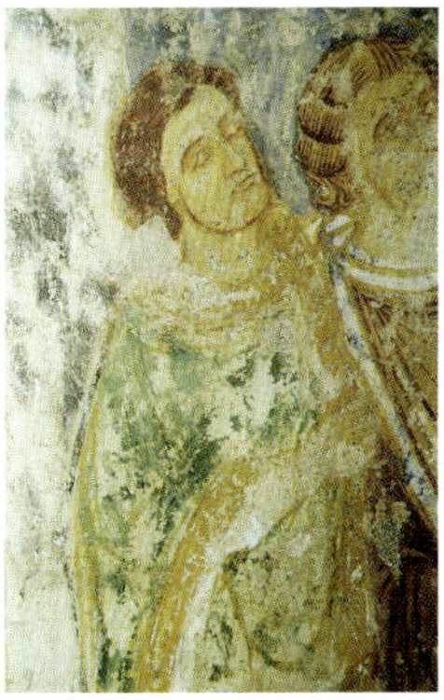

Plate 99 Norwich Cathedral: detail of the 'Losinga' cycle of paintings (c1190), showing the green mantle of the western figure on the west segment of the vault. A copper chloride green pigment, combined with lime white and applied over a lime ground, was employed in this case. (Photograph: H. Howard; © Courtauld Institute, Conservation of Wall Painting Department)

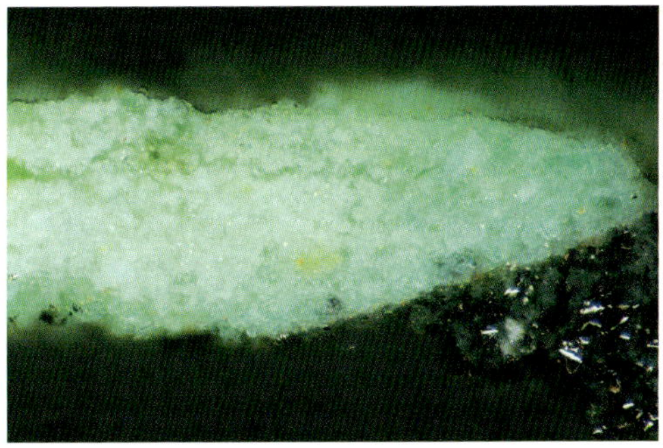

Plate 100 Norwich Cathedral, 'Losinga' cycle. Cross-section of a sample from the brilliant green roof of the cathedral in the lowest roundel on the soffit of the arch, showing the copper chloride green applied over dark grey underpaint of charcoal black combined with lime white. The grey underpaint would have increased the covering power of the copper chloride green and thereby enhanced its green appearance, original magnification 250×. (Photomicrograph: H. Howard; © Courtauld Institute, Conservation of Wall Painting Department)

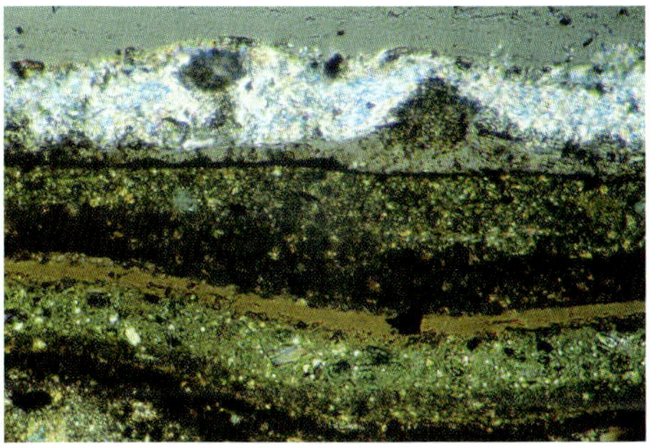

Plate 101 Malachite has been identified by instrumental analysis in only one instance in English Romanesque wall painting – in the excavated fragments of c1180 from the chapter house of Sherborne Abbey (Dorset). Similarly, in the Gothic period, identification has usually only been made by polarised light microscopy. This thin section, photographed in transmitted light with semi-crossed polars shows the green copper-containing pigment – probably malachite – employed for the original background of St. William of York (c1302–08) in the feretory of St. Albans Cathedral (Plate 45). The original green background was overpainted later in the medieval period with verdigris. A thick wax coating is visible on top of the paint sample, original magnification 200×. (Photomicrograph: H. Howard; © Courtauld Institute, Conservation of Wall Painting Department)

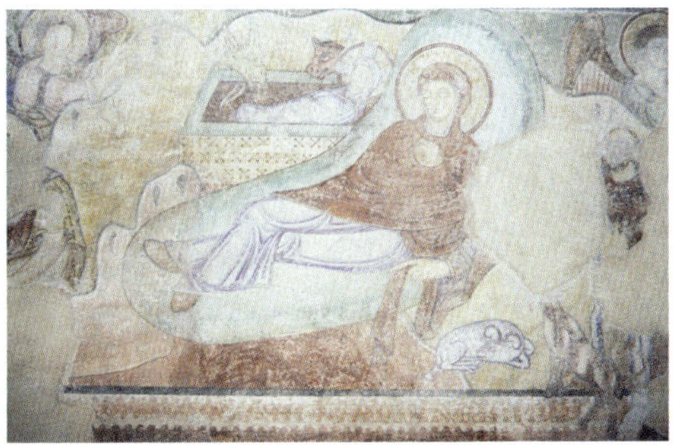

Plate 102 Detail of the Nativity *in the scheme of c1165 in the Johanneskapelle, Pürgg (Austria). Both malachite and a green made from combining charcoal black and yellow earth have been identified in this scheme. (Photograph: H. Howard; © Courtauld Institute, Conservation of Wall Painting Department)*

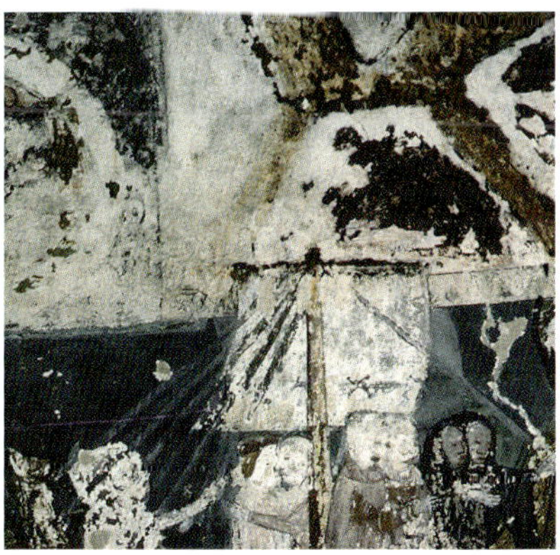

Plate 103 Horsham St. Faith Priory (Norfolk): detail of the Return to England *from the* Foundation Narrative *on the east wall of the refectory (c1250, but repainted in the 15th century). The presence of verdigris combined with prepolymerised linseed oil was confirmed in a sample taken from the repainting in the area above the boat. (© Crown copyright. NMR)*

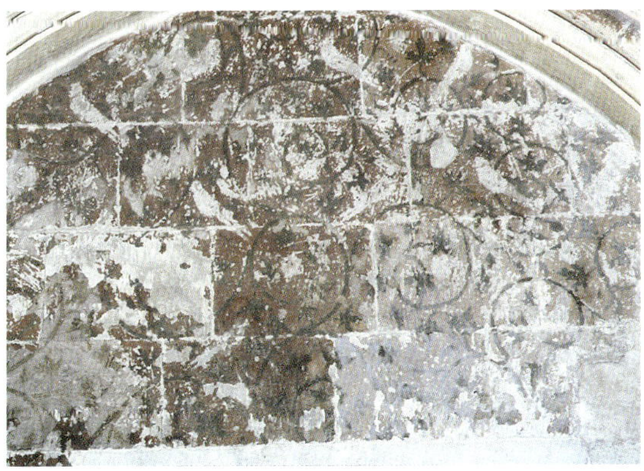

Plate 104 Rochester Cathedral: detail of the painting of c1340 on the back wall of the tomb of William of Perth. Analysis of the vine leaves confirmed the presence of verdigris bound with walnut oil and subsequently glazed with copper resinate. (© Courtauld Institute, Conservation of Wall Painting Department)

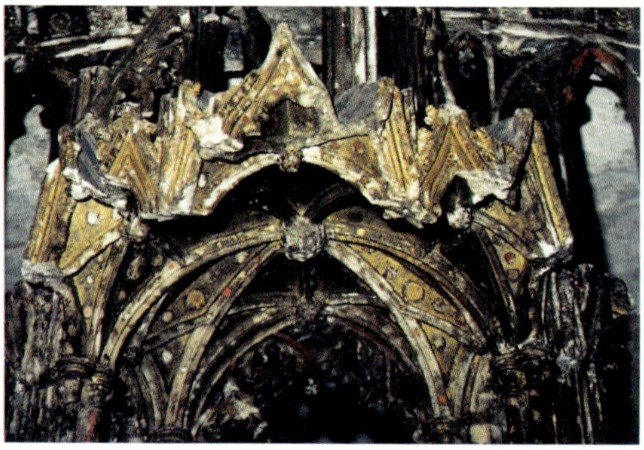

Plate 105 Canterbury Cathedral, Chapel of Our Lady Undercroft: detail of the central canopy of the reredos (c1370). (© Canterbury Cathedral Wall Painting Workshop)

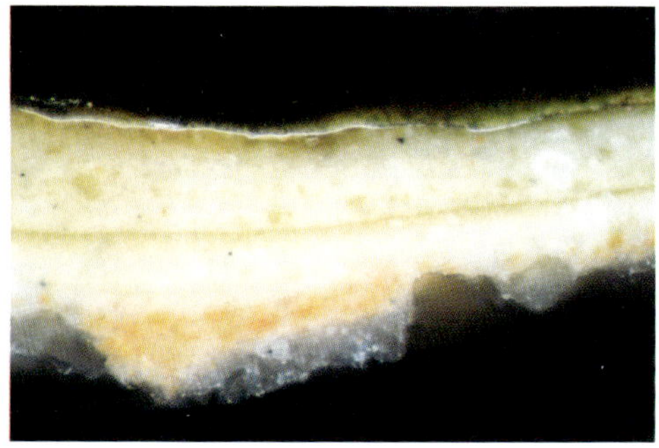

Plate 106 Cross-section of a sample from one of the raised decorative elements on the canopy shown in Plate 105, original magnification 400×. A luminous effect was produced by applying verdigris in an oil medium over a layer of silver leaf. (Photomicrograph: H. Howard; © Courtauld Institute, Conservation of Wall Painting Department)

Plate 107 Windsor Castle, King John's Tower: detail of the recently discovered 14th-century wall paintings. Accounts for 1365–66 list 12lb of verdigris and other pigments including 7lb asure de bys *at 3s per lb, purchased for these paintings. Preliminary analysis of the scheme indicated the presence of both verdigris and red lead. (© English Heritage Photographic Library)*

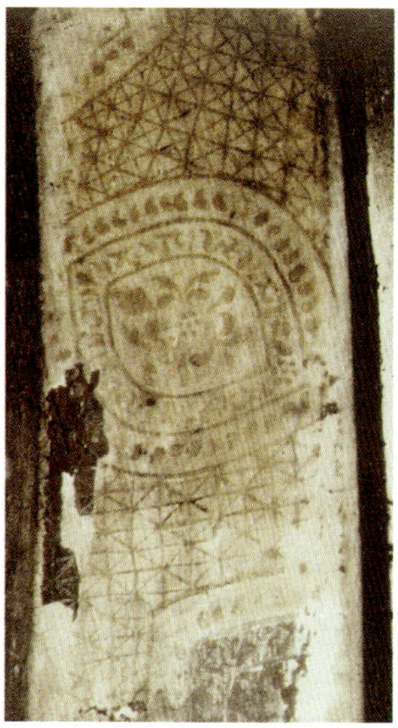

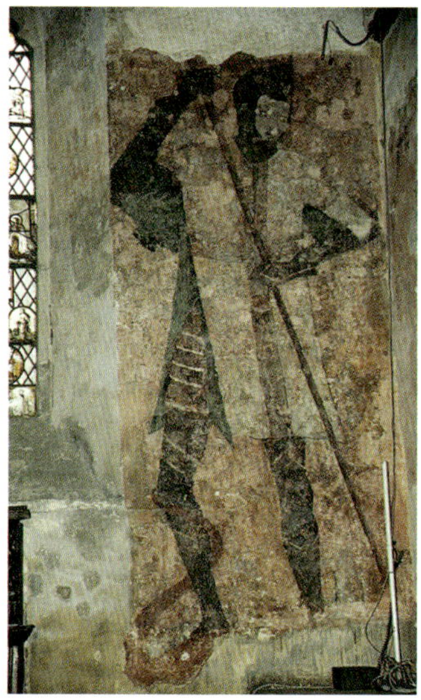

Plate 108 Farleigh Hungerford Castle, St. Leonard's Chapel: detail of the early 15th-century St. George *on the east wall. (© Courtauld Institute, Conservation of Wall Painting Department)*

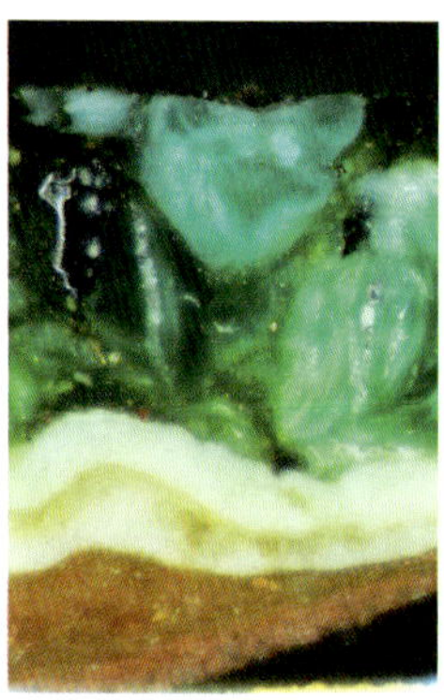

Plate 109 Farleigh Hungerford Castle, St. Leonard's Chapel. Cross-section of a sample from the green lining of the saint's cloak, original magnification 320×, showing a layer of verdigris combined with lead white and charcoal black applied in an oil medium over lead white. The particles of verdigris show signs of alteration to a pale blue copper chloride-rich product. (Photomicrograph: H. Howard; © Courtauld Institute, Conservation of Wall Painting Department)

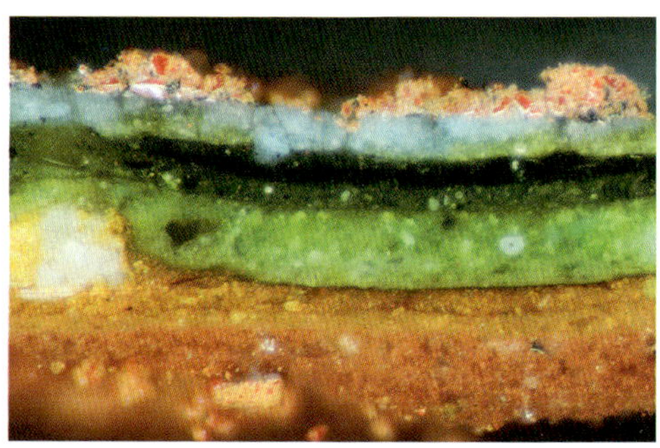

Plate 110 Farleigh Hungerford Castle, St. Leonard's Chapel. Cross-section of a sample from the contemporary Kneeling Knight *on the south wall of the chapel, original magnification 240×. Here multiple layers of verdigris – combined with varying quantities of lead white – have been applied over the red preparatory layers. A thin layer of vermilion is also visible on the surface of the sample. The uppermost verdigris layer has also partially altered to form a blue copper chloride-rich product. (Photomicrograph: H. Howard; © Courtauld Institute, Conservation of Wall Painting Department)*

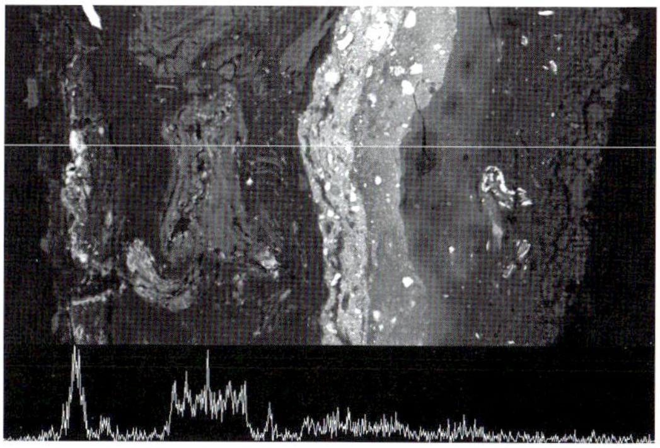

Plate 111 *Backscatter SEM micrograph of the sample shown in Plate 109. The line scan for chlorine indicates a higher concentration of the element in the partially altered verdigris particle and at the surface of the sample. (SEM micrograph: H. Howard; © Courtauld Institute, Conservation of Wall Painting Department)*

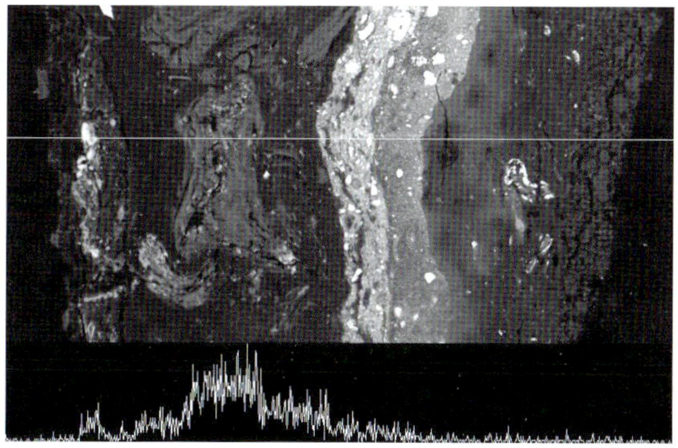

Plate 112 *Backscatter SEM micrograph of the sample shown in Plate 109. The line scan for copper indicates a higher concentration of the element in the blue particle and at the surface of the sample, though copper is present throughout the green paint layer. Significantly, the transformation of verdigris to form a copper chloride-rich alteration product seems to be associated with contact between the pigment and the overlying plaster. It appears to occur initially at the interface between the two layers, or where cracks in the green paint have provided a passage for further penetration of the lime-based plaster. (SEM micrograph: H. Howard; © Courtauld Institute, Conservation of Wall Painting Department)*

113

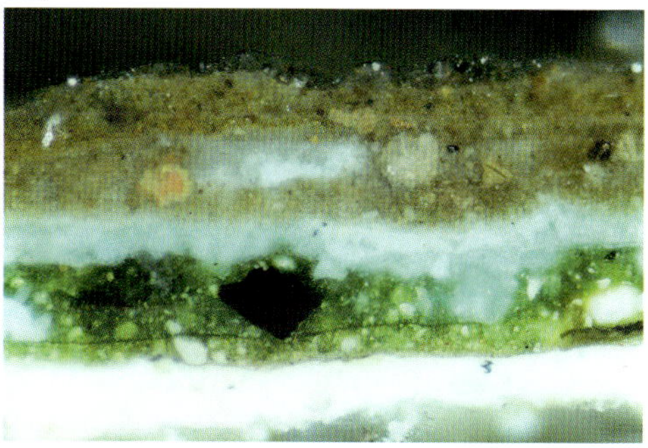

114

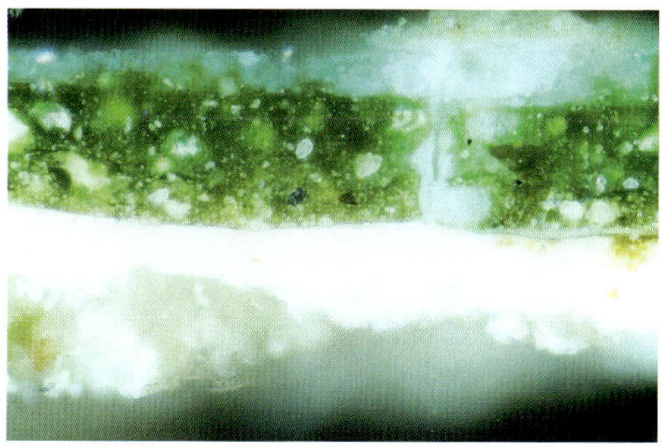

115

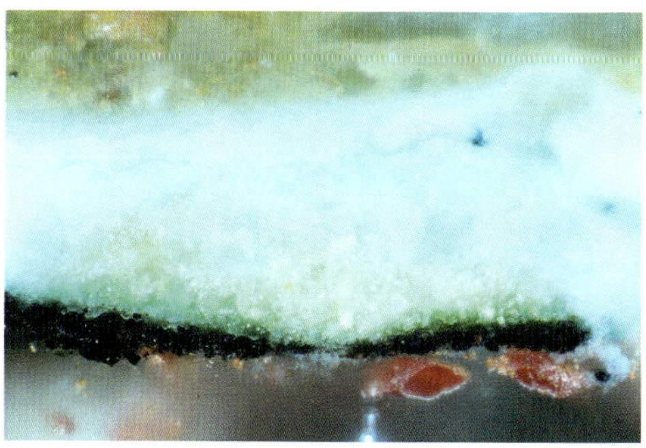

116

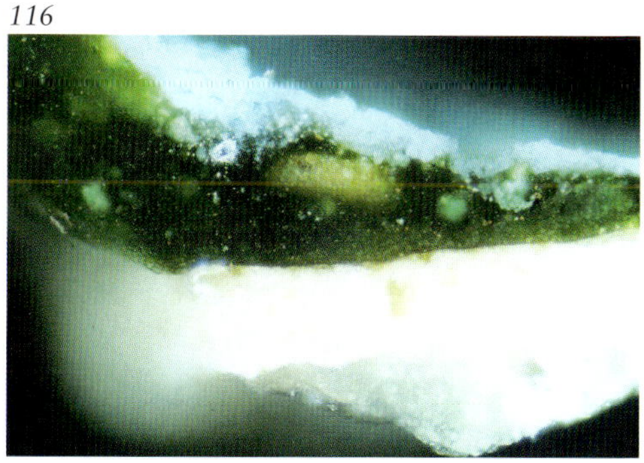

Plates 113–116 *Ingham (Norfolk), tomb of Sir Oliver de Ingham (c1344): cross-section (original magnification 324×) of a sample from the now blue polychromy on the bascinet. A pale blue alteration of verdigris has formed at the interface of this pigment and the coloured limewash. Where a crack exists in the verdigris layer, allowing increased contact with the alkaline limewash layer (Plate 114), the alteration has progressed to a greater depth in the green paint layer. In other areas where the alteration is more advanced, only a tiny trace of the green colour visible at the base of the pale blue layer (Plate 115) provides an indication of the original colour and nature of the paint layer. The same alteration phenomenon occurs on the de Bois tomb (c1380) at Ingham (Plate 116). This sample, taken from one of the weepers on the tomb chest, shows the formation of the pale blue alteration product of verdigris at the surface of the paint layer (from which the covering limewash has now been removed). (Photomicrographs: H. Howard; © Courtauld Institute, Conservation of Wall Painting Department)*

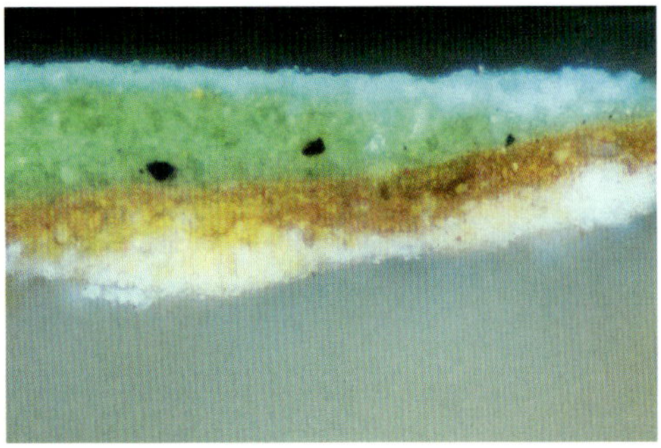

Plate 117 *The alteration of verdigris has also been noted in several Gothic wall paintings as here in the mid-14th-century scheme over the chancel arch at Ickleton (Cambridgeshire). The alteration product has formed in a layer 15µm thick over the original layer of verdigris combined with lead white. It is perhaps significant that the transformation product is not apparent where an original dark green glaze over the verdigris layer remains intact. (Photomicrograph: H. Howard; © Courtauld Institute, Conservation of Wall Painting Department)*

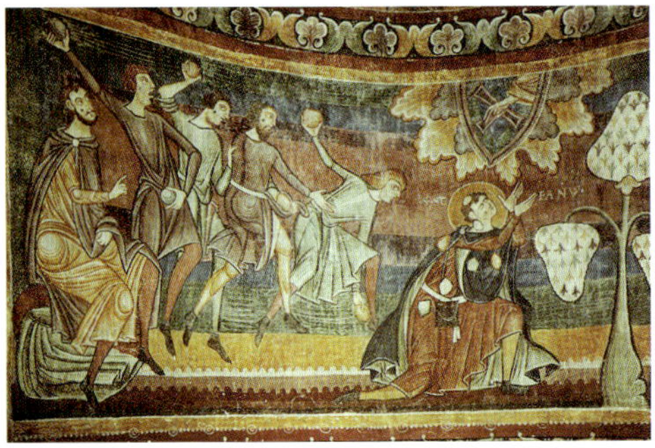

Plate 118 *Müstair (Switzerland): detail of the late 12th-century scheme showing the extensive use of various red pigments: vermilion, red lead and red earth. Vermilion was employed extensively in cathedral schemes in England in the Romanesque period, though there is little evidence for its use in parish churches at this time. (© Colorphoto Hans Hinz SWB)*

Plate 119 *Winchester Cathedral, Holy Sepulchre Chapel: detail of the Deposition (c1175) on the east wall of the chapel. Vermilion was used for numerous details such as the panelled background and drapery of St. John. (© Courtauld Institute, Conservation of Wall Painting Department)*

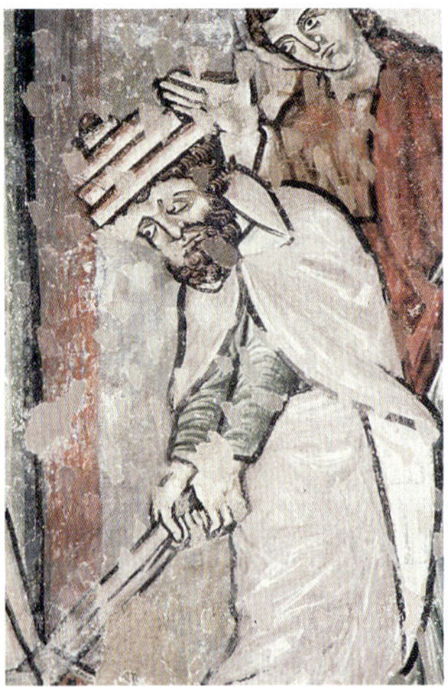

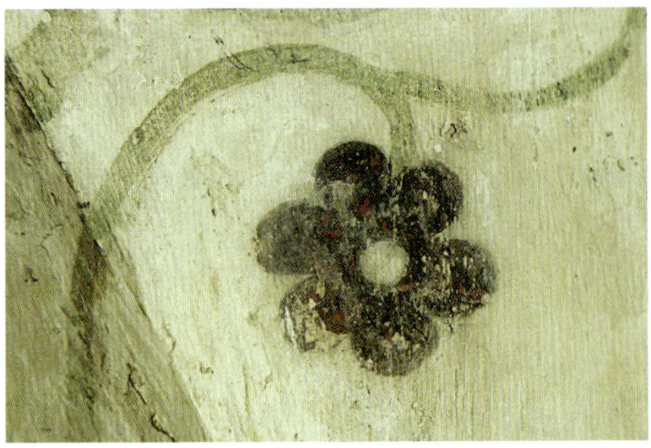

Plate 120 *Vermilion has sometimes altered, or partially altered, to its black form, metacinnabar. In the 13th-century scheme in the Guardian Angels' Chapel, Winchester Cathedral, the once brilliant red paint layers now appear almost black as in this rosette on the west segment of the vault. (© Courtauld Institute, Conservation of Wall Painting Department)*

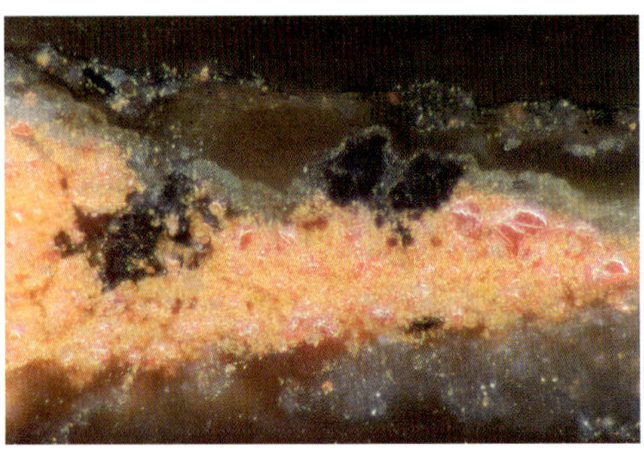

Plate 121 *Cross-section (original magnification 400×) of a sample from the rosette shown in Plate 120. Here the particles of darkened vermilion can be seen at the top of the otherwise brilliant red paint layer. A sulphated crust is visible on the surface of the paint layer. (Photomicrograph: H. Howard; © Courtauld Institute, Conservation of Wall Painting Department)*

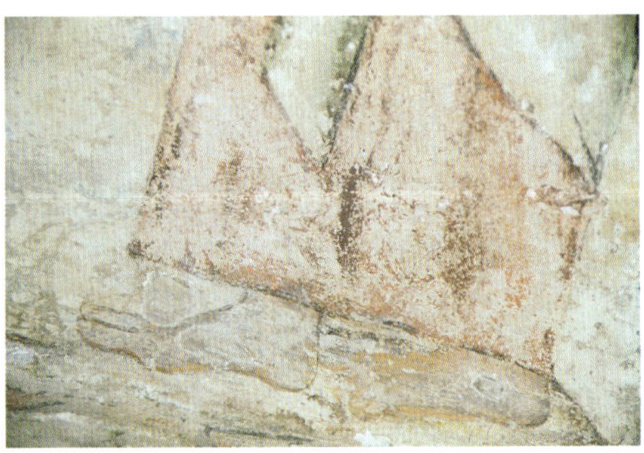

Plate 122 *Norwich Cathedral, Ante-Reliquary Chapel: detail of St. Paul's red drapery (c1300). This has a rather patchy red-brown appearance, again due to the partial alteration of vermilion. (Photograph: H. Howard; © Courtauld Institute, Conservation of Wall Painting Department)*

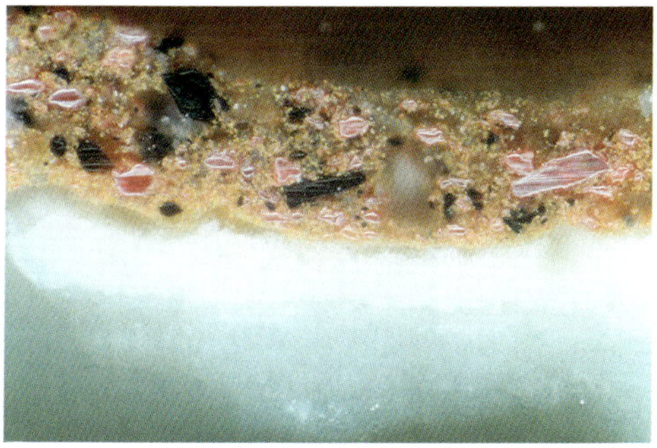

Plate 123 *Norwich Cathedral, Ante-Reliquary Chapel. Cross-section of a sample taken from St. Paul's drapery, original magnification 775×. Red particles of vermilion are mixed throughout the paint layer with those which have darkened. (Photomicrograph: H. Howard; © Courtauld Institute, Conservation of Wall Painting Department)*

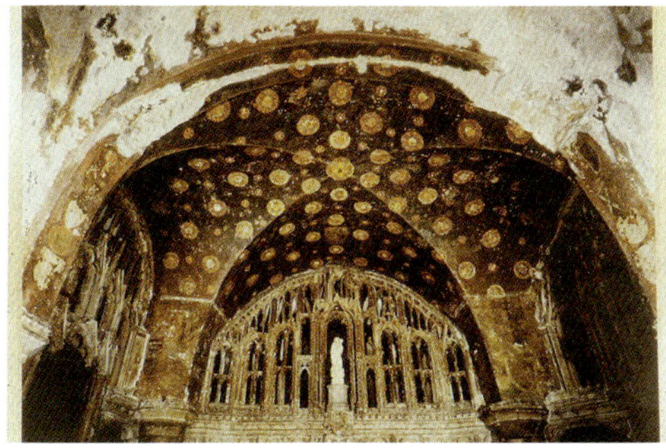

Plate 124 *Canterbury Cathedral, Chapel of Our Lady Undercroft: detail of the painting on the eastern bay of the vault before cleaning (c1370). (© Canterbury Cathedral Wall Painting Workshop)*

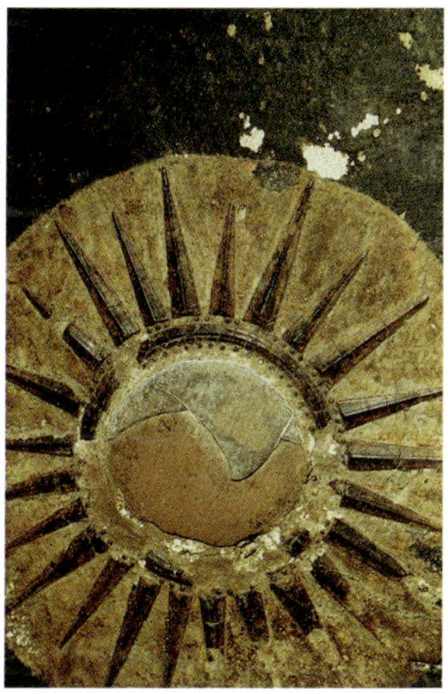

Plate 125 *Canterbury Cathedral, Chapel of Our Lady Undercroft. Detail of one of the attached suns which once had a reflective centre and gilded rays. (© Canterbury Cathedral Wall Painting Workshop)*

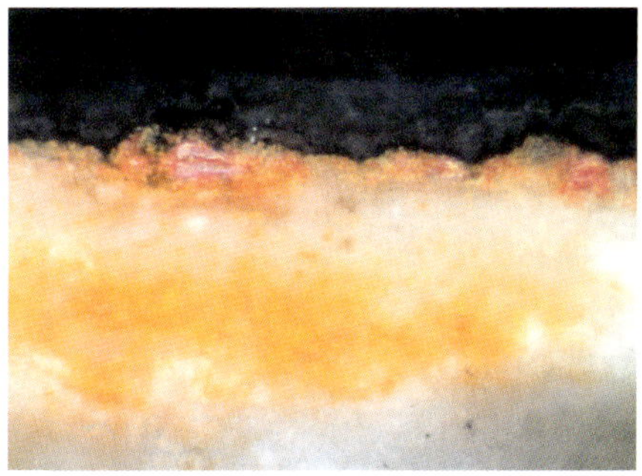

Plate 126 *Canterbury Cathedral, Chapel of Our Lady Undercroft. Cross-section (original magnification 400×) of a sample from the dark background of the vault before cleaning. Although it was thought that the background was originally blue, examination of the cross-section confirmed the presence of vermilion applied over a layer of red lead to produce a rich orange-red effect. (Photomicrograph: H. Howard; © Courtauld Institute, Conservation of Wall Painting Department)*

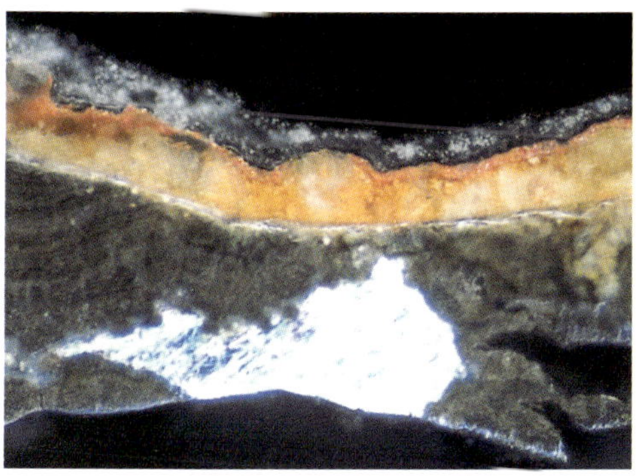

Plate 127 *Canterbury Cathedral, Chapel of Our Lady Undercroft. Cross-section of a sample taken from the base of one of the attached stars. A layer of partially oxidised tin leaf is present at the base of the sample. This was originally gilded and then the orange-red background was painted in the areas between the raised rays. The layers of vermilion over red lead which formed the red background are again clearly visible. (Photomicrograph: H. Howard; © Courtauld Institute, Conservation of Wall Painting Department)*

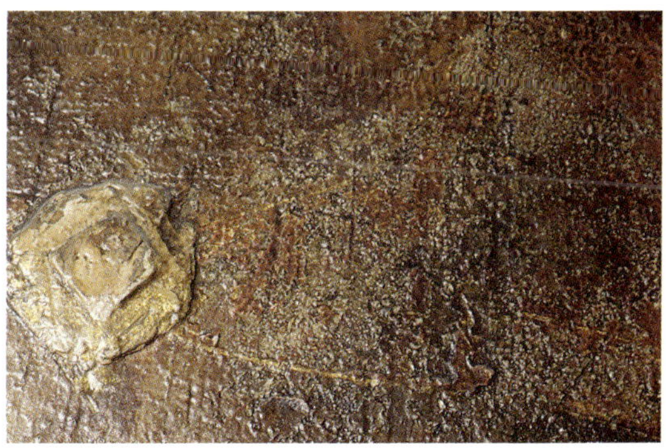

Plate 128 *Westminster Abbey, chapter house: detail of Christ's red drapery in the* Majesty, *showing the attached 'brooch' and linear details in gold. (Photograph: Howard/Kakoulli; © Courtauld Institute, Conservation of Wall Painting Department)*

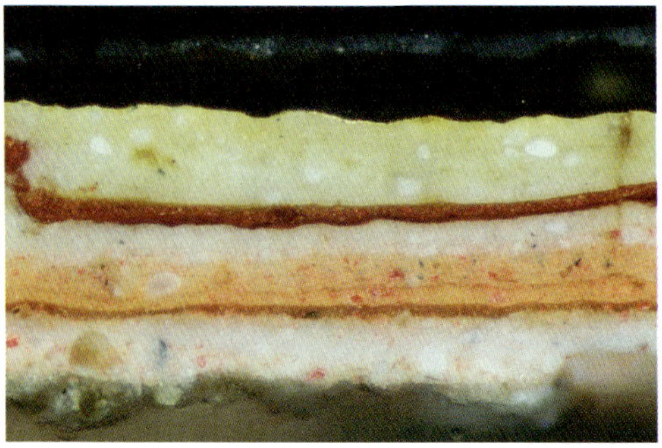

Plate 129 *Westminster Abbey, chapter house. Cross-section of a sample taken from Christ's brilliant red drapery, original magnification 430×. The modelling of the drapery has been built up with numerous layers of vermilion combined with varying quantities of lead white. (Photomicrograph: H. Howard; © Courtauld Institute, Conservation of Wall Painting Department)*

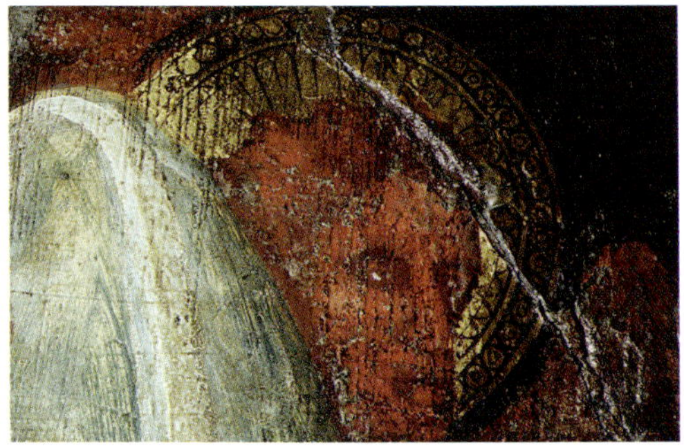

Plate 130 *Westminster Abbey, chapter house: detail of one of the angels in the* Judgement, *showing the brilliant red flesh tone which consists of vermilion applied in a thin layer (10µm) over red earth. (Photograph: Howard/Kakoulli; © Courtauld Institute, Conservation of Wall Painting Department)*

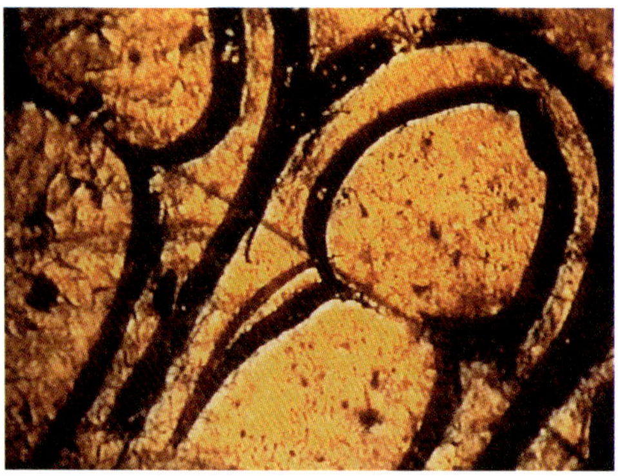

Plate 131 *Westminster Abbey, chapter house: detail of the nimbus of one of the angels in the* Judgement. *Lines used to set out the design were incised into the red preparatory layer. The fine black linear detail was applied with a brush on the surface of the gold leaf. (Videomicroscopy: Heritage/Howard; © Courtauld Institute, Conservation of Wall Painting Department)*

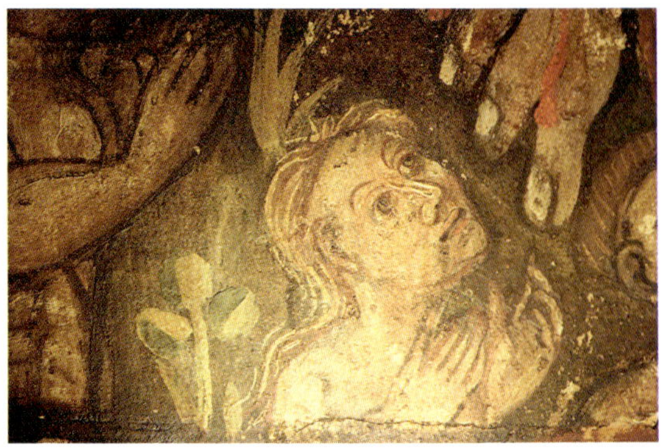

Plate 132 *Coventry, Holy Trinity: detail of the* Last Judgement *(c1435–60) over the chancel arch. (Photograph: H. Howard; © Courtauld Institute, Conservation of Wall Painting Department)*

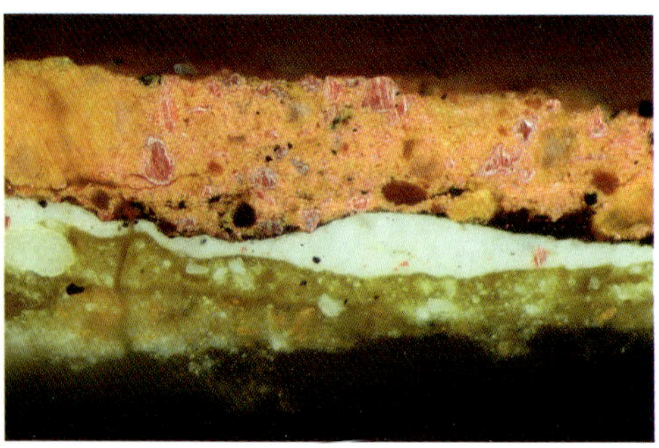

Plate 133 *Coventry, Holy Trinity. Cross-section of a sample from one of the large drips of blood on Christ's foot, original magnification 338×. For the rendering of fresh blood a combination of vermilion with red lead and red lake has been applied over a thin layer of carbon black on the flesh tone. (Photomicrograph: H. Howard; © Courtauld Institute, Conservation of Wall Painting Department)*

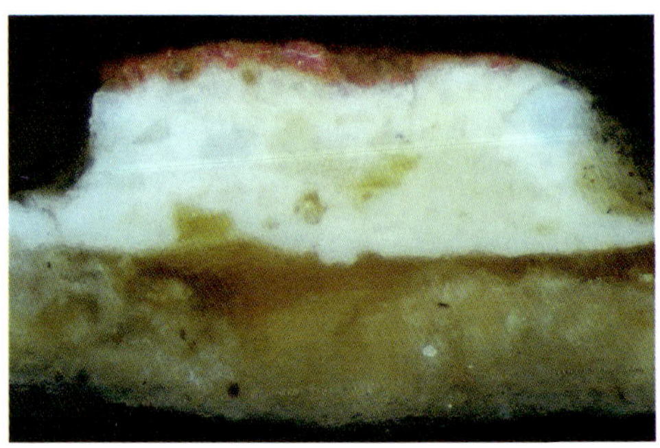

Plate 134 *Bristol, St. Mark's: cross-section (original magnification 500×) of a sample from the angel's wing in the* Nativity *(c1510), showing vermilion applied over a layer of lead white in oil. Beneath the lead white ground, an oily sealant layer on a layer of oil bulked with chalk and calcium sulphate is clearly visible. (Photomicrograph: H. Howard; © Courtauld Institute, Conservation of Wall Painting Department)*

Cost

The nomenclature used for lake pigments generally in English medieval accounts includes *cynopre*, *cinopel*, *cinople*, *cynople*, *synople* or *synapo lake* (Erskine 1981: 144; Salzman 1952: 168). Records of the purchase of pigments for St. Stephen's Chapel, Westminster Palace in 1351–54 indicate that red lakes were among the most expensive. For example, 1lb of *cynople* was priced at 30s, while 2lb of vermilion – the most costly of all the other red pigments – was 3s 4d, or 1/20th of the cost (Salzman 1952: 168). This evidence also suggests that several different types of red lake pigment were purchased for the scheme. One *cynople* was priced at 17s 3d for 1½lb, another *cynopre* 10s for ½ lb, while 2lb *cynopre of Montpellier* (de Monte Pessulano) was 16s (Salzman 1952: 168).[53] This price differential, with one of the lakes twice the price of the others, must be due to a substantial difference in quality or to the lake having been prepared using a more expensive variety of dyestuff, such as kermes. Unfortunately this suggestion cannot be substantiated since analysis of the dyestuff from the surviving fragments of wall painting from St. Stephen's Chapel has not yet been undertaken.[54] However, recent research comparing 15th- and 16th-century pharmacy price lists in Germany has established that madder lake was considerably cheaper than lake pigments prepared from scale insects, for example entries for 1505 suggest that madder was forty times cheaper than *grana* (Burmester and Krekel 1998: 102). Indeed it is clear that madder was cheaper than lake prepared from scale insects throughout northern Europe.[55]

Optical, chemical and physical properties

The optical characteristics of red lake pigments are largely dependent on the substrate employed. However, for madder on a hydrated alumina ($Al(OH)_3$) substrate the refractive index is given as approximately 1.64. In plane polarised transmitted light this pigment appears as large, rounded, gel-like flakes of translucent crimson-red colour.[56] There are no crystal faces and it is isotropic (McCrone *et al.* 1979: 1408, 1412). The type of substrate – hydrated alumina, calcium carbonate, calcium sulphate, lead white or kaolin – has a profound effect on the relative translucency of lake pigments. That a hydrated alumina support produced the highest translucency was well understood in the Middle Ages is suggested by the fact that Jehan le Begue stresses that a lake precipitated with alum may only be used for shading. To provide body (opacity) to the organic colorant he suggests the addition of gypsum or lead white (Merrifield 1967: 1: 292).

Owing to the conformity of their refractive indices, the combination of a hydrated alumina substrate and an oil medium provided the greatest translucency and allowed the red organic pigments to be used as glazes. At 1.50–1.56 the refractive index for hydrated alumina is only slightly higher than that for fresh linseed oil. Since hydrated alumina has a

53. Eastlake (1960: 1, 117–18) associates *Montpellier cynople* with madder lake from Montpellier. An even more exaggerated price difference is recorded in the Exeter Cathedral accounts for 1321–22, which probably relate to polychrome sculpture. Here one lake is almost four times the price of another; see Erskine 1981: 144–5. A qualitative difference in the type of red lake pigments available is also suggested by the French treasury accounts of the Countess of Artois, which in 1304 record '*de boin sinnoble*' employed at Hesdin Castle; Richard 1887: 331.
54. A crimson-coloured lake pigment was identified; van Geersdaele and Goldsworthy 1978: 11.
55. I am grateful to J. Kirby for this information.
56. Traditional early madder lakes may appear more orange-red in colour.

high oil absorption, a large quantity of oil would have been necessary to produce a workable paint layer. This high medium:pigment ratio also served to increase the translucency of the paint layer (Kirby and White 1996: 57).[57] Although red lakes were combined with other organic binding media – for example, Cennini suggests tempering fine lac with yolk of egg for use on walls – the translucency is reduced in these instances (Cennini 1960: 55). Since red lake pigments are affected by alkalinity, they are incompatible in *buon fresco* painting.

The principal disadvantage of these red pigments is their tendency to fade with exposure to light (of which the UV component is the most damaging) (Saunders and Kirby 1994: 93). While the binding medium and substrate may also affect the rate of fading, lake pigments mixed with white, applied over a pale-coloured or white ground, and unprotected by a glaze are particularly vulnerable to internal reflection (Johnston-Feller and Bailie 1982). Recent investigations have also demonstrated that, for lakes prepared on the same substrate, madder is the least fugitive, while brazilwood lakes are the most photosensitive of the red lake pigments available in the medieval period (Saunders and Kirby 1994: 86).

Preparation

Examination of extant artists' treatises suggests that before the 13th century red lakes for painting were commonly made from madder and generally precipitated onto a calcium carbonate substrate.[58] Eraclius (1967: 250–51) describes the preparation of a lake pigment by boiling madder roots (*Rubia tinctorum* L.) or wall flower (*viola lutea*), and grinding the strained liquid in a mortar with chalk (calcium carbonate). However, he also describes the preparation of a red lake – suitable for use on walls – made from *brasilium* (brazilwood), urine, alum and quicklime (calcium hydroxide) (Eraclius 1967: 234–5). The recommendation for the use of brazil lake on walls is unusual since this pigment is extremely photosensitive and therefore was not generally considered suitable for anything but manuscripts where it would receive relatively little light. A third variant, 'beloved of painters', was apparently made from the fresh resinous exudate of ivy.[59] Although initially the appearance of the sap was not red, on standing it is said to have assumed a blood red colour (Eraclius 1967: 190–93).[60] Theophilus mentions a pigment called carmine for which no preparation method is discussed; he may well be referring to a scale-insect lake such as kermes, and is certainly not describing madder since this is mentioned elsewhere in relation to staining bone and ivory. He suggests that for painting wood, and also for manuscript illumination, carmine should be combined with glair (Theophilus 1979: 33, 38, 40, 188–9).

Anglo-Norman instructions in the late 12th-century manuscript BL Cotton Titus D. XXIV include one of the earliest instructions for the preparation of brazilwood for use in

57. In this context, it is noteworthy that Cennini (1960: 26) comments that the fatness of red lakes is due to the alum.

58. Brazilwood lake is also frequently mentioned in early sources, but generally those relating to manuscript illumination. The late 12th-century manuscript Cotton Titus D. XXIV lists pigments suitable for illumination, including *lac* and *carmin* (probably lake from the kermes insect); Petzold 1995: 63.

59. J. Kirby of the National Gallery is very doubtful that a red lake pigment was made from this. Her practical experiments have proved unsuccessful, and she is inclined to believe that this entry is due to a misunderstanding of the nature of stick lac.

60. Theophilus (1979: 188–9) describes staining bones with madder, although there is no indication that they were subsequently ground up for use as a pigment.

manuscript illumination: 'for brazil, let it be scraped into an eggshell and glair added. After a day add good alum and leave it for three days to mature. And do not let it get wet' (Petzold 1995; Hunt 1995: 209). The brazilwood would have been finely cut or rasped into the egg shell in order to break up its structure, thus maximising the surface area for oxidation and thereby increasing the yield of brasilein ($C_{16}H_{12}O_5$) which is the red colorant.[61]

By the end of the 14th century one of the most important methods for the preparation of red lake pigments involved using clippings of dyed cloth. The red colour was extracted from the cloth by boiling with aqueous alkali, usually lye (potassium carbonate), which in the presence of alum (potassium aluminium sulphate) forms a precipitate of hydrated alumina to which the dye is attached.[62]

In some instances, various sorts of clippings were probably combined in the dyebath, making the precise nature of the organic colorant in the resulting pigment difficult to assess; for example, some of the cloth clippings may have been originally dyed with madder, while more expensive cuts were coloured with kermes.[63] The pigment resulting from the re-extraction of the dyes from these materials would therefore be a composite of the two colorants. Kirby and White (1996: 67–8) have found this combination in a number of instances; for example, kermes and madder dyestuffs occur in the same pigment employed for a robe in a French panel painting of *Saint Giles and the Hind* of *c*1500. In addition, their research indicates that in later works artists may have combined lake pigments to achieve a precise colour. This notion is confirmed by sources such as the treatise of Jehan le Begue, which suggests combining lac lake with brazil for use in manuscript illumination (Merrifield 1967: 1: 294–6). Pigment composition would also depend on the age of the dye vat and the number of times it had been used.[64]

Artists' treatises also suggest that various sources of red organic colorants were sometimes combined during the preparation of red lake pigments. Two of the three recipes of Peter of St. Omer, writing in *c*1300, describe such mixtures: madder with red lac apparently made from gum of ivy,[65] and brazilwood with gum of ivy distempered with urine. In each case the dyestuff was extracted by boiling – in water for the madder, and in red wine for the brazilwood – while the addition of alum provided a substrate for the lake pigment (Merrifield 1967: 1: 144–6).

Use in English medieval wall painting

Whether a single dyestuff or complex combinations of dyes are present in a lake pigment, these organic colorants have rarely been identified.[66] Consequently, although red lakes have been detected in many wall paintings in England, the dyestuff itself has only been identi-

61. The precursor, brasilin, is colourless (Mills and White 1994: 144). See also Wallert 1986: 53.
62. Aluminium hydroxide itself is colourless but is able to form stable colour-enhancing complexes.
63. The dyestuff present in a single textile nominally dyed with a particular dye may not be completely uniform. For example, a dyebath at the end of its use may have been boosted with another colorant; see Kirby 1999: 2.
64. Kirby and White (1996: 67) also note that the presence of mixed dyestuffs in a single pigment confirms the connection between textiles and lake pigments.
65. J. Kirby believes that stick lac rather than ivy gum was employed to produce the red lac described here.
66. Given the small sample size usually available and – particularly with wall paintings – the frequently low concentration of dyestuff, HPLC or fluorescence spectrometry have proved to be the most accurate methods of analysis. For the analysis and uses of red lake pigments, see Kirby and White 1996.

fied in three instances: madder lake has been detected in the scheme of *c*1300 in the Ante-Reliquary Chapel, Norwich Cathedral, and in the paintings of *c*1302–08 in the feretory of St. Albans Abbey, while the presence of kermes and brazilwood lakes has now been confirmed in the late 14th-century wall paintings in the chapter house of Westminster Abbey.[67]

Although lakes were common in sophisticated wall paintings from the 13th century onwards, when they were highly prized for their translucency in oil media, these pigments were not often employed in English Romanesque wall painting. Given the emphasis on *opaque* primary colours in the Romanesque palette, it is not surprising that in one of the two instances in which a red lake has been identified in this context – in the 12th-century paintings in the Holy Sepulchre Chapel at Winchester – the organic pigment forms part of an opaque paint layer. Here, the unusual pink colour of Nicodemus' robe in the *Deposition* was achieved by combining a red lake pigment with a tiny trace of ultramarine in a calcium carbonate matrix (Plates 160 and 161). Given the ultramarine close to Nicodemus in the background of the scene, it is possible that the blue pigment may have been an accidental inclusion, although it seems more likely that it was combined with the red lake to produce a slightly purple hue. This paint layer, 100μm thick, was applied directly over the plaster substrate which contains a few inclusions of charcoal black. The greatest concentration of such inclusions in the plaster generally was found beneath blue paint layers, and it seems likely that the charcoal black was incorporated to reduce the light scatter from the white substrate, thereby increasing the covering power of blue mineral pigments.[68]

A red lake pigment has also been identified in the wall painting fragments of *c*1190 recently excavated from the crypt of Glasgow Cathedral. By contrast with the Winchester example, the pigment here was applied over a layer of vermilion, presumably to intensify and saturate the appearance of the underlying red layer, thus functioning as a glaze (Plates 163 and 164).[69] Analysis also indicated that red lake was painted over a layer of green earth on another fragment. Since this would tend to produce a neutral grey colour it seems likely that details in red were originally applied over green underpaint, as in the Winchester paintings where *impasto* motifs in red lead decorate the green drapery of the Magdalene.

Unfortunately, only a tiny quantity of the lake pigment is present in the samples from Winchester and Glasgow, preventing identification of the source of the organic colorant. However SEM/EDX analysis of the sample from the Holy Sepulchre Chapel suggested that the substrate for the red dye may have been calcium sulphate, while calcium carbonate was found to be the substrate in the Glasgow sample (Howard 1995a: 101, 1998a; Park 1998). This result is interesting in the light of Laurie's suggestion that lakes precipitated onto calcium sulphate or calcium carbonate bases tend to be much less translucent than those precipitated with alum which was generally recommended in the Gothic period (Laurie 1913: 318).[70] Recent analysis also suggests that the presence of a calcium salt in a red lake glaze may produce a slightly cloudy effect, and may also make the organic colorant more liable to fading (Campbell *et al.* 1997: 37–8).

67. Analysis of the madder lakes was undertaken by Dr A. Wallert using spectrofluometry and DT-MS, and identification of kermes and brazilwood lakes was carried out by J. Kirby using HPLC.
68. Even a slightly grey underpaint beneath the pink of Nicodemus' robe would have had the additional benefit of reducing the fading of the lake pigment.
69. For the dating and technique of these fragments, see Howard 1998a and Park 1998.
70. Depending on the dyestuff, the colour may also be slightly affected; J. Kirby, pers. comm.

Despite their known susceptibility to fading,[71] red lakes were used increasingly in the Gothic period – particularly in high-quality schemes where translucent effects were achieved by application of these pigments as a glaze in an oil medium. In this period, red lakes have thus far been identified in the mid-13th-century paintings at Horsham St. Faith Priory (Norfolk) (Babington and James 1988); the scheme of *c*1300 in the Ante-Reliquary Chapel, Norwich Cathedral (Howard 1995c: 5; Park and Howard 1996: 396); the paintings of *c*1302–08 in the feretory of St. Albans Cathedral (Howard 1993a: 35); the mid-14th-century painting in the recess of the so-called 'Tomb of William of Perth' in the north-east transept of Rochester Cathedral (Rickerby 1988: 11); the mid-14th-century paintings of St. Stephen's Chapel, Westminster (van Geersdaele and Goldsworthy 1978: 11); the late 14th-century paintings in the chapter house of Westminster Abbey (Hluvko 1991: 55–6),[72] and in the Byward Tower of the Tower of London (Hluvko 1991: 55–7); the early 15th-century scheme in St. Leonard's Chapel at Farleigh Hungerford Castle (Howard *et al.* 1998: 60); the *Last Judgement* of *c*1435–60 over the chancel arch of Holy Trinity, Coventry (Howard 1996b: 5); the late 15th-century paintings in Eton College Chapel (Plummer 1986: 36); and in a shell palette from the Carmelite Friary at Coventry (Plates 175 and 176) (Howard 1997a: 2, pl. 4).

Application

Over a white ground or directly on the plaster support

The earliest identified use of a red lake pigment in English Gothic wall painting is in the splendid scheme of *c*1250 at Horsham St. Faith. For the border below the *Foundation Narrative* on the east wall of the refectory, red lake was combined with lead white and applied over a white ground (Babington and James 1988: 15–16).[73]

Analysis by fluorescence spectroscopy confirmed the use of madder lake in the scheme of *c*1300 in the Ante-Reliquary Chapel at Norwich (Plates 165 and 166).[74] This is one of only three instances in which the dyestuff of a red lake pigment has been identified in English medieval wall painting.[75] The madder lake was employed extensively to achieve translucent effects by the application of pigment mixtures and glazes over a reflective white ground of lead white combined with calcium carbonate in oil. A pure layer of the red lake pigment, 50μm thick, was applied over the white ground to produce the sumptuous red of the apostles' drapery on the western arch of the chapel (Howard 1995c: App. II).

Red lakes applied over a white ground have also been detected in the early 15th-century wall paintings in the chapel of Farleigh Hungerford Castle. For St. George's

71. Cennini (1960: 27) comments of red lake that 'the air is its undoing'.

72. Howard 2000 and forthcoming.

73. See also unpublished notes by Babington, held at the Courtauld Institute, on the examination of paint samples from these paintings. Although in other samples it seems likely – from the complex stratigraphy and the presence of intermediate glaze or varnish layers – that a red lake was applied as part of the 15th-century repainting of the scheme, the sample from the border (which also has a trace of carbon black on the surface) probably consists of the 13th-century paint layers alone.

74. For the principles of spectrofluometry, see Wallert 1986b.

75. This identification of madder is of particular interest since a madder market existed in Norwich from at least the 14th century; see Spence and Kent 1970. The market was certainly present in the 14th century, since the name Madermarket was employed in 1361, though the spelling Madyrmarket was used in 1368; Duxson 1933: 2–3.

surcoat, red lake in a white matrix was applied over a ground consisting of lead white with the addition of a little calcium carbonate in oil (Howard *et al.* 1998: 60).

Over coloured underpaint

Although red lake glazes were often applied over a white ground to enhance the luminosity of the organic colorant, they were also frequently applied over a layer of lead white combined with lake or another pigment. This method of application was employed in the mid-14th-century wall painting of St. Stephen's Chapel, Westminster, where the red lake was combined with lead white and then glazed with a translucent crimson lake in an oil medium (van Geersdaele and Goldsworthy 1978: 11).

Lake pigments were also applied over a pure red underpaint to enhance and enrich the red colour. For example, the rich red background of the vinescroll and popinjays painted in the recess of the 'Tomb of William of Perth' (*c*1340) in Rochester Cathedral was achieved by applying a red lake glaze over red lead on a ground of red earth (Rickerby 1988: 11). Similarly, in the *Majesty* painting of *c*1375–1404 in the chapter house of Westminster Abbey, a red lake glaze was applied over vermilion combined with red lead, while for St. John's robe in the *Apocalypse* cycle a crimson-coloured glaze covers a layer of vermilion on a chalk ground (Hvluko 1991: 55–8). Elsewhere in this cycle of late 14th-century wall paintings in the chapter house, a red lake, identified as kermes, was combined with lead white and applied over red lead combined with vermilion to form a brilliant pink highlight over red drapery.[76]

Underpaint of other colours was also employed. Thus, in a sample taken from one of the *Foundation* scenes at Horsham St. Faith Priory, the organic pigment was applied as a glaze over a grey layer of white combined with a trace of carbon black. Since the stratigraphy of this sample is complex, with at least three layers – apparently including a glaze – below the grey underpaint, the upper lake glaze was probably applied during the 15th-century partial repainting of these mid-13th-century paintings (Babington and James 1988: 15–16).

Red lake pigments were extensively used in the wall paintings of *c*1302–08 in the feretory of St. Albans Cathedral and on the shrine of St. Alban itself. The technique of the painting of the red chasuble of St. William, on the north side of the feretory, is particularly interesting since three separate types of layer structure were identified, each associated with the different degree of shading desired. For the darkest tones (generally corresponding to areas of deep folds delineated with black lines) a layer of red earth with natural vermilion scumbled across the top was applied directly to the lime ground. For the middle tones (now a soft, slightly dusky pink) a secondary ground of yellow earth was used, and over this a layer of red lake with charcoal particles was applied. Finally, for the lightest areas, a layer of red lake was applied directly to the lime ground. A sample taken from an area where the red lake was applied over the lime ground indicated that substantial fading has occurred.

Other notable occurrences of red lake include that employed in conjunction with orpiment in the scheme of *c*1420 in the chapel of Farleigh Hungerford Castle. In the shield above St. George, red lake in a white matrix was applied over a layer of red lake combined with vermilion over orpiment (Plate 189) (Howard *et al.* 1998: 62). Red lake

76. Sample 44/2582 from the lateral group, arch 3, south-east bay; Howard 2000 and forthcoming. The kermes lake was identified by J. Kirby using HPLC, see Kirby 1999: 2–3.

has also been identified in the *Last Judgement* of *c*1435–60 over the chancel arch of Holy Trinity Church, Coventry. In this instance the pigment has a clay-rich chalk substrate (Howard 1996b: 6).

Mixed with other pigments

Red lake pigments were frequently combined with azurite to produce a purple colour. In the Ante-Reliquary Chapel at Norwich, for example, a combination of azurite and madder was employed for the rich purple draperies of St. Peter and St. Andrew on the vault, and of one of the apostles on the western arch. For the more opaque, pale purple drapery of the *Christ in Majesty* on the vault, lead white was added to the azurite-madder mix.

A combination of madder lake and azurite applied over a lead white ground was identified in a sample taken from the purple chasuble of the standing bishop saint of *c*1302–08 on the south side of the feretory of St. Albans Cathedral. The source of the dyestuff has been identified in English medieval wall painting in only two other instances.[77]

Kermes lake with a trace of brazilwood lake combined with lead white has been identified in a sample from the angel's pink drapery in *The Mighty Angel* (*c*1375–1404) in the chapter house of Westminster Abbey. The quantity of brazilwood lake is minute, and its presence is probably due to the process of manufacture of the lake pigment from a dyebath of cloth clippings.[78] Elsewhere in this *Apocalypse* cycle in the chapter house, azurite combined with red lake and lead white was applied over tin leaf for the tiny mandorla in the scene of *St. John Rebuked by the Angel* in the second arch of the south-west bay (Plates 51 and 52).[79] Red lake and azurite were also employed for St. Michael's robe in the paintings of *c*1395 in the Byward Tower of the Tower of London, although in this instance the purple layer was applied over lead white (Plates 167 and 168) (Hvluko 1991: 55–6).

Applied as a glaze over gold leaf

Analysis of a sample from the lateral group of late 14th-century wall paintings in the chapter house of Westminster Abbey confirmed the presence of kermes lake applied over gold leaf to produce the luminous red appearance of the brooch of the lower left figure.[80]

Binding media

Preliminary analysis of the red lake glaze applied over green earth in the wall painting fragment of *c*1190 from the crypt of Glasgow Cathedral indicated that the layer was particularly medium rich. Further analysis undertaken using GC-MS confirmed the presence of protein, and suggested that casein or casein with animal glue may have been employed as the original binding medium.[81]

The very translucency of many red lake layers employed in the Gothic period suggests the use of an oil medium, although little instrumental analysis has been undertaken.

77. I am grateful to Dr Arie Wallert for identifying the madder lake using DT-MS.
78. Howard 2000 and forthcoming. The analysis of the kermes and brazilwood lakes was kindly undertaken by J. Kirby using HPLC, see Kirby 1999: 1–2.
79. Howard 2000 and forthcoming.
80. Sample 52/2590, Howard 2000 and forthcoming; see also Kirby 1999: 4.
81. I am grateful to Dr P. Colombini for undertaking the GC-MS analysis for me.

However, in various cases such analysis of layers containing other pigments in the same scheme has confirmed the presence of an oil medium. Thus, oil has been identified in the paintings of *c*1250 at Horsham St. Faith Priory; the scheme of *c*1300 in the Ante-Reliquary Chapel, Norwich Cathedral; the painting of *c*1340 in the recess of 'Tomb of William of Perth' at Rochester; the mid-14th-century wall painting fragments from St. Stephen's Chapel, Westminster; the late 14th-century *Apocalypse* cycle in the chapter house of Westminster Abbey; the early 15th-century paintings at Farleigh Hungerford Castle; the *Last Judgement* of *c*1435–60 at Holy Trinity, Coventry; and in the late 15th-century scheme in the chapel of Eton College.

Comparison with its use in other media and in continental wall painting

Wall painting

The red lake in the Holy Sepulchre Chapel, Winchester Cathedral, is the earliest identification of this pigment in English wall painting, although at Müstair (Switzerland, Plate 162) a red lake pigment of purple hue – possibly madder – was employed in the Carolingian scheme of *c*800 for glazes in the flesh tones and for Christ's mandorla in the *Ascension*. Mairinger and Schreiner suggest that madder was common at this period, and note that cultivation of madder (*warentia*) is recommended in one of Charlemagne's decrees, the *Capitulare de villis*.[82]

In the Gothic period, a red lake pigment has been detected in the Saint Maurille cycle of *c*1270–80 in the choir of Angers Cathedral. For a red robe in the banquet scene, a layer of red lake combined with lead white was applied over a lead white ground and subsequently glazed with a layer of red lake, while for a grey-blue vestment, red lake was combined with a copper blue pigment and overlaid with copper blue and lead white (Subes-Picot 1992: 89; Demailly *et al.* 1998: 11 and pls. 5 and 6).

Analysis of the red organic colorant employed for the paintings of *c*1332–49 on the inner face of the choir screens at Cologne Cathedral indicated the presence of brazilwood (*Caesalpinia sappan* L.) (Jägers and Schulze-Senger 1989: 187–8; Maul 1992). Given the supreme quality of these paintings – among the most sumptuous in northern Europe in the 14th century – the use of brazil lake seems incongruous since it was considered a poor quality red lake, perhaps because of susceptibility to fading. It is probably for this reason that brazil was not generally employed for high-quality panel or wall painting, although it was frequently used in manuscript illumination where it would be less exposed.[83] It was also employed where longevity was unimportant; for example 45s was spent on purchasing brazil with other colours for painting tents in England in 1350–52 (Staniland 1986: 238). However, despite their impermanence, brazilwood lakes are a beautiful red if adsorbed on a hydrated alumina substrate, and a brilliant pink on calcium carbonate or calcium sulphate substrates. Thus, they may have been selected on the basis of their colour for the Cologne scheme.

82. TLC analysis suggested the presence of madder, although the results were not considered completely convincing; Mairinger and Schreiner 1986: 196. For reference to the *Capitulare de villis*, in which kermes lake (*vermiculo*) is also mentioned, see Schweppe and Winter 1997: 109, and Schweppe and Roosen-Runge 1986: 259.

83. Brazil is frequently mentioned in early sources relating to manuscript illumination; Kirby and White 1996: 56.

Brazil lake was extensively applied over gold leaf in the Cologne paintings.[84] This would have enhanced the colour's translucency, but would also have increased susceptibility to fading due to intense internal reflection from the metallic leaf.[85] In other areas the red lake was employed over opaque paint layers to enrich the underlying colour. Although the varied methods of application may account for the different hues of red lake apparent across the surface of the painting, it is also possible – as indicated by the documentary evidence for the contemporary scheme in St. Stephen's Chapel, Westminster Palace – that several different dyestuffs were employed; applied either singly or in combination to produce particular coloristic effects as found in various panel paintings (Kirby and White 1996: 67). Alternatively, as determined by recent analysis of the wall paintings in the chapter house of Westminster Abbey, the brazilwood may be one of several dyestuffs present, probably due to manufacture of the red lake pigment from a dyebath of cloth clippings.[86]

The Westminster and Cologne schemes are the only instances where brazilwood lake has been identified in medieval wall painting, although documentary evidence indicates that the dye was purchased for decorating the Chapel Royal at Stirling (Scotland) in 1501.[87] Documentary sources also record the purchase of brazil among other pigments for painting in northern France in the 14th century: 'Brésil, teinture rouge obtenue à l'aide d'un bois étranger' (Richard 1887: 323). In 1339, the same accounts record the purchase of both brazil and alum.[88]

Cochineal lake appears to have been identified in only one scheme of wall painting: that in the Landkartengalerie, Salzburg, dating from the late 16th century.[89]

Polychrome sculpture

Red lake pigments have been frequently identified in English Gothic polychrome sculpture, as on the Shrine of St. Alban of *c*1302–08 (Howard 1993a); on early 14th-century roof bosses of Exeter Cathedral;[90] on the effigy of Sir Oliver de Ingham of *c*1344 at Ingham, Norfolk (Howard 1993d); in the 14th- and 15th-century polychromy on the west front of Exeter Cathedral, where tiny traces of the pigment remain on draperies and mouldings (Sinclair 1991: 122, 1992: 10); on the approximately contemporary Great Screen at Winchester Cathedral, where azurite and red lake were combined to produce a deep purple colour for the Virgin's cloak (Brodrick and Darrah 1989: 615–16); and on the late 15th-century Fitzalan effigies at Arundel (Sussex) where a red lake glaze was applied over gilded raised wax appliqué (Brodrick and Darrah 1986: 72, 91).

Instrumental analysis of the lake pigment in a sample from the appliqué decoration on the camail of Sir Oliver de Ingham confirmed the presence of madder lake, applied in a thick translucent layer over gold leaf (Plate 174).[91] Analysis of the binding medium in

84. It has also been suggested that for areas of the background lead 'leaf' was glazed with a red lake pigment; Jägers and Schulze-Senger 1989.
85. Cennini 1960: 88 describes a method for rendering drapery by painting the pattern of the cloth in vermilion over silver leaf and then glazing these layers with lake in an oil medium.
86. J. Kirby, pers. comm.
87. '8 oz bresell at 6d per oz'; Apted and Hannabus 1978: 133.
88. 'Pour brezil et pour alun viid'; Richard 1887: 341.
89. H. Leitner, pers. comm.
90. See analysis by J. Darrah and A. Roy in Hulbert 1991.
91. My thanks are due to Dr Arie Wallert for identifying the madder lake using DT-MS. For the analysis of this tomb, see Howard 1993d.

another sample from the tomb where red lake was applied as a glaze over vermilion suggested the presence of oil in the red lake layer, while GC-MS analysis of the red lead layer below the vermilion confirmed the presence of glue as the binding medium (Plates 172 and 173).[92]

Red lakes have also been identified in polychromy elsewhere in northern Europe: on a 12th-century Virgin and Child from Auvergne (Kargère 2002: 508–9); on the south portal of Lausanne Cathedral (*c*1216–20) (Furlano *et al.* 1981: Brodrick 1993: 20), on the '*Portail de la Mère Dieu*' (*c*1220–60) of Amiens Cathedral (Weeks 1998: 102), and on the 13th-century tomb of Count Palatine Henry II at Maria Laach (Germany) where madder lake was identified (Oellermann 1993: 237).

Panels

Red organic pigments have been identified in only two 13th-century English panel paintings: the ceiling paterae of *c*1263 from the Painted Chamber of Westminster Palace; and the Westminster Retable of *c*1270. On the retable, a red lake on a hydrated alumina substrate was applied as a pure layer over an opaque layer of red lake combined with lead white.[93]

Red lakes were employed extensively on the Thornham Parva Retable of *c*1335 where, for example, a mixture of red lake combined with azurite was employed for St. John's cloak. Further investigation suggested that two different types of lake pigment were employed on this panel, since for St. Catherine's drapery there are two layers of red lake glaze; while the upper layer is thicker, it contains smaller lake particles (2–5μm) compared with particles of 20μm in the lower layer. Of particular interest is that the dyestuff from one of these red lake pigments was identified as Polish cochineal beetle lake (*Porphyrophora polonica* L.).[94]

Madder lakes have been identified in several panel paintings in the National Gallery, London; for example, in *The Exhumation of Saint Hubert* of *c*1440 by Rogier van der Weyden and workshop; Dieric Bouts' *The Virgin and Child with Saints Peter and Paul* dated to the 1460s; the workshop of Dieric Bouts' *Christ Crowned with Thorns c*1475–1500 (Kirby and White 1996: 72). This red organic colorant has also been detected on the late 14th-century Estouteville triptych (Villers *et al.* 1997: 674), the Herlin altarpiece of *c*1466 (Broekman-Bolstijn *et al.* 1970: 370), and the Zittau altarpiece wing of *c*1510–20 (Heydenreich 1994: 34).

Manuscripts

Red lake pigments have been identified in manuscripts from the British Library dating from the 8th century onwards such as the Canterbury Psalter (Vesp. A. I.) of *c*700, the early 11th-century Canterbury Psalter (Arundel 155), and the Winchester Psalter (Nero C.IV) which is some twenty years earlier than the Holy Sepulchre Chapel paintings. Laurie describes the lake as rather dull in colour (Laurie 1913: 328–9 and table 1).

92. I am grateful to Dr M. P. Colombini for undertaking the GC-MS analysis.

93. Dr A. Roy, pers. comm.

94. Analysis of the Polish cochineal lake was undertaken by J. Kirby of the National Gallery using HPLC. I am grateful to Dr S. Bucklow of the Hamilton Kerr Institute for discussing the results of his examination of the retable with me; see also Goldsmith and Bucklow 1998: 43.

Discussion

Red lakes were rarely employed in Romanesque wall painting in England; moreover, in one of the only two instances where they have been detected, the pigment was used as a component of an opaque paint layer. By contrast, red lakes were frequently employed in high-quality Gothic wall painting, and the fact that such expensive pigments were often applied over large areas – as for drapery on the vault and on the western arch of the Ante-Reliquary Chapel, Norwich – underlines the importance that was attached to these schemes. Remarkably, these costly photosensitive pigments have also been detected in the external polychromy of the west front of Exeter Cathedral, on the south portal of Lausanne Cathedral, and on the '*Portail de la Mère Dieu*' of Amiens Cathedral.[95]

Unfortunately, little analysis of the constituent dyestuffs has been undertaken for English wall painting, but it may be significant that in two of the three cases where such investigations have been undertaken – in the scheme of *c*1300 in the Ante-Reliquary Chapel, Norwich, and in the feretory paintings of *c*1302–08 at St. Albans Cathedral – madder lake has been detected. Since madder lakes were considerably cheaper than lakes from scale insects, and since madder is less photosensitive than the other red lakes then available, it is perhaps unsurprising that it was selected for these monumental schemes of painting.[96]

A strong orange fluorescence in UV light may indicate the presence of madder. The observation of this phenomenon has suggested the presence of this lake pigment in a few instances, as in the early 15th-century scheme at Farleigh Hungerford Castle, and in the slightly later paintings at Holy Trinity, Coventry.[97] Fluorescence in UV light cannot however be regarded as conclusive evidence for the presence of a madder lake since, when partially faded, kermes and lac lakes may also fluoresce.[98]

However, it is significant that kermes lake has now been identified in the late 14th-century wall paintings in the chapter house of Westminster Abbey. This is the first time that this organic colorant has been identified in English medieval wall painting, or indeed in any medieval wall painting in northern Europe. In one of the samples the kermes appears to have been combined with a trace of brazil lake, and as discussed above this is likely to result from the method of preparation of the lake pigment from a dyebath of cloth clippings.

More analytical data exist for panel painting than for wall painting. Kirby and White have shown that madder lake was generally employed for northern European panel painting between 1400 and 1600, with kermes and lakes produced from other scale insects occurring only at the end of the period. In Italy, however, lakes prepared from scale insects were generally used throughout this period. Owing to their poor stability,

95. For Exeter, see analysis by Roy in Sinclair 1991: 122, and Sinclair 1992: 10; for Lausanne, see Furlano *et al.* 1981; and for Amiens, see Weeks 1998: 102.
96. Whether artists were aware of the relative photosensitivity of various lakes at this period is a matter of conjecture. Certainly the use of brazilwood was often forbidden, although whether this was due to its rate of fading is not stated.
97. At Farleigh Hungerford SEM/EDX analysis indicated that the red lake has an aluminium-rich substrate, while at Coventry a clay-rich chalk substrate was employed; see Howard *et al.* 1998 and Howard 1996b.
98. In these instances the fluorescence is much less strong although still a pale orange colour. In addition, madder lakes may not always fluoresce; Kirby *et al.* 1997: 68.

brazilwood lakes were seldom employed in panel painting.[99] It is therefore all the more striking that brazilwood lake alone has been identified on the superb choir screen paintings in Cologne Cathedral.

Kermes has been identified in other forms of polychromy in England in the medieval period, and so it not surprising that it has now been detected in wall painting. The lake pigment has previously been identified, combined with bone black and a proteinaceous binder to produce a purple colour, on a pilgrim's staff from Worcester (Lubin *et al.* 1990: 14). A similar pigment combination was also detected on the bust of Henry VII of *c*1510 now in the Victoria and Albert Museum, although here the crimson lake was combined with carbon black (Lindley and Galvin 1995: 173–7). In addition, documentary evidence shows that cloth dyed with kermes was available in England in the medieval period, since the Royal Wardrobe accounts for 1212 record that '1400 ells of scarlate and 243 ells of blanket dyed in grain [kermes]' were purchased by officers of King John (Hluvko 1991: 50). However, whether the kermes – or cloth clippings dyed with kermes – was imported and the dyeing actually undertaken in England, or whether the dyed cloth itself was imported is unclear.

Given the clear connection between dyeing of cloth and the manufacture of lake pigments in the medieval period, and since the connection between the techniques of sculptural polychromy and wall painting is now so firmly established, it is perhaps only a matter of time, and application of the appropriate technology and technical expertise, before kermes and other lakes are identified in other medieval wall paintings in England and northern Europe.

Pigment alterations

Exposure to light is the principal reason for the fading of lake pigments, although other factors may substantially affect the rate of colour loss (Saunders and Kirby 1994: 93). Although UV light is the most damaging component of the spectrum, the greater total exposure to light, the greater the colour loss.[100] Of the red lake pigments employed in the medieval period, Saunders and Kirby (1994: 86–93) have shown that for lakes prepared on the same substrate, brazilwood lakes are the most fugitive, turning from red to brown on fading (and then sometimes to completely colourless), while madder is the least photosensitive. The substrate itself has also been shown to affect the rate of fading, with calcium salts – particularly calcium carbonate substrates – producing the most rapid fading. Particle size has also been implicated in the rate of fading of lake pigments, with larger particles proving most resistant. The quantity of organic binding medium may also affect the rate of fading, thus the leaner the paint film the less protection from fading. In addition, the type of binding medium is also a factor, with oil protecting lake pigments more effec-

99. It has been suggested that the supply of brazil lakes was disrupted from the mid-15th century (by the fall of Constantinople in 1453), thus stopping supply of brazilwood from the East (*Caesalpinia sappan* L.) until New World species (*Caesalpinia brasiliensis* L.; *C. Crista* L. and *C. Echinata* Lam) were imported from Pernambuco in northern South America in 1500: Kirby and White 1996: 64. Analysis of textile samples indicates that approximately 80% were dyed with brazilwood before 1450, and that scale insect dyestuffs were more frequently employed until New World redwood was discovered; Hofenk-de Graaff and Roelofs 1972: 25. Wallert (1991b: 81) suggests that a similar trend may have occurred in manuscript illumination.

100. To protect lake pigments from fading, Saunders and Kirby (1994: 93) recommend limiting the overall light exposure and, where possible, eliminating UV light.

tively than egg tempera (Saunders and Kirby 1994: 92–3). The extenders combined with lake pigments and the grounds over which they are applied also play a part, and lake pigments mixed with white, applied over a pale-coloured or white ground and unprotected by a glaze, are particularly vulnerable to internal reflection (light reflected within the paint layer from the surfaces of the white particles) and therefore fade more rapidly.[101] Lakes prepared directly from the natural dyestuffs have also been found to fade less rapidly than those prepared from cloth clippings (Saunders and Kirby 1994: 86–93).

The fading of the organic red lake – and in particular the variation in colour loss depending on the stratigraphy of the paint – is a striking feature of the 'St. William' at St. Albans (Plate 169). Thus the palest zones of the drapery, where the lake was applied directly over the lime ground, have faded substantially more than those areas of the pigment combined with charcoal black and applied over yellow underpaint (Plates 170 and 171). The red lake combined with azurite in the purple chasuble of the companion saint at St. Albans does not appear to have faded significantly, probably partly because of its combination with the blue mineral pigment, and partly because its location is better protected from daylight (Howard 1993a: 38).

The condition of the lake pigment identified in the *Last Judgement* of *c*1435–60 at Holy Trinity, Coventry, varies considerably across the painting. Where it was applied over gold leaf, internal reflection has exacerbated fading of the pigment, but where applied over dark underpaint and combined with vermilion and red lead the organic colorant has survived well (Howard 1996b: 6). In the slightly earlier scheme in St. Leonard's Chapel, Farleigh Hungerford Castle – where red lake in a white matrix was applied over a layer of red lake combined with vermilion over orpiment – fading of the pigment is evident in areas where it was combined with lead white and calcium carbonate in the uppermost portion of the stratigraphy (Plate 189) (Howard *et al.* 1998: 62). Some degree of fading is also evident in the red lake employed in the important scheme of *c*1479–87 in the chapel of Eton College. Here the pigment was employed to pick out important details such as the precious stones in the Virgin's crown (Plummer 1986: 36).

Red lead

The pigment red lead (Pb_3O_4) is the synthetic equivalent of the mineral minium. There is little evidence to suggest that the natural mineral was ever used as a pigment, but from the earliest times the synthetic pigment was well known and often employed. Pliny remarks that second-quality minium (red lead) was frequently used to adulterate – and thereby extend – the more expensive first-quality minium (vermilion).[102] He describes the manufacture of red lead by smelting lead-rich stones found in silver and lead mines (Pliny 1984: 91).

101. The acceleration of the rate of fading over time, due to an increased proportion of white, has been demonstrated experimentally for a lake mixed with white, and for a lake used as a glaze over a white ground (Johnston-Feller and Bailie 1982). It has also been shown that for paint layers of lake applied as a glaze over a white ground or for lake combined with white, an increasing proportion of white accelerates the rate of fading over time (Johnston-Feller 1986).
102. In antiquity *cinnabaris* and *minium* were used synonymously, while red lead was properly called *minium secundarium;* Pliny 1984: 84–5.

Cost

It is clear from the 14th-century accounts of Ely Cathedral that red lead was considerably cheaper than red lake pigments – indeed at 4d per lb compared with red lake at 16s per lb, it was some forty-eight times cheaper – and also two and a half times less expensive than vermilion at 10d per lb (Chapman 1907: 2: 121). Slightly later in the 14th century, accounts for St. Stephen's Chapel, Westminster Palace, also provide important information regarding the relative cost of red lead. Here, red lead at 4d per lb is only slightly more expensive than red earth which cost 2d for the same quantity, while 1s 8d a lb was paid for *vermelon* (Salzman 1952: 168).

Optical, chemical and physical properties

In plane polarised transmitted light the tiny irregular aggregates of red lead appear a bright orange-red colour with high refractive indices, α and $\gamma = 2.42$. In transmitted light with crossed polars, red lead is birefringent, with anomalous blue-green interference colours and distinct extinction. The pigment is soluble in dilute acids, and on heating strongly it turns to litharge (PbO).[103]

An important property of red lead is its drying action on oils. This action is enhanced by the presence of litharge (PbO), the principal impurity found within red lead, which is not added intentionally but results from the manufacturing process (Fitzhugh 1986: 122). Medieval accounts provide interesting information which may relate to the addition of lead pigments as driers to oil. Although their function is not specified, these materials are frequently purchased together as at Norwich in 1316–17 when red lead, white lead, oil and whitewash were bought for painting images in the choir (Park and Howard 1996: 401–2).

Although readily available and relatively inexpensive, red lead is unstable in alkaline conditions and therefore unsuitable for use in *buon fresco* techniques, or over fresh lime wash or combined with lime white. It was generally considered unsuitable for use on walls, and indeed artists' treatises warn against the use of the pigment in wall painting; for example, Cennini states that 'if you use it [red lead] on the wall it soon turns black on exposure to air and loses its colour'. He also lists it among colours which may not be used in fresco painting, indicating that it was considered to be reactive under alkaline conditions as well as in contact with air (Cennini 1960: 25, 50). In addition, its presumed incompatibility with pigments such as orpiment (arsenic sulphide) was well documented, as in the *Coloribus et Artibus Romanorum* by Eraclius (1967: 252–5). This probably related to fear of the formation of black lead sulphide caused by contact of a sulphide and a lead-based pigment. However, this is difficult to explain in the light of the apparent compatibility of red lead with vermilion (mercury sulphide), with which it was frequently combined. However, treatises such as the 14th-century *Liber de Coloribus* – relating to manuscript illumination – indicate that minium should not be mixed with any other colour *except* vermilion and white (Thompson 1926: 283).

Preparation

Red lead was generally prepared by heating lead white in air. While the most efficient transformation occurs between 425 and 430 °C (Fitzhugh 1986: 122), it seems likely

103. For a detailed discussion of red lead and its characteristics, see Fitzhugh 1986: 122–5.

that in the medieval period heat was simply applied until sufficient alteration had occurred. The *Mappae Clavicula* (1974: 27) describes its preparation by treating metallic lead with strong vinegar to form lead white, and then heating the white material until the red pigment is formed. The fundamental principles of this process are repeated by other authors, sometimes with minor alterations, such as Theophilus' suggestion that hot vinegar or urine be used for the initial production of lead white from strips of metallic lead. He also indicates that the lead white should be ground with a stone without water before being heated in small pots over hot coals and stirred periodically (Theophilus 1979: 41–2).

Use in English medieval wall painting

Red lead varies in colour from pale to brilliant orange-red depending on its preparation. The only other bright orange pigment available to the medieval artist was realgar (orange arsenic sulphide), but this seems not to have been employed in wall painting and rarely elsewhere. Nor do mixed oranges generally appear to have been used in the Romanesque period, although occasionally layers of red and yellow were applied to produce a slightly orange-red appearance, as in St. Gabriel's Chapel in Canterbury Cathedral; here, yellow iron oxide was applied over red iron oxide for a border on one of the apsidal arches. An intense orange colour, produced by mixing red and yellow earths, was also employed in the wall painting fragments of *c*1190 recently excavated from Glasgow Cathedral.[104] Theophilus (1979: 41–2) suggests a combination of orpiment and cinnabar to produce orange drapery, although not in wall painting. Although orange was a secondary colour in the Romanesque artist's palette, which relied so heavily on the play of bold blues and reds, it appears that red lead was employed wherever a brilliant orange colour was required.

Despite warnings against its use, red lead was widely employed in wall painting.[105] Although much more common in the Gothic period, it has nevertheless been identified in four English Romanesque wall paintings. Three of these are cathedral schemes, where the pigment was employed to provide a brilliant orange hue for minor details, and in conjunction with other pigments – particularly vermilion – to produce intense shades of red in larger areas. Red lead has also been identified in the scheme of *c*1110 in the parish church of Witley (Surrey), where it has darkened and now appears a brownish-black colour (Howard 1992a). In addition, red lead was employed to provide a bright orange colour in the paintings of *c*1170 at Cormac's Chapel, Cashel (Ireland).[106]

Use of red lead increases sharply in the Gothic period, and Hluvko details various paintings in which the pigment has been identified including the mid-13th-century *Wheel of Fortune* in the choir of Rochester Cathedral (Hluvko 1991: 26; see also Darrah 1992) and the *Foundation Narrative* at Horsham St. Faith Priory (Hluvko 1991: 26; see also Babington and James 1988); the early 14th-century scheme at Bradwell-juxta-Coggeshall (Essex) (Hluvko 1991: 26);[107] the mid-14th-century painting in the tomb recess in the north-east transept of Rochester Cathedral (Hluvko 1991: 26; see also Rickerby 1988:

104. This pigment combination was used for fine linear details on one fragment; see Howard 1998a and Park 1998.
105. Fitzhugh (1986: 132–5) lists numerous occurrences in wall painting.
106. A sample taken from a capital to the south of the altar was identified by XRD; McGrath 1987: 58.
107. See also analysis by H. Howard in Cather and Park 1998.

10–11); the mid-14th-century wall paintings from St. Stephen's Chapel, Westminster Palace (Hluvko 1991: 26; see also van Geersdaele and Goldsworthy 1978); both the late 14th- and early 16th-century schemes in the chapter house of Westminster Abbey; the paintings of *c*1395 in the Byward Tower, Tower of London; the late 15th-century Legend of St. Eustace at Canterbury;[108] the paintings of *c*1479–87 in the chapel of Eton College (Plummer 1986: 36); and the early 16th-century scheme of domestic wall painting at 7 Silver Street, Ely (Hluvko 1991: 27).

More recent analysis has established the pigment's use in the mid-13th-century painting in the Ante-Reliquary Chapel of Norwich Cathedral (Howard 1995c: 4); the early 14th-century paintings in the Feretory of St. Albans Cathedral (Howard 1993a: 37); and the important cycle of domestic wall paintings of *c*1330 at Longthorpe Tower, Cambridgeshire.[109] Other occurrences in major schemes include the 13th-century schemes in the Guardian Angels' Chapel at Winchester Cathedral,[110] and the vault paintings (dated to the 1370s) of the Chapel of Our Lady Undercroft, Canterbury Cathedral (Howard 1993b). Red lead has also been identified in the 14th-century paintings at Harbledown (Kent),[111] and in the 15th-century paintings at Farleigh Hungerford Castle (Somerset) (Howard *et al.* 1998: 59–60), at Willingham (Cambridgeshire) (Curteis 1995a: 9), Kingston (Cambridgeshire) (Curteis 1995b: 5), and Berry Pomeroy Castle (Devon).[112] The pigment has also been identified in a 13th-century scallop shell palette excavated from Clarendon Palace (Hughes and Lewis 1988: 258–60), which seems likely to have been used for one of the many wall painting schemes undertaken there for Henry III.

It is clear from the instances listed above, and from other examples in Table 3 (see Appendix 3), that the use of red lead becomes increasingly common in the Gothic period, and that this cheap and intense orange-red pigment was then widely employed in wall paintings of all types. It is likely that the increased use of organic binding media during the Gothic period facilitated the incorporation of this alkaline-sensitive pigment in the palette. Moreover, the siccative properties of red lead would have provided a compelling rationale for its use with an oil medium.[113]

Application

Red lead in priming layers

The pigment was frequently employed in combination with lead white and oil as a priming layer. This is evident particularly where an ashlar substrate is present, as in the mid-14th-century wall paintings from St. Stephen's Chapel, Westminster, where the priming was applied in two layers: first red lead in oil, over which a layer of red lead combined with lead white and oil was applied (Plate 159) (van Geersdaele and Goldsworthy 1978: 9, 11–12). In the early 15th-century scheme at Farleigh Hungerford Castle a complex stratigraphy exists in the preparatory layers. For the figures of St.

108. For the chapter house, Byward Tower and St. Eustace, see Hluvko 1991: 27.
109. Unpublished research by H. Howard.
110. Unpublished research by H. Howard.
111. For analysis by Howard, see Organ 1992.
112. Analysis by H. Howard in Gill *et al.* 1996.
113. The good physical and mechanical properties of red lead in linseed oil may be attributed to the chemical formation of lead soaps: the resultant films are both tough and elastic; Fitzhugh 1986: 114.

George and the kneeling knight, a layer of red earth combined with calcium carbonate was applied in large rectangular zones, doubtless to provide particular emphasis within the overall decorative scheme. Over this dark red layer, and throughout the remainder of the scheme, another red ground – comprising red lead combined with oil and a little red iron oxide – was applied to provide a unifying substructure (Plate 183) (Howard *et al.* 1998: 59–60). Similarly, a unifying red lead preparatory layer was employed for the late 15th-century Legend of St. Eustace at Canterbury, although in this instance red lead was combined with lead white and a little chalk (Hluvko 1991: 28). Such layers would have provided a smooth and compact surface for painting, and additionally acted as a sealant, ensuring that the translucency of the polychromy would not be compromised by leaching of the paint medium into the porous support.

Red as a preparatory layer or mordant beneath gold or silver leaf
For St. Eustace's spur in the 15th-century painting at Canterbury, a thick layer of red lead with a little vermilion, chalk and lead white was applied in preparation for an oil mordant and a layer of gold leaf (Hluvko 1991: 28). This bright orange layer would certainly have influenced the coloristic properties of the overlying gold leaf.

Over a white ground or directly over the plaster support
Traces of red lead applied over a lime white ground survive in a shield depicted on one of the wall painting fragments of *c*1190, recently excavated from the crypt of Glasgow Cathedral (Plates 148 and 149) (Howard 1998a; Park 1998). For the brilliant orange on the collar of the best-surviving angel in the scheme of *c*1250 in the Ante-Reliquary Chapel, Norwich Cathedral, red lead was applied in a layer 55µm thick over a ground of lead white combined with calcium carbonate on the stone support (Plates 150 and 151) (Howard 1995c: 4 and App. 1). Similarly, in the 13th-century scheme in the Guardian Angels' Chapel, Winchester Cathedral, a sample from one of the angel's wings showed that red lead was applied over a lime white ground.[114]

Over coloured underpaint
In the 12th-century scheme in the Holy Sepulchre Chapel, Winchester Cathedral, the pigment was used for its distinctive coloristic qualities, as for the brilliant orange details on the pale green of the Magdalene's drapery in the *Deposition* (Hluvko 1991: 27). Red lead, combined with lead white and calcium carbonate, has also been identified in the intense orange motifs applied in *impasto* over a thin layer of carbon black in the geometric border dividing the two tiers on the east wall (Howard 1995a: 101).

Red lead glazed with red lake
The high refractive index and consequent good covering power of red lead, particularly when applied in an oil medium, produced dense and opaque paint films. The intense orange-red layer would also have been highly reflective, and thus produced rich saturated reds when glazed with a red lake pigment. This type of stratigraphy was employed for the mid-14th-century painting in the recess of the 'Tomb of William of Perth' in Rochester Cathedral. For the background of the vinescroll and popinjays a layer of brilliant red lead was applied over a ground of red ochre and subsequently glazed with a red lake pigment (Rickerby 1988: 10–11).

114. Unpublished analysis by H. Howard.

Mixed with other pigments

A combination of red lead and vermilion was extensively employed in the paintings of *c*1130 in St. Gabriel's Chapel, Canterbury Cathedral (Plate 140). Thus, for the red drapery in the *Birth of the Baptist* scene, for the stars on the soffit of the altar arch, and for other details, vermilion was combined with red lead and applied over a white, lead-containing underpaint. In other areas, such as the haloes on the altar arch, the same pigment mixture was applied directly over the lime plaster substrate (Cather and Howard 1994: 145; Howard 1992b: App.).

In the scheme of *c*1175 in the Holy Sepulchre Chapel at Winchester, red lead combined with vermilion was employed for the brilliant red details on Nicodemus' right leg in the *Deposition* (Hluvko 1991: 27).

Numerous instances of the combination or layering of red lead with vermilion in the Gothic period – for both coloristic and economic reasons – are cited in the vermilion chapter. Perhaps the most notable is in the late 14th-century decoration on the vault of the Chapel of Our Lady Undercroft in Canterbury Cathedral, where a thick layer of red lead was applied over the entire surface in preparation for a final layer of vermilion (Howard 1993b: 32). Examples of red lead combined with vermilion include the wings and crimson robe of St. Michael in the scheme of *c*1395 in the Byward Tower, Tower of London (Hluvko 1991: 28); the pink background of the rose decoration beneath the Bestiary scenes of *c*1510 at dado level in the chapter house of Westminster Abbey (Hluvko 1991: 28),[115] and the rosettes forming part of the 14th-century decoration on the nave south wall of Willingham (Curteis 1995a: 9).

Red lead combined with red earth and applied over charcoal black underpaint was employed for details of an angel's wing on the vault of St. Gabriel's Chapel, Canterbury Cathedral (Cather and Howard 1994: 145; Howard 1992b: App.). Red lead was combined with both red earth and lead white and applied over a red earth ground in the domestic wall paintings of *c*1530 at 7 Silver Street, Ely (Hluvko 1991: 28).

The intense orange colour of the maniple of the unidentified archbishop in the early 14th-century paintings in the feretory of St. Albans Cathedral was achieved by combining red lead with a small amount of charcoal black, and applying it in a thick layer over a lead white ground (Plates 152 and 153) (Howard 1993a: 38). Similarly, red lead and carbon black were combined for the angels' wings in the late 15th-century scheme at Kingston, where the red lead has now darkened producing a dense black appearance (Curteis 1995b: 5). Red lead was also applied over a layer of charcoal black combined with red iron oxide for the outline of St. Gabriel's sleeve in the Annunciation scene in the14th-century scheme at Harbledown.[116]

Binding media for red lead

In addition to the warnings against employing red lead in wall painting given by Cennini and others, some authors comment on the type of binding media suitable for use with this pigment. Theophilus suggests the use of cherry or plum tree gum with all pigments except red lead, lead white and vermilion, which he suggests should be ground with egg white (Theophilus 1979: 32–3). By contrast, the treatise of Peter of St. Omer

115. Howard 2000 and forthcoming.
116. For analysis by Howard, see Organ 1992.

recommends that, 'in laying it [red lead] upon walls it is ground with gum-water but never egg' (Merrifield 1967: 1: 140).[117]

Although gum has not as yet been identified as a binding medium for red lead in English medieval wall painting, in several instances analysis has determined the presence of a proteinaceous component in the medium. Examination of the paintings of *c*1175 in the Holy Sepulchre Chapel at Winchester Cathedral has confirmed that, although the carbonation of lime is the principal binding mechanism, a proteinaceous component is also sometimes present; for example, in a paint layer consisting of red lead, lead white and calcium carbonate in a sample taken from the *impasto* decoration of the geometric border dividing the tiers on the east wall of the chapel (Plates 145–147). A proteinaceous component has been identified in a layer of calcium carbonate used for the raised decoration on the borders of the Magdalene's drapery (Howard 1995a: 96), while a ninhydrin test also suggested the presence of glue distemper or casein in a sample of vermilion combined with red lead taken from a highlight on Nicodemus' right leg (White 1998). Similarly, analysis by the National Gallery, London, indicated that the red lead preparation of the paintings of *c*1352–63 from St. Stephen's Chapel, Westminster Palace, includes a proportion of egg tempera in the binder.[118]

Given the strong drying action which red lead exerts on oils, often it would probably have been combined with an oil medium. FTIR analysis of the secondary red lead preparation for the early 15th-century scheme at Farleigh Hungerford Castle indicated that it was indeed bound with an oil medium. This scheme once covered most of the east end of the chapel thus using extremely large quantities of both red lead and oil. The secondary red ground, which provided a unifying substructure throughout, would have acted as a sealant, ensuring that the translucency of the polychromy was not compromised by the leaching of the paint medium into the plaster support (Howard *et al.* 1998: 59). Similarly, for the late 15th-century *Legend of St. Eustace* at Canterbury Cathedral, red lead was combined with lead white and a little chalk in an oil medium for the ground layer (Hluvko 1991: 28). Walnut oil was identified as the binding medium in the 14th-century wall painting in the recess of the 'Tomb of William of Perth' at Rochester Cathedral, although whether red lead here was combined with this medium is not specified (Rickerby 1988).

At present there is little analytical evidence for the binding media used with red lead in medieval wall painting elsewhere in northern Europe. However, both protein and a small quantity of oil have been identified as the binding media for layers of red lead in the Carolingian paintings at Müstair (Mairinger and Schreiner 1986: 196), and organic binding media were also identified in some of the upper paint layers of the Romanesque scheme incorporating red lead in the same church (Emmenegger 1986: 198). In the early 12th-century paintings at Berzé-le-Ville, Burgundy, it has been suggested that red lead was combined with an organic material, perhaps glue. Such an application *a secco* may explain why motifs applied in red lead survive only in protected areas of the paintings.

117. The *Liber de Coloribus* suggests that red lead should be combined with egg white for use in manuscript illumination; Thompson 1926: 283.

118. For analysis by Plesters and Mills, see van Geersdaele and Goldsworthy 1978: 11–12. Recent experiments by Higgitt *et al.* (2002) indicate that the reaction between red lead and an oil medium to produce lead-fatty acid salts and lead carbonate, results in a low azelate peak in GC-MS analysis. Low azelate levels are characteristic of egg tempera, but in this case, a reinterpretation of the GC-MS results may suggest an oil-based medium.

In polychrome sculpture, linseed oil has been identified as the binding medium of the pigments – including red lead – in the original polychromy of the frieze from the west front of Lincoln Cathedral (*c*1140),[119] and an oil medium has likewise been detected in the polychromy – including both red and white lead – of the 11th-century Hørning plank from a Danish stave church.[120] On the de Ingham tomb (*c*1344) at Ingham, Norfolk, instrumental analysis of a red lead layer below vermilion confirmed the presence of glue as the binding medium.[121] By contrast, on the slightly later 14th-century de Bois tomb at Ingham, one of the remaining pink 'jewels' on Sir Roger's helm was found to consist of a complex mixture of red lead, red earth, lead white, calcite and azurite in an oil medium applied in thick *impasto* (350µm) over a layer of silver leaf. Here, the red lead and lead white pigments would have functioned as driers (Howard 1993d: pl. 8). GC-MS analysis confirmed the presence of linseed oil, which had been partially heat-bodied to promote drying (White 1998: 1–2).

Comparison with its use in other media and in continental wall painting

In the Romanesque scheme at Müstair, the intrinsic coloristic qualities of red lead and vermilion are cleverly juxtaposed, with cheek patches painted in the more orange-red lead while lips are picked out in vermilion. Red lead was also used for larger areas such as borders, although the palette also includes red earth and vermilion (Emmenegger 1986: 198). Red lead, used in conjunction with red ochre and vermilion, was also used for a detail of orange-red drapery on a fragment of wall painting from the abbey of Cluny.[122] Gothic examples include the important mid-13th-century wall paintings in the choir of Angers Cathedral, where red lead was combined with lead white to form the ground in some of the bays (Demailly *et al.* 1998: 11 and pl. 6). The pigment has also been identified in medieval wall paintings in the Briefkapelle of St. Mary's Church, Lübeck, Germany, and the scheme of *c*1330–50 at Waltensburg in Germany (Fitzhugh 1986: 132).

Polychrome sculpture

A combination of red lead and vermilion has been identified on the late 12th-century figures from St. Mary's Abbey, York, where the red of St. John's pallium consists of these pigments with the addition of lead white (Brodrick 1993: 23). Red lead and lead white priming layers have also been identified in other examples of polychrome sculpture, as on the Great Screen of *c*1475–90 at Winchester Cathedral, where the proportion of red and white lead in the pigment mixture varies considerably across the surface (Brodrick and Darrah 1989: 616–17), and on the west front of Exeter Cathedral where red lead was incorporated in priming layers of the 14th- and 15th-century polychromy (Sinclair 1991: 122, 1992: 10). Red lead was combined with lead white in a preparation layer for the 14th-century polychromy in the Lady Chapel of Ely Cathedral (Hluvko 1991: 28).

Where red lead was incorporated in an oil mordant, it would not only have accelerated drying but would also have had an impact on colour, giving a warmer cast to the

119. It has been suggested that traces of red lead found in the hair of several figures indicate that these areas were originally gilded: Larson 1992: 188, 1993: 34.
120. Mads. Christensen and J. Glastrup, poster at ICOM 1996.
121. I am grateful to Dr P. Colombini for undertaking this analysis.
122. Unfortunately it has not been possible to date this fragment securely to the Romanesque period; Rollier-Hanselmann 1997: 63.

metallic leaf. Red lead was employed in this way on both the 14th-century tombs at Ingham, where it was combined with lead white and applied beneath silver leaf (Howard 1993d: App. 5 and 6). For the Martyrdom Doorway (*c*1236) at Canterbury Cathedral a similar pigment combination was employed beneath gold leaf (Hluvko 1991: 28), while red lead alone was applied beneath gold leaf on the Great Screen at Winchester Cathedral (Brodrick and Darrah 1989: 615–17).

Mixtures of red lead and vermilion have also been identified on Romanesque polychrome sculpture including the 12th-century lion capital of Vossestrand, Norway (Plahter 1990). On Gothic sculpture, red lead has been detected in the polychromy of *c*1220–60 on the '*Portail de la Mère Dieu*' at Amiens Cathedral, where it was combined with lead white for the flesh tints (Weeks 1998: 102); on the approximately contemporary south portal of Lausanne Cathedral (Furlano *et al.* 1981); on the 13th-century tomb of Pfalzgraf Heinrich II at Maria Laach, Germany (Oellermann 1993: 264); and on the portal sculptures dating from the 14th century of the Holy Cross Minster in Schwäbisch Gmünd (Hauff 1988: 715).

Panel painting

Red lead occurs in the original paint layers of *c*1220 on the nave ceiling of Peterborough Cathedral. Here the pigment was combined with both calcium carbonate and charcoal black and used as an underpaint beneath vermilion for the draperies of St. Paul and St. Peter (Howard 1997b: tables II and III). The pigment was applied over a calcium sulphate preparation in the reredos of *c*1180 now at Adisham, Kent (Howard 1993c: App.).[123] Later in the 13th century red lead was employed on the ceiling paterae from the Painted Chamber of Westminster Palace (Liversidge *et al.* 1995; Lynn 1997), while an extremely unusual combination of red lead with ultramarine and lead white was employed for the orange lining of the draperies of one of the figures in the Westminster Retable of *c*1270.[124]

Red lead has also been identified in the Thornham Parva Retable of *c*1335, where it was combined with lead white to model St. Peter's undergarment. Over this modelled underpaint, red lake and lead white were applied for the final layers (Goldsmith and Bucklow 1998: 43). The pigment also occurs in the early 14th-century screen from Kingston Lacy, Dorset, where the bright orange chasuble of the bishop on the right side was painted in red lead shaded with crimson lake (Tracy 1997: 28).

Red lead has frequently been identified on Norwegian altar frontals (Plahter 1995: 114–15). However, it was used as an underpainting for vermilion in only a small number of cases, as on the frontals from Skaun, Hauge (before *c*1300), and Dale I (after *c*1300) (Plahter 1995: 122–3).

Manuscript illumination

Red lead was commonly used in manuscript illumination; indeed the word 'miniature' derives from the Latin *miniare*, to write with minium (Fitzhugh 1986: 113). Laurie has identified the pigment in English manuscripts from the 8th to the 12th centuries (1913: 325, table 1), and it was also commonly employed in the Gothic period.

123. The reredos has recently been redated to *c*1180 by C. Tracy on the basis of dendrochronological analysis.
124. A. Roy, pers. comm.

Discussion

Although red lead has been identified in Romanesque wall painting and sculptural polychromy in England and elsewhere in northern Europe, it was much more commonly employed in the Gothic period. Since the cost of this synthetic pigment was low, the paucity of its use in the 12th century is likely to have been for technical rather than for economic reasons. Certainly, the incompatibility of red lead with alkaline conditions would have made the pigment a much less attractive option when lime binders were so commonly employed. It is significant that where analysis of the original binding medium has been undertaken, proteinaceous and oily components to red lead layers have been identified in Romanesque wall painting.

The increased use of oil in the Gothic period would have made red lead an extremely popular choice, since the pigment had the additional benefit of acting as a drier. In late Gothic wall painting, red lead in an oil medium was frequently employed as a preparatory layer on both stone and plaster supports. Such layers would have been selected not only for their coloristic properties but also to seal the porous support and prevent leaching of the binding medium from paint layers applied subsequently.

With so few brilliant orange-red pigments available in the medieval period, red lead was an important component of the palette, and was applied both as a pure colour and in pigment mixtures. The characteristic intense orange colour of a pure layer of red lead can be seen in the scheme of *c*1250 in the Ante-Reliquary Chapel of Norwich Cathedral where it was employed for the collar of the angel on the south side of the eastern arch, and in the maniple of the early 14th-century archbishop on the south side of the feretory of St. Albans Cathedral (Plates 150–153). Red lead was commonly combined with vermilion, or used as an underpaint beneath it, presumably to extend this more expensive red pigment. Red lead glazed with vermilion was employed for the late 14th-century vault painting of the Chapel of Our Lady Undercroft in Canterbury Cathedral, while a combination of the two red pigments was employed throughout the medieval period, as in the 12th-century scheme in St. Gabriel's Chapel at Canterbury and in the late 14th-century scheme in the Byward Tower, Tower of London.

Pigment alterations

The darkening of red lead

The darkening of red lead has long been recognised and has been frequently detected in wall painting.[125] Although the darkening of red lead and lead white has generally been presumed to be through a process of oxidation (Matteini and Moles 1981; Petushkova and Lyalikova 1986: Giovannoni *et al.* 1990), the mechanism is still not completely understood.[126] The alteration of lead white has been the subject of several detailed studies which suggest that alteration is due to high alkalinity, humidity and the presence of an oxidising agent (Matteini and Moles 1981; Petushkova and Lyalikova 1986: Giovannoni *et al.* 1990; Koller *et al.* 1990). An investigation of the alteration of lead pigments in

125. Pliny (1984: 91) noted that the shine of second quality cinnabar – i.e. minium (red lead) – is affected by 'rust' in wall painting. Fitzhugh (1986: 109–39) discusses the alteration of red lead and provides numerous examples of its darkening.

126. However, if lead sulphide were formed as the alteration product, no oxidation would have taken place.

English medieval wall painting showed that in all cases the product of alteration was lead dioxide (plattnerite) (Welford 1991).

High humidity and light have also been implicated in the darkening of red lead (Gettens and Stout 1966; Weber 1923). The effect of binding media on the darkening of the pigment has also been the subject of several studies which indicate that glue may protect red lead in environments rich in oxygen and carbon dioxide, although not in nitrogen (Kenjo and Toishi 1964), while oil and synthetic binding media appear to protect the pigment from darkening (Fitzhugh 1986: 118). A recent study (Aze *et al.* 2002) reported that the blackening of red lead (bound by the carbonation of lime) occurred after five years in a sheltered external wall painting. The authors determined that darkening was due to the formation of plattnerite, and that alteration occurred on the surface of the paint layer and was particularly prevalent in any hollows in the surface. They suggest that the oxidation of red lead to plattnerite is influenced by condensation.

Although darkening does not appear to have occurred in the two identified examples in English Romanesque cathedral schemes, the pigment has clearly darkened in the paintings of *c*1110 in the parish church of Witley (Surrey) (Plates 137 and 138). Lead dioxide (plattnerite) is the product formed in the darkening of both red lead and lead white, and the pigments are indistinguishable once the conversion has taken place. At Witley, however, it appears that both red lead and lead white were employed. White must have been the colour originally intended as a background to the red of the cross nimbus of Christ in the *Harrowing of Hell* on the west wall, but this area now appears red on the surface of the plaster and dark brown-black in the interstices, and a sample indicated the presence of dark red-brown pigment particles in the interstices of the coarse plaster. EDX analysis confirmed the presence of lead in these particles. In the scene of the *Feast of Dives* on the south wall, the collar of Dives now appears dark brown-black, and again analysis indicated the presence of lead dioxide (Plates 138 and 139). Examination of this area suggested that the original intention was for vertical bands of red and white to create the effect of a high, ornate collar. The red appears to be the same colour as the overall background, executed in a red earth pigment, so it seems likely that vertical strips of lead white, painted over red, have altered to produce the dark effect. A sample taken from the neck of a dog in the scene to the left of the Feast suggests that the now-darkened lead pigment was combined with red earth. Originally, either red lead may have been combined with a red earth pigment to produce a strong red colour, or lead white mixed with the red earth would have produced a pale red.

Examples of the alteration of red lead in the Gothic period include the scheme of *c*1220 in the Holy Sepulchre Chapel, Winchester Cathedral, where a darkened lead pigment has been detected: for example, in the border of a prophet roundel on the south segment of the east bay of the vault (Howard 1995a: 102); the 13th-century scheme on the vault of the Guardian Angels' Chapel in the same cathedral;[127] the 14th-century *Christ in Majesty* at Farringdon, Hampshire, where darkening has occurred;[128] the scheme of *c*1330 at Bradwell-juxta-Coggeshall (Essex), where darkened red lead is apparent;[129] and in the 15th-century scheme at Berry Pomeroy Castle, Devon, where both darkening and lightening appear to have occurred.[130]

127. Unpublished analysis by H. Howard.
128. Analysis by J. Darrah.
129. Analysis by H. Howard in Cather and Park 1998.
130. Analysis by H. Howard in Gill *et al.* 1996.

Other instances of the phenomenon have been recorded elsewhere: for example XRD analysis of darkened red lead in a sample from a wall painting fragment from Kizil in Chinese Turkestan (dated from the 5th to the 8th centuries, and now in the Freer Gallery of Art, Washington) (Gettens 1938b; Fitzhugh 1986: 115–17), and in Swiss wall paintings of the 13th to 17th centuries[131] confirmed the presence of lead dioxide. Darkened red lead has also been detected in the Carolingian scheme at Müstair (Switzerland), where the pigment only remained unaltered where covered by other opaque pigments. Here, XRD analysis also confirmed the alteration product of both the red lead and lead white pigments to be dark lead dioxide, and media analysis suggested the presence of both protein and oil (Mairinger and Schreiner 1986: 195–6). Other significant occurrences of darkened red lead in continental wall painting include the 12th-century crypt paintings in S. M. del Piano in Ausonia, Italy, where the red lead is likely to be a *secco* addition over the *buon fresco* painting beneath (Bensi 1990); and the Legend of the True Cross in S. Franceso, Arezzo, by Piero della Francesca, where red lead was applied in an oily medium over a ground of lime white and carbon black (Matteini *et al.* 1989: 246–7).

It has also been suggested that the alteration of red lead may have produced some of the red-brown tonalities in the early 12th-century scheme at Berzé-le-Ville, as on the drapery of a figure in the scene of the Martyrdom of St. Vincent.[132] Darkened red lead has also been identified in the Romanesque paintings in the crypt of Auxerre Cathedral (France), in this case over red iron oxide underpaint. The red lead here may represent a repainting of the original Romanesque scheme, although given the number of other instances of the use of the pigment at this period it seems more likely that it is part of the original palette (Cadet 1995: 35).

In manuscript illumination darkened red lead has been reported in a number of instances, including the 8th-century Lindisfarne Gospels (Fitzhugh 1986: 115). The type of red brown appearance described above – particularly if found in areas of otherwise convincing flesh tones – may suggest a partial alteration of red lead to form dark lead dioxide. However, the presence of dark flesh tones does not always signal the alteration of a lead pigment. Indeed dark red-brown flesh tones are sometimes used in a curious convention – signifying hellfire, damnation and love – for angels in scenes of holy women at the tomb of Christ. Examples of this iconography have been found in manuscripts from the 11th to the 16th centuries.[133] Dark red flesh tones have also been employed for angels in the 14th-century scheme of wall paintings in the eastern bay of the chapter house of Westminster Abbey, and here analysis indicates that vermilion was applied over a layer of red earth.[134]

Lightening of red lead

The lightening of red lead to form a white product has, until recently, been less well studied than the darkening phenomenon, although it has been observed by a number of authors who implicate a variety of factors: carbon dioxide, humidity and sunlight (Eikhoff 1973 :51), light (Gettens and Stout 1966: 153), carbon dioxide and air (*Paint*

131. Kühn (1976) pers. comm. to Fitzhugh.
132. It has also been suggested that the lead pigment was added during a phase of repainting in the Gothic period; Rollier-Hanselmann 1997: 63.
133. Examples of this convention, which is based on a verse from St. Matthew 28,3, *his face was like lightning*, are provided by Petzold 1992: 155.
134. Howard 2000 and forthcoming.

Technology Manual 1966: 194). Other experiments to investigate the lightening phenomenon indicated that RH in excess of 70% accelerates the process (Zuixiong 1990: 44–6). Zuixiong (1989: 254) also observed that the white product developed generally across the surface of the paint layer and suggested that this may be related to a plentiful supply of carbon dioxide.[135] Lightening of the pigment was noted by Augusti in his study of Pompeiian pigments, 'but I have not found it [red lead oxide] free among the Pompeiian pigments. The reason is that it was all transformed into [white] lead carbonate and altered' (Augusti 1965: 68). Gettens and Stout comment that red lead in oil, when exposed out of doors may eventually turn pink or white because of the formation of lead carbonate (1966: 153), while Wehlte attributes the lightening of red lead to the formation of white lead sulphate (1975: 104). Duval attributes the lightening of red lead in an oil-bound ground layer to a reaction with the drying oil (Duval 1992).

As already noted, artists' treatises warn against the use of red lead in wall painting. Cennini's remark that 'questo color è solo buono a lavorare in tavola, ché se l'adoperi in muro, come vede l'aria diventa subito nero e perde suo color' (Cennini 1991: 48), 'this colour [red lead] is good only for working on panels, for if you use it on the wall it soon turns black, on exposure to the air and loses its colour' (Cennini 1960: 25) has traditionally been interpreted to imply that loss of colour is by darkening alone. An alternative interpretation might be that loss of colour is by darkening or lightening

The lightening of red lead was first identified in English wall painting in the early 12th-century scheme in the apse of St. Gabriel's Chapel, Canterbury Cathedral. Here, red lead was so extensively employed that its lightening has fundamentally affected the appearance of the paintings (Plate 140). The alteration is most evident where the pigment was combined with vermilion, or used as a layer beneath it. However, red lead was also applied in a number of other ways in this scheme, in some areas directly over the lime plaster and in others over a white, lead-containing underpaint. In one instance, where it was combined with red iron oxide and applied over charcoal black for an angel's wing in the western segment of the vault, no alteration was detected (Cather and Howard 1994: 145, 149).

One of the most striking aspects of the current condition of these paintings is that over many of the red areas there is now a grey veil which fluoresces orange in UV light. This surface phenomenon must be a relatively recent occurrence, since in 1880 Scott Robertson described now affected areas as 'rosy' (Scott Robertson 1880: 74, 67). However the alteration is already clearly visible in colour photographs published by Demus (1970: pls. 234–5).

Examination of cross-sections and analysis by SEM/EDX indicated that this grey veil consists of a white lead-containing material. Particles of the veil and of the underlying red paint layer were analysed separately by XRD, confirming the presence of cerussite and red lead respectively.[136] It is particularly significant that cerussite (lead carbonate, $PbCO_3$)

135. This contrasts with the dark oxidation product of lead pigments (plattnerite) which forms in patches across the surface (Welford 1991: 17). Zuixiong (1990: 69) concluded that the white alteration product of red lead is lead white. He also suggested that this may be converted in the presence of oxalic acid from microbes to lead oxalate which may in turn convert to lead dioxide (PbO_2) by some unknown reaction.

136. I am grateful to J. Webb of the Ancient Monuments Laboratory for carrying out initial examination by XRD, and to Dr Ashok Roy of the National Gallery for confirming the presence of cerussite using a Derbye-Scherrer camera with copper radiation and nickel filter. Experimental values agree with Joint Committee on Powder Diffraction Standards (JCPDS) file 5/417 (1953) which gives

rather than lead white – basic lead carbonate ($2PbCO_3 \cdot Pb[OH]_2$) – was found.[137] Although the manufacture of lead white was common in the medieval period, cerussite was not synthesised, confirming that it represents an alteration product of the red lead pigment layers rather than a distinct paint layer or glaze which was originally applied to the surface.[138]

Cross-sections taken from the affected areas each show slightly different aspects of the alteration phenomenon. A sample (Plates 141 and 142) from an area of red drapery shows cerussite – converted from the red lead layer beneath – bursting through the upper natural vermilion layer to form the grey crust on the surface. Another sample (Plate 144) shows the conversion of a red lead highlight applied over the natural vermilion of the drapery; here the crust almost appears to have 'grown' on the surface. Red lead has a specific gravity of 8.73 while that of cerussite is lower at 6.55. This reduction in specific gravity and associated increase in volume in the transformation process results in the formation of surface crusts such as that visible in Plate 144. In Plate 143, a larger proportion of the transformation product is apparent towards the upper part of the paint layer, a ratio typical of an alteration product and indicative of a process brought about by an external episode at, or treatment applied to, the surface. In addition, the vermilion particles in the red lead matrix are present in the same proportion in the as yet unaffected paint layer and in the 'crust', indicating an alteration of the upper portion of the paint layer. All the samples taken from the affected areas show a very open, porous structure – likely to encourage chemical reactions to take place – and a higher ratio of cerussite towards the surface.

Similar grey surface crusts associated with a red lead paint layer, as well as orange fluorescence in UV light, have been detected in other medieval wall paintings including the late Romanesque scheme in the Holy Sepulchre Chapel, Winchester Cathedral, and the early 14th-century paintings at Bradwell-juxta-Coggeshall, Essex, although the alteration product has not been identified.[139]

Examination of paint cross-sections from the late 14th-century *Apocalypse* cycle of wall paintings in the chapter house of Westminster Abbey clearly indicated that the red lead has undergone partial transformation to form a white, lead-containing alteration product (Plates 154–158). Of particular interest is the volume change associated with the transformation, and the disruption this produces within the paint layers (Plate 155). At high magnification, a yellow zone – probably due to incomplete alteration – is visible at the centre of some of the white particles (Plate 155) while the outer disrupted portion

strong d(Å) lines at 3.593, 3.498, 3.074, 2.522, 2.487, 2.081, 1.933 and 1.859. Raman spectroscopy, undertaken by Dr R. Withnall at University College, London, was also employed to confirm the presence of cerussite in a number of samples (confirmatory peaks at 62.1, 78.7, 109.9, 554.2, 844.2, 849.6 and 1055.8 wavenumbers).

137. Cerussite crystals are of varied habit: tabular, granular crystalline aggregates, and fibrous; see Klein and Hurlbut 1985: 338–9.

138. For the manufacture of lead white in the Middle Ages, see Gettens *et al.* 1993a: 68.

139. Unpublished analysis by H. Howard. In addition, tantalising information from conservators suggests that the alteration phenomenon may be more widespread than has been supposed. For example, Ballantyne (1992: 39) reported 'a pale grey which seems at one time to have been a pinkish hue' in relation to the 14th-century Passion cycle at White Notley, Essex. Cerussite has also been identified on a 13th-century painted fragment excavated from St. Augustine's Abbey, Canterbury (Sherlock and Woods 1988: 124). Although the presence of cerussite has not been linked to the alteration of red lead in this case, further investigation of these samples may prove fruitful in this respect.

shows a pattern of radial cracks, and perpendicular to these, wavy striations (Plate 158). Elemental analysis of the altered particles by SEM/EDX confirmed the presence of lead and oxygen, while examination of the backscatter image suggested that the central portion is of higher atomic number (or more dense, or both) than the disrupted outer area (Plates 157 and 158).[140] Alteration products of red lead with similar characteristics have been noted in a number of panel paintings, and in these cases instrumental analysis has confirmed that the transformation is due to a reaction between the red pigment and an oil medium to form lead-fatty acid salts and lead carbonate.[141] GC-MS analysis of the red lead in a sample taken from the tiny angels above the Old Testament figures in the *Apocalypse* wall painting cycle confirmed the presence of linseed oil and perhaps a trace of pine resin. In addition, FTIR analysis suggested that the white material is lead carboxylate, indicating soap formation due to a reaction between the original oil medium and the lead pigment as in the panel painting samples investigated.[142]

In addition to the lightening of red lead due to a reaction with an oil medium, it is clear that red lead in any binding medium can lighten in conditions of high humidity. An important new study examining test samples of red lead bound in gum, linseed oil, casein and acrylic media subjected to accelerated ageing under varying degrees of RH and varied atmospheric gases (O_2 and CO_2), has provided significant new findings regarding the lightening of this red pigment (Saunders *et al.* 2002). The authors confirm that: the primary alteration product is basic lead(II) carbonate (lead white); the higher the RH the greater the degree of lightening; samples kept at 75% RH or above altered in the absence of light; lightening is more rapid in aqueous binding media than in linseed oil; and leanly bound films alter more rapidly indicating that penetration of light and moisture into the paint layer is an important factor in the lightening process. Interestingly, at low RH (less than 51%) initial darkening of the red lead was detected in samples containing residual PbO from the pigment preparation. When the RH was subsequently raised above 75% lightening occurred. The reaction of red lead with oil reported by Higgitt *et al.* (2002) did not occur in the samples investigated, and Saunders *et al.* (2002) suggest that reaction to form lead-fatty acid salts does not take place where red lead is more rapidly converted to lead carbonate in the presence of high humidity and light.

It is interesting that the white product of alteration in the Romanesque wall paintings in St. Gabriel's Chapel has been identified as cerussite ($PbCO_3$) rather than lead white ($2PbCO_3 \cdot Pb[OH]_2$), the primary product of alteration detected in the laboratory tests by Saunders *et al.* (2002). However, since these wall paintings have been subjected to a wide range of environmental conditions and invasive treatments, including lime-watering, other reaction sequences and alteration products might be expected. Clearly further work in this area is a priority.

Perhaps the most important conclusion regarding wall paintings that can be drawn from the various studies is that since high humidity is implicated in the lightening of red lead, consolidation by lime-watering – still a standard treatment in English wall painting

140. Howard 2000 and forthcoming.

141. This reaction has also been identified in paint films containing lead-tin yellow in an oil medium, and has significant implications where medium analysis has been undertaken by GC-MS since the reaction between red lead or lead-tin yellow and an oil medium produces a low azelate peak and fatty acid ratios indicative of an egg tempera medium; see Higgitt *et al.* 2002.

142. This analysis was kindly undertaken by Marika Spring and Catherine Higgitt of the National Gallery's Scientific Department.

conservation involving multiple saturations of a wall painting – is likely to be highly inadvisable where lead pigments are present. It is also clear that, wherever possible, stable low RH (below 51%) should be maintained for wall paintings containing lead pigments. Analysis of the formation of cerussite in St. Gabriel's Chapel also indicated that hygroscopic milk proteins such as casein, which are also often used for consolidation, may accelerate the alteration process.[143]

143. FTIR and GC-MS analysis undertaken by Raymond White of the lead carbonate (cerussite) 'crust' on two samples from the paintings of *c*1130 in St. Gabriel's Chapel confirmed the presence of protein. Moreover, moristate levels indicated milk lipids from casein or skimmed milk, which may have been applied during conservation treatment. The milk protein may contribute to the alteration process, and its hydrophilic character would certainly promote high RH which is a necessary condition for its occurrence.

❦ SIX

Red and yellow earth pigments

Red and yellow earth pigments are tinted by the oxides and oxyhydroxides of iron – the red colour is usually due to the presence of haematite (Fe_3O_4), while yellow is due to goethite – but other iron-rich minerals such as maghaemite, lepidocrocite and akaganeite may also be present. Iron-rich minerals such as haematite and goethite have a high pigmenting strength and although they provide the colour of the earth pigment, they may be present in relatively low concentrations combined with a number of accessory minerals such as quartz, kaolin and other clays. The colour of the oxides and oxyhydroxides of iron varies enormously, from dark red purples – usually containing a high concentration of a very pure form of haematite – to rich red earths which may contain oxides and oxyhydroxides of iron and varying quantities of aluminium silicates. The yellow and yellow-brown forms contain goethite ($\alpha FeO^{-}OH$) together with various quantities of clays and quartz, while umbers also contain manganese dioxide.

Although these earth pigments were frequently obtained from local sources, particularly fine quality earth pigments were also traded; for instance, accounts for 1522 record the purchase of one stone of *Flanderis ocker* (Apted and Hannabus 1978: 143). A number of good sources in England are known, such as those in the Lake District, Forest of Dean and in the Mendips. A particularly important source is between Bristol and the Mendips, near the village of Winford.[1] In antiquity the finest red earth, purchased from the Pontine city of Sinope, was sold under seal to maintain its quality and was known as *sealed Sinope*.[2] In the Middle Ages the term *sinope* was used to describe a variety of red earths, and in Italy gave its name to the stage of drawing undertaken on the penultimate layer of plaster, usually in red earth or *sinoper*. In medieval accounts, earth pigments are also referred to as *rodel* and *hetre*, and the more brown-coloured earths simply as *brun*. Care should be taken not to confuse references to red earth (*sinope* or *sinoper*) with *synople*, *sinopis* and *cynople* which denote red lake pigments (Thompson 1956: 98, 124).

1. I am grateful to Brian Young of the British Geological Survey for this information.
2. Vitruvius VII, vii in Granger 1962: 113, and Thompson 1956: 98.

Cost

The cost of earth pigments would clearly vary enormously depending on whether they were imported or a local source was exploited. Generally, however, they appear to have been extremely cheap; for example, in 1275 *ochre* was purchased for Westminster at 1d per lb, while the mid-14th-century accounts for St. Stephen's Chapel, Westminster, record *oker* at 2d per lb (Salzman 1952: 168). Both red and yellow *oker* are included in an early 16th-century list of colours purchased for Christ's College, Cambridge, and that vast quantities of these pigments were employed is confirmed by an account of 1376 which itemises half a hundred weight of ochre bought in London for 3s (subsequently employed for ochreing a chapel at Havering). Similarly, in Winchester in 1532, 30lb of *yolowe oker* was purchased for colouring vaults although their precise location is not specified (Salzman 1952: 159).

The Westminster Palace accounts for 1275 also list *brun* at 3½d, which may have been a burnt ochre or a naturally occurring yellow-brown earth pigment (Salzman 1952: 168). The purchase of brun d'Auxerre at 3d per lb is also mentioned in the accounts of the Countess of Artois at the beginning of the 14th century, and it is likely that, where available, good sources of particularly attractive shades of ochre would have been sought out and exploited commercially (Richard 1887: 323, 333).

Optical, chemical and physical properties

Red ochre and haematite appear bright red in transmitted plane polarised light and have refractive indices of $\alpha = 2.95$, $\gamma = 2.74$. With crossed polars they display red-red birefringence and distinct extinction. Red ochre consists of very finely ground particles of haematite often with a significant quantity of other mineral impurities. The particle size for red ochre is typically 1–3µm, while that of haematite is approximately 1–40µm. Yellow ochre ranges in colour from bright to dull yellow in plane polarised transmitted light, with refractive indices of $\alpha = 2.26$, $\beta = 2.39$, $\gamma = 2.40$ (McCrone *et al.* 1979: table XXXI). The particles may appear as fine needle-like aggregates. With crossed polars the pigment is birefringent and displays distinct extinction.

Red and yellow earth pigments are generally very stable and durable, and are suitable for use both in alkaline conditions and combined with a wide range of organic binding media. However, they will dissolve in hot concentrated acids and on prolonged heating the pigments will oxidise. Yellow ochres are susceptible to more profound changes at lower temperatures, and a colour change from yellow to red due to loss of water from the hydrous iron oxides typically occurs in the temperature range 200–400 °C, resulting in the formation of haematite (Rickerby 1991: 42).

Some red earths which contain particularly high concentrations of aluminium silicates are capable of taking a high polish. They were therefore well suited for use as a bole beneath layers of burnished gold leaf. In the medieval period the red clay (*bole*) from Armenia was well known in this respect, but presumably other fine red clays would also have been employed (Gettens and Stout 1966: 98).

Preparation

Earth pigments were typically prepared by washing, levigating and grinding the source material to produce as pure as possible a form of the iron-rich minerals haematite or geothite. In addition, Theophilus describes the production of a red earth pigment, *prasinus*, by burning yellow ochres (1979: 17, 29–30).[3]

Use in English medieval wall painting

The stability, low cost and ready availability of these pigments in a wide range of hues – from rich red-purple to pale pinks, yellows and browns – made them ideal for use in wall painting and polychrome sculpture. There are innumerable examples of their use in England and elsewhere in Europe throughout the Middle Ages (see Appendix 3, Tables 3 and 4), and consequently only a few of the more interesting examples will be discussed here.

One of the most important uses of these earth pigments was as part of the overall process of setting out a wall painting scheme *in situ*. The most common form of setting out in medieval wall painting is the use of a preliminary drawing in a red or yellow earth pigment on the uppermost plaster layer or ground. However, true *sinopie* have also been detected in a few instances in Romanesque wall painting in England.

Application

Use of red and yellow earth pigments for sinopie

A *sinopia* or sketch, usually in red iron oxide, on the penultimate layer of plaster has clear implications for painting technique and is usually associated with the application of at least the initial layers of paint in *buon fresco*. Cennini describes the use of *sinopie*, suggesting that refinement to a charcoal sketch on the dry *arriccio* or levelling plaster be made in sinoper.[4] Since the *sinopia* was always intended to be covered with a new sketch provided to guide the painter on the uppermost plaster layer, its principal function was to map out the overall design within its architectural framework.

In England, *sinopie* are found only in the Romanesque period, and then extremely infrequently. The most celebrated example was discovered during the detachment of a portion of the 12th-century scheme from the south wall of the Holy Sepulchre Chapel in Winchester Cathedral in the 1960s, when a *sinopia* for three scenes set beneath architectural canopies was revealed (Plate 177).[5] The discovery of such an extensive *sinopia* is so far

3. This reference to a red pigment called *prasinus* is interesting since this term – deriving from the Greek for leek, *prason* – is usually taken to mean green earth (Grissom 1986: 142–3). Recent research by Rickerby (1991: 40–42) shows that although red ochre is produced on heating yellow ochre to temperatures in the range of 300–600 °C, a green hue is produced at 900 °C. In the light of these results, it is interesting to speculate whether Theophilus is indeed referring to a green pigment made from burning yellow ochre.
4. Once the fresh plaster patch or *intonaco* for a day's work has been applied over the *sinopia*, Cennini describes in detail how to set out the design on this uppermost layer, from the use of snapped lines to locate the composition to the precise pigment mixture – black, dark [yellow] ochre, lime white and pale red ochre combined with water and applied with an almost dry, pointed brush – to accomplish the initial drawing. Refinements to this final stage of the drawing are made in terre-verte or verdaccio; Cennini 1960: 42–5. *Sinopie* are particularly common in Italian trecento wall painting, and although not discussed here, numerous examples are given by Borsook 1980 and Arts Council 1969. Mora *et al.* (1984: 139) describe in detail the initial setting out in charcoal which was refined and corrected by the *sinopia* in trecento Italy.
5. The *sinopia* on the east wall of the chapel – traces of which are visible through tiny losses in the *intonaco* – is discussed and illustrated by Park (1983: 39–41 and pl. XVc). Evidence from the

unparalleled in England, although examination of two lacunae in the *intonaco* on the chancel south wall at Hardham (West Sussex) (*c*1100) has revealed red vertical lines suggesting the presence of a *sinopia* on the levelling plaster (Sawdy *et al.* 1998: 34–5). However, the tiny traces currently visible at Hardham make it impossible to be sure whether a detailed *sinopia* or simply a few setting-out lines are present beneath the uppermost plaster layer.

A preparatory sketch in red depicting the Virgin has also been found on the penultimate layer of plaster of the Romanesque scheme on the north wall of the chancel at Henham, Essex.[6] Traces of a yellow earth pigment were found on the penultimate plaster layer on fragments of *c*1090 excavated from Lewes Priory (Sussex), and may also represent the remains of a *sinopia* (unusually executed in yellow) before the thin uppermost plaster layer was applied (Howard and Park 1997).

Examination of late 12th-century painted plaster excavated from the crypt of Glasgow Cathedral revealed tiny traces of a red line at the interface of two plaster layers on one fragment (No. 124), while extensive areas of red pigment are apparent on the lower plaster layer of fragment No. 120 (Plate 178). Samples taken from the red paint on the latter fragment confirmed the presence of both vermilion and red earth. While red earth alone is typically employed for *sinopie*, vermilion combined with red iron oxide was also detected in the *sinopia* at Winchester. Although it is possible that the traces of red paint on the penultimate plaster layer at Glasgow represent a *sinopia*, it is also possible that two distinct decorative schemes are present. It is striking, for example, that the lower plaster layer on fragment No. 120 is similar in both thickness and texture to that employed for the masonry pattern on another fragment, and it may be that a very simple scheme in red was applied shortly before the more complex overall decoration including figurative elements was executed (Howard 1998a; Park 1998).

A preparatory drawing in red below a tinted lime ground was employed to set out the mid-12th-century painting of *St. Paul and the Viper* in Canterbury Cathedral (Gärtner and Organ 1989: 8, 11).[7] Similarly, examination of paint samples from the chasuble and blue fringe of the maniple of the archbishop saint of *c*1302–08 in the feretory of St. Albans Cathedral indicated the presence of a preparatory drawing in red earth on the lime substrate, but in this case below a lead white ground. Although preparatory drawing in red is extremely common in wall painting, it is highly unusual – indeed, perhaps unique – for it to be carried out below a thick opaque lead white ground layer (Howard 1993a: 38).

Use of red and yellow earth pigments for preliminary drawing

Preliminary drawings were typically carried out on the uppermost plaster or ground layer – that is, the same level on which the painting was executed.[8] In the scheme of *c*1110 at Witley (Surrey), the intention of the preliminary drawing in red seems to have been to map

stratigraphy of the plaster, and from the examination of cross-sections, indicates a trace of red pigment (red earth with some vermilion) beneath approximately 4mm of plaster and suggests that the *sinopia* was applied either directly onto the ashlar support or on the levelling plaster.

6. For the Henham painting, see Ballantyne 1989, Cather *et al.* 1990: xiv, and Park 1990b: 239–40 who dates the scheme to the late 11th century. Ballantyne indicates that three small zones where paint survives on the final plaster layer can be seen at the left side of the *sinopia*. She also confirms that the plaster layers for the *sinopia* and for the final paint layers are of identical appearance.
7. Winfield (1977: 11) suggests that the first version of St. Paul's halo was painted in yellow ochre.
8. Preliminary drawing in red or yellow is equally common in medieval wall painting elsewhere in northern Europe, as in the Romanesque wall paintings at Müstair (Emmenegger 1986: 198), the scheme of *c*1260 in Gurk Cathedral, Austria, and the roughly contemporary paintings at St. Georgen

Plate 135 Vermilion has been identified in external polychromy as on the west front of Wells Cathedral where the pigment was applied in an egg tempera medium over a lead white ground. This plate shows details of a roof delineated in vermilion in the Naming of St. John *on the north side of the apse of St. Gabriel's Chapel, Canterbury Cathedral. (© Courtauld Institute, Conservation of Wall Painting Department)*

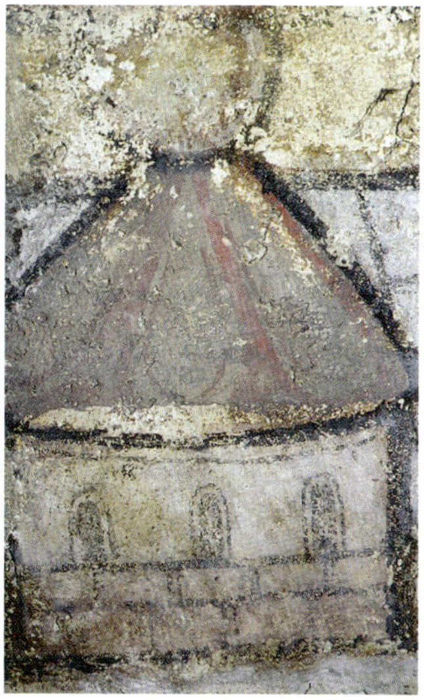

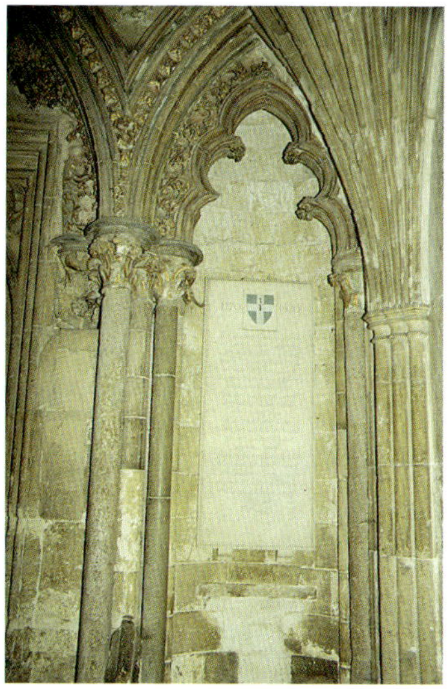

Plate 136 Canterbury Cathedral, cloister: detail showing part of the Martyrdom Doorway *of c1236. Here vermilion has been identified, combined with a little red lead and applied over a lead white ground. (© Courtauld Institute, Conservation of Wall Painting Department)*

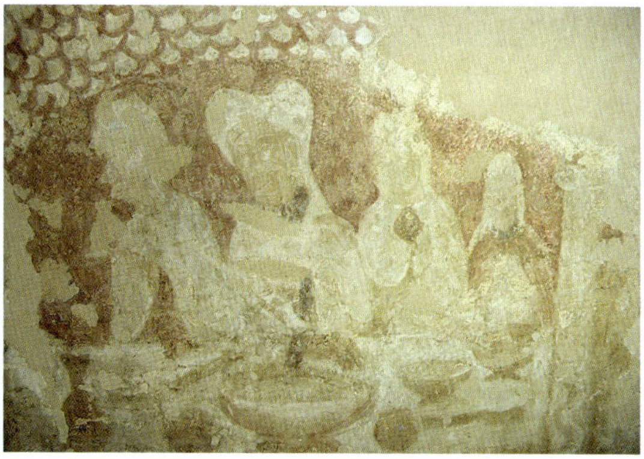

Plate 137 Witley (Surrey): detail of the Feast of Dives *on the south wall of the nave (c1110). The presence of an altered lead pigment in this scheme was signalled by the dark zones which occur in the collar and cuffs of Dives, and in the brooch and collar of two other figures. (© Courtauld Institute, Conservation of Wall Painting Department)*

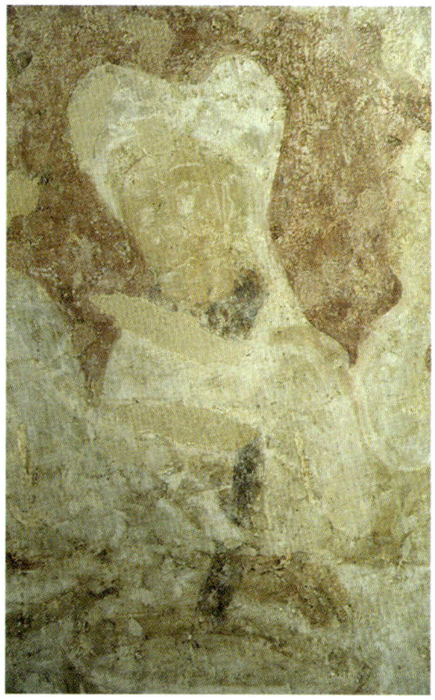

Plate 138 Witley (Surrey). Detail of the figure of Dives showing the alternating red and white ornament on his collar, with grey-black areas characteristic of altered lead. (© Courtauld Institute, Conservation of Wall Painting Department)

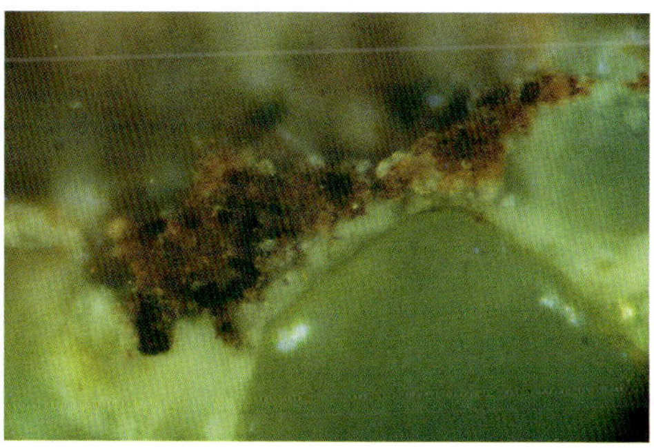

Plate 139 Witley (Surrey). Cross-section of a sample taken from Dives' collar, showing the abraded paint layer surviving in the interstices of the uneven plaster. SEM/EDX analysis confirmed the presence of lead in the paint layer. (Photomicrograph: H. Howard; © Courtauld Institute, Conservation of Wall Painting Department)

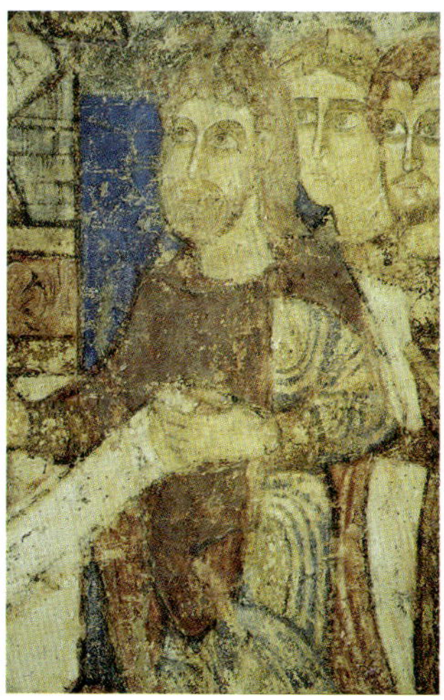

Plate 140 Canterbury Cathedral, St. Gabriel's Chapel: detail of the Birth of the Baptist *on the north side of the apse (c1130). The grey 'crust' over the red drapery occurs across many areas of the paintings and is associated with the alteration of red lead. (© Courtauld Institute, Conservation of Wall Painting Department)*

141

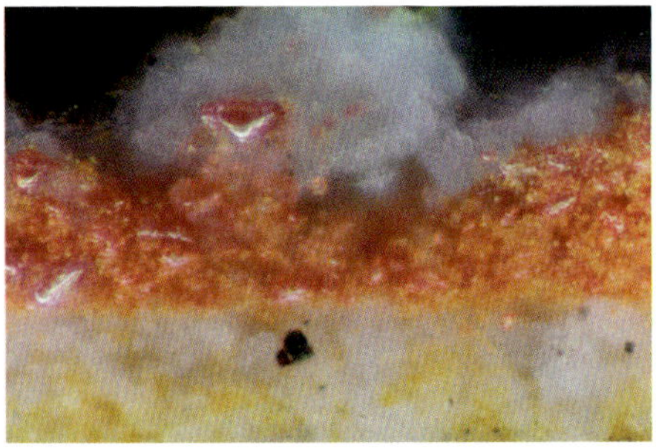

142

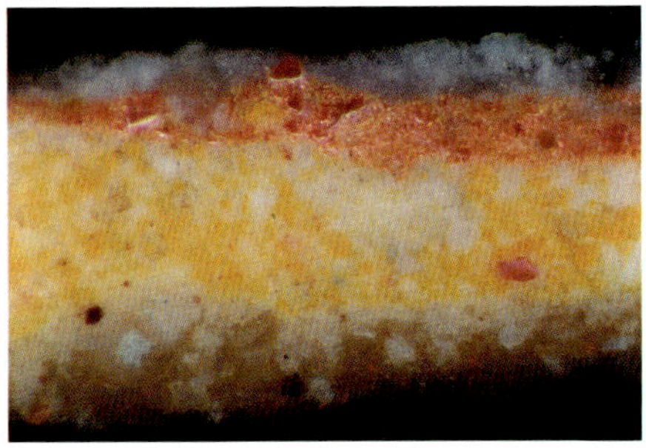

143

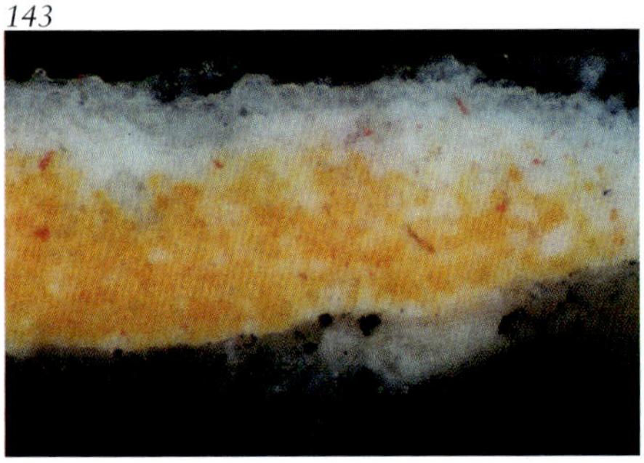

144

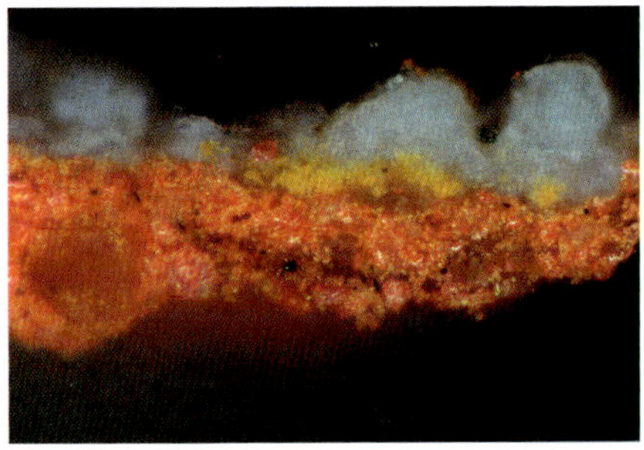

Plates 141–144 *Cross-section of a sample (141) from the red drapery shown in Plate 140, original magnification 1000×. Particles of cerussite – converted from the red lead layer beneath – can be seen bursting through the upper layer of vermilion and forming a grey crust on the surface. Similarly, in a sample from the tree on the north buttress (Plate 142), the white alteration product of red lead has pushed through the vermilion layer and formed a crust on the surface. In another sample from the apse paintings, a layer of cerussite has formed on the surface of the original layer of red lead combined with a little vermilion (Plate 143). In this sample (Plate 144) from a detail in red lead over the vermilion drapery, the crust forms a particularly uneven layer on the surface. (Photomicrographs: H. Howard; © Courtauld Institute, Conservation of Wall Painting Department)*

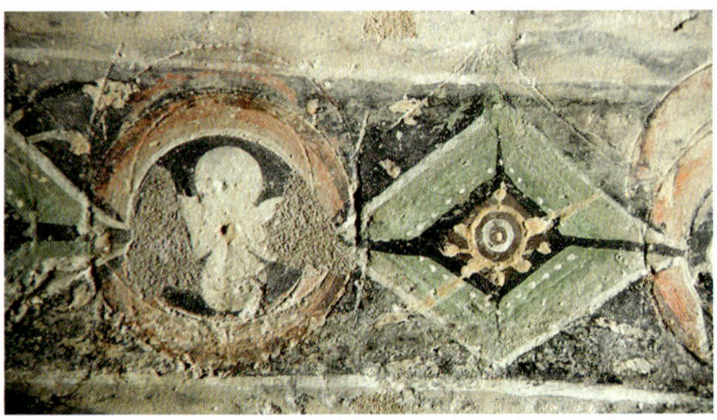

Plate 145 *Winchester Cathedral, Holy Sepulchre Chapel: detail of the paintings of c1175 on the east wall, showing the border dividing the narrative tiers. The geometric elements within the border were initially incised into the wet plaster with the aid of a compass; when the border was moved to a slightly lower position new incisions were made, but a different quality of line is evident since the plaster was clearly no longer fresh. Details added in impasto in red lead are visible at the right. (© Courtauld Institute, Conservation of Wall Painting Department)*

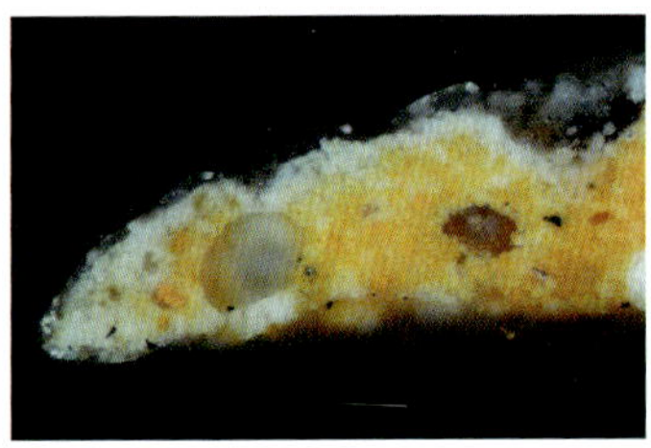

Plate 146 *Cross-section (original magnification 302×) of a sample from the raised detail in red lead shown in Plate 145. The red lead is combined with lead white and calcium carbonate, and analysis of the binding medium suggested the presence of a proteinaceous material. (Photomicrograph: H. Howard; © Courtauld Institute, Conservation of Wall Painting Department)*

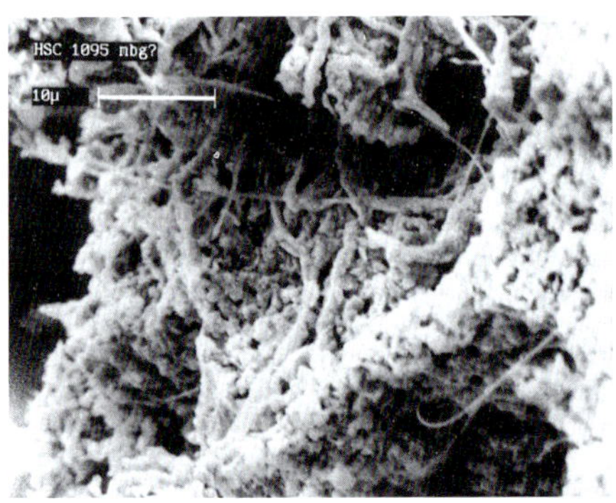

Plate 147 *SEM micrograph of the sample in Plate 146, showing what appear to be hyphae within the paint layer. The presence of microbiological growth may herald an alteration in the lead-based paint layer, since humidity has been implicated in the alteration of lead pigments. (SEM micrograph: H. Howard; © Courtauld Institute, Conservation of Wall Painting Department)*

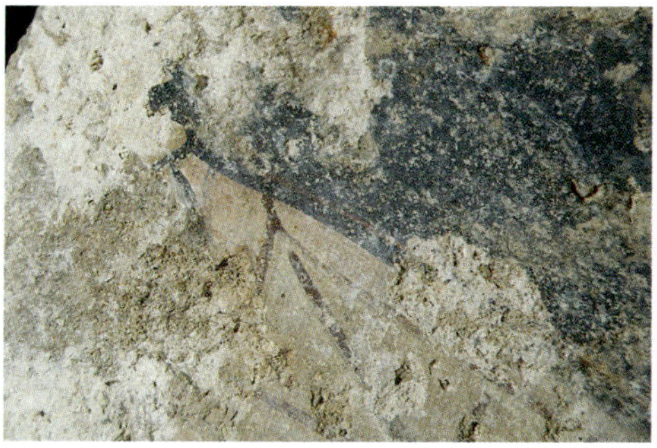

Plate 148 Glasgow Cathedral: detail of a fragment of wall painting excavated from the crypt (c1190). (Photograph: H. Howard; © Courtauld Institute, Conservation of Wall Painting Department)

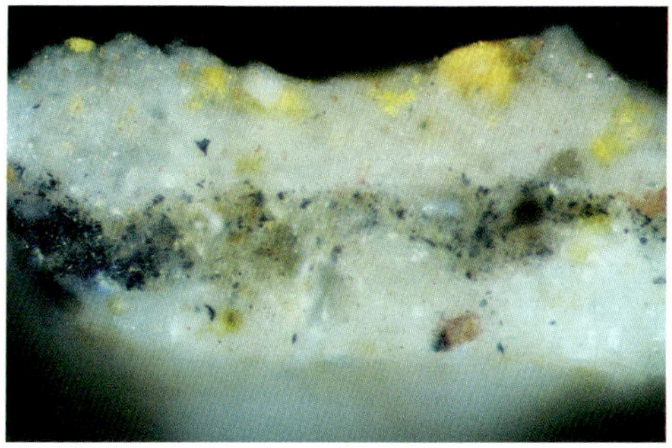

Plate 149 Cross-section of a sample taken from the shield in Plate 148, original magnification 446×. For the orange colour at the tip of the shield, red lead was applied over a local lime white ground on a layer of charcoal black combined with lime white and a little red lead. (Photomicrograph: H. Howard; © Courtauld Institute, Conservation of Wall Painting Department)

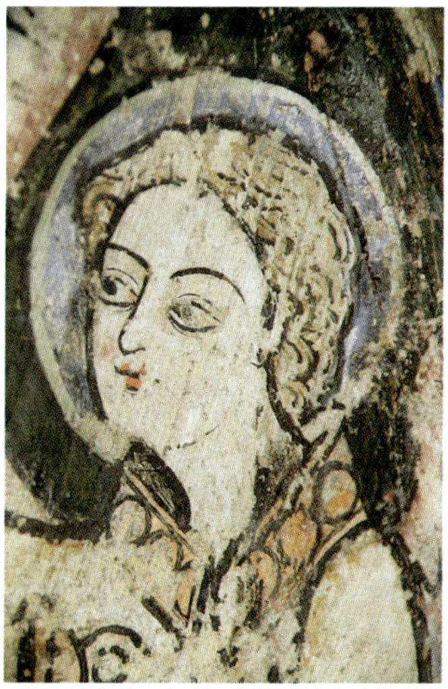

Plate 150 Norwich Cathedral, Ante-Reliquary Chapel: detail of an angel beside the eastern arch (c1250). (Photograph: H. Howard; © Courtauld Institute, Conservation of Wall Painting Department)

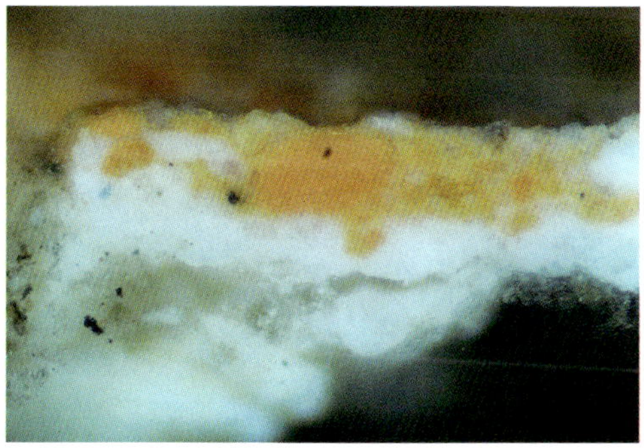

Plate 151 Norwich Cathedral, Ante-Reliquary Chapel. Cross-section of a sample from the angel's collar, showing a layer of red lead applied over lead white on the lime plaster, original magnification 500×. (Photomicrograph: H. Howard; © Courtauld Institute, Conservation of Wall Painting Department)

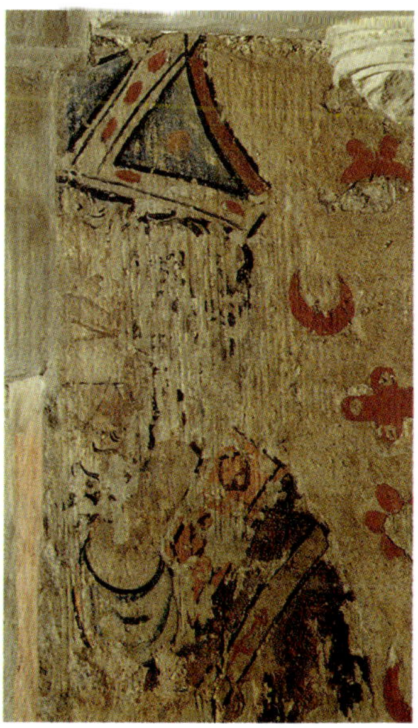

Plate 152 St. Albans Cathedral, feretory: detail of the unidentified archbishop on the south-east pier (c1302–08), after recent cleaning. (© The Perry Lithgow Partnership)

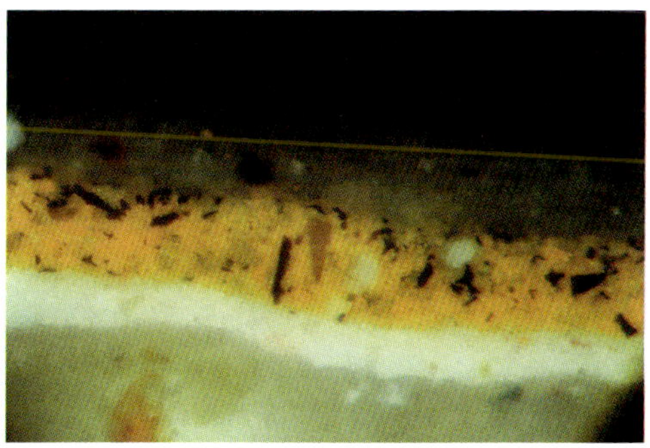

Plate 153 St. Albans Cathedral, feretory. Cross-section of a sample from the archbishop's orange maniple, original magnification 400×. A lead white ground covers the lime substrate used to level the tooled ashlar. The paint layer consists of red lead combined with charcoal black and is covered with a wax coating which was removed during recent conservation work. (Photomicrograph: H. Howard; © Courtauld Institute, Conservation of Wall Painting Department)

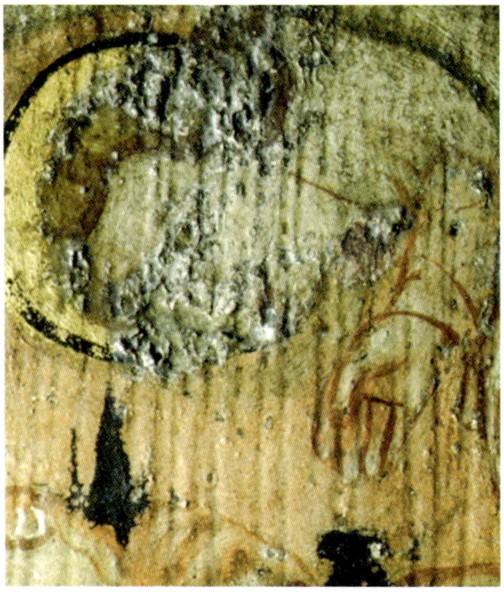

Plate 154 *Westminster Abbey, chapter house: detail of* Babylon is Fallen *from the* Apocalypse *cycle of c1375–1404. The tool marks of the ashlar support are clearly visible. The red background of the scene consists of red lead applied over a chalk ground. (Photograph: Howard/Kakoulli; © Courtauld Institute, Conservation of Wall Painting Department)*

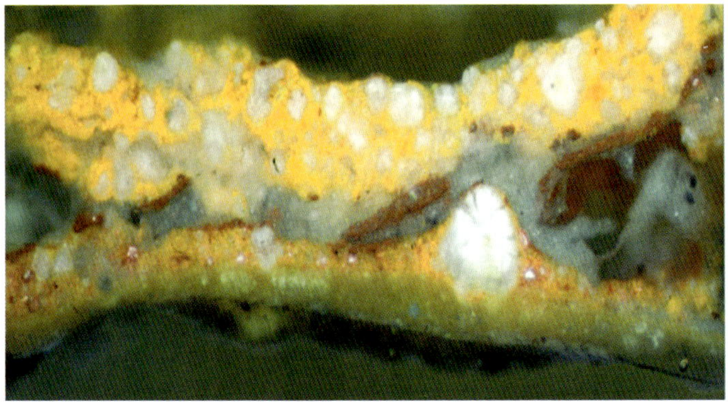

Plate 155 *Westminster Abbey, chapter house. Cross-section of a sample from arch 3 of the south bay of the* Apocalypse *cycle, original magnification 422×. The upper orange-red paint layer consists of red lead with white, lead-containing inclusions which represent an alteration of the red pigment. Below this is a thin dark red layer of red earth which was applied over a third red layer consisting of vermilion combined with red lead and large lead-containing inclusions, which are also due to the alteration of red lead. A yellow zone – probably due to incomplete alteration – is visible at the centre of some of the white particles. Of particular interest is the volume change associated with this transformation and the consequent disruption of the paint layers. (Photomicrograph: H. Howard; © Courtauld Institute, Conservation of Wall Painting Department)*

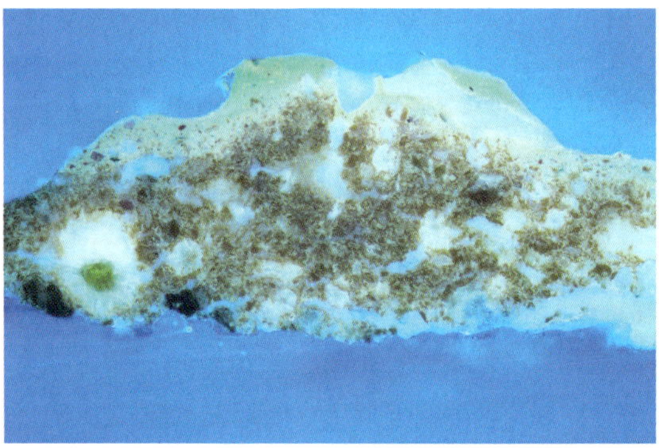

Plate 156 *Westminster Abbey, chapter house. This sample from the south-east bay shows a similar alteration of the red lead layer as in Plate 155. In UV light it produces a slightly yellow fluorescence, while the coating on the surface has a milky appearance. (Photomicrograph: H. Howard; © Courtauld Institute, Conservation of Wall Painting Department)*

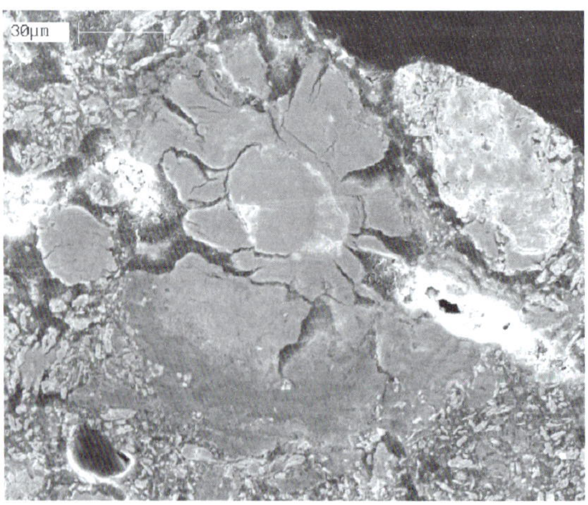

Plate 157 *SEM micrograph showing one of the particles of altered red lead from the sample in Plate 156. (SEM micrograph: H. Howard; © Courtauld Institute, Conservation of Wall Painting Department)*

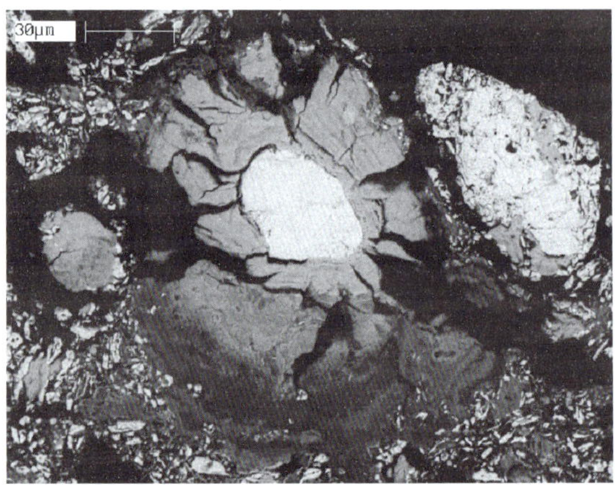

Plate 158 *Backscatter SEM micrograph of the particle shown in Plate 157. The backscatter image provides information relating to atomic number – the higher the atomic number the lighter the portion of the image. Thus it is clear that the centre of the altered particle (which appears yellow in normal light, as in Plate 155), is of higher atomic number (or more dense, or both) than the rather disrupted outer portion. Elemental analysis by SEM/EDX confirmed the presence of only lead and oxygen in both of these zones. The disrupted portion of the particle shows a pattern of radial cracks, and perpendicular to these, slightly wavy striations. The transformation of red lead here is due to a reaction between the pigment and an oil medium. (SEM micrograph: H. Howard; © Courtauld Institute, Conservation of Wall Painting Department)*

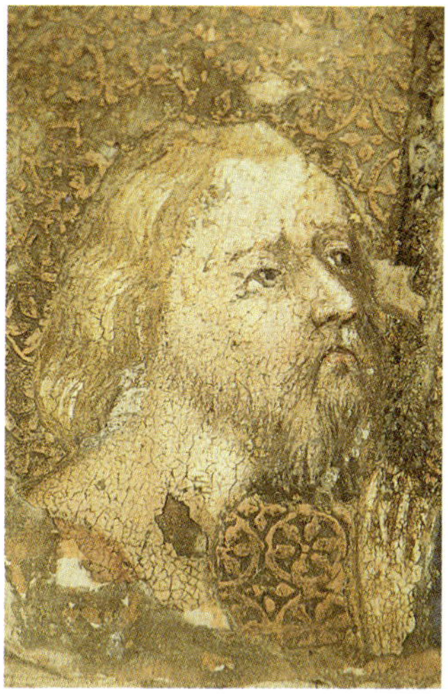

Plate 159 Red lead was employed extensively in the mid-14th-century wall paintings of St. Stephen's Chapel, Westminster Palace. The red lead also employed for the raised appliqué decoration is clearly visible where the uppermost paint layers have been lost. (© Copyright The British Museum)

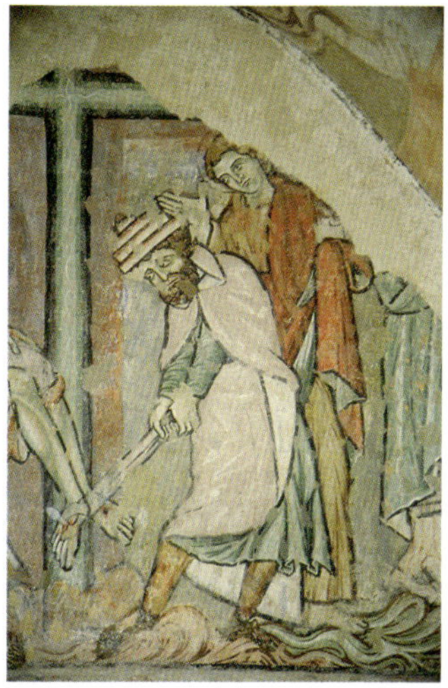

Plate 160 Winchester Cathedral, Holy Sepulchre Chapel: detail of the Deposition on the east wall (c1175), showing Nicodemus' pale pink robe. (© Courtauld Institute, Conservation of Wall Painting Department)

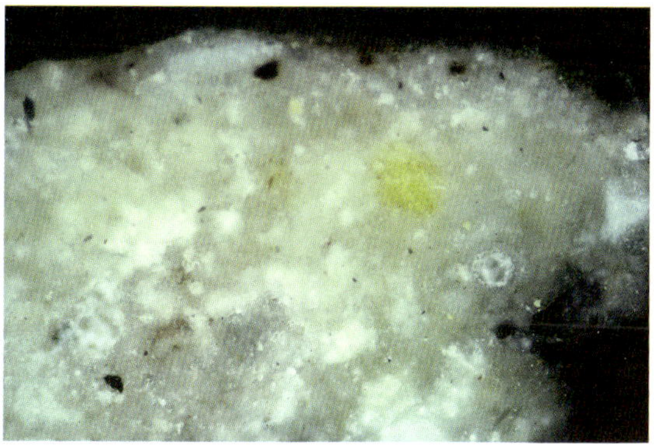

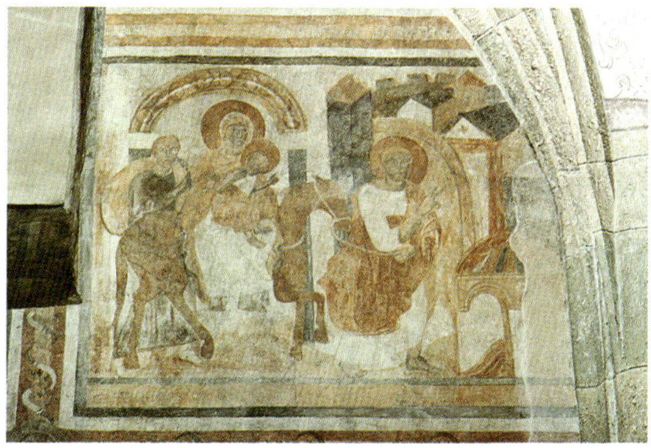

Plates 161 and 162 Winchester Cathedral, Holy Sepulchre Chapel. Cross-section (original magnification 266×) of a sample from Nicodemus' robe, showing how the unusual pink colour was achieved by combining a red lake pigment with a tiny trace of ultramarine in a calcium carbonate matrix. (Photomicrograph: H. Howard; © Courtauld Institute, Conservation of Wall Painting Department) This is the earliest identification of red lake in English medieval wall painting, although at Müstair (Switzerland) it was employed in the Carolingian scheme of c800, of which a detail is shown in Plate 162. (© Colorphoto Hans Hinz SWB)

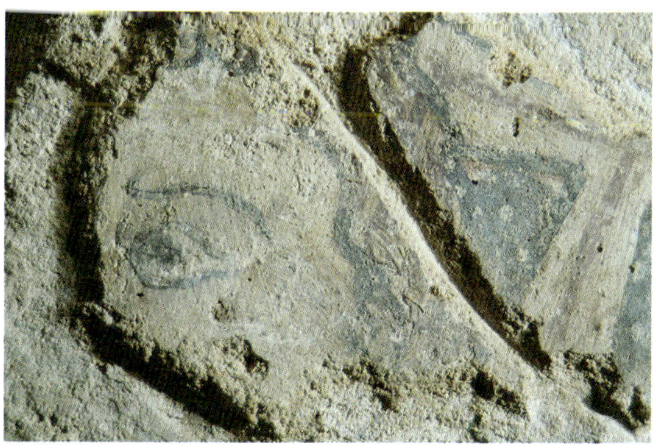

Plate 163 Glasgow Cathedral: detail of a fragment of wall painting (c1190) excavated from the crypt, showing a male head. (Photograph: H. Howard; © Courtauld Institute, Conservation of Wall Painting Department)

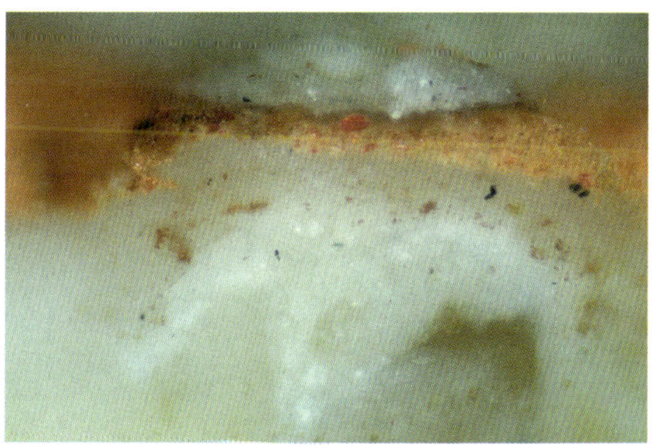

Plate 164 Tiny traces of a red lake pigment were detected in another of the fragments from Glasgow Cathedral. The cross-section shows that it was applied over a layer of vermilion, presumably to saturate its appearance, original magnification 500×. This is the first identified instance of the use of red lake as a glaze in Romanesque wall painting in Britain. (Photomicrograph: H. Howard; © Courtauld Institute, Conservation of Wall Painting Department)

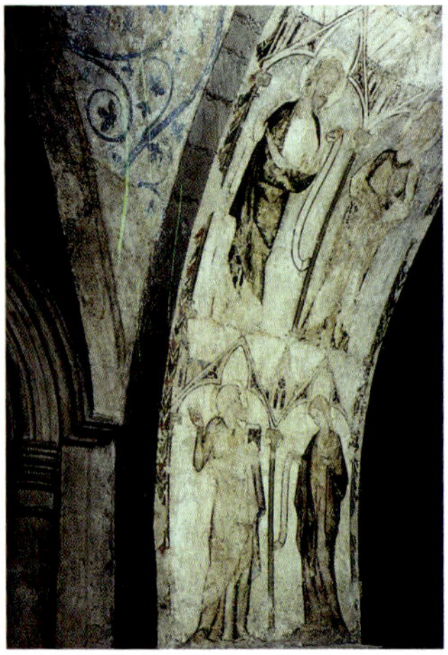

Plate 165 *Norwich Cathedral, Ante-Reliquary Chapel: detail of the paintings on the western arch (c1300). (© Crown copyright. NMR)*

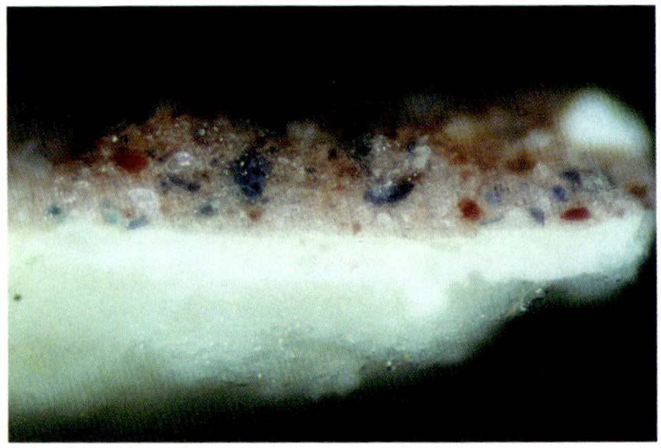

Plate 166 *Norwich Cathedral, Ante-Reliquary Chapel. Cross-section (original magnification 500×) of a sample from the drapery of one of the apostles on the arch. Red lake was combined with natural azurite and applied over a lead white ground to produce the rich purple colour. A combination of azurite and red lake was also employed for the purple draperies of St. Peter and St. Andrew on the vault of the chapel. Analysis by fluorescence spectroscopy of the dyestuff employed in this scheme confirmed the presence of a madder lake. (Photomicrograph: H. Howard; © Courtauld Institute, Conservation of Wall Painting Department)*

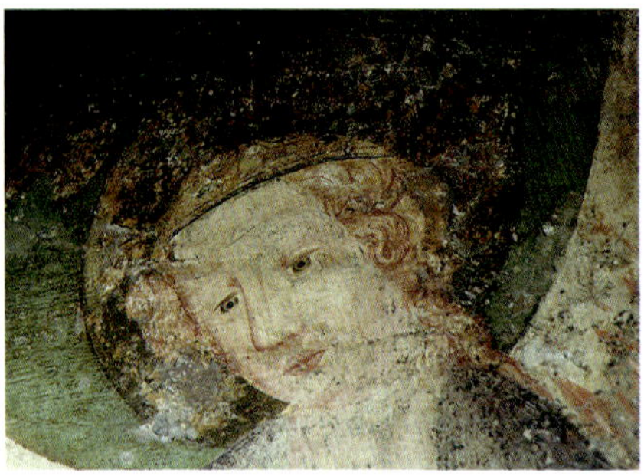

Plate 167 *Tower of London, Byward Tower: detail of St. Michael (c1395). (© Courtauld Institute, Conservation of Wall Painting Department)*

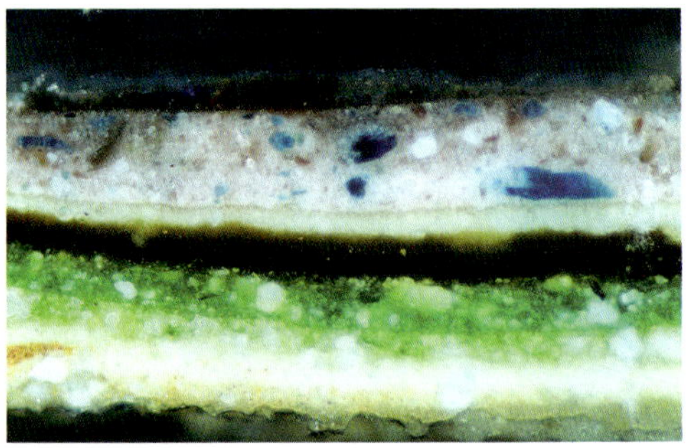

Plate 168 *Tower of London, Byward Tower. Cross-section of a sample from St. Michael's robe, showing the mixture of red lake, azurite and lead white employed to achieve the soft purple colour. The purple robe was painted over the green background of the scheme, for which verdigris was combined with lead white. (© Courtauld Institute, Conservation of Wall Painting Department)*

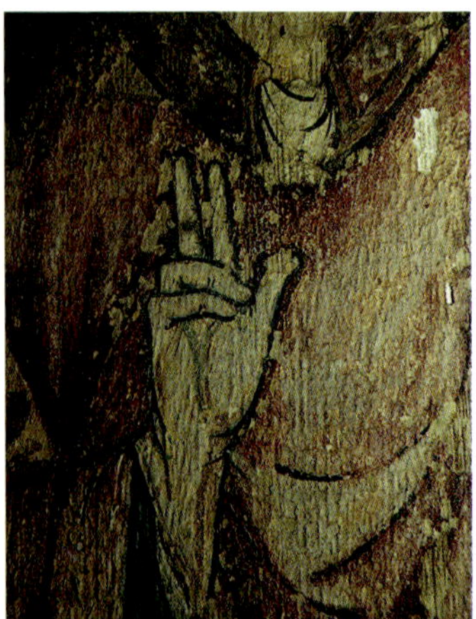

Plate 169 *St. Albans Cathedral, feretory: detail of St. William of York on the north-east pier (c1302–08). The highlights of the chasuble show considerable signs of fading. (Photograph: H. Howard; © Courtauld Institute, Conservation of Wall Painting Department)*

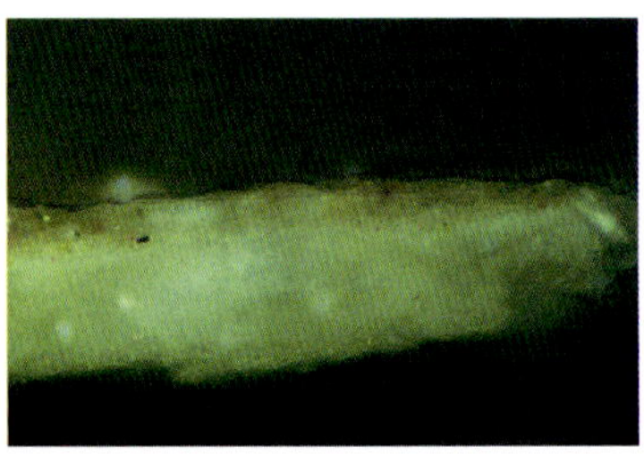

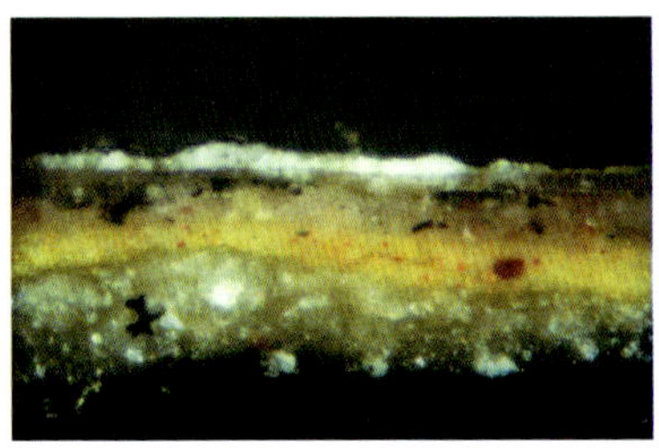

Plates 170 and 171 *Cross-section of a sample from the faded area of drapery shown in Plate 169. Here the red lake has been applied in a matrix of lime white and lead white over the lime ground, making it more susceptible to internal reflection. In these areas it has faded almost entirely. By contrast, where it has been applied over coloured underpaint and mixed with charcoal black for an area of deep pink shading (Plate 171) there is less internal reflection and consequently the colorant shows less sign of fading. Madder lake was identified in a sample taken from the corresponding archbishop on the south-east pier. (Photomicrographs: H. Howard; © Courtauld Institute, Conservation of Wall Painting Department)*

172

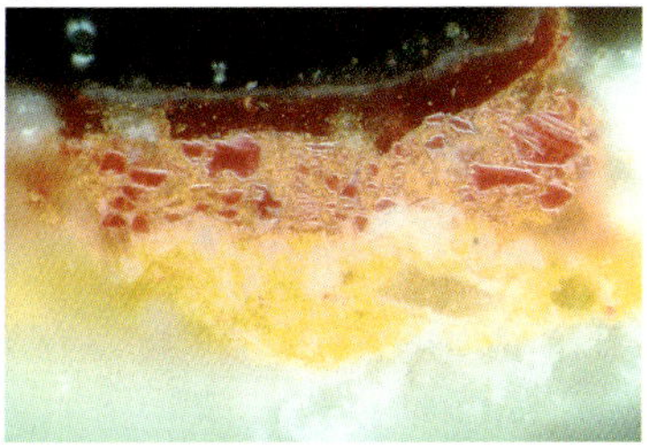

173

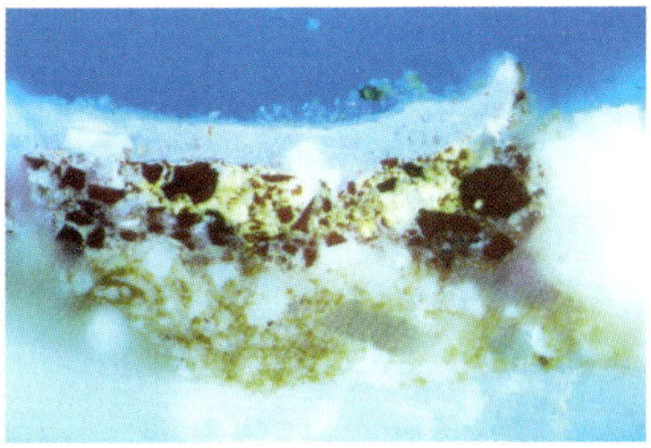

174

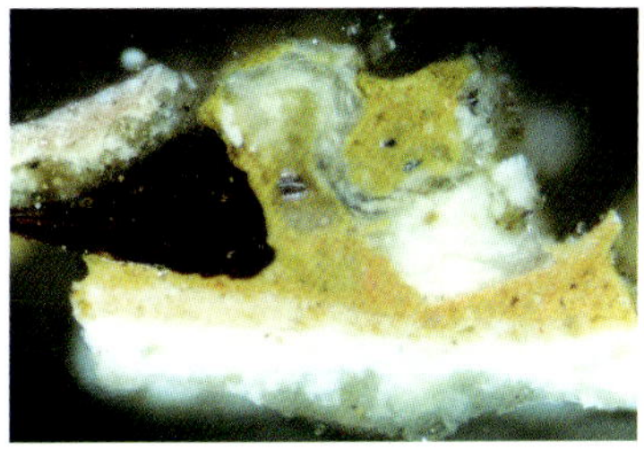

Plates 172–174 *Red lakes have frequently been identified in English medieval polychrome sculpture as on the tomb of Sir Oliver de Ingham (c1344) at Ingham, Norfolk. This sample (172), taken from a fragment probably from the tomb canopy, shows a layer of red lake applied as a glaze over vermilion. The same sample, stained for protein with FITC and viewed in UV light (Plate 173), indicates that while the vermilion layer was applied in a proteinaceous binding medium, the red lake layer is in an oil medium. In a sample from the appliqué decoration on Sir Oliver's camail (Plate 174), a thick red lake glaze was applied over gold leaf. Instrumental analysis of the constituent dyestuff confirmed the presence of madder lake. (Photomicrographs: H. Howard; © Courtauld Institute, Conservation of Wall Painting Department)*

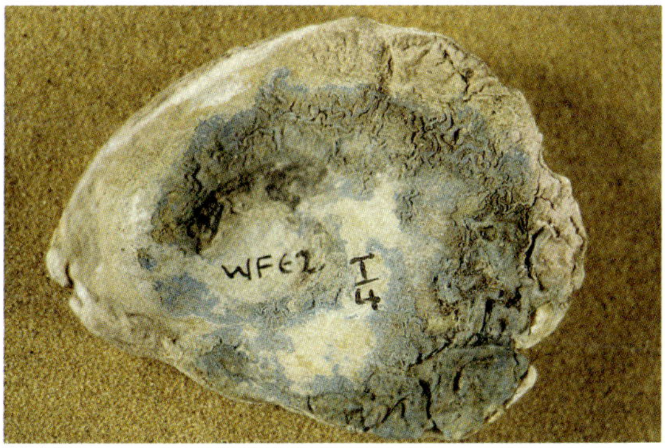

Plate 175 *Detail of an oyster-shell palette, dating from between the mid-14th and the mid-16th centuries, excavated from the Carmelite Friary at Coventry. (Photograph: H. Howard; © Courtauld Institute, Conservation of Wall Painting Department)*

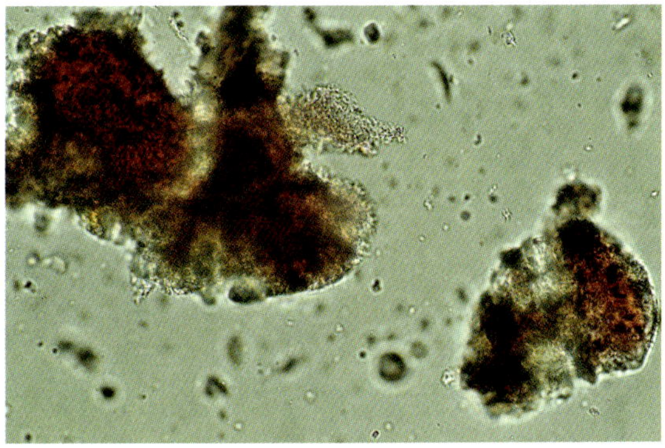

Plate 176 *Analysis of the pink paint around the perimeter of the shell palette shown in Plate 175 confirmed the presence of a red lake pigment. The photomicrograph shows particles of the lake pigment in plane-polarised transmitted light, original magnification 900×. The blue pigment in the centre of the palette is natural azurite. (Photomicrograph: H. Howard; © Courtauld Institute, Conservation of Wall Painting Department)*

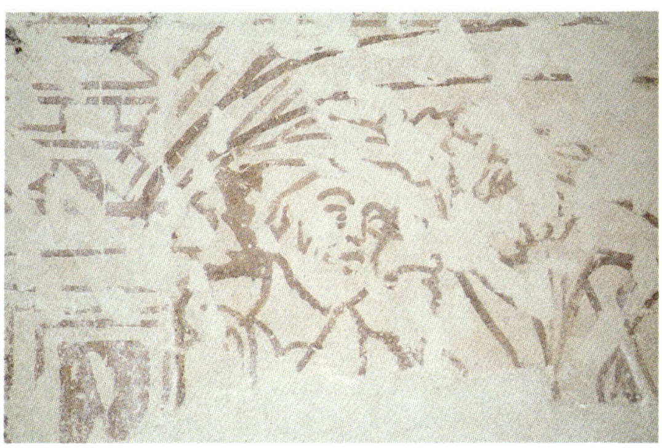

Plate 177 *Winchester Cathedral, Holy Sepulchre Chapel: detail of the* sinopia *for three scenes set beneath architectural canopies (c1175). While red earth is typically employed for* sinopie, *vermilion combined with red earth was used in this case. (© Courtauld Institute, Conservation of Wall Painting Department)*

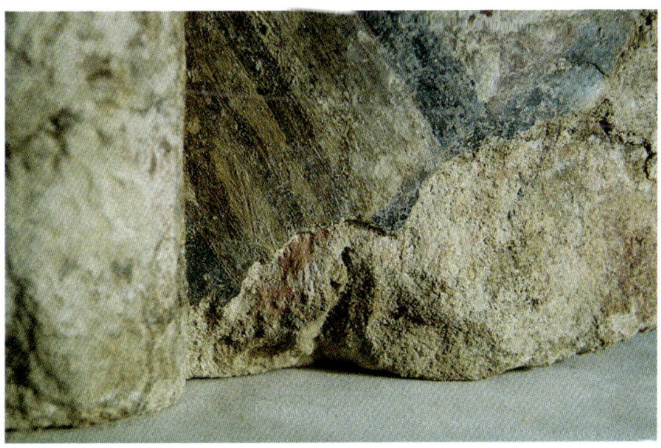

Plate 178 *Glasgow Cathedral: detail of a fragment of wall painting (c1190) excavated from the crypt. Extensive areas of red paint are present on the penultimate plaster layer of this fragment – visible at lower left centre – which may represent either a* sinopia *or the presence of two distinct decorative schemes. (Photograph: H. Howard; © Courtauld Institute, Conservation of Wall Painting Department)*

out the basic geometric forms and place the main pictorial elements, such as the figures and their architectural context, within the visual field. Evidence of setting out in red earth is particularly clear in the very abraded areas on the north side of the nave and in the middle tier of the south wall, where a few preparatory lines principally mapping out architectural features are evident. Preliminary drawing for the figure subjects is most conspicuous in the *Miraculous Draught of Fishes* on the south wall, where alterations are particularly evident in the drawing of Simon Peter's hand (Plate 179) (Howard 1992a: 9).[9]

A preliminary drawing in yellow iron oxide was employed for setting out on the uppermost plaster layer of the 12th-century scheme in the Holy Sepulchre Chapel of Winchester Cathedral. Of particular interest in this case are the changes made by the painter at an advanced stage of the painting process. These are especially evident in the *Entombment* where, to the left of the Virgin's head, the remains of another head and a broad-brimmed hat can just be seen, indicating the original position of the figure anointing Christ's body with oil (in the final version placed to the right of the Virgin).[10] Such sweeping compositional changes made at an advanced stage of the painting process suggest that initial and continued planning were largely undertaken on the wall itself at this period. Many other alterations are also visible. For instance, the figure of St. John in the *Deposition* was finally painted with his hand gesturing towards Christ, whereas the yellow preliminary drawing indicates that he was originally conceived with his hand held to his face in grief (Howard 1995a: 94).

Close examination of the late Romanesque 'Losinga' cycle of paintings in Norwich Cathedral has shown that different colours were employed for the underdrawing on the vault (yellow) and soffit (red). This use of different colours for setting out in distinct parts of the surviving painting, and the more precise and assured nature of the drawing on the soffit, has suggestive implications concerning the division of labour in the workshop organisation: perhaps the yellow underdrawing on the vault was carried out by an assistant, allowing for alterations and refinements in red by the master if necessary (Howard and Gasol 1996). Since so little of the overall programme survives, however, this must remain speculative.

For the setting out of the apse paintings of *c*1130 in St. Gabriel's Chapel, Canterbury Cathedral, initial underdrawing in yellow earth is visible wherever the subsequent paint layers have been lost.[11] This pale colour facilitated correction, and *pentimenti* are evident particularly in the lower tier on the north wall, as in the repositioning of the inscriptions held by Elizabeth and the figure facing her. By contrast with Witley, where the initial outline in red earth was strengthened and refined using the same pigment, in the St. Gabriel's paintings the yellow preliminary sketch was finalised in red earth (Plate 180). This can be seen particularly clearly in the setting out of the meander pattern and busts on the entrance arch soffits.[12]

---

ob Judenburg, Austria (Leitner 1989). Preliminary drawing in the early 12th-century paintings at Berzé-la-Ville (Burgundy) is illustrated by Demus (1970, especially pls.120 and 121).

9. In the Carolingian wall paintings at Müstair the sketch on the *intonaco* was likewise generally drawn in red earth; see Emmenegger 1986: 197.
10. The icongraphic significance of this and other alterations are discussed by Park 1983: 43–5.
11. Recent examination of the scheme at Idensen (Germany) provides a very close comparison for the use of a preliminary sketch in yellow (see Drescher *et al.* 1990: 80–81).
12. For Witley, see Howard 1992a:8, and for St. Gabriel's Chapel, see Cather and Howard 1994:143–4 and Howard 1992b. In the approximately contemporary scheme at Copford (Essex), a preliminary sketch in yellow was likewise refined in red iron oxide; see Organ *et al.* 1991.

Unusually, refinement in green of a yellow preliminary drawing occurs in the painting of *c*1140–50 in the north transept of Chichester Cathedral. This green underdrawing was also used to model areas such as the feet of the Apostles, and it can be seen through the semi-translucent yellow flesh tone applied over it.[13]

In early Gothic painting, preliminary drawing in red earth has been detected in the scheme of *c*1220 in St. George's Chapel, Ely Cathedral (Howard and Pathak 1995: 3), while refinement in yellow of an initial sketch in red was employed in the approximately contemporary paintings in the refectory of the Eastbridge Hospital, Canterbury. In the latter, both the shape and position of Christ's cushion have been altered, and the detail of the design developed in the second phase of drawing (Howard 1994).

Preliminary drawing in red, refined and finalised in the same pigment, occurs in the 14th-century scheme at Bradwell-juxta-Coggeshall, Essex (Cather and Park 1998). Although often less extensive, and exhibiting fewer subsequent alterations, preliminary drawing executed directly on the wall continues throughout the later Middle Ages. Drawing in red was employed to set out the main compositional elements in the *Last Judgement* of *c*1435–60 over the chancel arch at Holy Trinity, Coventry (Howard 1996b: 5), while preliminary drawing in a brownish earth pigment was used to set out the design and to indicate much of the final detail in the splendid paintings of *c*1470 in the Deanery at Durham. Drawing in brown was also used for setting out the approximately contemporary *Legend of St. Eustace* in Canterbury Cathedral,[14] while drawing in red is evident in the scheme of *c*1510 at St. Mark's Bristol (Gill and Howard 1997: 99).

Use of red and yellow earth pigments within a limited palette

Red and yellow ochres and lime white are the dominant colours in the paintings of *c*1110 at Witley. Examination and analysis of the recently uncovered section of this scheme on the west wall of the nave indicated that full advantage had been made of this limited 'egg and bacon' palette by layering and mixing the pigments. Following completion of the preliminary drawing, the basic design was articulated in lime white and red and yellow earth pigments. Over the basic mid-tones, details were built up in a range of tonalities with lime white highlights added finally in *impasto* on the surface. In some areas, no basic background colour appears to have been employed but the *impasto* paint layers applied directly to the buff-coloured plaster; this occurs in a detail of Adam's hand in the *Harrowing of Hell*, where the plaster provides the mid-tone to set off the white highlight. For areas of flesh painting, a pink colour – red earth combined with lime white – was applied to the plaster, and then linear details in red and yellow earth pigments (again combined with lime white) were applied in *impasto*. Highlights in pure lime white were added last. Similarly, the draperies were rendered with an initial base colour, with folds and highlights added in darker and lighter tones; thus the basic pink tone of Christ's garment was articulated with deep pink shadows and lime *impasto* (Howard 1992a).

In the slightly earlier scheme at Hardham, yellow and red earth were also employed for the flesh tones. For the Virgin's hand in the *Annunciation*, red earth combined with lime white was applied over yellow ochre combined with lime white, with a highlight of

13. The green pigment employed here is thought to be malachite, although no analysis was undertaken; Gärtner 1988.
14. For the Durham scheme, see Allardyce *et al.* 1983: 8, and for the St. Eustace painting, Allardyce 1984: 125.

pure lime white added in *impasto* on the surface.[15] In the scheme of *c*1130 in St. Gabriel's Chapel, Canterbury Cathedral, yellow earth was used extensively for drapery and borders, and also – combined with varying amounts of lime white – to provide the basic flesh tones over which shadows or highlights were applied depending on the tonality required. In addition, black underpainting was used for some areas of yellow, as in a sample from the drapery of *Christ in Majesty* (Cather and Howard 1994: 144–5).

In the Gothic period, earth pigments continued to be used to full effect in the context of a limited palette, particularly in parish church schemes such as that of the early 13th century at Duxford, Cambridgeshire (Courtauld Institute of Art 1987). However, a similarly restricted palette was also occasionally used in grander contexts, as in the splendid paintings of *c*1270 in the south-east corner of the south transept of Winchester Cathedral. Here the exquisite nimbed figure in a Benedictine habit, and traces of his companion, were delicately executed in red and yellow earth pigments, lime white and carbon black alone (Curteis *et al.* 1994: 27).

Accessory mineral inclusions

Additional mineral inclusions are frequently detected in red and yellow earth pigments. Thus, in the early 12th-century scheme at Witley, the yellow flesh colour of one of the souls in Hell was achieved by combining quartz-rich red and yellow earth pigments with lime white (Plate 181). The large quartz particles in the paint layer may be seen clearly in Plate 182 owing to their high birefringence in transmitted light with crossed polars. The Hardham paintings share this use of a quartz-rich red earth pigment, and in some red earth paint layers there are large silica inclusions. However, the red particles themselves are extremely small, often less than 1μm, and analysis by SEM/EDX also indicates that the red earth pigment is clay-rich, and that a few of the red particles have a high titanium content.[16]

High concentrations of quartz minerals have also been found in a sample of yellow earth taken from the west wall of the apse of St. Gabriel's Chapel, Canterbury Cathedral. Here the yellow quartz-rich mixture was employed as an underpainting beneath red iron oxide. By contrast, a yellow earth pigment without the large quartz inclusions was employed for the preliminary drawing and to block in areas of the basic background forms of the composition (Cather and Howard 1994: 145).

Analysis of a yellowish pigment from the earliest plaster layer on the south side of the chancel arch of Stoke Orchard Church, Gloucestershire, suggested that it also contained clay and other accessory minerals, and that it may have been obtained from a local source (Rouse and Baker 1967: 116–17).

Use as an underpaint

Yellow and red earths were often employed as a local ground or underpaint beneath other pigments. At Hardham, for example, red and yellow earth pigments were combined with lime white and used as local grounds beneath a copper chloride green: red earth as an underpaint for the green nimbus of the *Agnus Dei* above the chancel arch, and

15. Analysis by Howard, in Sawdy *et al.* 1998.
16. See analysis by Howard in Sawdy *et al.* 1998. Correspondence with the British Geological Survey has, to date, unearthed no information on titanium-rich red earth deposits in England. A trace of titanium was also found in a yellow sample from a 17th-century layer at Stoke Orchard Church, Gloucestershire; Rouse and Baker 1967: 117.

yellow beneath the green nimbus of one of the flanking censing angels.[17] Yellow earth was applied beneath vermilion to produce a brilliant red colour in Joseph of Arimathea's robe in the scheme of *c*1175 in the Holy Sepulchre Chapel at Winchester (Howard 1995a: 101). In these cases, the red earth underpaint probably served both to extend and to enhance the brilliant and more expensive red pigment. At St. Gabriel's Chapel, Canterbury Cathedral and at Cormac's Chapel, Cashel (*c*1170) red iron oxide was combined with charcoal black to produce a dark purple underpaint in preparation for the application of a layer of ultramarine (Howard 1991; Cather and Howard 1994).

In the feretory of St. Albans Cathedral, red and yellow earth pigments were used as secondary grounds to great effect for the figure of St. William of York (*c*1302–08). A local ground of red earth was used below his face, hair, and the blue drapery of his dalmatic; this produced a warmer tone in the fair hair and the flesh painting, while below the blue drapery it resulted in a more purple effect. An underpaint of yellow earth mixed with a little red was employed below the original green background, and below certain areas of the red drapery of the chasuble. This yellow added warmth to the translucent red of the chasuble and enhanced the warm green of the background. Below the gilding of the crozier, yellow earth was combined with vermilion (Howard 1993a: 35).

Over a white ground or directly over the plaster support
Orange-red ochre from the flesh of one of the apostles in the *Washing of the Feet* at Hardham consists of red and orange-red earth particles intimately combined with the lime white ground and applied directly to the plaster substrate, indicating that the earth pigment was bound by the carbonation of lime (Hluvko 1991: 14).

Combined with other pigments
The bright orange-red colour of the background of the serpent in the *Temptation* at Hardham was achieved by combining red ochre with a few particles of yellow ochre in a lime white matrix, and applying the mixture in a layer 50μm thick over the lime plaster substrate (Sawdy *et al.* 1998: 34–5). For Nicodemus' flesh tone in the late 12th-century scheme in the Holy Sepulchre Chapel at Winchester, particles of red and yellow were combined with carbon black and lime white, while for Christ's cloak deep maroon and orange-red earths were combined with a few carbon black particles (Hluvko 1991: 15–16). Examination of fragments of 12th-century painted plaster from Sherborne Abbey, Dorset, indicated that red and yellow earth pigments were extensively employed, while instrumental analysis confirmed that haematite was combined with chalk and oil for the pink preparation for gilding in some of the fragments (Howard and Park forthcoming a). Red lead combined with red earth and then overlaid with vermilion was employed for painted fragments from the niche adjacent to the Martyrdom Doorway at Canterbury Cathedral. These fragments may be contemporary with the doorway itself (*c*1236), or may have been painted later (Hluvko 1991: 15, n. 98).

Mixed with lead white
Red and yellow earth pigments are frequently found combined with lead white in the Gothic period. For example, in the scheme of *c*1220 in the Eastbridge Hospital, Canterbury, yellow earth combined with lead white was employed for the feathers of the symbol

17. Analysis by Howard in Sawdy *et al.* 1998: 34–5.

of St. John (Howard 1994: App. I). A similar pigment mixture, applied beneath a layer of pure yellow earth, was used for the yellow drapery of the angel of *c*1250 to the south of the eastern arch of the Ante-Reliquary Chapel in Norwich Cathedral. The same pigment mixture, with the addition of a little calcium carbonate, occurs again in the slightly later scheme in the same chapel in the lining of St. Peter's drapery on the vault (Park and Howard 1996; Howard 1995c: App. I–III).

Brown ochres

Shades of deep yellow and yellow-brown ochres seem to be evident in the draperies of the figure of St. Thomas Becket in the north transept at Stow, Lincolnshire (*c*1200). The technique and style of this painting is reminiscent of the monumental figures of St. Cuthbert and St. Oswald in the Galilee Chapel at Durham, although at Stow the palette appears to be particularly rich in yellow, browns and brown-reds. Brown iron oxides have been visually identified in the mid-13th-century *Wheel of Fortune* in Rochester Cathedral, and on the Romanesque frieze of the west front of Lincoln Cathedral (Darrah 1992; Larson 1992: 188). In the Rochester painting a rather translucent purple-brown ochre was detected at the bottom edge of the wheel, combined with chalk and applied directly on the plaster support (Darrah 1992). A range of brown and orange-brown earth pigments have also been identified in the late Romanesque fragments from Glasgow Cathedral (Howard 1998a).

In the late 12th-century scheme in the Holy Sepulchre Chapel, Winchester Cathedral, SEM/EDX analysis of a sample from the dark shadow on Christ's yellow loin cloth in the *Entombment* indicated the presence of manganese in addition to iron, suggesting the use of an umber (which contains small amounts of manganese oxides or oxyhydroxides). Further examination suggests that a single layer of yellow iron oxide combined with umber and a few charcoal black particles, just 30μm thick, was applied directly to the white plaster substrate (Howard 1995a: 96, 101). Umber has also been detected in the painting of *c*1220 at Eastbridge Hospital, Canterbury, where the pigment was employed for a detail on the neckband of Christ's drapery (Howard 1994: 4–5, App. 1).

Binding media

In the Romanesque period earth pigments were frequently bound by the carbonation of calcium hydroxide in some form. For example, in the Holy Sepulchre Chapel at Winchester it seems likely that the preliminary drawing and initial pigment layers were applied *a fresco*. However, the stratigraphy of the final painting is remarkably complex and varies substantially across the pictorial surface, from thin single layers applied directly on the lime plaster substrate to paint applied in considerable *impasto*, often over coloured underpaint. In some samples, a single rather thin paint layer is evident applied directly to the plaster substrate, and is likely to have been applied *a fresco*. By contrast, for areas of flesh painting, multiple layers were applied to produce complex effects of modelling. The mid-dark flesh tone of Christ's foot in the *Deposition*, for example, was achieved by the application of five different layers of earth pigments – green, red and yellow – applied either singly or in combination, often with the addition of lime white and charcoal black (Plate 160). Here, carbonation of the lime white pigment provided additional binding capacity within the complex layer structure (Howard 1995a: 96). The intrinsic stability of earth pigments is complemented through binding by the carbonation of calcium hydroxide, and the resultant paint layers tend to be extremely durable. By contrast, where earth pigments – particularly red earths – are combined with organic binding media, the paint layers are often very powdery and friable. This is probably due to the very fine particle size of these pigments and the high binder:pigment ratio which is necessary for adequate binding.

Comparison with their use in other media and in continental wall painting

Wall painting

Earth pigments were as widely employed in other northern European wall painting as in England, and important examples include the Romanesque schemes in France at Berzé-la-Ville and in the crypt of Auxerre Cathedral (Rollier-Hanselmann 1997; Cadet 1995); and the scheme of *c*1260 in the choir of Angers Cathedral, where a wide range of other red and yellow pigments were included in the palette, such as red lake, vermilion, orpiment and lead-tin yellow (Demailly *et al.* 1998: 11). Although yellow ochre was detected in the splendid mid-14th-century scheme in the choir of Cologne Cathedral – indeed it was the only yellow pigment identified – red lake and vermilion were used in preference to red earth pigments (Jägers and Schulze-Senger 1989: 187–8).

Examination of the scheme of *c*1165 at Pürgg, Austria, indicated not only that the preliminary drawing in yellow earth was undertaken while the plaster was still fresh, but that in some areas the drawing was refined and finalised in red earth, while in others a mixture of green, umber and yellow earth was used (Howard 1995d). Red iron oxide was employed as an underpaint for vermilion in the 12th-century scheme at Müstair (Switzerland) (Mairinger and Schreiner 1986), and for Christ's drapery in the *Maiestas Domini* of *c*1130 at Idensen (Germany), although in the latter instance the red iron oxide-based underpaint was combined with lime white (Matteini and Moles 1994: 83, pl. 9.3c). This application follows closely the recommendation in Theophilus' treatise for red mixed with lime as a local ground beneath cinnabar on walls (1979: 25). At Idensen, grey underpainting for areas of yellow earth was also employed as seen in samples taken from areas of drapery (Matteini and Moles 1994: pl. 10.5, 85). This practice mirrors that in the approximately contemporary scheme in St. Gabriel's Chapel, Canterbury. Red ochre was also combined with black for dark tones in the Idensen scheme (Matteini and Moles 1994: 83).

Polychrome sculpture

Earth pigments were also used for polychrome sculpture throughout the medieval period. Red earth combined with lime white and applied over a white ground has recently been found on the capitals of *c*1180 in the ante-room of the medieval chapel in the keep of Guildford Castle (Howard 1998c). Other important examples of the use of earth pigments include 13th-century polychromy on the west front of Wells Cathedral (Sampson 1998: 118–19; Roy and White 1982); the mid-13th-century polychromy on the west front of Salisbury Cathedral, where traces of finely divided red earth have been identified (Howard 1996d: 2–3); the effigy of Sir Oliver de Ingham (*c*1344) at Ingham, Norfolk, where red and yellow earth pigments were incorporated in raised appliqué elements such as his camail (Howard 1993d: App. 2); the de Bois tomb of *c*1380, also at Ingham, where yellow earth was combined with lead white and applied as a ground on the tomb chest (Howard 1993d: App. 1); the early 14th-century bosses in the choir of Exeter Cathedral, where ochre was incorporated in a gold size (Hulbert 1998: 35–6); the 14th- and 15th-century polychromy on the west front of Exeter Cathedral, where both red and yellow ochre were extensively employed (Sinclair 1991: 122–4, 1992: 10); and in the late 15th-century polychromy of the Great Screen at Winchester Cathedral, where size for gilding was tinted by incorporating yellow ochre (Brodrick and Darrah 1989: 616).

Continental examples include the 12th-century lion capital from Vossestrand (Norway) on which red earth with a high silica content has been identified (Plahter 1990: 280); the '*Portail de la Mère Dieu*' of *c*1220–60 at Amiens Cathedral, where red, pink, orange and yellow earth pigments were detected, and examination also indicated that gold leaf was applied over red and yellow ochres (Weeks 1998: 102); and the early 13th-century polychromy on the south portal of Lausanne Cathedral (Switzerland), where red, yellow and brown ochres have been identified (Furlano *et al.* 1981).

Panels

Earth pigments have also been identified in a number of important English medieval panel paintings, as on the Westminster Retable of *c*1270 where the flesh tones consist of red earth combined with vermilion.[18] Red earth has been identified in the original paint layers on the nave ceiling of *c*1220 at Peterborough Cathedral, where the pigment was combined with lead white and carbon black to produce the purple colour of the Psaltery Player's instrument (Howard 1997b: table 1). Both red and yellow earths have been detected in the Thornham Parva Retable of *c*1335, where they were employed for the modelling of hair, beards and drapery (Goldsmith and Bucklow 1998: 43). Similarly, a purple-red earth and yellow earth have been identified in the early 14th-century screen from Kingston Lacy, Dorset. In this instance the yellow ochre was employed in the mordant for all gilded areas (Tracy 1997: 28).

Discussion

Earth pigments are among the most widely employed pigments in all types of large-scale painting throughout the medieval period. Their low cost and ready availability in a wide variety of colours would be reason enough for this popularity, but their stability in both alkaline conditions and in organic binding media made them particularly suited to the wall painter's palette.

While a range of reds in addition to red earth – red lead, vermilion and red lake – were available and used throughout the medieval period, the choice of yellows was more limited. Orpiment has not been detected in English wall painting in the Romanesque period, and only rarely in Gothic schemes. Lead-tin yellow has been found in a few instances in the 15th and early 16th centuries, but yellow lakes have yet to be identified. It is therefore not surprising that in a number of high-quality schemes yellow earth is the only yellow pigment present, whereas a variety of reds are included in the palette. Thus, in the elaborate scheme of *c*1300 in the Ante-Reliquary Chapel in Norwich Cathedral, red earth is excluded from the palette in favour of madder lake and vermilion, whereas yellow earth is the only yellow pigment employed (Howard 1995c: App. I–III). In the costly paintings in the choir of Cologne Cathedral a similar situation occurs, with a yellow earth pigment juxtaposed with red lakes and vermilion (Jägers and Schulze-Senger 1989: 187–8). Similarly, yellow earth is the only yellow pigment identified in the late 14th-century scheme in the Chapel of Our Lady Undercroft in Canterbury Cathe-

18. A white pigment was also combined in this mixture. I am grateful to Dr A. Roy for this information.

dral, and in the 15th-century painting at Berry Pomeroy Castle, Devon. In these instances red earth is included in the palette together with vermilion and red lead.[19] By contrast, in the 14th- and 15th-century polychromy on the west front of Exeter Cathedral, yellow ochre is used in conjunction with orpiment and lead-tin yellow, and with red lakes, vermilion, red lead and red earth (Sinclair 1991:123–4, 1992: 10), while in the mid-13th-century wall paintings in the choir of Angers Cathedral yellow earth is employed in the palette together with lead-tin yellow and orpiment (Demailly *et al.* 1998: 11).

Pigment alterations

Earth pigments are extremely durable and stable, although yellow earths in particular undergo alteration to form haematite on heating. In his study of heat alterations to pigments painted in fresco techniques, Rickerby found that yellow ochre changed to a slightly red colour at 300 °C and was completely altered to red at 350 °C. He also noted that haematite was only formed between 400 and 700 °C, and that there was an unexpected alteration towards green at 700–900 °C. Red earth pigments are generally considered to be stable at red heat – 500–1000 °C – only decomposing to form magnetite Fe_3O_4 at temperatures in excess of 1300 °C. However, Rickerby recorded this transformation at a lower temperature – 1000 °C – than previously recorded (Rickerby 1991: 42).

The Romanesque cycle of wall paintings in the chapter house of Santa Maria de Sigena (Spain) have been compared stylistically and iconographically with late 12th-century paintings at Winchester, and may have been painted by English artists. These paintings are particularly interesting with respect to the heat alterations of pigments since they were partially destroyed by fire at the beginning of the Spanish Civil War. Many of the brownish tones visible throughout the remains of the cycle are thought to be due to the alteration of yellow and red earth pigments.[20]

19. For the Canterbury chapel, see Howard 1993b; for Berry Pomeroy, see Gill *et al.* 1996.
20. For the recent technical examination of the paintings, see Gasol Fargas 1999; and for issues of style, iconography and dating, see Pächt 1961, Oakeshott 1972 and Schuler 1995.

❦ SEVEN

Yellow pigments

Orpiment

Orpiment, the yellow trisulphide of arsenic (As_2S_3), was valued in the medieval period for its intense pale yellow colour and micaceous sparkle, properties which led to its use for imitating gold in painting.[1] Indeed its Latin name, *auripigmentum*, quite literally means pigment the colour of gold or gold paint (Fitzhugh 1997).

Mineral orpiment occurs in subvolcanic conditions: as a volcanic sublimate, as a hot spring deposit, and in mine fires (Wallert 1984: 45; Fitzhugh 1997: 54). It also frequently occurs as an alteration product of other arsenic ores, such as realgar.[2] Orpiment was also sometimes excavated from seams in gold mines, and in these instances often contained small quantities of gold as an impurity. Sublimation of the mineral thus sometimes produced metallic gold – a discovery that led alchemists to believe that gold was actually formed during the process of its distillation.[3]

The source of the orpiment employed in medieval painting is uncertain, although it has been suggested that it was imported from Asia Minor, Hungary and Macedonia (Wallert 1984: 45). Although the mineral could be ground for use as a pigment, it was also manufactured artificially by the sublimation of sulphur and arsenic ores.

Cost

The first documentary evidence for the use of orpiment in English painting is in the Westminster Palace accounts for 1265, where 2lb of orpiment and ochre are recorded with

1. Mosaic gold (stannic sulphide), also employed as a substitute for gold, has been found in English manuscripts of *c*1265–80 (Morgan 1988: 33), although it has not been identified in wall painting.
2. Red arsenic sulphide, realgar (AsS or As_4S_4), has been found combined with mineral orpiment as a pigment, as in an Icelandic manuscript of *c*1360 (Best *et al.* 1995: 36–7). Realgar has not been identified in wall painting but it has been detected in medieval icons from Bulgaria (Fitzhugh 1997).
3. Pliny (1984: 61–3) describes the avarice of Emperor Caligula in ordering the smelting of orpiment in the hope of producing gold. Although high-quality gold *was* produced, the quantity was so small that no financial gain was made.

other pigments purchased by William Page (Colvin 1971: 399). The Westminster accounts for 1275 also itemise orpiment, and show that at 6d per lb it was six times the price of yellow ochre (Salzman 1952: 168).[4] The same price is given in the Sacrist's Rolls of Ely Cathedral in 1337 (Chapman 1907: II: 98). Unfortunately, the precise nature of the painting to be undertaken with the pigment is not specified in any of these accounts, although the quantity purchased in 1265 certainly seems to indicate large-scale polychromy.

Unfortunately, no accounts survive from the Romanesque period, when it is more likely that mineral rather than synthetic orpiment would have been employed. It is clear, however, that in the Gothic period orpiment was not particularly expensive, but that its price was similar to that of other synthetically produced pigments such as lead white. For example, in the Great Wardrobe accounts of 1350–52, 15lb of *orpyment* at 6d per lb and lead white at 8d per lb were purchased for streamers and standards for Edward III.[5] A sharp increase in the price of the pigment in the later Middle Ages may be indicated by accounts of 1501, which include 'half ane pund *orpement*' at 8s per lb, for colours to be used in the Chapel Royal at Stirling (Apted and Hannabuss 1978: 145), although this high price may reflect the use of mineral orpiment rather than its synthetic equivalent.

Optical, chemical and physical properties

In plane polarised transmitted light, particles of both mineral orpiment and its dry-process synthetic equivalent are pale yellow to lemon in colour, with high relief and refractive indices of $\alpha = 2.4$, $\beta = 2.81$ and $\gamma = 3.02$. The micaceous structure of the pigment is readily apparent. With crossed polars, orpiment is strongly birefringent with anomalous blue-green polarisation colours (Fitzhugh 1997: 55–6).

Orpiment initially darkens on heating, but will burn in air forming white arsenic trioxide (As_2O_3). Both orpiment and realgar lose colour with light – realgar alters to form para-realgar, while orpiment oxidises to form translucent or white oxides of arsenic. The pigment decomposes slowly in water and is also soluble in both acids and alkalis (Fitzhugh 1997: 51–2).

Owing to its micaceous structure, grinding of the pigment was clearly problematic in the medieval period, and a number of authors recommend the addition of other materials to facilitate the process (Wallert 1984: 48–9). The *Mappae Clavicula* provides a recipe for grinding orpiment with saffron and glass to imitate gold letters (Wallert 1984: 46). Eraclius (1967: 238–9) describes grinding the pigment with water and a little calcined bone on a marble slab, while Cennini (1960: 29) suggests grinding it on a stone with a little glass and working it up for as long as possible. Glass may have had a purely mechanical function in the grinding process, however with its siccative properties it would also have acted as a drier for an oily medium."[6] Since orpiment dried poorly in oil, the addition of siccatives such as red lead or glass was essential (Wallert 1984: 48–9).

Orpiment is said to be incompatible with lead and copper pigments because of a reaction between the sulphur from the pigment and copper or lead ions (Weber 1923:

4. Later, in 1299, orpiment at 8d per ½ lb is listed among pigments and media brought from Arras to Hesdin; Richard 1887: 331.
5. Other colours purchased included azure, and indigo at 18d per lb; Staniland 1986: 240.
6. Medieval glass often contained lead oxides which would have acted as siccatives.

96; Thompson 1956: 178). This incompatibility is noted in the *Mappae Clavicula* (1974: 27–8), and Eraclius (1967: 252) suggests that the pigment should not be combined with green (presumably verdigris), minium (red lead) or folium. Verdigris and lead white, in particular, were known to be incompatible with orpiment (Thompson 1956: 169, 178).

Since orpiment is also unstable in alkaline conditions, it is unsuitable for use in *buon fresco* or combined with lime milk or the pigment lime white. Several artists' treatises warn against its use in wall painting; for example, Theophilus indicates that orpiment and *anything mixed with it* is not good on a wall (1979: 22). Cennini provides contradictory advice, on the one hand insisting that orpiment should not be applied either in *fresco* or *secco* to a wall since it will turn black on exposure to air,[7] while on the other suggesting a combination of orpiment and indigo for painting rooms in *secco* (Cennini 1960: 29, 32, 33, 50).

Perhaps owing to the high toxicity of the pigment, a few authors provide recipes for yellow colours to simulate orpiment; thus Eraclius describes *auripigmento simili,* which is made from the liquor of fish gall mixed with a little vinegar and some white clay on a marble slab (1967: 198–9). Jehan le Begue suggests combining orpiment with sulphur vivum to provide a yellow colour, and although this would not have reduced the overall toxicity, it would have served to extend the poisonous yellow (Merrifield 1967: 1: 96).

Preparation

Although the mineral may be coarsely ground for use as a pigment, it was also produced synthetically, and Cennini describes the pigment as both extremely poisonous and made by alchemy (1960: 28).[8] The early 15th-century French treatise by Jehan le Begue describes the production of orpiment by sublimation from sulphur and realgar (Merrifield 1967: 1: 78–9; Wallert 1984: 45).

Use in English medieval wall painting

To date there are no known examples of orpiment in English Romanesque wall painting and only six in the Gothic period: in early 13th-century fragments of wall painting excavated from Worcester Cathedral;[9] in the mid-13th-century *Wheel of Fortune* in the choir of Rochester Cathedral (Darrah 1992); in the 14th-century *Christ in Majesty* at Farringdon, Hampshire;[10] in the *Apocalypse* scheme of *c*1375–1404 in the chapter house of Westminster Abbey;[11] in the scheme of *c*1420 at Farleigh Hungerford Castle, Somerset (Howard *et al.* 1998: 61–2); and for the heraldic devices on the north wall at Ashby St. Ledgers, Northamptonshire, dating from before the end of the 15th century (Curteis 1994a).

7. Such blackening is more likely to be due to a chemical reaction between orpiment and calcium hydroxide.
8. I am grateful to J. Kirby for the information that she has found a recipe for heating an arsenic compound with vermilion to make orpiment. This method should not work chemically, but it may be indicative of the broad experimentation of alchemists in the medieval period.
9. SEM/EDX analysis of the pigment was kindly undertaken by Dr A. Roy; see Park 1997: cat. 4, figs 4.17, 4.22 and 4.23.
10. Analysis by J. Darrah in Ballantyne 1988.
11. Howard 2000 and forthcoming.

Application

Over coloured underpaint

In the Rochester *Wheel of Fortune*, orpiment was employed for Fortune's brooch, and applied over indigo combined with chalk to produce the green colour at the base of the wheel.

Mixed with other pigments

Analysis of a sample from the early 13th-century wall paintings excavated from the crypt of Worcester Cathedral suggested the presence of orpiment combined with lead white used for drapery, and in the fragment of a face which is only 6.5cm high. The tiny and relatively uniform particle size – 2–5µm – suggested that this orpiment was produced artificially, probably by sublimation.[12] This is the earliest identified example of orpiment in English wall painting and a very early example of artificial orpiment.

Recent analysis of the late 14th-century *Apocalypse* cycle in the chapter house of Westminster Abbey has confirmed the presence of orpiment employed for the flesh tone of a tiny nimbed figure in scene 60 in the first arch of the south bay. Here the flesh tone consists of orpiment combined with yellow and red earth pigments, glazed with red lake. The pigment was also employed beneath a layer of tin leaf for the trumpet of the *Musical Angel* in the second arch of the south-east bay, and as a component of an oil-based mordant with inclusions of white, red and black pigments for a layer of tin leaf used for the window tracery in the scene of the *Seventh Vial* (Plates 187 and 188).[13]

Unusual combinations of orpiment – with lead-tin yellow, vermilion and lead white – have been identified in the paintings of *c*1420 at Farleigh Hungerford Castle. For St. George's garter a combination of orpiment, vermilion and lead white was applied over gold leaf to intensify the pinkish-orange colour of the metal, while the gold colour for the plate armour under the garter appears to have been produced by mixing orpiment and lead-tin yellow. Extensive use of orpiment applied on its own is also found in areas such as the shield above St. George (Plate 189) (Howard *et al.* 1998: 61–2).

A sample taken from Fortune's sleeve at Rochester Cathedral confirmed that orpiment and indigo were also combined to produce a pale blue-green colour (Darrah 1992: 1–2).[14] A mixed green – of orpiment and an unidentified blue pigment – was also employed for the heraldic devices on the north wall at Ashby St. Ledgers. Here, foliage on a layer beneath the late 15th-century fleur-de-lys motif was painted with the orpiment mixture (Curteis 1994a).

Binding media

Some medieval treatises propose particular binding media for the use of orpiment in wall painting. Eraclius suggests a combination of orpiment and egg (Merrifield 1967: 1: 238–9), while Jehan le Begue recommends fig milk (Merrifield 1967: 1: 96–7), and Cennini prescribes nothing but size (1960: 29).

Instrumental analysis of the original binding medium has only been undertaken at Farleigh Hungerford, where it was identified as oil. However, the Rochester *Wheel of Fortune* is also presumed to be in an oil medium, and given the complex stratigraphy in

12. There is also a tiny amount of vermilion present (less than 2% of the sample).
13. Howard 2000 and forthcoming.
14. Darrah 1992:1–2. A combination of orpiment and indigo was suggested by Theophilus (1979: 22) for painting drapery on panels, and by Cennini (1960: 32) for painting rooms in *secco*.

all the other instances and the instability of orpiment in alkaline conditions, it is likely that an organic binder is invariably present.

Comparison with its use in other media and in continental wall painting

Wall painting

Orpiment has been detected in bay two of the scheme of *c*1270–80 in the choir of Angers Cathedral in France, while lead-tin yellow and yellow ochre have been employed elsewhere (Demailly *et al.* 1998: 11).

Polychrome sculpture

Although orpiment has not yet been identified in early medieval wall painting, early examples of its use have been confirmed in other media: on fragments of carved wood of *c*960 from the Grave of the King of Gorm, Jelling, Jutland; on a plank from the 11th-century Danish stave church at Hørning;[15] on a 12th-century Virgin and Child from Auvergne (Kargère 2002); and on the mid-13th-century painted wooden ceiling at St. Michael's Church, Hildesheim (Denninger 1969).

Documentary sources provide evidence for the use of orpiment for English polychrome sculpture in the Gothic period. For example, colours purchased for the restoration of the Shrine of St. William in Norwich Cathedral in 1325 include 40 leaves of gold leaf, 12lb of white lead, vermilion and *orfainment* at 2s 2d, together with oil for painting (Saunders 1930: 111). In 1334, 2s 2d was paid for 3lb of 'gold colour' – probably orpiment – for the vaulting of the new belfry at Ely (Salzman 1952: 168). Moreover, the analysis of the polychromy on the west front of Exeter Cathedral has confirmed the presence of orpiment in both the 14th- and 15th-century phases of decoration. The pigment was usually applied pure, in a thin layer, and was employed for details such as crowns and hats (Sinclair 1991: 123, 1992: 10).

Panels

A combination of orpiment and indigo – now much faded – has been identified in the Wilton Diptych (*c*1395), in the grey-green foliage in the foreground of the *Virgin and Child with Angels* and of the *White Hart* (Roy 1997: 133–4). The same pigment mixture has also been identified on a Norwegian altar frontal from Eid, dated after *c*1300 (Plahter 1995: 120), while a mixed green of orpiment and charcoal black was employed on another Norwegian frontal – from Kinsarvik (dated before *c*1300) – although in this instance the orpiment has now 'faded' (Plahter 1995: 120). The pigment also occurs on all three altarpieces from Tingelstad (dated to 1250–1350); in each case the binding medium for the orpiment appears to be egg tempera, while the pigment remains intact only in areas where it has been covered by another paint layer. The largest particle detected in the Tingelstad I altarpiece (*c*1260–1300) is some 5mm in length, and here the coarsely ground bright yellow material, with a pearly lustre and lamella cleavages, is thought to be ground mineral orpiment (Plate 186) (Plahter *et al.* 1974: 96).

Manuscript illumination

Orpiment combined with indigo has been identified in the Book of Kells (*c*800) (Fuchs and Oltrogge 1994: 136). Documentary sources also indicate that orpiment was

15. Mads. Christensen and J. Glastrup, poster at ICOM 1996.

employed from the early Middle Ages, and it is listed among pigments suitable for manuscript illumination in a mid-12th-century manuscript, probably from Salzburg (Petzold 1995: 62–3).

Discussion

Despite its instability in alkaline conditions, and in combination with lead and copper pigments, orpiment has been detected in a number of Gothic wall paintings. Although the pigment was employed in other media in the Romanesque period, it has not yet been identified in wall painting. Its instability in alkaline conditions would be sufficient reason for the preference for other yellow pigments in Romanesque wall painting, where the carbonation of lime is the most important component of the binding process. However, it also seems likely that the pigment would have been more expensive before the methodology for the manufacture of dry-process synthetic orpiment was widely available in the Gothic period.

The date of the earliest confirmed identification of the pigment in English wall painting – in the early 13th-century excavated fragment of wall painting from Worcester Cathedral – accords reasonably well with the first documentary evidence for its use in painting, in the Westminster Palace accounts of 1265. Elsewhere in northern Europe, the only other use of orpiment in wall painting known to the author is in the choir scheme of *c*1270–80 in Angers Cathedral.

Pigment alterations

Darkening of orpiment

The incompatibility of orpiment with lead and copper pigments is well documented, and is probably due to a reaction between the sulphur from the orpiment and copper or lead ions, to produce copper or lead sulphides (Weber 1923: 96; Thompson 1956: 178).

In recent experiments, lead white, red lead, verdigris and lead acetate were combined with orpiment in a number of organic binding media and subjected to varying environmental conditions. Where stand oil was employed as a medium there was no sign of alteration, while similar pigment mixtures in a medium of egg yolk, gum arabic or cherry gum darkened in a reaction accelerated by increased humidity. In addition to the role of the binding medium in the darkening process, Wallert also noted that alteration was particularly marked with pigment mixtures including verdigris and lead white (both prepared with acetic acid). He concluded that excess acetic acid may initiate the darkening reaction by decomposing the orpiment, and that an aqueous binding medium and humid conditions also appear to be necessary for the formation of arsenic sulphides (Wallert 1984: 50–51).

Scant evidence, which *may* implicate orpiment in the darkening of both a lead and a copper pigment, exists for one medieval wall painting in England. Analysis of paint samples from the 14th-century *Christ in Majesty* at Farringdon (Hampshire) confirmed the presence of orpiment as well as a discoloured lead pigment – presumed to be red lead – and a partially altered copper blue pigment.[16] However insufficient detail is reported to determine whether the orpiment is immediately adjacent to, or combined with the lead

16. Analysis by J. Darrah.

or copper pigments (Ballantyne 1988). Since alteration of lead and copper pigments is commonly reported in wall paintings where orpiment is not present, the yellow pigment cannot be firmly implicated in the pigment transformations which have occurred at Farringdon. In both other cases where orpiment has been identified in combination with a lead pigment – at Worcester Cathedral with lead white, and at Farleigh Hungerford Castle with both lead white and lead-tin yellow – no darkening is apparent. In the Farleigh Hungerford painting (the only instance where organic analysis has been undertaken) it seems likely that the oil medium provides sufficient isolation of the pigment particles to prevent reaction between the sulphur and lead.

Lightening

Loss of colour in the orpiment particles is, however, evident at Farleigh Hungerford, particularly in the uppermost glaze layer of a sample from St. George's garter, where some of the pigment particles have altered to form a white or translucent material. Both orpiment and realgar lose colour on exposure to light: realgar alters to form para-realgar, while orpiment oxidises to form translucent or white oxides of arsenic.

The 'fading' of orpiment on exposure to light has been described by a number of 20th-century authors (Laurie 1967: 90; Church 1915: 184; Weber 1923: 96). More recent experiments have indicated that although orpiment is generally stable to visible and UV light when dry, increased humidity *and* light induces the oxidation of the pigment to form white arsenic trioxide and hydrogen sulphide.[17] As with the darkening of the pigment – which may occur in mixtures with copper and lead pigments, resulting in the formation of black copper and lead sulphides – orpiment appears to be more susceptible to decolorisation in paint films with an aqueous binding media.[18] Although the original binding medium at Farleigh Hungerford is oil, it is possible that leaching of the medium may have occurred – particularly from the upper portion of the paint layers – during recent programmes of invasive conservation and cleaning of the paintings (Howard *et al.* 1998: 60).

Decolorised orpiment has been identified in a number of other instances, as on the Kinsarvik altar frontal (of before *c*1300) (Plahter 1995: 120), and on the 12th-century lion capital from Vossestrand where a greyish white material following the line of the carved brow was identified as decomposed orpiment. In other areas on the capital, where it was protected from light, the pigment was found to be better preserved (Plahter 1990: 279). Examination of the *Book of Kells* (*c*800) also revealed that the orpiment had sometimes lost its brilliant yellow hue, appearing white or pale brown in colour. Here the alteration is thought to be due to light, humidity and dirt (Fuchs and Oltrogge 1994: 134–5).

Lead-tin yellow

There are two forms of lead-tin yellow distinguishable by XRD, Types I and II, the more common being Type I (Pb_2SnO_4). Type II is also lead-tin oxide, but additionally may contain free tin oxide and additional silicon (Si).

17. Orpiment has also shown severe colour loss on exposure to ozone in the absence of light; Fitzhugh 1997: 51.
18. Dr A. Roy, pers. comm.

There has been widespread confusion about the identity of this pigment over the centuries. *Giallorino* (or *giallolino*) means a pale yellow colour. Merrifield (1967: 1: clviii) asserted that *massicot* in the northern manuscripts was identical with *giallolino* mentioned in Italian manuscripts. *Giallolino* was translated as *massicot* or *general* by Haydocke (1598), and in another 16th-century English source *massicot* and *general* are listed separately, indicating that they were regarded as two distinct pigments. However, it is now generally accepted that in the medieval period *giallorino* (*giallolino*) was the term employed in southern Europe to describe lead-tin yellow, while *massicot* was employed to describe this pigment in northern texts.[19] Thus, where massicot is mentioned in documentary sources – as at York and Westminster – lead-tin yellow is indicated (Salzman 1952:168–9). However, today the terms massicot and litharge are used for the yellow monoxide of lead (PbO) which is often found as an impurity with red lead as a result of the process of manufacture.

Yellow lead monoxide (PbO) was also known in the medieval period, and Jehan le Begue describes yellow *sandaraca* (clearly lead oxide) which is made by heating lead white, and which will turn into red lead if heating continues (Merrifield 1967: 1: 314). It is clear that this yellow pigment was employed for painting since he suggests that it should be combined with water for masonry, egg for parchment and oil for wood (Merrifield 1967: 1: 314).[20]

Cost

Evidence for the use of lead-tin yellow in 15th-century England is provided by an entry in the Fabric Rolls for York Minster for 1483, which documents the purchase of 4lb *masticote* together with verdigris, vermilion, red lead, lead white and indigo for the central tower vault. Later, the Westminster Palace accounts of 1532 specify *mastecote* at 2s 8d per lb. The expensive nature of this pigment is clear given that the same accounts reveal that verdigris cost only 1s per lb and vermilion 1s 2d (Salzman 1952: 168).

Optical, chemical and physical properties

In plane polarised transmitted light both types of lead-tin yellow appear as finely divided very pale yellow particles, sometimes displaying conchoidal fracture. The refractive indices of both types are high (greater than 2.0). Type II lead-tin yellow is isotropic, while Type I has a tetragonal crystal system and displays birefraction with crossed polars (refractive indices of $\alpha = 2.29$ and $\gamma = 2.31$).

19. For a full discussion of the obsolete terminology, see Kühn 1993b: 83–6.
20. In the only two instances in which instrumental analysis has confirmed the presence of yellow lead monoxide in English medieval polychromy – in the scheme of *c*1220 on the nave ceiling of Peterborough Cathedral (Howard 1997b: 6), and in the 16th-century heraldic wall paintings on the south pier of the crossing at Holy Trinity, Coventry (Howard 1996b: pl. 57.) – only a single particle appears to be present in the samples examined, and it is likely that these represent an impurity formed in the production of red lead. However, since litharge – lead monoxide – also acts as a powerful drier for oils (Fitzhugh 1986: 122), it may have been added intentionally in these instances for this reason. In a few other cases the pigment has been identified on the basis of PLM and a microchemical test for lead alone, but this provides insufficient analytical evidence to differentiate between massicot (yellow lead monoxide), lead-tin yellow and lead antimonate yellow.

Due to the high refractive indices of lead-tin yellow, both types have good hiding power when combined in an oil medium. In addition, these lead-based pigments have siccative properties, accelerating the drying rate of oil media, and forming lead salts of fatty acids (Kühn 1993b).

Both types of lead tin-yellow are stable in acidic conditions unless boiled with concentrated acids for prolonged periods. The pigment is also stable with respect to light, although heating above 900 °C will induce decomposition with the formation of free tin oxide. Exposure to soluble sulphides or hydrogen sulphide may induce darkening, with the formation of black lead sulphide. Where an oil or oil-resin medium is employed such a reaction is extremely unlikely, although an aqueous medium may facilitate such an alteration (Kühn 1993b: 91–4).

Lead-tin yellow is extremely stable in alkaline conditions, and Cennini recommends it for painting in both *fresco* and *secco*, writing that 'it lasts forever, ground in clear water'. He goes on to describe *giallorino* at length, indicating that the pigment is synthetic, solid and rather difficult to grind. He recommends it for producing mixed greens (particularly for foliage), and for representing silk cloth and shot drapery (Cennini 1960: 28, 50, 53, 56, 90).

Preparation

Lead-tin oxides were employed as opacifiers for glass long before the earliest recipe for the manufacture of the pigment *giallorino* (lead-tin yellow) is given in a Bolognese manuscript dated to the early 15th century. This recipe recommends combining calcined lead and tin, and heating them with minium and sand. The pigment obtained by following these instructions is lead-tin yellow Type II. Type I is made by heating three parts of lead monoxide, minium or lead dioxide with one part of tin dioxide to between 650 and 800 °C ((Kühn 1993b: 89–91).

Use in English medieval wall painting

The few instances of the use of lead-tin yellow that have been confirmed in English wall painting are generally of 15th- or early 16th-century date, and include the early 15th-century scheme in St. Leonard's Chapel, Farleigh Hungerford Castle (Howard *et al.* 1998); 15th-century paintings in the presbytery of St. Albans Cathedral (Perry 1998: 6); the Eton College paintings of *c*1479–87 (Plummer 1986: 36), and the paintings from St. Mark's, Bristol, of *c*1510 (Gill and Howard 1997: 99). The pigment has also been identified on an excavated fragment of (?) 14th-century date from the Bedern at York.[21] Preliminary identification of the pigment has also been made in an early 15th-century wall painting in the south choir aisle of Lichfield Cathedral.[22]

In the case of the Farleigh Hungerford paintings, XRD analysis confirmed the presence of lead-tin yellow Type I.[23] Examination of the SEM/EDX spectrum from a sample

21. The fragment may have been part of the original decoration of the chapel (*c*1340s), or associated with a late 14th-century phase of rebuilding; see analysis by Howard in Wong forthcoming. Whether lead-tin yellow Type I or II is present in this instance has yet to be determined by XRD.
22. Unpublished analysis by J. Darrah.
23. This analysis was generously undertaken by Dr A Roy of the National Gallery using a Debye-Scherrer camera with copper radiation (40 KeV and 35 milli-amps for 3½ hours).

of the pigment employed in the Bristol paintings confirmed the presence of a significant quantity of silicon (Si), and it is possible that Type II lead-tin yellow is present although XRD analysis would be necessary to prove this.

Application

Over coloured underpaints

In the scheme at St. Mark's, Bristol, a layer of lead-tin yellow 35µm thick was applied over a complex system of pale pink, white and grey underpaint for the halo of the female figure in the *Noli me Tangere* (Plates 86 and 190). The precise function of the pale pink layer immediately beneath the yellow is unclear, but it may have served to warm the slightly acid-yellow colour of the lead-tin yellow (Gill and Howard 1997: 101, pl. XXXIId).

Mixed with other pigments

At Farleigh Hungerford, lead-tin yellow was employed in various pigment combinations to achieve different coloristic effects. It was combined with lead white and applied over an underpaint of verdigris and lead white in oil for the pale yellow heraldic motif immediately above St. George on the east wall of the chapel. In addition, a complex mixture of lead-tin yellow, verdigris, red lake and lead white was detected in another part of this motif. Examination of a sample from St. George's garter suggested that a gold colour for the plate armour under the garter may have been achieved by mixing orpiment and lead-tin yellow (Howard *et al.* 1998: 60–62).

Binding media

In both the Farleigh Hungerford and Bristol paintings instrumental analysis indicated that oil was employed as the binding medium, while at Eton linseed oil was identified. Owing to its high refractive index, lead-tin yellow has extremely good covering power in oil and oily-resinous media, and will also act as a siccative.

Comparison with its use in other media and in continental wall painting

Lead-tin yellow has been identified in the fifth and seventh bays of the wall paintings of *c*1270–80 in the choir of Angers Cathedral (Demailly *et al.* 1998: 11). This is one of the earliest identifications of the pigment in any work of art, but unfortunately few details of its application are provided. The pigment has also been found in 15th-century wall paintings at Ghent and elsewhere on the continent, though again no details of the application method are provided (Kühn 1993b: 106–7). The pigment has also been identified in one instance in Danish wall painting,[24] and on two wall paintings of *c*1500 from Sweden (Nord and Tronner 2000: 99).

In English polychrome sculpture, the smallest traces of lead-tin yellow have been identified in the 14th- and 15th-century polychromy on the west front of Exeter Cathedral,[25] and on the late 15th-century effigies at Arundel (Sussex) (Brodrick and Darrah 1986). An oyster-shell palette dated to the 13th century and containing lead-tin yellow was found in

24. K. Trampedach, pers. comm.
25. The pigment was identified on a rosette on the armour of a knight (B19) dated to the 14th century, and on a 15th-century supporter angel of figure C17; see Sinclair 1991: 123–4 and 1992: 10.

excavations at St. Augustine's Abbey, Canterbury, although unfortunately the type of painting for which the pigment was employed is unknown (Sherlock and Woods 1988: 200).

The pigment has also been identified in a few instances in polychrome sculpture elsewhere in northern Europe. The earliest instance is on a *Virgin* of *c*1350 now in the Diocesan Museum, Freising (Germany) (Kühn 1993b: 106). Other examples include the sandstone *Moerlin Epitaph* of 1497 (Kühn 1993b: 106); and fragments of late Gothic sandstone sculpture excavated in the vicinity of Berne Minster (Switzerland) (Gutscher *et al.* 1989: 117–18).

Lead-tin yellow seems to occur first in panel paintings in *c*1300. The pigment was used throughout the 14th century, principally as Type II. From the 15th century, numerous instances of its use have been identified (Kühn 1993b: 86, 99–111). One notable occurrence is on the late 14th-century Estouteville triptych from Germany (now in the Gallery of the Courtauld Institute of Art, London), where the pigment was combined with verdigris to produce a range of greens in the groups of Apostles (Villers *et al.* 1997: 674).

Discussion

Despite its stability in alkaline conditions and with organic binding media, lead-tin yellow has been identified in only a small number of medieval wall paintings. All identified examples of this comparatively expensive pigment in English wall painting date from the 14th to 16th centuries. The few identified examples of the use of the pigment in English polychrome sculpture also date from the 14th and 15th centuries, while on the continent the earliest identified example in wall painting is *c*1270–80, and in sculptural polychromy of *c*1350. Although lead-tin yellow first occurs in panel paintings of *c*1300, it again remains fairly uncommon until the 15th century.

In the Farleigh Hungerford paintings, XRD analysis confirmed the presence of lead-tin yellow Type I. On the continent, XRD analysis has determined the presence of Type I lead-tin yellow in 15th-century wall paintings at Freiberg (Germany).[26] So few samples from wall paintings have been subjected to this type of analysis that it is impossible to draw any conclusions regarding the use of Type I in these instances. However, evidence from samples of the pigment employed in easel paintings suggests that, of the two crystalline types, Type II is generally the earlier pigment, with notable examples in Florentine and Bohemian paintings of the 14th-century. Research – albeit based on SEM/EDX analysis and SEM imaging rather than XRD analysis – has also indicated that, in Italy, Type I replaced Type II during the second quarter of the 15th century (Martin and Duval 1990). In any event, the identification of lead-tin yellow Type I in the early 15th-century scheme at Farleigh Hungerford Castle, although unusual for wall painting, conforms to the expected crystalline type of the pigment for contemporary northern European painting.

Pigment alterations

The pigment is generally considered to be stable, and no darkening has been detected in the samples identified.

26. For these and other notable occurrences, see Kühn 1993b: 99–111.

EIGHT

White pigments

Lime white

The lime white pigment *bianco di San Giovanni* ($CaCO_3.Ca(OH)_n$) was an important component of the wall painter's palette; indeed for the Romanesque artist it played a crucial role in painting technology. Unlike other calcium carbonate whites, lime white retains a considerable proportion of unreacted calcium hydroxide and therefore, in addition to its role as a white pigment, functions as a binding medium.

Cost

Lime is often mentioned in English medieval accounts, but it is clear from the quantities involved that lime putty for building rather than the pigment lime white was being purchased. For example, for works at Westminster Abbey in 1253, Agnes the lime burner received 12s 6d for two and a half hundredweight of lime (Colvin 1971: 261). Although, in the Exeter Cathedral accounts for 1323–24, lime appears in a list with various painting materials including a black colour, silver foils and verdigris, it is still not clear whether the three quarters of lime at 15d were employed as a painting or building material since timbers are also mentioned (Erskine 1981: 146–7).

Optical, chemical and physical properties

Particles of lime white are often extremely finely divided, displaying no characteristic form even at high magnification. Once fully carbonated, however, in common with other forms of calcium carbonate, lime white has two principal refractive indices $\alpha = 1.658$ and $\gamma = 1.486$. In transmitted light with crossed polars, the pigment is highly birefracting (Gettens *et al.* 1993b: 208–13).

The important distinction between the pigment lime white and pure slaked lime (lime putty) is that while the latter is semi-translucent and has a high coefficient of shrinkage on drying, lime white is an opaque white pigment which shrinks little on drying. In addition, when combining other pigments with lime white it is possible to determine the desired hue and tone during the mixing process, whereas with slaked lime the resulting colour cannot be determined until the mixture has completely dried. However, the most important characteristic of lime white is that it retains a high proportion (approximately 30%) of unreacted calcium hydroxide which, on carbonation, functions as a binding medium.[1]

Since the pigment retains unreacted calcium hydroxide it remains highly alkaline for a considerable time; indeed, Denninger found the pH of a saturated solution of *bianco di San Giovanni* to be 13 (1974: 186). This residual alkalinity means that the pigment may be combined with casein or egg white to form extremely stable calcium caseinates and calcium albumin compounds, and it is likely that this property was both known and utilised by the medieval painter.

The pigment is stable in alkaline conditions, but contact with even dilute acids results in the evolution of gaseous carbon dioxide. Lime white is stable to light, but heating to temperatures above 500 °C will initiate decomposition, with the evolution of carbon dioxide. Temperatures between 800 and 1000 °C will result in complete calcination and the formation of calcium oxide (CaO) (Rickerby 1991: 42). Lime white is stable in contact with sulphide-based pigments, although it is not suitable for use with alkali-sensitive pigments such as red lead or lead white (Gettens *et al.* 1993b: 210). Contact with sulphur dioxide and high humidity results in the formation of calcium sulphate which causes serious damage due to the greater volume and deliquescence of calcium sulphate.[2]

Preparation

Cennini describes the process for manufacturing lime white: combining air-slaked lime[3] with water and stirring the mixture for eight days, before moulding it into small cakes and allowing them to dry in the sun. This process was repeated as necessary. For painting on the wall the cakes were ground thoroughly in water. Cennini indicates that absolutely no other binding medium is necessary, and that the pigment is essential for flesh tones – combined with finest, lightest *sinopia* – and for mixing with other pigments (Cennini 1960: 23, 34).

Use in English medieval wall painting

Throughout the Romanesque period lime white was universally employed as a white pigment. Even where lead white was incorporated in the palette, it was used in conjunction with lime white. Lime white was used not only for highlights but combined with other pigments in order to obtain the desired tone, and frequently also functioned to provide additional binding capacity within a complex stratigraphy.

1. Experiments to determine the proportion of unreacted calcium hydroxide present in lime white were undertaken by Denninger (1974: 186).
2. Calcium sulphate has an equilibrium RH (RHeq) of 99.96% and can therefore be regarded as a deliquescent salt. For a study of the kinetics of deliquescence and crystallisation of soluble salts in wall painting, see Sawdy 1995.
3. Slaked lime (calcium hydroxide) is obtained through a double transformation of calcareous rock: (a) calcination at 1000 °C in the lime kiln produces calcium oxide or quicklime ($CaCO_3 \rightarrow CaO + CO_2$); (b) water is added to produce slaked lime ($CaO + H_2O \rightarrow Ca(OH)_2$ + heat). Depending on the precise amount of water and conditions involved in the latter process the product may be a fine white powder or lime putty of the type employed in lime mortars. However, if quicklime (CaO) is allowed to absorb water and carbon dioxide from the air, this is called air-slaking, and it is this process which Cennini employed for the production of *bianco di San Giovanni*.

With the increased use of lead white and a greater desire for translucent effects, lime white was less suited to the Gothic wall painter's palette.[4] Since organic binding media were also increasingly common in this period, the function of lime white as a binding medium was correspondingly less important. Nevertheless, the pigment was still widely employed in the Gothic period, particularly in parish church schemes.

Application

Use as a ground

In the scheme of *c*1100 at Hardham (West Sussex) it is clear that the intention was to paint largely in a *buon fresco* technique, and the primary mechanism for binding the pigments was the carbonation of calcium hydroxide from the lime plaster. However, local lime grounds – applied in one or two layers – were also employed in some areas. The use of this white pigment as a local ground, and in various pigment mixtures providing additional binding capacity, is not only consistent with the *fresco* process, but is a crucial element of the technique.[5]

From the first half of the 13th century lime grounds were frequently employed for wall painting. In the refectory of the Eastbridge Hospital, Canterbury – where the scheme of *c*1220 is executed over an earlier painting – a calcium carbonate ground, approximately 1–2μm thick, was applied with a broad brush in several layers in preparation for the painting. Similarly, in the approximately contemporary decoration on the entrance arch of St. George's Chapel, Ely Cathedral, a lime ground was used to provide a compact white surface for painting. The ground here also serves to enhance the visual unity of the diverse levels of the support, which vary from the cross-bedded ashlar blocks of the voussoirs and quoins to the rubble core of the arch.[6]

Lime white combined with other pigments

At Hardham, lime white was frequently combined with other pigments, not only to produce tonal variation and particular coloristic effects but also to provide additional binding capacity. For example, it was combined with the copper green pigment atacamite for the nimbus of the *Agnus Dei* over the chancel arch, and with red and yellow earth pigments for the flesh tones of figures. Additionally, it was combined with charcoal black to achieve a false blue colour in the dado and elsewhere.[7] In the paintings of *c*1170 at Cormac's Chapel, Cashel (Ireland), lime white was also combined with many of the pigments, for example, with vermilion to produce a brilliant red on the western segment of the vault, and with ultramarine over a local ground of charcoal black mixed with lime white (Howard 1991: 2).

Similarly, the pigment played an important role in more sophisticated paintings, such as that on the east wall of the Holy Sepulchre Chapel at Winchester Cathedral (*c*1175). Here, multiple layers were applied to produce complex effects of modelling in areas of flesh painting. For example, the mid-dark flesh tone was produced by the application of

4. Cennini (1960: 59) states that lime white may not be combined with oil.
5. Analysis by Howard in Sawdy *et al.* 1998: 34–5.
6. For the Canterbury paintings, see Howard 1994; and for St. George's Chapel, see Howard and Pathak 1995.
7. For this false blue pigment, see chapter nine on black pigments, and for the pigment analysis at Hardham by Howard, see Sawdy *et al.* 1998: 34–5.

five different layers of earth pigments – green, red and yellow – applied either singly or in combination, usually with the addition of lime white and charcoal black. The carbonation of the lime white pigment here would have provided additional binding capacity within the complex layer structure (Howard 1995a: 101).

Theophilus frequently refers to lime white for use on walls, generally combined with other pigments; for example, he suggests that for laying in a robe, lime white should be combined with ochre (1979: 23). This methodology is roughly adhered to in the scheme of *c*1110 at Witley, Surrey. Here, once the preliminary drawing was complete, the main features of the design, including drapery and architecture, were painted in mixtures of lime white and red and yellow earth pigments. Details were built up in a range of tonalities over the basic mid-tones, with lime white highlights added finally in *impasto* on the surface (Howard 1992a: 10).

Lime white was still widely employed in the Gothic period. For example, in the paintings of the first quarter of the 14th century at Little Wenham (Suffolk), a synthetic copper blue pigment was combined with lime white and applied over a lime ground for the blue draperies of the female saints on the chancel east wall. A similar pigment mixture was employed for the background of the fictive reredos of *c*1335 on the chancel east wall at neighbouring Brent Eleigh (Howard 1988: 34, App. II). In a cathedral context, lime white was incorporated in the palette of the early 13th-century scheme in the Holy Sepulchre Chapel at Winchester, where charcoal black combined with lime white was used as an underpaint beneath vivianite for Joseph of Arimathea's leg in the Deposition (Howard 1995a: 102). Lime white was also identified in the 13th-century painted coffin-lid of Archbishop Walter de Gray in York Minster (Ramm *et al.* 1971: 122).

Impasto *highlights*

Since lime white pigment retains a high proportion of unreacted calcium hydroxide and therefore requires no additional binding medium, it is particularly suitable for highlights (Denninger 1974: 185–7). For this purpose, the pigment was used to particularly good effect in the scheme of *c*1100 at Coombes (West Sussex), and also in the paintings of *c*1130 in St. Gabriel's Chapel, Canterbury Cathedral. In the latter, details such as the flesh highlights and the imitation pearls in the jewelled borders were built up using pure lime white. In the flesh painting, layers of varying combinations of yellow earth and lime white were applied, with the lime white highlights then applied in thick *impasto* still being clearly visible in UV light (Plate 191).

Lime white highlights applied in considerable *impasto* were also employed to great effect in the 'Losinga' cycle of paintings of *c*1190 in Norwich Cathedral. Many of these highlights are still clearly visible, as on the foliate decoration between the roundels on the soffit. Others – such as the details of the collars and clasps of the standing figures on the vault – are much more difficult to detect, and now are only clearly visible in UV light (Plate 192).

Binding media

In the late 12th-century 'Losinga' cycle at Norwich, lime white was used extensively to provide additional binding capacity for the complex stratigraphy and *impasto* of the upper paint layers. In some areas of the painting, as well as functioning as a binding medium, the pigment radically influences the tone of certain colours; thus, for the very pale blue pigment on the east reveal of the western arch, it was combined with natural ultramarine. In other areas, where a substantial alteration of tone may have been considered undesirable – as for the brilliant ultramarine sky behind the cathedral in the lowest roundel on the soffit – samples show that the quantities of lime white were considerably reduced, and indeed that lime in some other form, such as lime-milk or lime-water, may have been employed to ensure that the intensity of the blue colour was maintained (Howard and Gasol 1996: 20–21). Instrumental analysis to determine the presence of any additional organic binding medium in this scheme was undertaken on two samples – one of vermilion combined with lime white from the central surviving roundel on the soffit, and the other, of atacamite in a lime matrix (applied over charcoal black in a lime matrix) from the roundel below – but in neither case was there evidence of original organic binding material.[8]

Analysis of fragments of *c*1190 excavated from the crypt of Glasgow Cathedral confirmed that although the carbonation of lime is the principal binding mechanism – particularly where a multiple layer structure is evident – a proteinaceous component is also present in some samples. For example, FTIR analysis of a green earth paint layer confirmed the presence of a trace of protein in addition to calcium carbonate and the green pigment. The red lake glaze on the surface of the green layer in this sample was found to be particularly medium-rich, and preliminary GC-MS analysis confirmed the presence of protein, possibly casein or casein with animal glue (Howard 1998a). Whether the reaction between lime white and casein to form a stable calcium caseinate was employed in this instance would require further analytical investigation. However, in other instances the combination of these materials has been identified; for example, lime combined with casein was detected in the scheme of *c*1240 at St. Georgen, Austria (Plate 193), in the preparatory layers of bays two and three of the scheme of *c*1270–80 in the choir of Angers Cathedral (France), and in the early 12th-century polychromed stone sculpture in the apse of the Church of St. Michele, Val Susa, Italy where the stone was coated with a layer of red iron oxide combined with calcium caseinate in preparation for painting.[9] A lime-casein medium has also been identified in the mid-13th-century painted wooden ceiling at Hildesheim, Germany (Denninger 1969: 92–3).

Comparison with its use in other media and in continental wall painting

Lime white was as commonly employed in Romanesque wall painting elsewhere in northern Europe as it was in England, and has been identified, for example, in the paintings of *c*1130 at Idensen (Germany) (Matteini and Moles 1994: 82); in the scheme of *c*1165 in the Johanneskapelle, Pürgg (Austria) (Howard 1995d: 4); and in the scheme of *c*1200 at Müstair (Switzerland) (Emmenegger 1986: 197–8).

Fewer examples have been identified in Gothic wall painting, and apparently none in medieval panel painting where chalk, lead white or calcium sulphate were employed. There are only a few identified examples of the use of lime white in polychrome sculpture, where again the use of lead white or chalk is more common. However, lime white has been identified in the 13th- and 14th-century polychromy on the west front of

8. I am grateful to R. White for undertaking organic analysis of these samples; White 1998: 1.
9. The analysis at St. Georgen was undertaken by Dr K. Petersen using antibody staining techniques. For the technique of these paintings, see Leitner 1989: 109–11. For the technique of the Angers wall paintings, see Demailly *et al.* 1998. The sculptor of the polychromed figures in the apse at Val Susa is Magister Nicholaus, known for his work at Ferrara Cathedral, Parma and Cremona. I am grateful to S. Pulga for information regarding the sculpture at Val Susa.

Salisbury Cathedral where it has undergone partial transformation to form calcium sulphate and calcium oxalate (Howard 1995e: 2, 1996d: 3), and in the preparatory layers of the 14th- and 15th-century architectural polychromy on the west front of Exeter Cathedral (Sinclair 1991: 120–21, 1992: 10).

Discussion

Lime white was supremely suited to the palette and technology of Romanesque wall painting, and the pigment was commonly employed in that period in England and elsewhere in northern Europe. In these paintings, lime white functioned both as a white pigment and as a binding medium for other pigments compatible in an alkaline environment.

With the increased use in the Gothic period of alkali-sensitive pigments, and the related increase in the proportion of organic binding media, lime white was less frequently employed. Its extensive use in Romanesque wall painting was matched by the equally extensive employment of lead white in the Gothic period. Even the *impasto* highlights in lime white, which are such a feature of many Romanesque schemes, were easily replaced by the use of lead white combined with heat-bodied oil in Gothic painting. The higher refractive index of lead white and consequently greater covering power in oil media also made it a more suitable choice in sophisticated Gothic wall paintings where the application of translucent layers of colour over a highly reflective white ground was often a priority. Lime white does not appear to have been used at all in medieval panel painting, and very little in polychrome sculpture.

Pigment alterations

One of the major causes of deterioration in wall paintings is the action of atmospheric sulphur – often the result of pollution from burning fossil fuels – on lime white, lime grounds and lime plasters, resulting in the formation of calcium sulphate (Torraca 1988: 41). This phenomenon has been detected in the scheme of *c*1220 in the Eastbridge Hospital, Canterbury (Howard 1994: 9–10). Here FTIR analysis confirmed the presence of varying quantities of calcium sulphate in samples taken from the original calcium carbonate ground layers. In this instance, the covering of the wall by a fireplace and chimney in the 17th century, and consequent combustion of coal and other organic material would certainly have provided a rich source of sulphate ions. Similarly, in the scheme of *c*1100 at Hardham, sulphation due to the use of coal-burning stoves in the past has caused considerable damage. Given the widespread use of the pigment lime white, and the lime plaster substrate, the entire stratigraphy of these important paintings is susceptible to damage from sulphation (Sawdy *et al.* 1998: 64–9).

Sulphation is damaging since the volume change on the alteration of calcium carbonate to calcium sulphate is detrimental to the structure of the painting, often causing spalling and flaking of the surface. In addition, calcium sulphate is slightly soluble, and consequently – particularly in Romanesque wall paintings where calcium carbonate forms the fundamental structure of the support, and is the primary mechanism for binding the paint layers – the entire structure is theoretically extremely susceptible to moisture damage. Even when calcium sulphate is no longer being formed, it may be activated during episodes of condensation, causing further damage through swelling, partial dissolution and subsequent recrystallisation.

Chalk

Where lime white was not required to function as a binding medium, there was little or no advantage in using this pigment in preference to calcium carbonate ($CaCO_3$) in some other form, such as chalk, pulverised limestone, or perhaps even ground marble. Indeed, given the high alkalinity of lime white, there was a positive advantage in employing such alternatives.

There are good accessible chalk deposits in the south and east of England, and across the Channel in France and Belgium, and in these regions, it is likely that the local material would have been widely employed. Where no natural chalk exists, as in Norway, research has shown that the material was imported (Perch-Nielsen and Plahter 1995).

Cost

A Westminster Palace account of 1351 records that 4 gallons of size and a bushel of chalk dust supplied by John Crepelgate was used for whitening the walls of a downstairs chamber (Salzman 1952: 157). Given the accessibility of deposits in southern England, it certainly seems likely that chalk would have been an inexpensive commodity. This is supported by accounts of 1501 for the Chapel Royal at Stirling Castle – to which the material presumably would have been transported a considerable distance – where 4 stones of chalk were purchased at only 1s per stone (Apted and Hannabuss 1978: 134).[10]

Optical, chemical and physical properties

Once lime white is fully carbonated it can be difficult to distinguish the small particles from other forms of finely ground calcium carbonate. However, chalk often displays characteristic features, including tiny nanofossils known as coccoliths, which can be detected at high magnification in transmitted light with crossed polars (Plate 211) (Gettens *et al.* 1993b: 204–5, 211–12). The tiny platelets are highly birefringent and the two principal refractive indices are $\alpha = 1.658$ and $\gamma = 1.486$ (Gettens *et al.* 1993b: 208–12).

Although much less so than lime white, chalk is still alkaline, with the pH of precipitated chalk given as 9.8. In common with all carbonates, chalk is rapidly decomposed by acids with the evolution of carbon dioxide. As with other forms of calcium carbonate, heating to around 500 °C will initiate the decomposition of chalk, although the optimum temperature for complete decomposition and transformation to calcium oxide (CaO) is usually given as 800–900 °C (Rickerby 1991: 42).

Preparation

Chalk is generally ground under water, washed and then allowed to settle out. The finest grade of the material is found at the surface, and this was probably the grade that would have been collected and dried for use as a pigment (Gettens *et al.* 1993b: 205).

10. Although this measure of weight varied with different commodities, it would have been equivalent to several pounds at the very least, thus suggesting that chalk was cheap.

Use in English medieval wall painting

The presence of chalk has been confirmed in a few Gothic wall paintings in England, as in the mid-13th-century *Wheel of Fortune* in Rochester Cathedral, where it was combined with indigo;[11] in the splendid cycle of domestic wall paintings of *c*1330 at Longthorpe Tower (Cambridgeshire), where it is a component of the uppermost paint layer in a sample taken from a white stencilled motif on a king's mantle in the *Three Living and Three Dead* (Plates 210 and 211);[12] and in the early 16th-century paintings from St. Mark's, Bristol, where it was combined with calcium sulphate to bulk out the oil-based sealant for the stone support; it was also mixed with lead white and calcium sulphate as a component of white paint layers in a sample taken from an area of sky (Gill and Howard 1997: 99–100). Recent analysis of the late 14th-century *Apocalypse* cycle in the chapter house of Westminster Abbey also confirmed that a chalk ground was applied over the ashlar support in preparation for painting (Plates 218–220).[13]

Binding media

In the late 12th-century *Deposition* on the east wall of the Holy Sepulchre Chapel, Winchester Cathedral, a layer of calcium carbonate – which appears to be in the form of chalk – combined with an organic binding medium was identified in a layer of raised decoration on the borders of the Magdalene's drapery. The same mixture was detected beneath an oil-based mordant for gold leaf in a sample from the halo of the angel in the *Entombment* (Plates 212 and 213) (Howard 1995a: 96). Similarly, calcium carbonate has been identified as a component of the mordant employed for gilding on fragments of *c*1180 from Sherborne Abbey, Dorset (Plates 214–217) (Howard and Park forthcoming a).

Comparison with its use in other media and in continental wall painting

Chalk has also been detected in a number of instances in sculptural polychromy, all dating from the Gothic period: as a component of the *pastiglia* on the de Bois tomb of *c*1380 at Ingham, Norfolk; in the 14th- and 15th-century polychromy on the west front of Exeter Cathedral where it was combined with red lead, lead white and red iron oxide in the initial priming layers; in the ground layers of some samples taken from a group of early 15th-century polychrome stone fragments from Tewkesbury Abbey, Gloucestershire; in the ground of the 15th-century Fitzalan effigies at Arundel; on the fragments of *c*1475–90 from the Great Screen of Winchester Cathedral, where it was used, for example, in the grey underpaint of the blue lining of the Virgin's cloak in *The Virgin and Child*; and as a ground on the late 15th-century Cartmel Fell figure of Christ.[14]

Northern European panel paintings typically have a chalk ground (Bomford *et al.* 1990: 17–19), and recently analysed examples include the ceiling paterae of *c*1263 from

11. In a few samples a chalk ground was also tentatively identified; Darrah 1992.
12. Preliminary unpublished analysis by H. Howard.
13. Howard 2000 and forthcoming.
14. For the de Bois tomb, see Howard 1993d: App. 5; the Exeter polychromy is discussed by Sinclair 1991: 120–21; and for the examination of paint samples from the Tewkesbury polychromy, see UCL 1996. The most important of the Tewkesbury remains – which may originate from an altarpiece – include more than twenty fragments of figures. For the Fitzalan effigies, see Brodrick and Darrah 1986: table II; for the Great Screen, see Brodrick and Darrah 1989: 615–17; and analysis by Plesters and Roy of the Cartmel Fell figure is published by Behrens 1982: 133.

the Painted Chamber of Westminster Palace, which have a chalk ground applied in three layers, and the Thornham Parva Retable of *c*1335.[15] Chalk grounds have also been detected on a number of 13th and 14th-century Norwegian panel paintings and polychrome sculptures, and analysis indicates that, from the mid-13th century to the early 14th century, the chalk employed was generally imported from the Channel region. It has also been suggested that the use of Channel region chalk at this period is associated with English stylistic influence on Norwegian painting (Perch-Nielsen and Plahter 1995).

Discussion

Chalk has been detected in only a few instances in English medieval wall painting, with the majority dating from the Gothic period. Where this material does appear to have been used in Romanesque wall painting at Winchester and Sherborne, it is employed as a component of an oil-based mordant, or in raised *pastiligia*-like additions applied with an organic binding medium on the surface. Although lime white is generally such a feature of Romanesque wall painting, in these instances its high alkalinity would have been disadvantageous, and the additional binding capacity superfluous.

Although chalk was commonly employed as a ground in medieval panel painting in northern Europe, it was not generally used as a preparatory layer for wall painting. Indeed, a chalk ground has been identified in only one instance, in the important late 14th-century *Apocalypse* cycle in the chapter house of Westminster Abbey. In general, it seems likely that, even where the carbonation of lime was not employed as a binding mechanism, the application of a limewash ground would have been more economic on a large scale, and that once fully carbonated, it would have functioned similarly to a chalk ground.

Pigment alterations

Both chalk and lime white are susceptible to damage through sulphation. A discussion of this process is included in the section on lime white.

Calcium sulphate whites

Naturally occurring mineral gypsum is calcium sulphate dihydrate ($CaSO_4{\cdot}2H_20$). Perhaps the best known form of the mineral is alabaster – which is fine-grained and often pure white in colour – although other forms such as rock gypsum, selinite and satin spar are also common. Calcium sulphate anhydrite ($CaSO_4$) also occurs naturally, often beneath the hydrated form, and the two forms are also found in combination.

Gypsum deposits at Nore Down in the Isle of Purbeck were known in the medieval period, and documentary evidence also confirms that 'plaster of Paris' was brought from Purbeck to Clarendon in 1288. Burnt plaster of Purbeck is also mentioned at Portchester in 1397 (Salzman 1952: 156). There were also good deposits of gypsum in Yorkshire,

15. For the Westminster panels, see Liversidge *et al.* 1995: 499; and for the technique of the Thornham Parva Retable, see Goldsmith and Bucklow 1998; Bucklow forthcoming.

while the alabaster from Nottingham is renowned, and the waste and inferior deposits of this would doubtless have been utilised.

Calcium sulphate combined with animal glue is commonly known as gesso, and this material was widely used as a ground for panel paintings, particularly in the Italian trecento. Two types of gesso are detailed by Cennini for use as grounds for panel painting: *gesso grosso* – pure anhydrite ($CaSO_4$), or a mixture of agglomerates of anhydrite and gypsum ($CaSO_4 \cdot 2H_20$); and *gesso sottile* – composed of very fine gypsum crystals. *Gesso sottile* formed the finer, upper portion of the ground layer, which was often covered with a bole in preparation for gilding.[16]

Cost

Plaster of Paris is itemised in the mid-14th-century accounts for Westminster Palace; for example, 3 bushels at 5s were purchased for mending the walls of St. Stephen's Chapel. Here, as in other cases, it is clear that it was not intended for use as a painting material, but rather for plastering or whitening walls (Salzman 1952: 156).

Optical, chemical and physical properties

The refractive indices of calcium sulphate dihydrate are low: $\alpha = 1.521$, $\beta = 1.523$ and $\gamma = 1.530$. In transmitted light with crossed polars the particles are birefringent and display oblique extinction (McCrone *et al.* 1979: 1407). The refractive indices of anhydrite ($CaSO_4$) are also low – $\alpha = 1.570$, $\beta = 1.575$ and $\gamma = 1.614$ – and these particles are highly birefringent and display distinct extinction.

Gypsum is slightly soluble in water (1.29×10^{-2} moles per litre at 25 °C) and therefore may undergo cycles of deliquescence and recrystallisation in the presence of humidity.[17] Calcium sulphate dihydrate is also soluble in dilute hydrochloric acid. On heating, gypsum will give off water: temperatures of around 128 °C result in the production of the hemihydrate ($CaSO_4 \cdot H_20$), while heating to between 130 and 160 °C results in a combination of the hemihydrate and anhydrite ($CaSO_4$).[18] However, at temperatures above 650 °C an insoluble form of anhydrite forms, while above 900 °C a mixture of insoluble anhdrite and calcium oxide is formed which will set very slowly on contact with water (Federspiel 1995: 59).

Preparation

For use as a pigment or as a component of a preparatory layer, mineral gypsum could simply be finely ground. Cennini, however, describes in detail the preparation process for the two types of gesso so frequently employed as a ground in panel painting. For *gesso grosso*, mineral gypsum was burned in a kiln to drive off the waters of crystallisation (Cennini 1960: 70–73). The resulting material would be pure anhydrite (soluble and insoluble), or a mixture of anhydrite and hydrated forms of calcium sulphate depending

16. For gesso grounds in Italian trecento panel painting, see Gettens and Mrose 1954, and Bomford *et al.* 1990: 17–19.
17. Calcium sulphate has an RHeq of 99.96%, and can be considered as deliquescent; Sawdy 1995:29.
18. This material is now commonly known as *plaster of Paris*, which on contact with water rapidly reverts to the dihydrate form and sets.

on the extent of the firing process. However, since Theophilus describes gypsum 'burned in the fashion of lime' for use as a ground on wood, it seems likely that temperatures of around 900 °C were employed in the process, well above the 650 °C necessary for the formation of insoluble anhydrite (Theophilus 1979: 27). For the production of *gesso sottile* the burnt gypsum was slaked in excess water for an extended period and then thoroughly ground after excess water was removed.

Use in English medieval wall painting

Where calcium sulphate (gypsum ($CaSO_4{\cdot}2H_20$)) has been detected in medieval wall painting and polychrome sculpture, it is often present as an alteration product of calcium carbonate in the presence of atmospheric sulphur and humidity.

Application

In only one instance to date has the presence of calcium sulphate as an original component of a paint layer been confirmed in English medieval wall painting. In the early 16th-century scheme in St. Mark's, Bristol, calcium sulphate was not only combined with chalk and oil as a sealant, but also employed in a mixture with lead white and chalk for the sky in the *Noli me Tangere* (Gill and Howard 1997: 99–100).

However, in a few cases, calcium sulphate-based plasters were employed for wall painting, as for the *Last Judgement* of *c*1435–60 over the chancel arch of Holy Trinity, Coventry. Here the soffit board and triangular infill panels in the upper corners of the painting were prepared with a layer of calcium sulphate with virtually no aggregate (Howard 1996b: 4–5). The presence of a gypsum-based plaster has also been confirmed in the early 16th-century paintings in the Dominican Friary at Beverley (Humberside).[19] For the magnificent scheme of *c*1370 in the Chapel of Our Lady Undercroft in Canterbury Cathedral, a calcium sulphate ground was employed to prepare the vault for painting, with layers of red lead and vermilion providing the intense red background for the attached suns and stars (Howard 1993b: 32–3). In these examples the calcium sulphate has often been resolubilised in the presence of humidity and has penetrated the paint layers.

Comparison with its use in other media and in continental wall painting

Although chalk was commonly employed as a ground in medieval panel painting in northern Europe, unusually the Adisham reredos (*c*1180) has a gesso (calcium sulphate) ground.[20] A calcium sulphate ground has also been identified on the most important surviving English medieval panel painting, the Westminster Retable (*c*1270) (Binski and Freestone 1995). Here there is not only a gesso ground, which has an Italian rather than a northern association, but there also appears to be a chalk ground – one of a sequence of layers – beneath the paint and gilding of the imitation enamels.[21]

Calcium sulphate has also been detected in a few zones of the preparatory layer on the 13th-century portal of the '*Portail de la Mère Dieu*' at Amiens Cathedral (Weeks 1998:

19. The plaster contains gypsum with a little siliceous matter; Hulbert 1986b: 1 and 1997.
20. C. Tracy has redated this reredos on the basis of dendrochronological analysis.
21. I am grateful to Dr Roy for discussing these findings with me.

102). However, in some examples where calcium sulphate has been detected, as on the Romanesque frieze at Lincoln or in the 13th-century polychromy on the west front of Wells Cathedral, it is still uncertain whether the calcium sulphate represents an original material or an alteration product.[22]

Discussion

Calcium sulphate appears to have been little used as a pigment in English medieval wall painting, and where it has been detected it is generally difficult to determine whether it represents part of the original palette or is an alteration product. However, it has been employed as a ground or plaster layer in a few instances, most notably in the late 14th-century scheme in the Chapel of Our Lady Undercroft, Canterbury Cathedral.

Lead white

The brilliant, dense and opaque white pigment, basic lead carbonate ($2PbCO_3.Pb(OH)_2$) – also known as *ceruse* – has been produced artificially since antiquity. Pliny indicates that although a natural source of ceruse earth was once known, synthetic lead white made from vinegar and lead was universally employed (1984: 289).

Cost

The accounts of Exeter Cathedral for 1320–21 provide some idea of the cost of lead white in relation to other pigments: at 3d per lb it was approximately three times cheaper than vermilion and verdigris, six times cheaper than indigo, and some fourteen times cheaper than the finest '*azure*' bought from London (Erskine 1981: 134). In general, it seems that throughout the Gothic period it was inexpensive, readily available and widely used.

The sheer quantity of lead white purchased for painting in the Gothic period is often astounding. Accounts detailing the materials purchased for painting the central tower of York Minster in 1473 list 319lb of lead white but just 2lb of vermilion and ochre, 6lb of red lead and 12lb of verdigris and other colours (Tatton-Brown 1989: 145). Such a large amount of lead white, particularly in relation to the much smaller quantities of other pigments purchased, suggest that in this instance it was employed as an overall ground.

Optical, chemical and physical properties

The tiny particles of lead white (1–2μm) appear pale and opaque in plane polarised transmitted light. The refractive indices are high at $\alpha = 2.09$ and $\gamma = 1.94$, and with crossed polars the particles display birefringence and parallel extinction.[23]

As a carbonate, lead white is soluble in dilute acids with the evolution of carbon dioxide gas. It is stable to light, but on heating to low temperatures the pigment turns

22. For the Lincoln frieze, see Larson 1992: 188; and for Wells, see Roy and White 1982.
23. For a description of the history, manufacture and properties of the pigment, see Gettens *et al.* 1993a: 68–71.

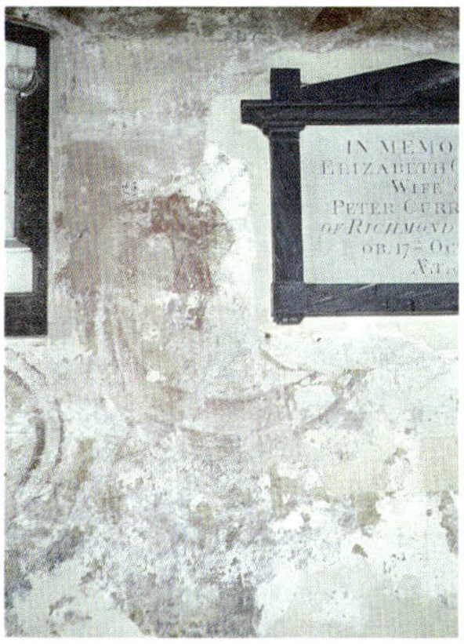

Plate 179 *Witley (Surrey): detail of the paintings on the nave south wall (c1110). Preliminary drawing in red earth was employed to place the main pictorial elements within the visual field. Preliminary drawing for figure subjects is particularly conspicuous in the* Miraculous Draught of Fishes *shown here. (© Courtauld Institute, Conservation of Wall Painting Department)*

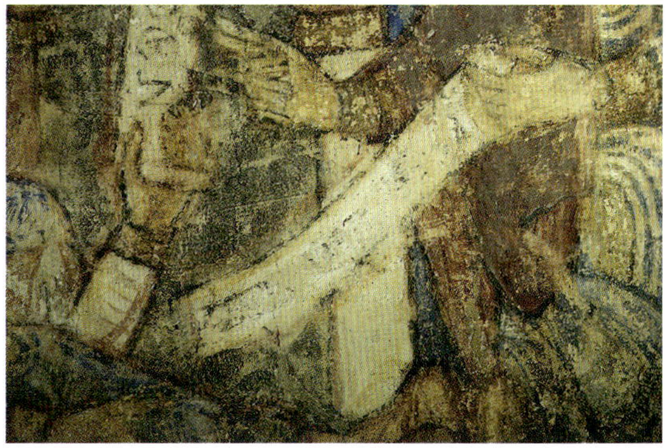

Plate 180 *Canterbury Cathedral, St. Gabriel's Chapel: detail of the* Birth of the Baptist *(c1130) on the north side of the apse. Here, preliminary drawing in yellow earth has been strengthened and refined in red earth. Pentimenti clearly show that the drawing was modified for a third time during the painting phase, as can be seen for the setting out of the scroll at the centre of the plate. (© Courtauld Institute, Conservation of Wall Painting Department)*

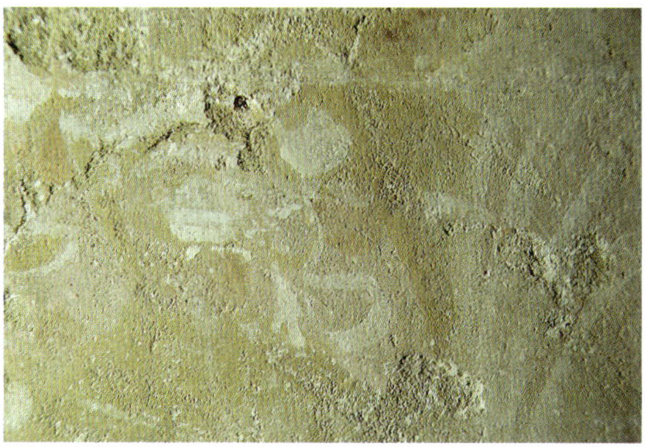

Plate 181 *Witley (Surrey): detail of the* Harrowing of Hell *on the west wall of the nave (c1110). Yellow earth, red earth and lime white are the dominant colours employed in this scheme. For areas of flesh painting, a pink colour – red earth combined with lime white – was applied directly to the plaster, then linear details in red or yellow earth combined with lime white were applied in impasto. Highlights in lime white were added last. (© Courtauld Institute, Conservation of Wall Painting Department)*

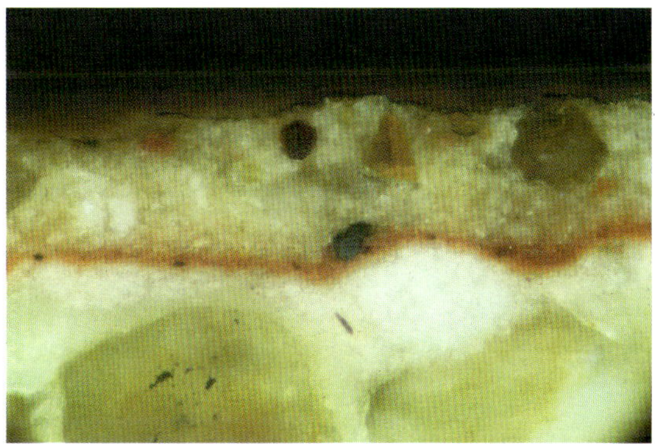

Plate 182 *Witley (Surrey). Cross-section of a sample from the flesh tone of one of the figures in the* Harrowing of Hell, *original magnification 340×. Red and yellow earth have been combined with quartz particles and lime white to produce a yellow highlight painted over a thin layer of red earth on the coarse lime plaster. (Photomicrograph: H. Howard; © Courtauld Institute, Conservation of Wall Painting Department)*

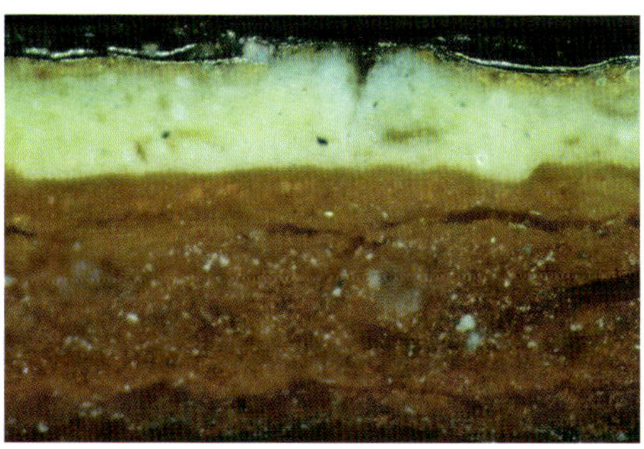

Plate 183 *Farleigh Hungerford Castle (Somerset) St. Leonard's Chapel: cross-section from the early 15th-century wall paintings, original magnification 500×. This sample, taken from St. George's helmet, shows the red preparatory layers for two layers of silver leaf. In preparation for the figures of St. George and the kneeling knight, a layer of red earth combined with calcium carbonate was applied in large rectangular zones in order to provide particular emphasis within the overall decorative scheme. Over this dark red layer – and throughout the remainder of the scheme – another red ground comprising red lead combined with oil and a little red iron oxide was applied to provide a unifying substructure throughout. This layer would also have acted as a sealant, ensuring that the translucency of the polychromy would not be compromised by leaching of the paint medium into the plaster support. (Photomicrograph: H. Howard; © Courtauld Institute, Conservation of Wall Painting Department)*

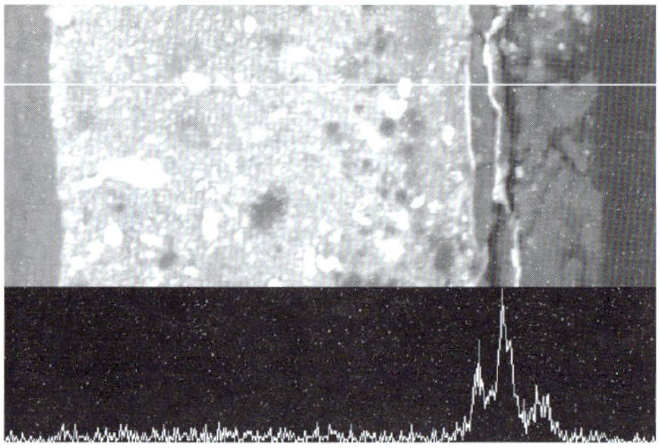

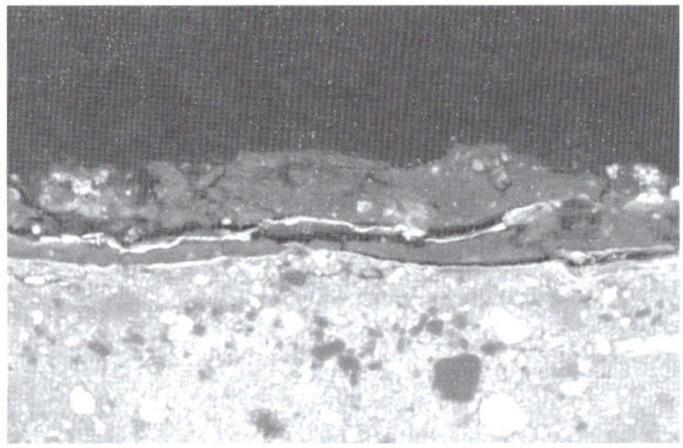

Plates 184 and 185 *Backscatter SEM micrograph with an X-ray line scan for silver confirming the presence of the two layers of the metal leaf in the cross-section shown in Plate 183. A gold effect was achieved by applying a yellow glaze over the lower layer of silver leaf, and further embellishment provided by slightly raised curvilinear details in silver leaf highlighted with a green glaze. Backscatter SEM micrograph (Plate 185) of the upper part of the sample shown in Plate 183. (SEM micrographs: H. Howard; © Courtauld Institute, Conservation of Wall Painting Department)*

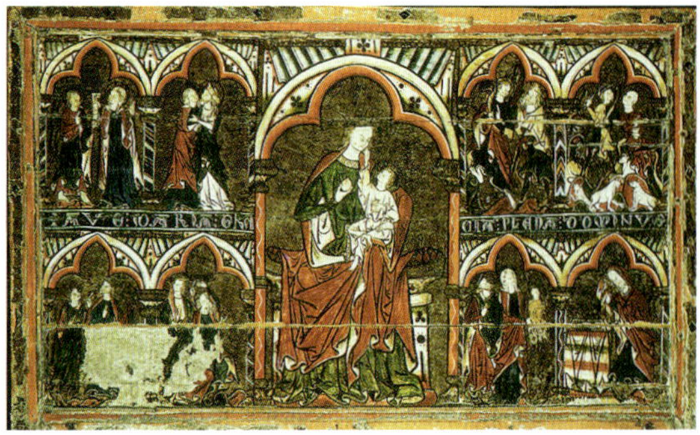

Plate 186 *Tingelstad I altarpiece of c1260–1300 (Norway). Orpiment occurs on all three of the altarpieces from Tingelstad. In this instance particularly large particles of the pigment were detected, some measuring 5μm in length. (Photograph: Jacques Lation; © University Museum of Cultural Heritage, University of Oslo)*

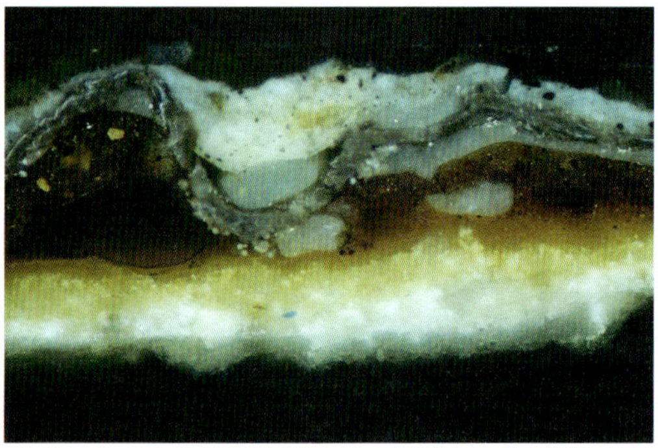

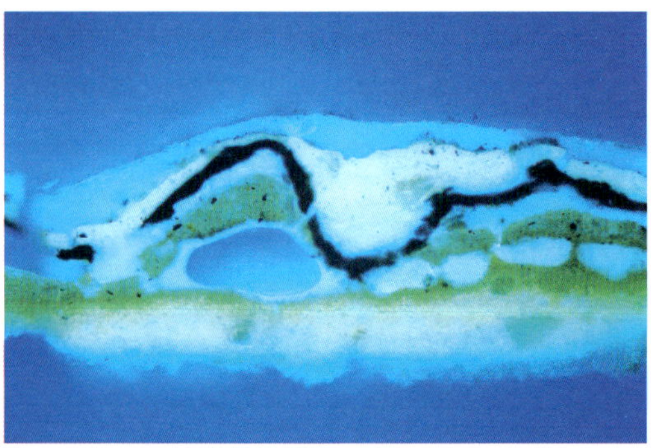

Plates 187 and 188 *Orpiment has only been detected in six English wall paintings from the Gothic period including the* Apocalypse *cycle of c1375–1404 in the chapter house of Westminster Abbey. This cross-section (top), taken from the window tracery in* The Seventh Vial, *shows particles of orpiment combined with lead white, and red and black inclusions in the mordant for the thick layer of tin leaf above. In UV light (Plate 188, bottom) the mordant fluoresces a warm yellow colour. (Photomicrographs: H. Howard; © Courtauld Institute, Conservation of Wall Painting Department)*

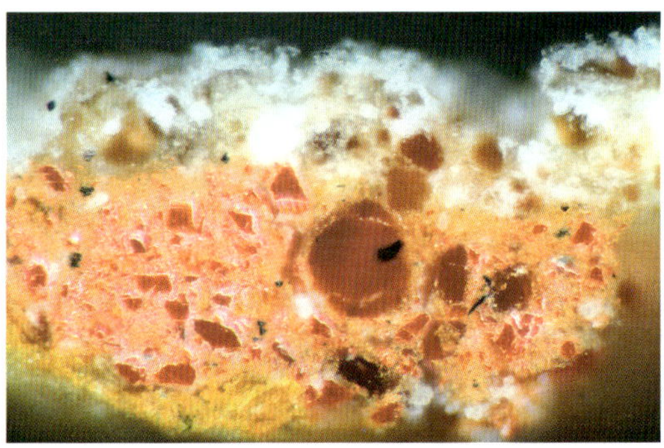

Plate 189 *Orpiment also occurs in the early 15th-century scheme at Farleigh Hungerford Castle (Somerset). This sample, from the shield above St. George (Plate 108), shows a thick layer of red lake applied over the pigment (at lower left of the sample). The lake shows considerable evidence of fading in the uppermost portion of the paint layer. (Photomicrograph: H. Howard; © Courtauld Institute, Conservation of Wall Painting Department)*

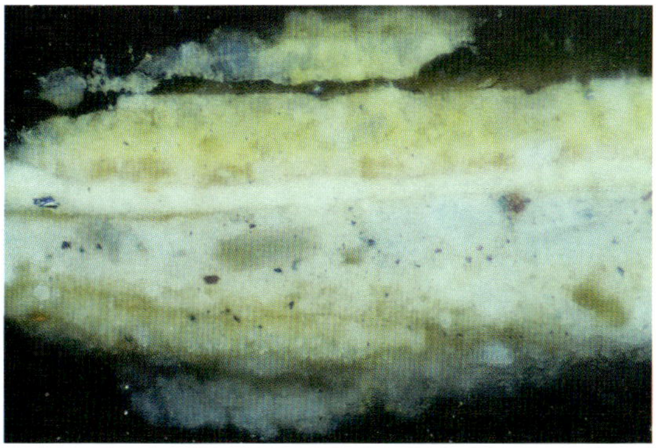

Plate 190 *Lead-tin yellow has only been identified in a few English wall paintings including those of c1510 in St. Mark's, Bristol. This cross-section shows the pigment applied over a complex system of pale pink, white and grey underpaints for the halo of the Virgin Mary (?) in the* Noli Me Tangere *(Plate 86). (Photomicrograph: H. Howard; © Courtauld Institute, Conservation of Wall Painting Department)*

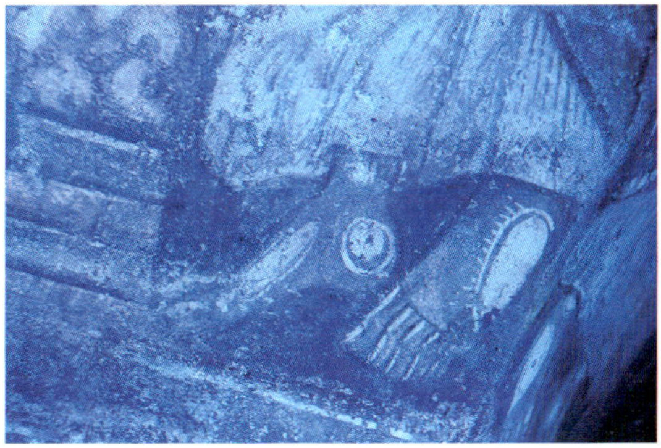

Plate 191 *Lime white was often used to particularly good effect for highlights of flesh painting and other details such as 'pearling' in fictive jewelled borders. The lime white used for highlights of the flesh in the apse paintings of St. Gabriel's Chapel, Canterbury Cathedral (c1130), shows up especially well in UV light. (© Courtauld Institute, Conservation of Wall Painting Department)*

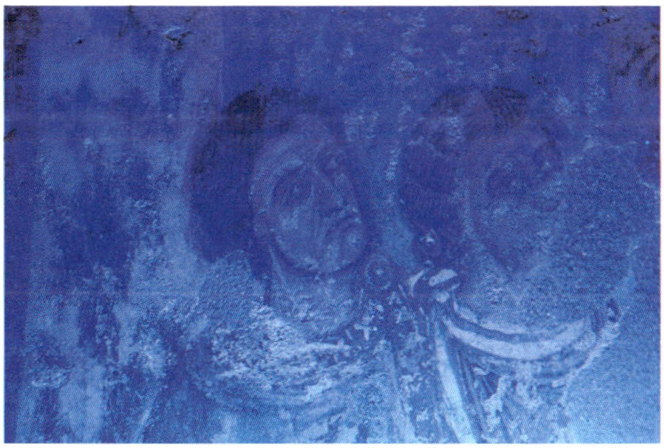

Plate 192 *Lime white highlights in considerable impasto were similarly employed for the 'Losinga' scheme of c1190 in Norwich Cathedral. Some details, such as the collars and clasps of the standing figures on the vault, are now difficult to detect and are only clearly visible in UV light. (Photograph: H. Howard; © Courtauld Institute, Conservation of Wall Painting Department)*

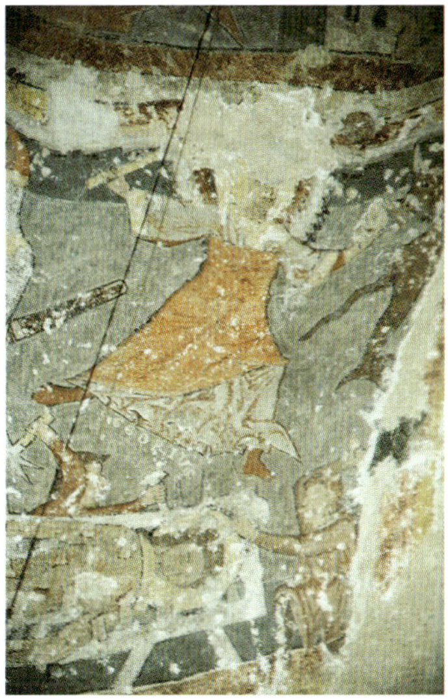

Plate 193 *Lime white reacts with casein to form a stable calcium caseinate, and this combination has been identified, for example, in the scheme of c1240 at St. Georgen-bei-Judenburg, Austria, a detail of which is shown here. (Photograph: H. Howard; © Courtauld Institute, Conservation of Wall Painting Department)*

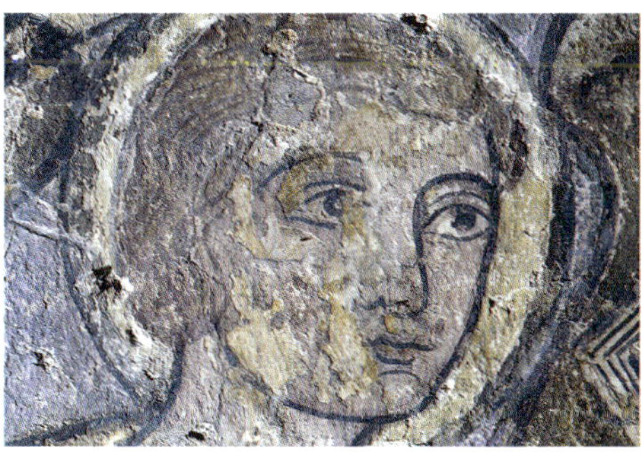

Plate 194 *Canterbury, Eastbridge Hospital: detail of the paintings of c1220, showing the angel symbol of St. Matthew. (Photograph: H. Howard; © Courtauld Institute, Conservation of Wall Painting Department)*

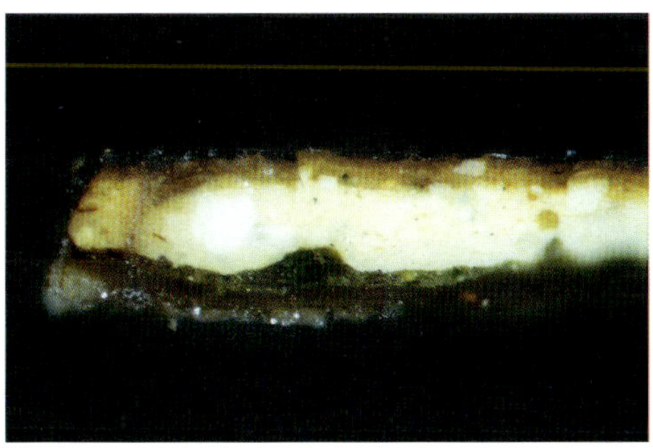

Plate 195 *Canterbury, Eastbridge Hospital. Cross-section of a sample from the white inscription 'MATHEVS', original magnification 228×. A layer of lead white has been applied in impasto over verdigris on the lime ground. Analysis by FTIR confirmed that both the lead white and verdigris layers have an oil medium. Yellow-brown surface coatings are evident at the top of the sample. (Photomicrograph: H. Howard; © Courtauld Institute, Conservation of Wall Painting Department)*

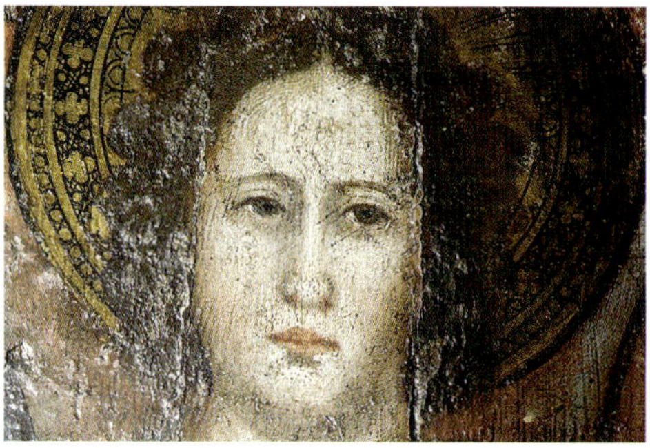

Plate 196 *Lead white was extensively employed in both the* Last Judgement *and the* Apocalypse *paintings in the chapter house of Westminster Abbey. This detail of the seraph in arch 2 of the east bay shows the subtle modelling of the flesh tones. To produce the pale flesh tones in the* Majesty, *layers of lead white were combined with varying quantities of vermilion and applied over a grey-green underpaint of charcoal black and yellow in a white matrix. (Photograph: Howard/Kakoulli; © Courtauld Institute, Conservation of Wall Painting Department)*

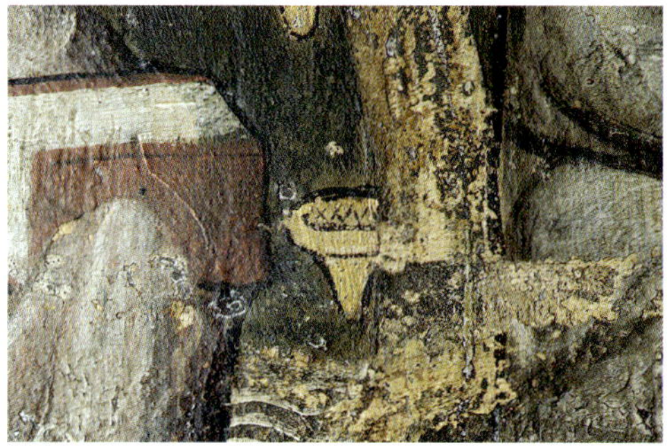

Plate 197 *Westminster Abbey, chapter house.* The Majesty and the Twenty-four Elders *in the north-west bay: for green details on the border of the mandorla, verdigris in a lead white matrix was applied over a secondary ground of lead white over tin leaf. (Photograph: Howard/Kakoulli; © Courtauld Institute, Conservation of Wall Painting Department)*

Plate 198 *Farleigh Hungerford Castle (Somerset), St. Leonard's Chapel: detail of the south wall of the nave showing the kneeling knight (early 15th-century). (© English Heritage Photographic Library)*

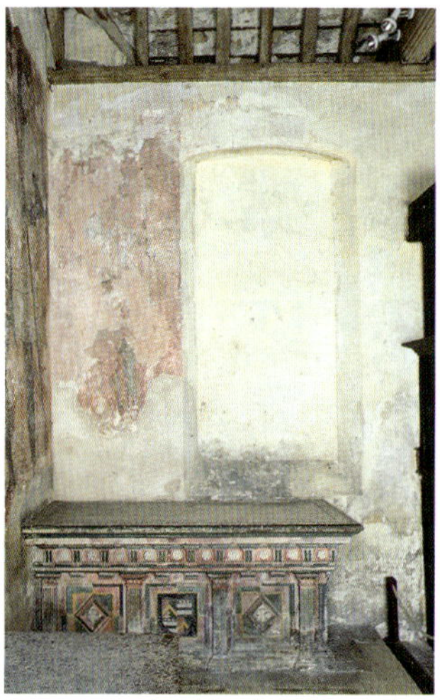

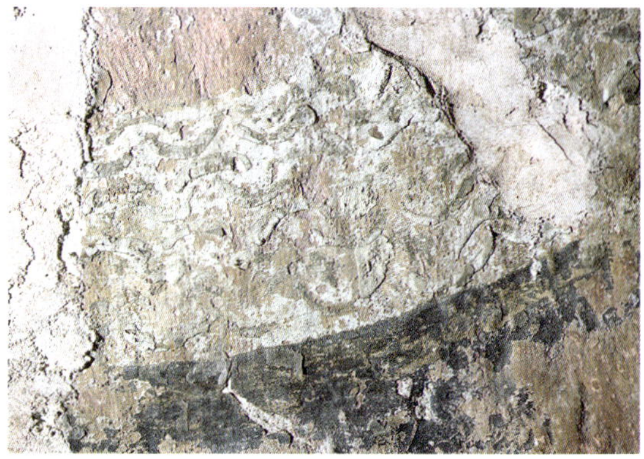

Plate 199 *Farleigh Hungerford Castle (Somerset), St. Leonard's Chapel. Detail of the impasto rendering of the mail around the neck of the knight. (© Courtauld Institute, Conservation of Wall Painting Department)*

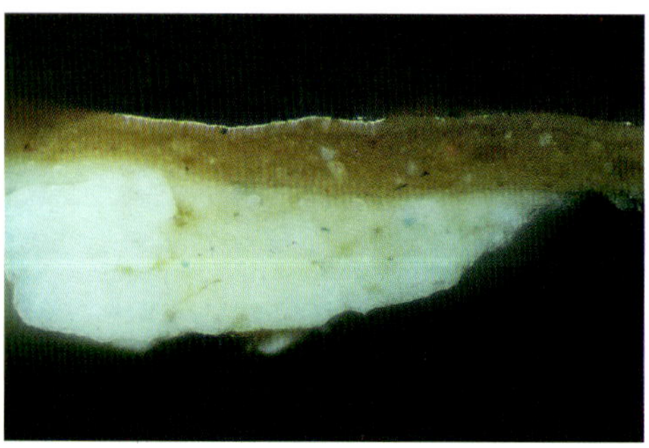

Plate 200 *Farleigh Hungerford Castle (Somerset), St. Leonard's Chapel. Cross-section of a sample from the mail, original magnification 500×. A thick paste of lead white in an oil medium was applied and the mail modelled* in situ. *The undulating surface was then covered with silver leaf, and although this would have been glazed or varnished to prevent tarnishing, no trace of this remains on the surface of the cross-section. (Photomicrograph: H. Howard; © Courtauld Institute, Conservation of Wall Painting Department)*

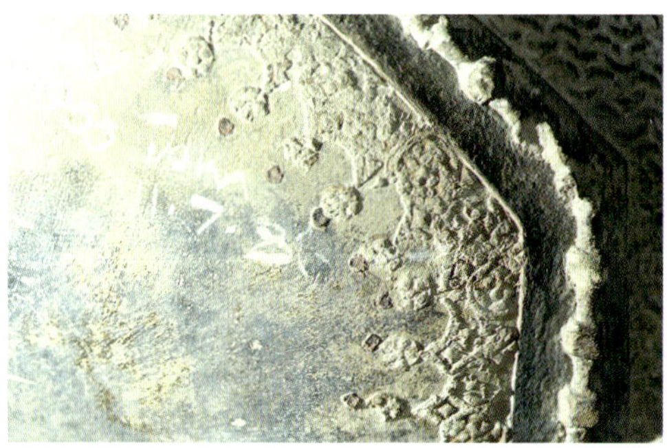

Plate 201 *Ingham (Norfolk), tomb of Sir Roger and Lady Margaret de Bois (c1380): detail of Sir Roger's helmet showing the pink 'jewels' which have been applied in considerable impasto on the surface. (© John Burbidge)*

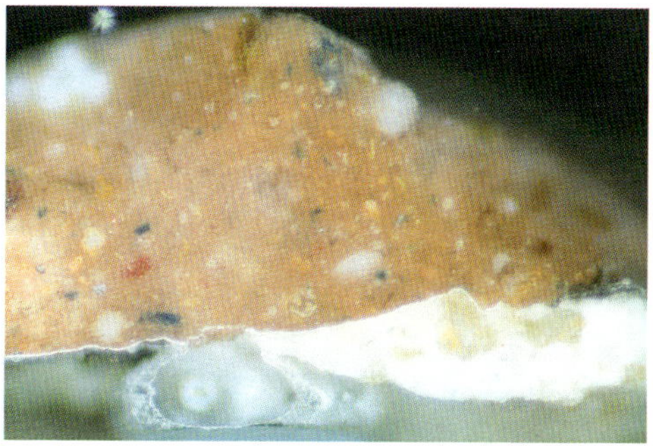

Plate 202 *Ingham (Norfolk), de Bois tomb. Cross-section of a sample from one of the fictive jewels, original magnification 250×. The impasto paint layer consists of lead white combined with red lead, red earth, calcium carbonate and azurite in an oil medium, applied over a layer of silver leaf. The lead and copper pigments here would have functioned as driers for the oil. GC-MS analysis confirmed the presence of linseed oil which was partially heat-bodied to promote drying. (Photomicrograph: H. Howard; © Courtauld Institute, Conservation of Wall Painting Department)*

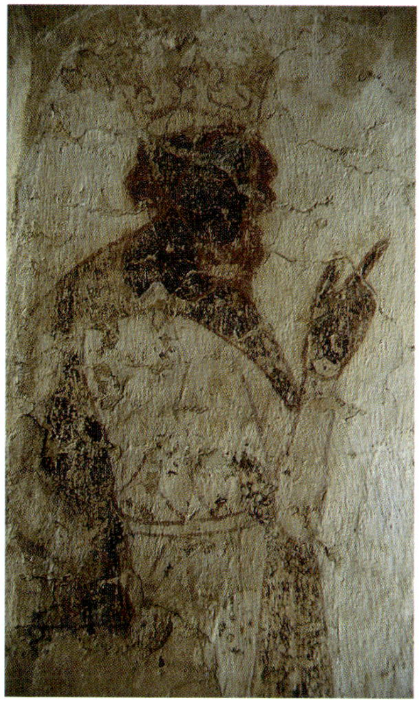

Plate 203
Longthorpe Tower (Cambridgeshire): detail of the Three Living and the Three Dead *in the early 14th-century cycle in the first floor solar. The darkening of the flesh tones in these and other paintings has been the subject of a detailed study (Welford 1991) which demonstrated that the phenomenon was due to the alteration of lead white to form dark lead dioxide (plattnerite). (© Courtauld Institute, Conservation of Wall Painting Department)*

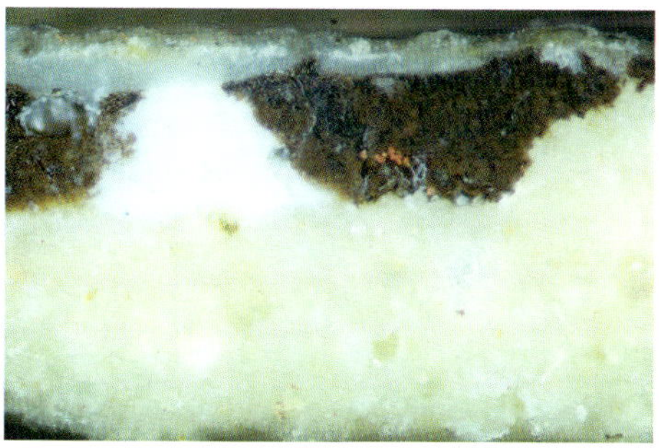

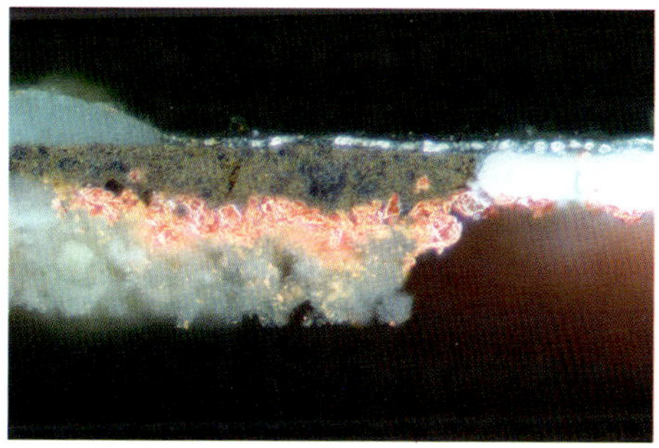

Plates 204 and 205 *Longthorpe Tower (Cambridgeshire). Cross-section (original magnification 268×) of a sample of darkened lead white from the flesh painting. This section (204) shows the curiously patchy nature of the darkening phenomenon, with zones of unaffected lead white juxtaposed with completely darkened areas. A similar partial alteration of lead white appears in a sample taken from the fictive tapestry on the south wall (Plate 205) where lead white was applied over vermilion. (Photomicrographs: H. Howard; © Courtauld Institute, Conservation of Wall Painting Department)*

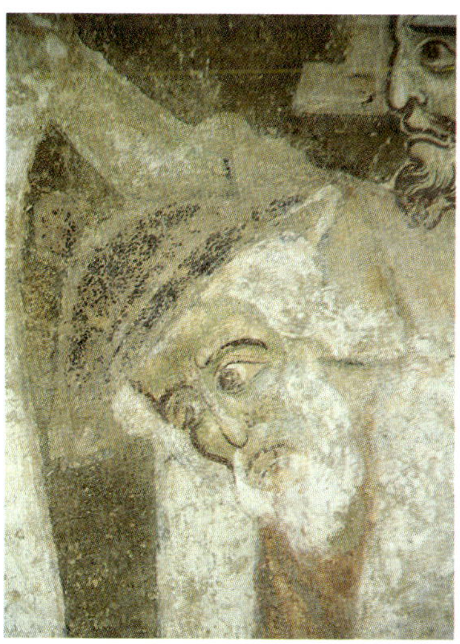

Plate 206 *Winchester Cathedral, Holy Sepulchre Chapel: detail of the* Deposition *of c1220. Alteration of a lead pigment is evident here in Nicodemus' hat which has a dark spotty appearance. (© Courtauld Institute, Conservation of Wall Painting Department)*

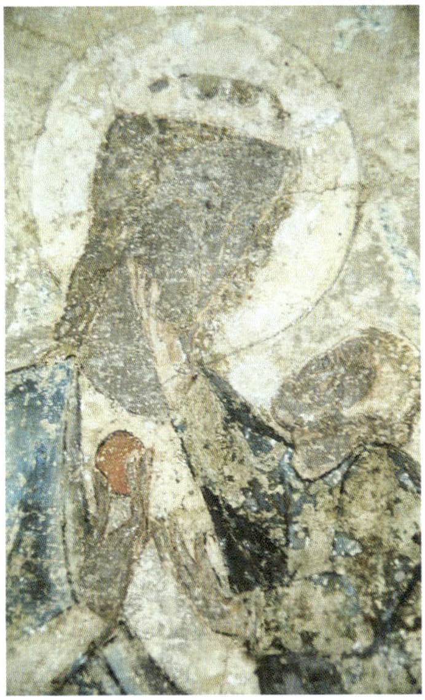

Plate 207 *Norwich Cathedral, Ante-Reliquary Chapel: detail of the* Virgin and Child *on the vault (c1300). Although dark flesh tones are frequently due to the alteration of lead white, XRD analysis of samples taken from the dark flesh areas in this scheme produced no evidence of altered lead pigments. (Photograph: H. Howard; © Courtauld Institute, Conservation of Wall Painting Department)*

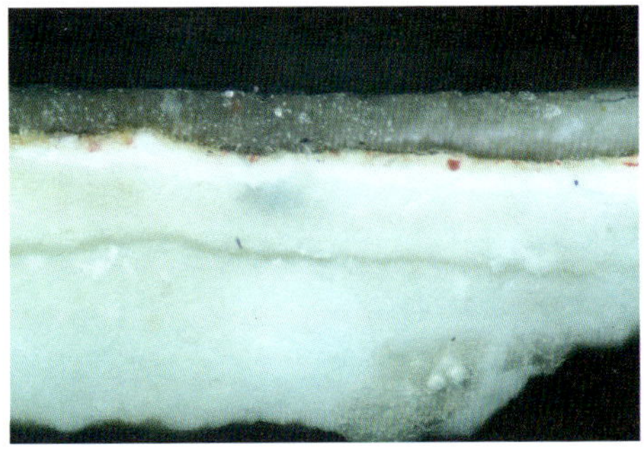

Plate 208 *Norwich Cathedral, Ante-Reliquary Chapel. The flesh tone of lead white combined with vermilion remains unaltered in this sample from the Norwich paintings, although wax and surface accretions are visible on the surface, original magnification 314×. Here, the patchy grey appearance of the flesh is simply due to the partial removal of past surface coatings and accretions. (Photomicrograph: H. Howard; © Courtauld Institute, Conservation of Wall Painting Department)*

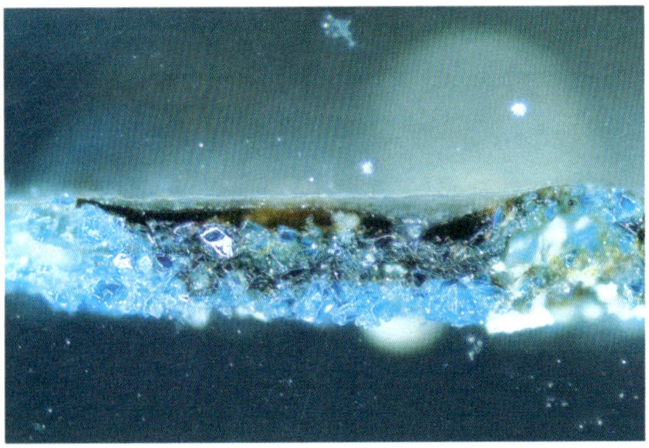

Plate 209 *Norwich Cathedral, Ante-Reliquary Chapel. This cross-section from the Virgin's robe also shows the effect of partial cleaning. A thin layer of wax remains on the surface of the cross-section, and below 19th-century resin varnish penetrates the azurite layers. The coatings have been removed from the high points in the paint layer, but have been left in the hollows producing the patchy blue-green-brown colour of the drapery. (Photomicrograph: H. Howard; © Courtauld Institute, Conservation of Wall Painting Department)*

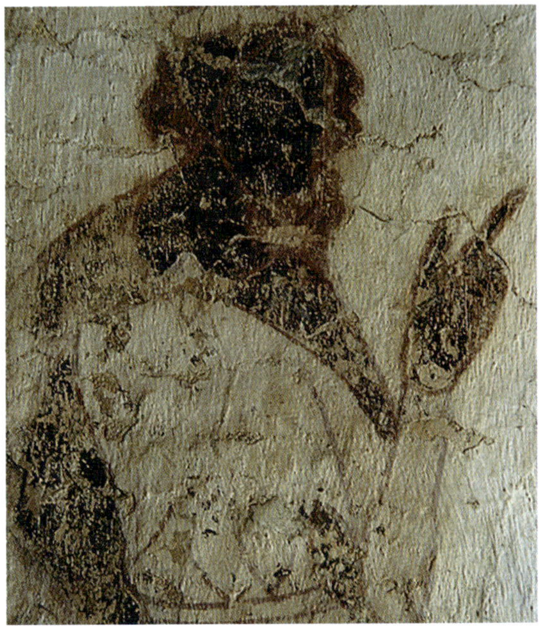

Plate 210 *Longthorpe Tower (Cambridgeshire): detail of the* Three Living and the Three Dead *(early 14th century). (© Courtauld Institute, Conservation of Wall Painting Department)*

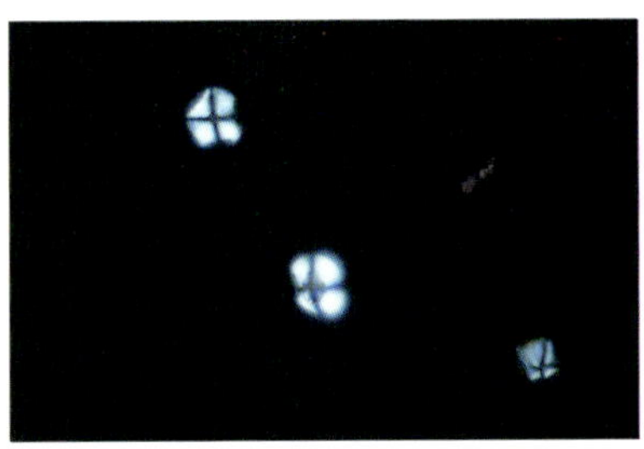

Plate 211 *Longthorpe Tower (Cambridgeshire). Dispersion of a sample from a white stencilled motif on one of the king's mantle, original magnification 940×. Tiny nanofossils known as coccoliths display characteristic crosses in transmitted light with crossed polars, indicating the presence of chalk. (Photomicrograph: H. Howard; © Courtauld Institute, Conservation of Wall Painting Department)*

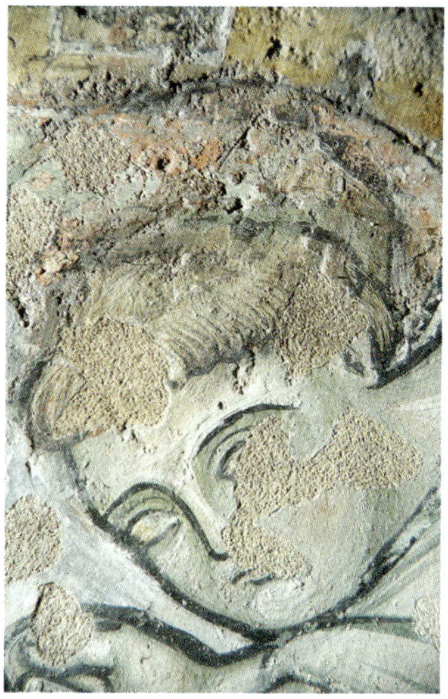

Plate 212 Winchester Cathedral, Holy Sepulchre Chapel: detail of the censing angel in the Entombment on the east wall (c1175). A tiny trace of the original gilding remains on the angel's nimbus. (© Courtauld Institute, Conservation of Wall Painting Department)

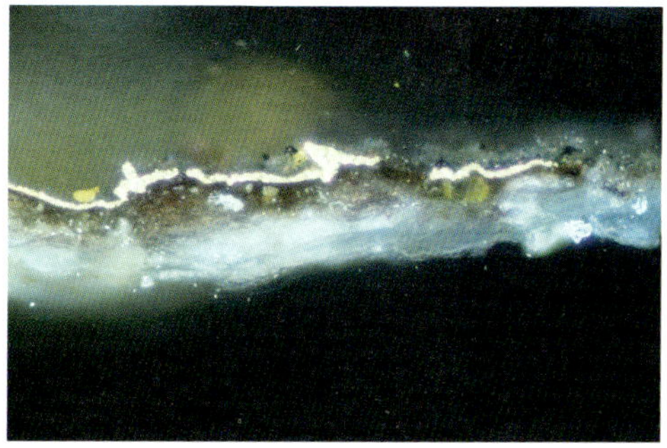

Plate 213 Winchester Cathedral, Holy Sepulchre Chapel. Cross-section of a sample from the gilding on the angel's nimbus, original magnification 500×. The oil-based mordant for the gold leaf has been bulked out with calcium carbonate, which appears to be in the form of chalk. (Photomicrograph: H. Howard; © Courtauld Institute, Conservation of Wall Painting Department)

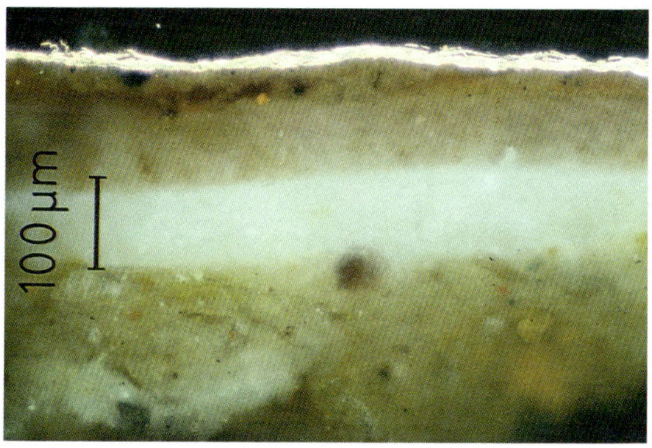

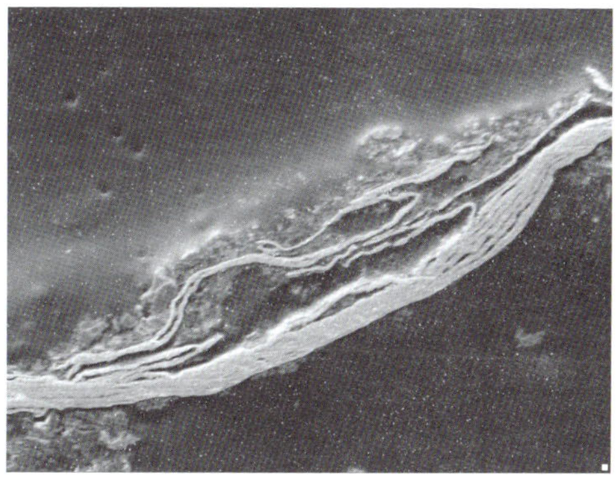

Plates 214 and 215 Cross-section of a sample from a wall painting fragment of c1180 excavated from the chapter house of Sherborne Abbey, original magnification 241×. Calcium carbonate – probably in the form of chalk – has been identified as a component of the oil-based mordant for the gold leaf. Multiple layers of gold leaf have been applied to the surface of the fragment, and these are particularly clearly visible in the SEM micrograph in Plate 215. (Photomicrograph and SEM micrograph: H. Howard; © Courtauld Institute, Conservation of Wall Painting Department)

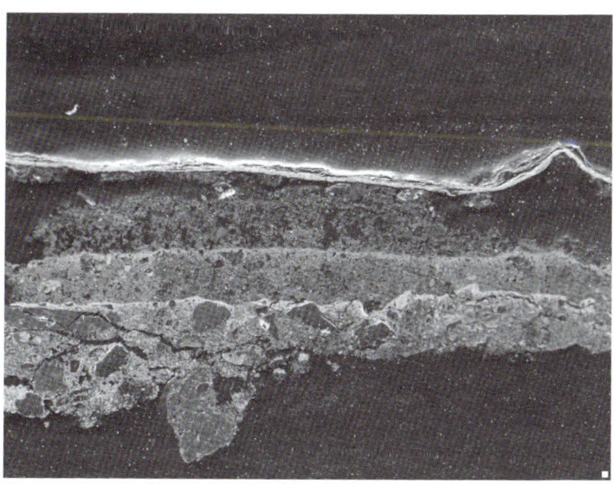

Plate 216 SEM micrograph of the sample in Plate 214, showing the coarse aggregate of the plaster support and the finer layers applied in preparation for gilding. (SEM micrograph: H. Howard; © Courtauld Institute, Conservation of Wall Painting Department)

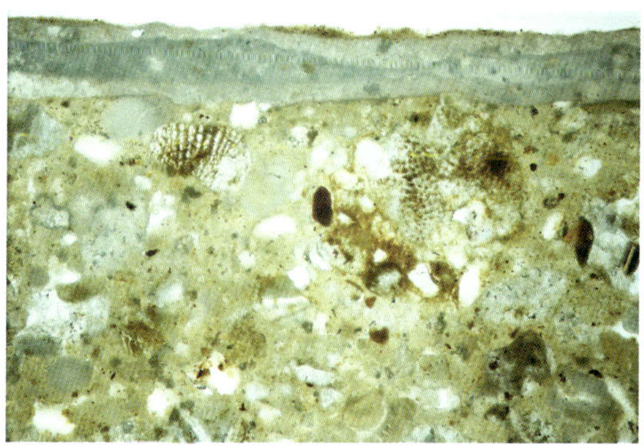

Plate 217 Thin section of another sample from the Sherborne fragments, photographed in plane-polarised transmitted light, original magnification 90×. The crushed fossil-rich limestone aggregate of the lime-based plaster is clearly visible. Since the Sherborne region is rich in Inferior Oolitic limestone, it is probable that this material is of local origin. (Photomicrograph: H. Howard; © Courtauld Institute, Conservation of Wall Painting Department)

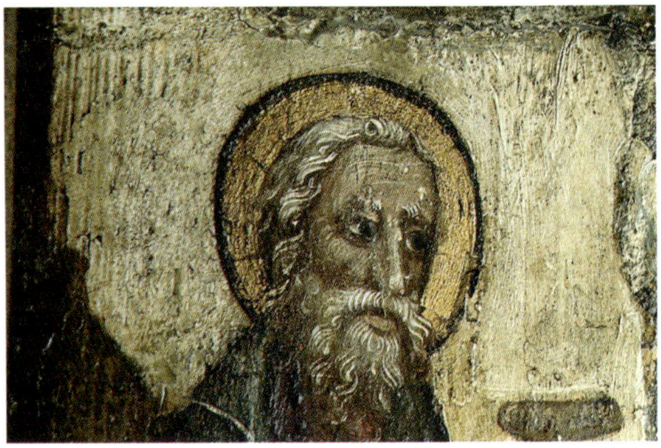

Plate 218 *Westminster Abbey, chapter house: detail of the* Celestial Host *in the* Apocalypse *cycle of c1374–1404. (Photograph: Howard/Kakoulli; © Courtauld Institute, Conservation of Wall Painting Department)*

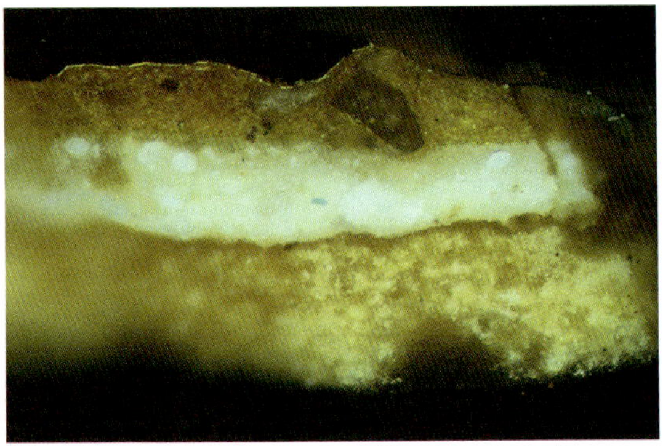

Plate 219 *Westminster Abbey, chapter house. Cross-section of a sample from the nimbus of St. John in the* Celestial Host, *original magnification 376×. The chalk ground is visible at the base of the sample and over this a secondary ground of lead white with a single azurite inclusion has been applied. The gold leaf has been adhered with an oil-based mordant. (Photomicrograph: H. Howard; © Courtauld Institute, Conservation of Wall Painting Department)*

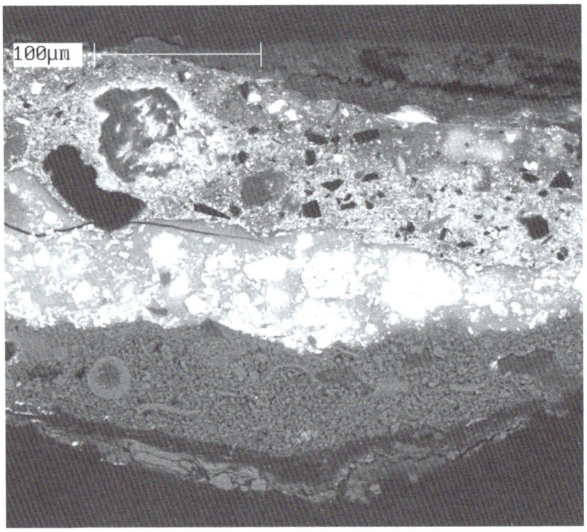

Plate 220 *Backscatter SEM micrograph of the sample shown in Plate 219. The coccoliths in the chalk ground are clearly visible at high magnification. (SEM micrograph: H. Howard; © Courtauld Institute, Conservation of Wall Painting Department)*

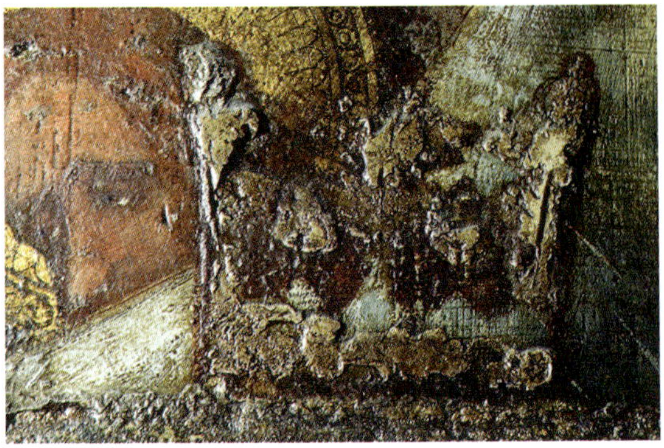

Plate 221 *Westminster Abbey, chapter house: detail of the* Last Judgement *(c1374–1404), showing the raised crown held by the seraph in arch 2 of the east bay in raking light. (Photograph: H. Howard; © Courtauld Institute, Conservation of Wall Painting Department)*

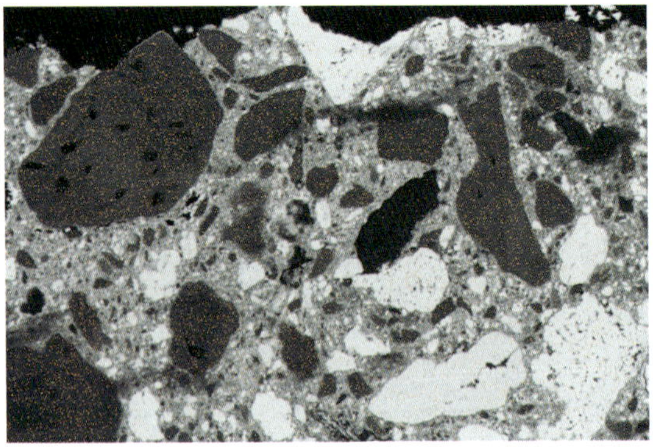

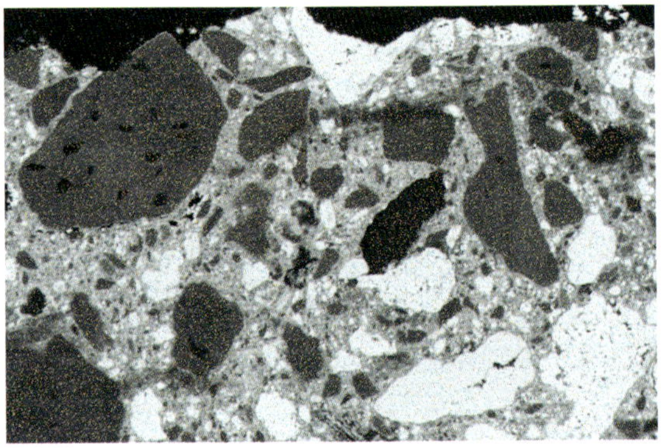

Plates 222 and 223 *Backscatter SEM micrograph of a sample from the raised crown in Plate 221, original magnification 532 000×, with an X-ray map for phosphorus (in orange). The dark particles in the cross-section clearly show the presence of phosphorus. Backscatter SEM micrograph (Plate 223) as above, but with X-ray dot maps for phosphorus (magenta), calcium (yellow), lead (cyan) and silica (orange). The dark particles in the cross-section are rich in calcium and phosphorus indicating the presence of bone white, while the matrix is lead-rich. (SEM micrographs: H. Howard; © Courtauld Institute, Conservation of Wall Painting Department)*

yellow through the formation of massicot (PbO), while further sustained heating will produce litharge, and finally red lead (Pb_3O_4).

On occasion lead white blackens to form lead dioxide, and although the mechanism for this darkening process is not fully understood, alkaline conditions, high humidity and the presence of an oxidising agent have been implicated in the process (Matteini and Moles 1981: n. 33; Petushkova and Lyalikova 1986: 65–9; Giovannoni *et al.* 1990: 22–3). The pigment is also sensitive to sulphides, again turning black, through the formation of lead sulphide. However, in many instances where lead white has been combined with vermilion (mercury sulphide) – as at Farleigh Hungerford Castle (Somerset) where this mixture was employed for the flesh tones of St. George – no alteration is observed (Howard *et al.* 1998: 60). This apparent stability may be due to the presence of an oil binder which envelops and protects the pigment particles.

The high refractive indices and consequent good covering power of lead white in oil mean that the pigment was often employed in this medium. Since it also functions as a drier for oil, forming lead salts of fatty acids, such paint layers are also extremely stable.

Preparation

The pigment was generally produced by exposing metallic lead to vapours of acetic acid in a warm environment rich in carbon dioxide (Gettens *et al.* 1993a: 68). Peter of St. Omer describes the process in detail, indicating that strips of lead should be suspended above vinegar and sealed in a vessel which was then placed in horse dung for thirty days (Merrifield 1967: 1: 120). By contrast, Theophilus suggests that thinned plates of lead should be covered with hot liquid vinegar and then sealed in a wooden chest for a month (1979: 41–2). Although Cennini provides no precise details regarding the preparation, he confirms that it was made from lead by alchemical processes. He also indicates that the more finely it was ground the more perfect the pigment became (Cennini 1960: 34).

Use in English medieval wall painting

Since red lead has been identified in both cathedral schemes and in more modest parish church paintings of the 12th century, it is not surprising that lead white also occurs in Romanesque wall painting. The earliest identified example of the use of both red lead and lead white in English medieval wall painting is in the scheme of *c*1110 in the parish church of Witley (Surrey) (Howard 1992a: 11–12). Both pigments have also been identified in the scheme of *c*1130 in St. Gabriel's Chapel, Canterbury Cathedral (Cather and Howard 1994: 145). Lead white has been identified in the apostle of *c*1130–40 in the nave of Little Easton (Essex), and in the slightly earlier figure of St. Margaret of Antioch at Binsted (West Sussex) (Welford 1991: 4–10). The pigment was employed in the wall paintings of *c*1170 in Cormac's Chapel, Cashel (Ireland), in a grey-blue sample taken from a recess to the south of the altar (McGrath 1987: 58).[24]

Lead white became increasingly common in the Gothic period (Appendix 3, Table 5), and was used as an overall ground from the middle of the 13th century. Important examples of its use include the paintings of *c*1220 in the Holy Sepulchre Chapel, Winchester

24. Lead white was also identified in a sample from the west side of the chancel arch where a trace of the pigment was detected in the lime ground, Howard 1991: 2.

Cathedral (Howard 1995a: 100); the mid-13th-century *Foundation Narrative* at Horsham St. Faith Priory (Norfolk) (Babington and James 1988: 13–14); the schemes of *c*1250 and 1300 in the Ante-Reliquary Chapel, Norwich Cathedral (Park and Howard 1996: 391, 396; Howard 1995c: 4–5); the paintings of *c*1302–08 in the feretory of St. Albans Cathedral (Howard 1993a: 36); the domestic scheme of *c*1330 at Longthorpe Tower (Cambridgeshire) (Welford 1991: 7–8); the mid-14th-century wall painting fragments from St. Stephen's Chapel, Westminster Palace (van Geersdaele and Goldsworthy 1978: 11–12); the late 14th-century wall paintings in the chapter house, Westminster Abbey;[25] and the late 15th-century scheme in Eton College Chapel (Plummer 1986: 36).

Application

Over a white ground or directly over the plaster support

Although the pigment has darkened and it is now impossible to determine its original colour, it seems likely that lead white rather than red lead would have been employed for the background of the red cross nimbus of Christ in the *Harrowing of Hell* on the west wall of the nave at Witley. The lead pigment employed in this area was applied directly to the plaster substrate, and the now darkened particles are visible in the interstices of the coarse plaster.[26] In another sample taken from the neck of a dog in a scene on the nave south wall, the now-darkened lead pigment was combined with red earth and applied directly over the plaster substrate. In this case lead white may originally have been combined with red earth to produce a pale pink colour, or red lead may have been mixed with the earth pigment to give a strong orange-red (Howard 1992a: 11, App.).

Use as a ground

Lead white was frequently employed in Gothic wall paintings to enhance translucency; as a brilliant white ground, it provided luminosity to subsequently applied layers. Even as early as *c*1130, in the paintings of St. Gabriel's Chapel in Canterbury Cathedral, lead white appears to have been used as a local ground for discrete areas of the scheme (Cather and Howard 1994: 144).[27] A local ground of this type seems to have been applied directly to the plaster substrate beneath the red background on the south face of the north buttress, and beneath an angel's draperies on the north side of the apse. A local lead white ground also seems to have been applied over yellow earth in an area of pink-grey in the *Birth of the Baptist* (Howard 1992b: 10–11, App.). This finding may be particularly significant since the use of local lead white grounds was previously unknown in English Romanesque wall painting.

Lead white appears to have been used as a ground for the *Foundation Narrative* scenes of *c*1250 (although substantially repainted in the 15th century) on the east wall of the refectory at Horsham St. Faith Priory (Babington and James 1998: 13–14). The pigment, combined with oil, was employed as an overall ground for the scheme of *c*1250 in the Ante-Reliquary Chapel, Norwich Cathedral (Howard 1995c: 3). Here, however, in

25. Howard 2000 and forthcoming.
26. EDX analysis confirmed the presence of lead in these particles. Howard 1992a: pls. 34–39.
27. It is, however, possible that the white, lead-containing material identified here is not the pigment lead white (basic lead carbonate) but cerussite (neutral lead carbonate), the alteration product of red lead.

common with most wall paintings of this period, the overlying paint layers are still generally rather opaque. By contrast, from about 1300 translucency became of paramount importance to the Gothic artist, and lead white grounds were frequently used to provide a compact reflective white surface to interact optically with the succeeding translucent paint layers. At St. Albans Cathedral, the unidentified archbishop of *c*1302–08 on the south side of the feretory was executed over a thick preparation of lead white in oil with a few inclusions of red lead. The painter of this figure was relying on strong reflections from the lead white ground through highly translucent paint layers, such as the red lake and azurite mixture employed for his chasuble.[28] Similarly, for the approximately contemporary second scheme in the Ante-Reliquary Chapel at Norwich Cathedral a lead white ground, applied in broad brushstrokes and giving a distinctive surface texture, was used to full effect. For example, the translucent greens, purples and reds employed for the drapery of the apostles on the western arch were laid over an intensely reflective lead white ground which provided a luminosity and brilliance to the overlying colours (Park and Howard 1996: 396). A thick lead white preparation also occurs in the paintings of *c*1510 at St. Mark's, Bristol, where it was applied in oil in layers up to 80μm thick (Gill and Howard 1997: 99).

Over coloured underpaint

In the scene of the *Feast of Dives* at Witley (*c*1110), the collar of Dives now appears dark brown-black. Analysis suggested the presence of lead dioxide, and examination of the area sampled suggested that the original intention was for vertical bands of red and white to create the effect of a high ornate collar. The red appears to be the same colour as the general background, executed in a red earth pigment, and so it seems likely that vertical strips of lead white were painted over red earth in this instance (Howard 1992a: 11, App.).

Mixed with other pigments

When applied as a ground, lead white served to enhance the *luminosity* of subsequently applied paint layers. By contrast, the high refractive index and consequent good covering power of the pigment also made it ideally suited for increasing *opacity* and lightening other colours when combined in pigment mixtures. In the Romanesque scheme in St. Gabriel's Chapel, Canterbury Cathedral, it was found combined in one area with ultramarine and charcoal black to form an underpaint beneath red lead and vermilion, and in another sample combined with verdigris (Howard 1992b: 10–11, App.).

In Gothic wall painting, lead white was incorporated in numerous pigment mixtures. For example, it was combined with verdigris to form the green background of the early 13th-century scheme in the Eastbridge Hospital, Canterbury (Howard 1994: App. 1), and with lead-tin yellow for the heraldic motif above St. George in the early 15th-century decoration of the chapel of Farleigh Hungerford Castle (Somerset). It also occurs in many other pigment mixtures in the Farleigh Hungerford paintings: for example with red lake for the border of St. George's surcoat, with verdigris for the lining of his cloak, and with vermilion for his flesh tones (Howard *et al.* 1998: 60–62). Lead white has been detected in a number of pigment mixtures in the late 14th-century wall paintings in the chapter house of Westminster Abbey. For example, to produce the pale flesh tones in the

28. A thick lead white ground layer has also been identified on the Shrine of St. Alban itself; see Howard 1993a: 37–8.

Majesty, layers of lead white were combined with varying quantities of vermilion and applied over a grey-green underpaint of charcoal black and yellow in a white matrix (Plate 196). For green details on the border of the mandorla in the *Majesty and the Twenty-four Elders*, verdigris combined with lead white was applied over a local ground of lead white over tin leaf (Plate 197). Isotope ratio studies of samples taken from the *Majesty* and *Apocalypse* cycles indicate that the lead used for the production of the lead white in the chapter house paintings came from Derbyshire, as has also recently been established for the lead pigments employed in the earlier 14th-century Thornham Parva Retable.[29] In the scheme of *c*1510 in St. Mark's, Bristol, lead white was combined variously with vermilion, indigo, carbon black and a copper chloride green (Gill and Howard 1997: 100). Lead white would have been incorporated to adjust the tonality of these pigments, but would also have functioned as a drier where – as at Eastbridge, Westminster, Farleigh and St. Mark's – an oily medium was employed.

Impasto *highlights and other details*

The density and brilliance of lead white also made it ideal as a pigment for *impasto* highlights. Lead white functioned as a siccative for drying oils, and so when combined with a heat-bodied oil could be applied in impressive *impasto*. In the painting of *c*1220 in the Eastbridge Hospital, Canterbury, it was applied in layers up to 100μm thick over the verdigris background to produce the inscriptions above the angel and eagle evangelist symbols (Plates 194 and 195) (Howard 1994: 5). Raised ornament imitating mail around the neck of the kneeling knight (early 15th century) on the south wall of the chapel at Farleigh Hungerford Castle was modelled *in situ* from a thick paste of lead white in an oil medium (Plates 198–200). The undulating surface of the mail was then covered with silver leaf, and although this would then have been glazed or varnished to prevent tarnishing, no trace of this remains on the surface of the cross-section (Plate 200).

Binding media

Lead white is generally stable when combined with a drying oil such as linseed oil, and Peter of St. Omer suggests this combination for use on walls (Merrifield 1967: 1: 120). Others, clearly aware of the tendency of the pigment to darken, strongly advise against its use on walls whatever the medium (Cennini 1960: 34; Dionysius 1974: 15), and it is significant in this respect that lead white has been shown to be unstable and darken in alkaline conditions (Matteini and Moles 1981; Giovannoni *et al.* 1990; Koller *et al.* 1990). In the St. Gabriel's Chapel scheme the unaltered state of the pigment indicates that it is likely to have been applied to dry plaster with an organic binding medium. Although linseed oil was identified as a component of the original binding media in these paintings, it was not possible to determine the particular medium combined with the lead white. However instrumental analysis has confirmed the presence of lead white in linseed oil, sometimes combined with carbon black, in the preparation for areas of gilding in the approximately contemporary scheme at Idensen (Germany) (Matteini and Moles 1994: 85).

Analysis of the paintings of *c*1175 in the Holy Sepulchre Chapel of Winchester Cathedral confirmed the presence of a proteinaceous component in the binding medium in some

29. For the lead isotope analysis of samples from the chapter house, see Stos-Gale 1999; for the Thornham Parva retable, see Goldsmith and Bucklow 1998: 40, and Bucklow forthcoming: appendix 4.

samples, as in a layer consisting of lead white, red lead and calcium carbonate in a sample from the *impasto* decoration of the border dividing the narrative tiers on the east wall.

Despite previous conservation treatments, FTIR analysis indicated that lead white combined with oil was used for the *impasto* details in the scheme of *c*1220 in the Eastbridge Hospital, Canterbury (Howard 1994: 5–6). Similarly, in the scheme of *c*1250 on the eastern arch of the Ante-Reliquary Chapel, Norwich Cathedral, FTIR analysis indicated that the secondary lead white ground was applied in an oil medium (Park and Howard 1996: 391–2; Howard 1995b: 3). Lead white was also applied in oil in the scheme of *c*1300 in the same chapel, and instrumental analysis by GC-MS confirmed the medium in this case as linseed oil.[30]

Use as a drier for oil

Accounts from the 13th and 14th centuries provide interesting information regarding the addition of driers to oil, particularly of lead-based driers. The two were frequently purchased together although their function is not specified. For example, lead white, oil and other colours were purchased for painting in the refectory of the Cathedral Priory at Norwich in 1290–91, while in 1316–17 red lead, white lead, oil and whitewash were bought for painting images in the choir of the cathedral itself. Items purchased for the decoration of the Shrine of St. William in 1325 included lead white, oil, orpiment and vermilion (Park and Howard 1996: 401–2). The use of lead and copper pigments as driers for oil are also indicated in medieval treatises. Cennini, for example, mentions a recipe for a mordant for use on walls involving linseed oil boiled with lead white and verdigris (1960: 96, 120–21).[31]

For the paintings of *c*1500 at St. Mark's, Bristol, the use of lead pigments as driers for areas of thick *impasto*, and visual evidence such as the craquelure, both suggested the use of an oil medium. Additional analysis using FTIR microspectroscopy indicated that oil was used both as a sealant – combined with lead white added as a drier – and as the binding medium for the paint (Gill and Howard 1997: 100).

Comparison with its use in other media and in continental wall painting

Lead white has been identified in Romanesque wall painting elsewhere in northern Europe, as in the scheme of *c*1130 at Idensen (Germany), where it was combined with linseed oil in the preparation for the gilding on Christ's mantle and on the small cross in his hand (Matteini and Moles 1994: 85).

Recent analysis of the wall paintings of *c*1270–80 in Angers Cathedral revealed a mixed ground of lead white sometimes combined with a little red lead. Lead white was also identified in other pigment mixtures; for example, a violet colour was achieved by combining lead white with a copper blue and red lake, and flesh tones by mixing vermilion with lead white (Demailly *et al.* 1988: 11). Further information on the use of lead white in French Gothic wall painting is provided by an account of 1320 in connection

30. GC-MS analysis detected linseed oil in a sample taken from the border above the head of the bishop on the south-east pier. This sample contained both 13th- and 14th-century phases of painting, although very little of the earlier scheme was present; see White 1998: 1.
31. Lead white was employed as a drier for oil from an early date: Eraclius (1967: 230–31) describes a method for painting stone plaques or for imitating marble on columns, in which the stone – previously dried in the sun – was prepared with a layer of lead white in oil.

with the decoration of Hesdin Castle, where it is stipulated that the finest lead white is to be used for the painting ground (Dumay 1874: 237–46). Lead white was also employed for the splendid mid-14th-century choir screen paintings of Cologne Cathedral (Jägers and Schulze-Senger 1989: 188).

Lead white grounds have also been identified in English Romanesque polychrome sculpture, as on the column figures of *c*1180–85 from St. Mary's Abbey, York. Here a lead white ground was employed for the figures of Moses, St. John, and Apostles. The pigment was also used in combination with ultramarine and charcoal black for the rich blue colour of Moses' robe (Brodrick 1993: 23–7).

The pigment has been more frequently identified, however, in Gothic polychrome sculpture. Analysis by Dr A. Roy of the National Gallery indicated that lead white and perhaps gypsum were employed as grounds for the exterior polychromy of Wells Cathedral of *c*1220–30 (Sinclair 1995: 110, n. 11), while a pure lead white ground has also been identified in the polychromy of *c*1236 on the Martyrdom Doorway in the cloister of Canterbury Cathedral (Hluvko 1991: 42). A lead white ground was identified on some fragments of early 15th-century polychrome sculpture at Tewkesbury Abbey (UCL 1996: 1), and lead white in oil was employed as a ground for the de Bois tomb of *c*1380 from Ingham, Norfolk.

Thick *impasto* modelling to create raised decorative elements, such as the mail at Farleigh Hungerford described above, is also found on polychrome sculpture. For example, lead white was fashioned into imitation jewels on the de Bois effigies of *c*1380 at Ingham. Lady Margaret's head-dress was decorated with fictive pearls apparently formed from lead white combined with heat-bodied oil, while a pink jewel on Sir Roger's helm consists of a complex mixture of lead white, red lead, red earth, calcite and azurite combined with linseed oil and applied in substantial *impasto* (350μm) over a layer of silver leaf (Plates 201 and 202) (Howard 1993d: pls. 5–8. In these instances the lead pigments would have functioned as driers, and GC-MS analysis of a sample confirmed the presence of linseed oil which had been partially heat-bodied to promote drying (White 1998: 1–2). Other examples of the use of lead white in late medieval painted sculpture in England include the 14th- and 15th-century polychromy on the west front of Exeter Cathedral, where the pigment was employed in the priming layers as an undercoat and in various pigment mixtures (Sinclair 1991: 120–21, 1992: 10); the late 15th-century Fitzalan effigies at Arundel (Brodrick and Darrah 1986: 91); and the late 15th-century polychromed sculptures from the Great Screen in Winchester Cathedral, where lead white was combined with red lead for the underpaint (Brodrick and Darrah 1989: 615–16).

Lead white applied in an oil medium has been identified in a number of continental Romanesque polychrome sculptures, as on the 11th-century Hørning plank, Denmark,[32] and on the Hemse crucifix of *c*1170–90 (Sweden) (Plahter 1984: 40–41).[33] In the latter case, lead white combined with a drying oil – probably linseed oil – was employed not only for the preparatory layers, but also for flesh tones, and in combination with ultramarine. Analysis also indicated that a high proportion of neutral lead carbonate

32. Analysis indicated that the medium is linseed oil, and that lead white is included in the palette together with orpiment, red lead and vermilion. Mads. Christensen and J. Glastrup, poster at ICOM 1996.

33. By contrast, pure unmixed lead white, in a medium likely to be egg tempera, has been identified on the lion capital from Vossestrand; Plahter 1990: 275.

(cerussite) was combined with the lead white (basic lead carbonate). Since neutral lead carbonate has a lower hiding power than its basic equivalent it has been suggested that its inclusion may help to account for the enamel-like appearance of the flesh tones (Plahter 1984: 40–41). Although neutral lead carbonate may occur as an impurity in basic lead carbonate (Gettens *et al.* 1993a: 67), it is significant that an alteration of red lead to form neutral lead carbonate (cerussite) has been confirmed in the Romanesque paintings of St. Gabriel's Chapel, Canterbury Cathedral (Cather and Howard 1994: 149), and it is possible that on the Hemse crucifix the paint layer originally consisted of a combination of lead white and red lead, which has now altered to cerussite.

Lead white grounds are commonly encountered in Gothic sculpture in northern Europe, and have been identified, for example, on the west portal of Chartres Cathedral of *c*1140–50 (Nonfarmale and Rossi Manaresi 1987: 270), and the external polychromy of *c*1160 at Bourges Cathedral (Rossi Manaresi 1987: 177). Later examples include the early 13th-century south portal of Lausanne Cathedral, where lead white was also employed as an overall ground (Furlano *et al.* 1981: table 2); the portal of the '*Portail de la Mère Dieu*' of *c*1220–60 at Amiens Cathedral, where a lead white preparation was employed and where the pigment was also combined with others (Weeks 1998: 102); the 13th-century tomb of Count Palatine Henry II at Maria Laach, Germany (Oellermann 1993: 264); and the mid-14th-century portal sculptures of the Holy Cross Minster in Schwäbisch Gmünd (Hauff 1988: 715).

A lead white *imprimatura* was often applied on top of a chalk ground in preparation for paint layers in northern European Gothic panel painting. One example which has recently undergone analysis is the Thornham Parva Retable of *c*1335. Lead isotope analysis of the lead white employed for this panel suggested that the lead ore from which the pigment was made came from Derbyshire (Goldsmith and Bucklow 1998: 40).[34] Other examples of lead white employed as a preparatory layer in English panel painting include the reredos of *c*1180 now in Adisham parish church, Kent, where a layer of lead white 45μm thick was applied over the gesso preparation (Howard 1993c: 8); the nave ceiling of *c*1220 at Peterborough Cathedral, where the pigment was combined with carbon black in some areas (Howard 1997b: 5–6); and the early 14th-century screen from Kingston Lacy, Dorset (Tracy 1997: 28; Plender and Marchant 2000).

Discussion

Lead white seems to have been employed relatively infrequently in Romanesque wall painting. Where the pigment has been identified in 12th-century schemes it was used either as a component of an opaque paint layer, or for specific areas of underpaint beneath such layers.

However, the use of lead pigments in the Romanesque palette is significant since it presupposes at least a partial use of an organic binding medium, and it is increasingly clear that mixed media were indeed employed in wall painting of this period. Where instrumental analysis of the original binding medium of lead white in Romanesque wall painting has been undertaken, an organic component has been identified; for example, linseed oil at Idensen. Furthermore, strong indications that organic binding media were used in paintings such as those of Witley, which were previously assumed to have been

34. Lead isotope analysis was undertaken by Dr Stos-Gale at the Isotrace Laboratory, Oxford.

bound by the carbonation of lime alone, suggests that originally there may well have been more substantial areas of lead and possibly other pigments than now survive. Indeed, although the areas of lead pigment first recognised were in the clothing of the opulently dressed figures in the *Feast of Dives* – a local use consistent with the iconography of the subject – they were subsequently identified in the background of the arcading on the north wall and on one of the dogs in the scene to the left of the Feast, indicating a very widespread use originally.

By contrast with the opaque colouring of the Romanesque period, translucency was increasingly desired in the Gothic period, and by *c*1300 lead white grounds were frequently employed to provide a compact reflective white surface to interact optically with the succeeding translucent paint layers. The incompatibility of lead white in alkaline conditions certainly helps to explain the relative paucity of its use in Romanesque wall painting where the carbonation of lime was often the major binding mechanism. However, lead white is supremely suited for use in an oil medium, and as the use of this medium steadily increased through the Gothic period – together with the desire for translucent effects – the use of lead white increased accordingly. The pigment was also cheap, and has been identified in wall paintings of all types in the Gothic period.

Pigment alterations

The alteration of lead pigments has been the subject of several detailed studies which suggest that it is due to high alkalinity, humidity and the presence of an oxidising agent (Weber 1923; Matteini and Moles 1981: n. 33; Petushkova and Lyalikova 1986: 65–9; Giovannoni *et al.* 1990: 22–3). An investigation of the darkening of lead pigments in English medieval wall painting showed that in all cases the product of alteration was lead dioxide (plattnerite) (Welford 1991). Lead dioxide is the product formed in the alteration of both red lead and lead white, although in neither case is the mechanism for the alteration completely understood.

Examples of the darkening of lead pigments in Romanesque wall paintings in England include the apostle of *c*1130–40 in the nave at Little Easton, Essex; the slightly earlier representation of St. Margaret at Binsted, West Sussex (Welford 1991); and the scheme of *c*1110 in the nave of Witley, Surrey (Howard 1992a). At both Little Easton and Binsted, the lead white pigment employed for flesh tones has partially altered to form dark lead dioxide, resulting in severe disfigurement of the image. There are many other examples of this phenomenon, particularly in the Gothic period, with notable examples in the scheme of *c*1300 in the Galilee Chapel, Durham Cathedral; the paintings of the first quarter of the 14th century at Little Wenham, Suffolk; and the early 14th-century scheme of Longthorpe Tower, Cambridgeshire (Plates 203–205).[35]

The dark appearance of flesh tones does not always indicate the alteration of a lead white layer to form plattnerite. In samples taken from dark flesh tones in the scheme of *c*1300 in the Ante-Reliquary Chapel, Norwich Cathedral, XRD analysis produced no evidence of the presence of plattnerite.[36] It seems likely that the patchy grey appearance here is due to the partial removal of past surface coatings and accretions (Plates 207–209).

35. For a full discussion of this phenomenon, see Welford 1991.
36. I am grateful to Christine Bläuer Böhm of the Fachhochschule Köln and Jilleen Nadolny for undertaking this analysis for me.

A different type of lead alteration seems to be present in the early Gothic wall paintings in the Holy Sepulchre Chapel at Winchester (Plate 206). Here, black spots evident on the paint surface throughout the scheme signalled the presence of this transformation, and analysis confirmed the presence of a dark red-brown lead-based material. SEM/EDX analysis of samples indicated a very low oxygen peak, which may suggest that a mixture of lead dioxide and lead sulphide is present in this instance (Howard 1995a: 100–101).

Bone white

Bone white ($Ca_3(PO_4)_2$)[37] has been identified in only one instance in English medieval wall painting: in the schemes of *c*1375–1404 in the chapter house of Westminster Abbey. Most notably, the white bone-ash pigment was combined with lead white and silica and employed for the raised crown held by an angel in the *Majesty* in the eastern bay (Plates 221–223). Bone white, combined with yellow earth, lead-based inclusions and organic material, was also identified beneath tin leaf in the scene of *St. John Rebuked by the Angel* in the *Apocalypse* cycle.

The presence of bone white has also been noted in two samples from the outer moulding of the tympanum super-arch of the porch of Salisbury Cathedral. Unfortunately, the date of the layers in which the pigment was identified is not specified and may range from the 13th to the 17th century (Sinclair 1997: 13).

37. Bone white is composed of tricalcium phosphate ($Ca_3(PO_4)_2$) although calcium carbonate and other minor constituents may also be present. To prepare the pigment, Cennini (1960: 5) recommends the selection of the second joint and wings of fowls or capons, putting them into the fire until they have turned whiter than ashes and then thoroughly grinding the residue.

❦ NINE

Carbon black pigments

The black pigments employed in the medieval period are known generically as carbon blacks. They are formed by the pyrolysis of organic matter and may be classified as crystalline or non-crystalline. The crystalline form of carbon black – graphite – occurs only in the form of black earth in medieval polychromy, where it is combined with accessory minerals such as quartz, and sometimes iron oxides which may contribute to the overall colour. The non-crystalline carbon blacks are characterised by Winter as flame carbons (those produced in a gaseous phase); cokes (with a liquid or plastic precursor[1]); chars (those having a solid precursor); and coals (formed by the natural carbonization of plant remains).[2] Charcoal black, probably the most common of the carbon black pigments employed in the medieval period, is classified as a char, often retaining the morphology of its precursor, wood, in considerable detail.

Cost

Charcoal black is included in the accounts for Westminster Abbey – for example, in 1253, 7s 4d was paid to a charcoal burner for a supply (Colvin 1971: 261, 283) – although unfortunately it is not clear whether this charcoal was intended as a painting material. 'Black chalk' and 'chalk of a black colour' are itemised among other pigments in the Westminster Palace accounts of 1532, but such vast quantities were purchased – 376lb of the black chalk – that it is hard to imagine its function (Salzman 1952: 168).[3] Although these entries do not provide any clear evidence of the cost of carbon blacks employed specifically for painting, given the low cost of the raw materials and their ease of preparation, it is likely that they would have been inexpensive.

1. Bone and ivory blacks are included among the cokes since the solid constituents, including gelatine and collagen, fuse on heating and are therefore in a liquid or plastic state immediately before carbonisation.
2. For a detailed discussion of the characterisation and properties of carbon blacks, see Winter 1983.
3. Black chalk may be classified as a black earth pigment. Cennini (1960: 20) refers to a soft black stone from Piedmont which may be sharpened and used for drawing, but given the quantity of 'black chalk' purchased for Westminster it is unlikely that it was used for this purpose.

Optical, chemical and physical properties

The presence of a carbon black may be determined by a simple microchemical test, but it is often less easy to identify the type of black that was originally employed.[4] However, in some instances non-carbon constituents may suggest the presence of a particular type of carbon black. Bone and ivory blacks, for example, consist not only of a coke but also of significant amounts of hydroxyapatite – $Ca_5(OH)(PO_4)_3$ – which may be identified by XRD, or by detecting the elemental constituents (phosphorus, calcium and sulphur) during SEM/EDX analysis. Inclusions of quartz and iron oxides may indicate the presence of a mineral carbon such as black earth.

Optical properties may also assist identification. The morphological features of the precursor may sometimes still be evident, as where large unpulverised pieces of charred wood are found in charcoal black. Flame carbons also have a characteristic morphology: the particles of lamp black are so tiny that they cannot be seen individually, and the minute spheres tend to form aggregates or chains.

Carbon blacks are generally highly opaque in transmitted light and are therefore seen only as a silhouette, which may be characteristic. For example most chars, such as wood and fruit-stone blacks, display distinctive elongated particles. The majority of chars and all flame carbons are isotropic, while in earth blacks particles of quartz are often visible, and graphite itself is anisotropic. Bone and ivory black particles tend to be irregular, with some 'lumpy' porous particles, and it may also be possible to see inorganic particles with carbon roughly covering the surface.

The difficulties in firmly identifying the specific form of carbon black results in the generic term being employed in many instances. Where further characterisation has been possible, charcoal blacks are the most commonly employed form in English medieval wall painting.

Preparation

The manufacture of charcoal for drawing is described in detail by Cennini who recommends tying bundles of willow twigs together, placing them in a sealed cooking pot and placing this in a baker's oven at the end of the day. In the morning the charcoal should be perfectly roasted; if overdone it would fail to hold together (Cennini 1960: 19). Cennini goes on to describe the various blacks which are available, remarking that charcoal black and almond or peach stone black are perfect colours. The process for producing lamp black is also outlined (1960: 22–3). Peter of St. Omer does not include details of the manufacturing process of black pigments, although for walls he indicates that charcoal black – made from leather, hay or any wood with the exception of oak – should be combined with water or egg. He also provides details of another 'excellent' black which appears to be a mixture of lamp black and the soot of rushes (Merrifield 1967: 1: 138).

Although various woods were probably employed for the production of charcoal in England, medieval authors such as Cennini refer to charcoal made from young shoots of grape vines as the *nigrum optimum* (Thompson 1956: 85; Cennini 1960: 22).[5] Preparation

4. For this destructive microchemical test a suitable particle is extracted from the sample and combined with a reagent comprising three parts concentrated nitric acid and one part concentrated hydrochloric acid. All but carbon black will go into solution.
5. A carbon black – thought to be vine black – was identified in the wall paintings of *c*1332–49 in the choirstalls of Cologne Cathedral; Maul 1992.

of the vine sprigs, which would have been less readily available in England, was carefully controlled to ensure thorough reduction to carbon, since otherwise a brown-black was produced. The sprigs were tightly packed, covered and sealed before being baked in a slow oven. The resulting charcoal was used in sticks for drawing, or powdered and ground for use as a pigment.

Use in English medieval wall painting

The majority of black pigments identified in medieval wall painting in England are chars; that is, those formed from the pyrolysis of wood or fruit-stones. Bone black – a coke black – has also been detected in a few instances, and although no flame blacks such as lamp black appear to have been employed, there is some evidence for the use of coal blacks.

Charcoal black

Major use of charcoal black: mixed with white to form a false blue colour

Charcoal black, either mixed with white or laid over a white ground in an attenuated layer,[6] produces a cool blue-grey effect and sometimes a positively blue appearance. The blue appearance of the surface is due to a phenomenon known as Rayleigh scattering: white pigments scatter the short wavelengths effectively and consequently blue light does not penetrate deeply into the paint film, whereas carbon black absorbs uniformly across the spectrum. In combination, white and carbon black appear blue-grey because the shorter average path-length of the short (blue) wavelengths allows less scope for absorption by carbon black, and therefore more blue light re-emerges as reflected light than the red portion of the incident light falling on the painting (Brill 1980: 87, 93–4). Although other carbon black pigments may sometimes produce a neutral or even cool grey colour, charcoal black is particularly suitable for the production of false blues. The stability of charcoal black in alkaline conditions, its ready availability and low cost would have made this an ideal pigment for the production of blue effects on the large scale required for wall paintings.[7]

The blue appearance of false blue can be enhanced by juxtaposition with warm colours such as ochres, a technique employed to great effect in the so-called 'Lewes Group' of wall paintings of *c*1100 in Sussex, where the palette consists predominantly of red and yellow ochres and lime white (Park 1984).[8] In the paintings at Coombes, for example, the warm orange-yellow – employed for areas such as the backgrounds and drapery in the narrative tiers – heightens the blue appearance of the fictive marbling at

6. The latter method of application was employed in the late 14th-century Byzantine frescoes of Sancta Sophia, Trebizond; see Plesters 1963: 133.
7. Theophilus (1979: 18, 20, 23 25) describes the pigment *veneda* as a mixture of black and white, and recommends it for underpainting azure, vegetable blue, and green. Jehan le Bègue, recommends that *venede* (or *veneda*) for use on walls be made from black and lime white; Merrifield 1967: 1: 310.
8. Analysis of slightly earlier fragments of wall plaster excavated from Colchester Castle (Essex) in which red and yellow ochres, lime white and charcoal black are mixed in various combinations to produce pinks and blue-greys is provided by Drury 1990: 111–22.

dado level (Plates 23 and 24).[9] Modelling effects have also been created by varying the amount of the lime white combined with charcoal black: mid-tones were applied first, followed by darker shadows and highlights in thick white *impasto* (Howard 1990: 195, 197–8). A cross-section taken from a dark shadow (Plate 25) clearly shows this stratigraphy, in which charcoal black was combined with lime white and applied in two layers – the upper with a higher proportion of carbon black – over the white plaster substrate.

False blue was also employed for a scene on the west wall of the chancel at Hardham (another of the Lewes Group churches), in which it appears that Adam and Eve are shown in water.[10] The strong blue effect here is enhanced by the proximity of the warm flesh tones of the naked figures. A sample taken from one of the 'waves' shows that charcoal black combined with lime white was applied in a single layer 75μm thick over the coarse plaster surface. The cross-section also reveals the presence of a single red iron oxide particle, although this is probably an incidental inclusion.

The rationale for the use of this false blue in wall painting was doubtless primarily that of cost and availability. The only mineral blue pigment generally employed in the Romanesque period was ultramarine, which was both imported and expensive making the use of an economic alternative easily explicable. However, false blue was also commonly employed in the Gothic period, when alternatives to ultramarine such as azurite, indigo and synthetic copper blues were readily available. It seems likely therefore that false blue was also selected for its particular coloristic qualities which were well understood and fully exploited by the medieval painter. The blue-grey appearance of false blue would have suited the Romanesque aesthetic where primary colours predominate particularly well, although it was generally employed extensively only where ultramarine was clearly not within the budget.

The cool blue-grey of false blue made it extremely useful as an underpaint beneath 'true' mineral blues, since it had the effect of enhancing their colour and increasing covering power by reducing light scatter from an otherwise white surface. This use of false blue as an underpaint for ultramarine has been detected in the scheme of *c*1170 at Cormac's Chapel, Cashel (Ireland), while examination of other samples from this scheme indicates that it was similarly employed beneath vermilion to produce a slightly purple appearance, and beneath yellow ochre mixed with charcoal black to enhance the resultant green (Howard 1991).

Blue-grey mixtures have also been identified later in the Romanesque period as well as in the Gothic period. For example, such a colour made from combining charcoal black and lime white has been identified in excavated fragments of *c*1180 from the chapter house of Sherborne Abbey, and the same pigment mixture also appears to have been employed for the *in situ* diaper pattern on the south wall (Howard and Park forthcoming a). A blue-black made from a mixture of charcoal black and lime white was also used on the *c*1255 painted coffin-lid of Archbishop Walter de Gray at York Minster, where

9. The lozenge pattern employed for the dado ornament at Coombes imitates the type of marble revetments found in late antique and early medieval churches, where the natural veining in the stone forms symmetrical patterns of chevrons and lozenges over the lower surfaces of walls and piers, as at San Vitale, Ravenna and San Vincenzo al Volturno (see Mitchell 1985: 129, 132, ns. 14–22 and 170 n.20).
10. For a discussion of this unusual subject, see Park 1984: 210–11, Howard 1990: 198–9 and Baker 1998: 217 and pl. 9.

analysis also indicated that the black background of the figure is painted in wood charcoal (Plate 26).[11]

False blues also occur in Romanesque wall painting elsewhere in northern Europe. For example, in the paintings of *c*1100 in the chapter house of the Abbey of La Trinité, Vendôme, the blue-grey colour of the sea in the best-preserved scene was achieved by mixing black, green yellow and orange pigments (Taralon 1986a: 409). Examination and analysis of the 9th–11th-century wall paintings in the crypt of Saint-Germain, Auxerre also indicated the presence of red inclusions combined with carbon black and lime white to produce false blue, but in this case analysis by Raman microspectrometry confirmed the presence of both red iron oxide and vermilion (Sapin and Couprey 1992: 85, 94–5).

Charcoal black combined with lime white was also employed to produce a blue-grey colour for the meander pattern at the east end of the nave north wall in the scheme of *c*1165 at Pürgg, Austria (Howard 1995d: App. 1). A mixture of charcoal black and lime white which may have been intended to imitate a blue pigment has also been identified in the Romanesque schemes in the crypt of Auxerre Cathedral (Cadet 1995: 38), and in the parish church of Vicq (Rollier-Hanselmann 1997: 66).

A mixture of charcoal black and lime white was likewise employed to produce a blue-grey colour on a 13th-century stone column in Lichfield Cathedral (Plate 27) (Howard 1988: pls. 70–71). Examples of the use of false blue have also been confirmed in other examples of polychromed sculpture, as on the 15th-century effigies in the Fitzalan Chapel at Arundel where an opaque blue-grey appearance was produced by combining charcoal black with lead white (Brodrick and Darrah 1986: 71).

Other uses of charcoal black

The extent to which charcoal blacks were utilised in the Romanesque palette is exemplified by the range of applications found in the scheme of *c*1130 in the apse of St. Gabriel's Chapel, Canterbury Cathedral. Here, the pigment was used extensively to produce a variety of effects by applying it in a number of ways: combined with ultramarine; combined with red earth as an underpaint for ultramarine; combined with lime white as a grey underpaint for ultramarine in the rich blue background of the narrative registers; as a local ground for green pigments; beneath red lead and red earth; to add shadows to flesh tones; and to finalise drawing. In addition, charcoal black was employed as an underpainting for areas of yellow, such as Christ's drapery on the vault (Cather and Howard 1994; Howard 1992b: 12, pl. 31).[12]

Final linear details in charcoal black are evident, for example, in a sample from the soffit of the south apsidal arch. Here the yellow preliminary drawing was refined and finalised in red earth before the final linear details were confirmed in charcoal black (Howard 1992b: pl. 11). The late 12th-century fragments of wall painting recently excavated from the crypt of Glasgow Cathedral are characterised by the use of free, confident brushstrokes which often follow the geometric designs, as seen clearly in the palmette motif on a fragment from the same group discovered earlier in the 20th century, with the final strokes applied boldly in charcoal black (Park 1998; Howard 1998a).

11. Analysis by J. Plesters in Ramm *et al.* 1971: 121–2 and 124.
12. This practice is mirrored in the contemporary scheme at Idensen, Germany; Drescher *et al.* 1990: 83.

Similarly, in the scheme of *c*1170 in Cormac's Chapel, Cashel, a range of pigment mixtures incorporating charcoal black have been identified: vermilion, charcoal black and red earth were combined to produce a dark purple colour; ultramarine was mixed with charcoal black to form a dark blue; and a dark maroon-red colour was produced by combining charcoal black with red earth (Howard 1991: App. 1). In the late 12th-century scheme in the Holy Sepulchre Chapel, Winchester Cathedral, charcoal black was mixed with the plaster itself specifically for zones to be painted blue, presumably to reduce the light scatter and thereby increase the covering power of the natural ultramarine. This can be seen, for example, in a sample taken from the background of the *Deposition* on the east wall of the chapel (Howard 1995a: 101).

The pigment continued to be used in Gothic wall painting albeit less extensively than previously. In the paintings of *c*1220 now transferred to the west wall of the Holy Sepulchre Chapel, charcoal black was employed as a local ground beneath vivianite (Howard 1995a: 102). A shadow in charcoal black was applied over the copper-green drapery of St. Margaret in the paintings of *c*1300 on the vault of the Ante-Reliquary Chapel, Norwich Cathedral, while for St. William's red chasuble in the feretory of St. Alban's Cathedral, charcoal black was combined with red lake to produce a darker richer red colour. Large particles of charcoal black combined with red lead are also evident in the maniple of the unidentified archbishop, also of *c*1302–08, on the south-east pier of the feretory.

Charcoal black has often been detected in wall paintings elsewhere in northern Europe, as for example in the Carolingian and Romanesque schemes at Müstair, Switzerland. In the latter scheme a mixture of charcoal black and lime white was employed as an underpaint beneath green borders, and beneath robes and backgrounds painted in ultramarine (Emmenegger 1986: 198). In the scheme of *c*1165 at Pürgg, Austria, the black pigment was combined with red earth and lime white to produce a purple underpaint for ultramarine, and was also mixed with lime white alone for a blue-grey layer beneath ultramarine (Howard 1995d: App. 1).

Charcoal black has also been identified in medieval polychrome sculpture. In the Romanesque period the pigment was employed on the late 12th-century figures excavated from St. Mary's Abbey, York. Here, charcoal black was combined with lead white and ultramarine for the drapery of Moses, and combined with lead white and red earth for St. John's eyebrow (Brodrick 1993: 23).

Carbon black for preliminary drawing

Carbon blacks were often employed for preliminary drawing in the Gothic period and have been identified in the paintings of *c*1395 in the Byward Tower, Tower of London, and in the 15th-century *Adoration of the Magi* at Berry Pomeroy Castle.[13]

13. These carbon blacks are generally of unspecified type. I am grateful to C. Babington for information on the Byward Tower painting. For Berry Pomeroy, see Gill *et al.* 1996. A charcoal black sketch enhanced in red ochre has been identified at Chippenham, Cambridgeshire (Hulbert 1992: 33), while in the 15th-century wall paintings at Cowlinge, Suffolk, preliminary drawing in black was finalised in red (Organ and Sawdy 1991: 2).

Bone black

Bone black has been identified by PLM in a number of English medieval wall paintings. The earliest identification to date is in the mid-13th-century *Wheel of Fortune* in the choir of Rochester Cathedral (Darrah 1992), while it has also been detected in the late 14th-century *Apocalypse* cycle in the chapter house of Westminster Abbey. Here a few bone black particles were detected in a sample of faded blue-black from the drapery of a figure in the south-east bay. Examination of the cross-section indicated that a few particles of bone black were combined with indigo in a lead white matrix and applied over a lead white ground. Since the number of bone black particles in this instance is rather small, their presence may be incidental; alternatively, they may have been added intentionally to produce a slightly darker blue colour (Howard 1988: 42).

Bone black was also employed in the paintings of *c*1470 in the Deanery of Durham Cathedral. Here the pale blue-grey armour of the sleeping soldier in the *Resurrection* was produced by the application of indigo combined with lead white over an extremely fine layer of bone black on a local ground of red earth (Howard 1988: pls. 63–4).

An 'animal black', presumably bone black, has also been identified in the splendid cycle of wall paintings of *c*1260 in the choir of Angers Cathedral (Demailly *et al.* 1998: 11).

In polychrome sculpture, bone black or bone brown has been identified on the 14th-century monument at Erwarton (Suffolk). Here, the pigment had the appearance of the partial pyrolysis of animal bones and was incorporated into the ground layer of the effigy. Elsewhere, charcoal black was employed, as in a sample taken from above the knight's left bicep (Roy 1984: 4).[14] Examination of paint samples from the 14th- and 15th-century polychromy on the west front of Exeter Cathedral also suggested that bone black (or possibly a coal or lamp black) may have been employed for details such as hair and to imitate Purbeck marble. This black pigment was also combined with vermilion and iron oxide to produce a rich brown, and with white and red iron oxide for a grey colour (Sinclair 1991: 124, 1992: 10).

Coal black

SEM/EDX analysis of the black pigment employed in the *Last Judgement* of *c*1435–60 over the chancel arch at Holy Trinity, Coventry, indicated a notably high sulphur content. Since no phosphorus or calcium was present, the use of bone black can be excluded, and it is possible that coal was the source of the black pigment (Howard 1996b: pls. 21–23).[15] Here, the black was employed as an underpaint for the green background, enhancing the intensity of the verdigris layer by reducing light scatter from the otherwise white ground. The black underpaint consists of the black pigment in an oil medium with lead white driers. A sulphur-rich black pigment has also been identified in the scheme of *c*1500 in the Lord Mayor's Chapel, Bristol (Howard 1995b: pl. 7). As at Coventry, a black underpaint was employed, though here beneath indigo. In this instance the reduction of

14. Bone black has also been identified on the 13th-century tomb of Count Palatine Henry II at Maria Laach, Germany; Oellermann 1993: 264.
15. Harley (1982: 157) provides references for the use of coal as a pigment in both oil and water-colour painting during the first part of the 17th century.

light scatter would have helped to reduce the fading of the organic pigment (Howard 1995b: pl. 7).[16]

Black earth pigments

Although brown earth pigments are discussed in detail in the section on earths, a few examples of black earth pigments,[17] identified by PLM in polychrome sculpture, are included here. Notable examples include the fragments from the late 15th-century Great Screen at Winchester Cathedral, where translucent particles of black or dark brown ochre were applied over an orange underpaint for the Virgin's right eyebrow in the *Virgin and Child* group. For the border of her gilded cloak, a black linear detail was also painted in black ochre over the gold leaf (Brodrick and Darrah 1989: 615–16).

Binding media

Where carbon blacks were combined with lime white to produce a false blue colour, unreacted calcium hydroxide in the white pigment would have functioned as the binding medium for the black pigment particles. However, since black pigments in an oil medium were notoriously poor at drying, siccatives such as verdigris or lead white were added. In order to retain the dark colour, verdigris was usually incorporated in preference to lead white since the latter, if introduced in any quantity, would have produced a grey colour.

Discussion

Carbon blacks were employed extensively for wall painting throughout the medieval period in England and elsewhere in Europe. In most cases the type of carbon black employed has not been identified and the generic term is given in many publications. Charcoal black does, however, appear to be the most common of the carbon blacks identified, although this may be due to its relative ease of identification owing to the characteristic remains of the precursor (wood). Perhaps the single most important use of charcoal black was in the production of a false blue colour. The use of this pigment mixture seems to have been particularly prevalent in Romanesque wall painting where it was an economic alternative to the costly mineral blues then available.

16. Salzman (1952: 170) suggests that the 'secole blake' included among other pigments used at Hampton Court in the 16th century may be sea coal, that is pit coal.
17. Black earths contain carbon, as well as considerable quantities of quartz and other inorganic minerals, often including iron compounds (Winter 1983: 57). Yellow and yellow-brown earths contain goethite ($\alpha FeO{\cdot}OH$) together with various quantities of clays and quartz, while umbers also contain manganese dioxide.

❦ TEN

Discussion

Thirty pigments have been identified in the present research, establishing beyond doubt the variety and sophistication of the palette of English medieval wall painting. While the use of most of these pigments was already known – from the documentary sources and from previous analytical evidence – four pigments have been identified for the first time. Furthermore, this work provides detailed evidence to illustrate the complexity of the evolving paint application methodologies in English medieval wall painting during the Romanesque and Gothic periods.

Pigments previously unidentified in English medieval wall painting

The four pigments identified for the first time in English medieval wall painting are kermes lake, madder lake, vivianite and salt green.

Kermes and madder lake pigments

While lake pigments generally have been identified in several English medieval wall paintings, as in the paintings of *c*1250 at Horsham St. Faith Priory and in the scheme of *c*1395 in the Byward Tower, Tower of London, the *type* of organic dyestuff employed to produce the pigment remained unknown until the identification of madder and kermes in the present research.

Medieval accounts sometimes record the purchase of lake pigments, as for the mid-14th-century wall paintings in St. Stephen's Chapel, Westminster Palace. However, the nomenclature employed for lake pigments in such documents is varied – including *cynopre*, *cinopel*, *cinople*, *cynople*, *synople* and *synapo lake* – and the particular type of lake purchased is therefore generally unclear. Since one type of lake purchased for St. Stephen's Chapel was twice the price of the others, it was probably prepared using a more expensive dyestuff, possibly kermes; however, analysis of the dyestuffs used in these wall paintings has yet to be undertaken. While there is plentiful documentary evidence that madder was cultivated in England in the medieval period, its use in English medieval wall painting has remained unconfirmed until now.

It is significant therefore that this work has identified the expensive organic red dyestuffs kermes and madder in several highly sophisticated paintings. Madder lake was extensively employed in the scheme of *c*1300 in the Ante-Reliquary Chapel, Norwich Cathedral and in the paintings of *c*1302–08 in the feretory of St. Albans Cathedral. Since madder lake is less photosensitive than the other red lakes available in the medieval period, and relatively less expensive than lakes made from scale insects, it is not surprising that this pigment would have been selected for wall painting. Much more expensive was the lake made from the scale insect *kermes vermilio Planch*, commonly known as kermes. This has recently been identified in the important late 14th-century wall paintings in the chapter house of Westminster Abbey. Sometimes as much as 40 times more costly than madder, its use underlines the opulence and extravagance of the Westminster paintings. In one sample from this scheme, a tiny trace of brazil wood lake (*Caesalpinia brasiliensis*) was found mixed with the kermes. This suggests that the source of the red dye employed to make the pigment was a dye bath of cloth clippings. From the end of the 14th century this was one of the main methods of preparing lake pigments, and such dye baths may have contained a variety of cloth clippings, including those originally dyed with kermes, madder and brazil.

Kermes was also employed for other forms of polychromy. For example, it has been identified on a pilgrim's staff from Worcester and on the bust of Henry VII of *c*1510 now in the Victoria and Albert Museum.

Vivianite

The identification of vivianite is of particular interest since it appears not to be mentioned in the contemporary documentary sources. This blue iron phosphate mineral has now been identified in both the Romanesque and Gothic schemes in the Holy Sepulchre Chapel, Winchester Cathedral, and in the 13th-century wall paintings in the Guardian Angels' Chapel of the same cathedral. As a locally available material it was employed together with the expensive minerals ultramarine and azurite, and appears to have been selected for its particular coloristic qualities. Since vivianite is generally stable in its blue form, it is significant that in the Holy Sepulchre Chapel paintings the particles have partially altered to a yellow colour, thus producing an overall green appearance in areas such as backgrounds which would originally have been blue. Significantly, the pigment has retained its blue colour only where it has been protected from a variety of invasive conservation treatments.

In only two other instances has vivianite been identified in wall painting in northern Europe: in the 13th-century paintings at Mästerby, Gotland (Sweden) and in the late 14th-century scheme at Urschalling (Germany). Analysis of polychrome sculpture has revealed its presence in a few cases – on an Anglo-Saxon fragment of stone sculpture from York, on the Romanesque lectern at Freudenstädter (Germany), on a seated Virgin Mary of *c*1250 (Germany); and on the 14th-century polychromed portals of Holy Cross Minster, Schwäbisch Gmünd (Germany) – and in two cases in panel painting, namely the 13th-century Norwegian altar frontal Tingelstad III, and on a 13th-century painting from Regensburg, Germany (see p. 37–8).

It is likely that vivianite was used as a pigment in the medieval period only where a local source of the mineral was accessible; its use in medieval polychromy has been confirmed most frequently in Germany and in England where mineral deposits are well known.

Copper chloride green pigments

The presence of salt greens (copper chloride green pigments) in English medieval wall painting has also been confirmed for the first time. These pigments have been detected in both Romanesque and Gothic wall paintings although they appear to have been most commonly employed in the 12th century. For example, the copper chloride green, atacamite, was employed for details such as the nimbi of the Virgin and other figures in the scheme of *c*1100 on the nave east wall at Hardham (West Sussex). The sophisticated technique of these paintings is underlined by a range of coloristic effects achieved by applying atacamite over a variety of differently coloured underpaints. Traces of an intense green colour have been detected in other areas of this important scheme, suggesting that the pigment was once used extensively throughout.

While atacamite was used to pick out only small details in the Hardham paintings, it was extensively employed for draperies and areas of background in the 'Losinga' cycle of paintings of *c*1190 at Norwich Cathedral. Here it was exploited in conjunction with green earth for its particular coloristic qualities, and was also applied over both light and dark underpaints to extend the subtle range of effects achieved.

A copper chloride green pigment has also been identified in the paintings of *c*1510 at St. Mark's, Bristol. Here it was combined with lead white and applied over a grey ground of carbon black combined with lead white to produce the rich green background in the *Trinity* scene.

Primary sources describe the preparation of salt green although not specifically in relation to wall painting. Copper chloride greens have frequently been identified in manuscript illumination, in panel painting and in Romanesque polychrome sculpture, and there are also confirmed instances of the use of the pigment in wall painting elsewhere in northern Europe, as in the scheme of *c*1130 at Idensen (Germany), and in 13th-century wall paintings from Gotland (Sweden) (see pp. 76–7).

In summary, the presence of original copper chloride greens has now been definitely established in English medieval wall painting. Considered with the evidence from elsewhere in northern Europe, it is likely that the pigment was more common in the Romanesque than the Gothic period, however further examples are needed to confirm this view.

Use of pigments and the changing nature of English medieval wall painting

Identification and use of lake pigments in Romanesque wall painting

Although some pigments identified in this book were known to have been employed in English Gothic wall painting, their earlier use was previously unknown. Thus, although red lake pigments were known to have been relatively common in sophisticated wall paintings from the 13th century onward, they have been identified in the present research in paintings as early as *c*1175. Examination of the unusual dull pink colour of Nicodemus' robe in the *Deposition* on the east wall of the Holy Sepulchre Chapel, Winchester Cathedral, revealed the presence of a red lake pigment combined with calcium carbonate and a tiny trace of ultramarine in an opaque paint layer. The substrate of the lake pigment in this instance was identified as calcium sulphate. This finding may be significant since lakes precipitated onto calcium carbonate or calcium sulphate tend to be less translucent than those precipitated with alum. This is consistent with artists' treatises which suggest that before the 13th century, red lakes for painting were generally precipitated onto a calcium carbonate substrate (see pp. 113–14). These lakes would have been entirely compatible with the opacity of Romanesque polychromy, while in the Gothic period – when translucency was so highly prized – alum-based lakes would have been more commonly employed.

A red lake has also been detected for the first time as part of a glaze layer in Romanesque wall painting in Britain. A red lake precipitated onto a calcium carbonate substrate and used as a glaze was identified in a fragment of the painting of *c*1190 excavated from the crypt of Glasgow Cathedral. Here, the pigment was applied over a layer of vermilion and would have saturated and enhanced the underlying red layer. Unfortunately, so little of the lake pigment survives in this instance that analysis of the constituent dyestuff was not possible.

While the earliest identified red lake in English wall painting is in the scheme of *c*1175 in the Holy Sepulchre Chapel at Winchester, elsewhere in northern Europe a red lake pigment has been identified in wall painting of as early as *c* 800 (see pp. 120–21). Such early use should not be surprising since the preparation of lake pigments is described in treatises and recipe books well before the Gothic period. Eraclius describes the preparation of lake pigments from madder roots, wall flowers, brazilwood and ivy, while Theophilus also mentions madder (albeit in relation to staining bones and ivory), and discusses the pigment *carmine* which may have been a scale-insect lake such as kermes (see pp. 114–15). Now that the presence of lake pigments has been established in early wall painting in England and elsewhere in northern Europe, further identifications of these pigments are likely.

Identification and use of lead pigments in Romanesque wall painting

A further significant finding is the cumulated evidence for the use of lead pigments in English Romanesque wall painting. Although red lead had already been detected in the wall paintings of *c*1170 at Cormac's Chapel, Cashel (Ireland), as well as in the Romanesque wall paintings in the Holy Sepulchre Chapel, Winchester Cathedral, its presence has now been confirmed in three other instances – at Witley (Surrey), St. Gabriel's Chapel, Canterbury Cathedral, and in the wall painting fragments excavated from Glasgow Cathedral. Similarly, the present research has confirmed the presence of lead white in two further Romanesque wall paintings schemes (see p. 177).

The use of lead pigments in England is mirrored by findings from Romanesque wall painting elsewhere in northern Europe. For example, lead white has been detected in the scheme of *c*1130 at Idensen (Germany), and red lead in both Carolingian and Romanesque schemes at Müstair (Switzerland). Similarly in polychrome sculpture, lead pigments have been identified both in England – as in the original polychromy of the frieze from the west front of Lincoln Cathedral (*c*1140) – and in a number of continental Romanesque polychrome sculptures (see pp. 132–3 and 181–3).

Although used less frequently in Romanesque wall painting than subsequently, evidence for the use of lead pigments in the 12th century is significant since it presupposes at least a partial use of an organic binding medium. Paintings such as those in St. Gabriel's Chapel were previously thought to have been painted purely in a *fresco* technique. However, lead pigments are unstable in alkaline conditions and therefore require an additional organic binding medium. Both linseed oil and a proteinaceous material have now been identified as part of the original materials employed in St. Gabriel's Chapel and this finding has implications both for the original technology of the painting and for its conservation. Furthermore, evidence that organic binding media were employed in paintings such as those of Witley, which were previously assumed to have been bound by the carbonation of lime alone, suggests that originally there may well have been more substantial areas of lead and possibly other pigments than now survive.

Evidence from continental paintings again supports these findings. Linseed oil was combined with lead white in the scheme at Idensen, and the binding media for layers of red lead in the Carolingian paintings at Müstair comprise a protein and a small quantity of oil. Likewise in polychrome sculpture, linseed oil has been identified as the binding medium of the pigments (including red lead) in the frieze from the west front of Lincoln Cathedral, and an oil medium has been detected in the polychromy (including both red and white lead) of an 11th-century plank from the stave church at Hørning (Denmark). A ground of lead white in egg was applied in preparation for the 12th-century polychromy of the *Montvianex Madonna* (Kargère 2002: 509)(see pp. 132–3 and 182–3).

Selection and use of blue mineral pigments

While ultramarine was almost invariably the blue mineral pigment employed throughout the Romanesque and into the Gothic period in both cathedral and parish church schemes, a range of substitutes, predominantly azurite, but also indigo and synthetic copper blues, become increasingly common from the middle of the 13th century.

The frequent use of ultramarine in English Romanesque wall painting has parallels elsewhere in northern Europe; for example, it was employed extensively in the scheme of *c*1130 at Idensen (Germany). Indeed, the mixture of charcoal black and white employed as a ground for ultramarine at Idensen mirrors that used in the approximately contemporary paintings in St. Gabriel's Chapel at Canterbury and in Cormac's Chapel at Cashel. Coloured grounds have also been detected elsewhere, as in the scheme of *c*1165 at Pürgg (Austria), where ultramarine was applied over a dark red ground to achieve a purple effect, and over grey where a dark blue colour was intended. Ultramarine has also been identified in English Romanesque polychrome sculpture, for example in the frieze from the west front of Lincoln Cathedral (*c*1140), and on the limestone figure of Moses of *c*1180–85 from St. Mary's Abbey, York (see p. 34).

The ascendancy of azurite over ultramarine in the Gothic period is intriguing for two reasons. First, azurite was clearly available in the Romanesque period, but in English wall painting its only confirmed use during this period is for the scheme of *c*1130 at Kempley (Gloucestershire); on the continent there are exceptional and scattered examples (see p. 43). Second, in the Gothic period ultramarine continued to be used in panel painting and for high-quality wall paintings such as those of *c*1350–60 in St. Stephen's Chapel, Westminster. Such examples are generally from important and expensive commissions. These observations suggest that ultramarine was considered the superior pigment and that there may have been a sharp rise in its cost in the early 13th century, making it prohibitively expensive for most large-scale painting thus leading to the use of cheaper alternatives.

Of the cheaper alternatives to the mineral blues, 'false' blue (a combination of charcoal black and lime white) was used extensively in English Romanesque wall painting, particularly in parish church schemes. Although the rationale for its use was primarily that of cost and availability, its continued use in the Gothic period – when indigo and synthetic copper blues were also commonly employed – indicates that it was also selected for its particular coloristic qualities. Analytical evidence suggests that the use of indigo increases substantially in England in the Gothic period and the pigment was clearly widely available at this time. Similarly, synthetic copper blues were commonly employed in Gothic wall painting; this may relate to improvements in the method of their manufacture in the 14th century.

Selection and use of green pigments

There is a chronological shift in the green pigments used for wall painting during the medieval period. The use of green earth declines at the end of the Romanesque period. It is likely that the soft warm colour of green earth was better suited to the Romanesque palette.

Although recipes for making verdigris are known from as early as the 8th century and recur in primary sources throughout the Middle Ages, the pigment was not generally used in wall painting until the 13th century. The paucity of its use in the Romanesque period is probably due to its instability in alkaline conditions. Verdigris was, however, the most favoured green pigment in Gothic wall painting, and in several instances analysis shows that it was employed in an oil medium.

Only once has the presence of malachite been confirmed by instrumental analysis in English medieval wall painting – on fragments of *c*1180 excavated from the chapter house of Sherborne Abbey. Previously, any copper green was simply presumed to be malachite, and it was therefore thought to be present at a number of sites. However, recent investigations have confirmed not only that many green copper pigments are heterogeneous in nature, but also that a number of copper-based materials such as chlorides and sulphates were also frequently employed as green pigments. Thus instrumental analysis of green copper-based pigments is mandatory if their precise nature is to be known.

Selection and use of red pigments

The selection of red pigments – vermilion, red lake, red lead and red earth – remains much the same throughout the medieval period although there is a marked increase in the use of red lakes and red lead in the Gothic palette. Despite the fact that vermilion was widely employed throughout the medieval period, the paucity of its use in parish church schemes in the Romanesque period is striking. One possible explanation is that the cheaper synthetic dry-process vermilion was not widely available at that time although recipes for its preparation are known from as early as the 8th century. Its scarcity would have inflated the price of the synthetic pigment, perhaps almost to that of imported mineral cinnabar. It is possible, therefore, that vermilion was too expensive during the Romanesque period for most modest parish church schemes.

Unfortunately, in only three instances has it been possible to analyse the constituent dyestuffs of the lake pigments identified in English medieval wall painting, but it may be significant that madder lake was detected in two of these.[1] In addition, kermes and brazil lake were detected in the important late 14th-century scheme in the chapter house of Westminster Abbey.

---

1. Madder lake has also been detected in contemporary English polychrome sculpture.

Selection and use of other pigments

While a number of pigments were known but not selected for wall painting until the Gothic period, some relatively new materials were also introduced at this time. Lead-tin yellow, an intense pale lemon yellow pigment, was first employed in panel painting from about *c*1300 but the first instance of its use in wall painting in England is not found until the (?)14th century, on a fragment excavated from the Bedern at York. All other examples of its use are in 15th- and 16th-century schemes of wall painting. Of the other yellow pigments available, yellow earths continue to predominate. Orpiment was rarely employed in English wall painting, and not at all until the Gothic period. The instability of orpiment in alkaline conditions would be sufficient reason to select alternatives in those Romanesque wall painting schemes where the carbonation of lime was the chief component of the binding process.

Lime white dominated the white section of the Romanesque wall painter's palette. The pigment was widely employed to vary the tone of pigment mixtures to provide additional binding capacity within the paint layer, as in St. Gabriel's Chapel, and for *impasto* highlights as in the 'Losinga' cycle of *c*1190 in Norwich Cathedral.

Lead white has been identified in only a small number of Romanesque wall paintings. However, it was pre-eminent among the whites of the Gothic period where its most important function was as a ground to provide a compact and highly reflective surface to interact optically with subsequently applied translucent paint layers. The pigment, often combined in an oil medium, was also used to vary the tone of pigment mixtures and to provide *impasto* highlights.

Carbon black pigments were extensively employed for wall painting throughout the medieval period. The relative ease of identification of charcoal black may to some extent account for its frequency of identification in comparison with other classes of carbon black.

Application methodology

In the past, the opaque and predominantly primary colours of the Romanesque palette were presumed to have been applied in a *buon fresco* technique. However, it is now clear that, although Romanesque wall painting technology is rooted in the *fresco* tradition and that calcium hydroxide (in a variety of forms) is the principal binding agent, overall the application methodology is much more complex.

Accumulated evidence indicates that plastering in large horizontal bands relating to scaffolding lifts was typical for English Romanesque wall painting and therefore, although the preliminary drawing and some of the initial paint layers were applied while the plaster was fresh and thus bound by its carbonation, the extremely large plaster areas preclude describing the overall technique as *buon fresco*. In addition, complex stratigraphies have been found with, for example, alkali-sensitive lead white underpaint used in some areas. In other instances, multiple paint layers were applied to produce complex effects of modelling. For example, as many as five paint layers – some incorporating combinations of two or three pigments – were employed to produce subtle shading of flesh tones. Local grounds were also used to influence the colour of subsequently applied paint layers, for example grey, red and white were used beneath blue. In these cases the uppermost layers clearly required additional binding media.

Furthermore, instrumental analysis has confirmed that organic binding media were commonly used as a component of the binding mechanism throughout the Romanesque period in both cathedral and parish church schemes. Indeed the range and application of organic binding media has been found to be both subtle and complex. Linseed oil and a proteinaceous material have been identified as original components of the binding mechanism as early as *c*1130 in the apse scheme of St. Gabriel's Chapel, Canterbury Cathedral. Only slightly later, in the paintings of *c*1175 in the Holy Sepulchre Chapel at Winchester Cathedral, linseed oil, modified by heating to promote drying, was used as a mordant for gilding indicating great sophistication in the use of this material. Significantly, the palette of both cathedral and parish church schemes has frequently been found to include pigments such as red lead and lead white which are incompatible with painting in a *fresco* technique. Thus it is clear that from at least the beginning of the Romanesque period, a variety of organic binding media were commonly used as a component of the original binding mechanism of wall paintings of all types (see pp. 5–6).

Gothic wall painting technique after about 1300 is governed by concepts of luminosity and translucency, perhaps influenced by the increasingly large areas dedicated to stained glass windows and in an effort to emulate their brilliant colour. There is a trend towards the use of an increasingly large proportion of organic binding media, particularly linseed oil, and a move towards the use of lead white grounds, often in combination with translucent glazes, to produce effects of luminosity. In addition, organic binding media are combined in increasingly complex combinations, frequently with different binding media used for particular layers in the stratigraphy such as vermilion in a proteinaceous medium glazed with red lake in oil, or even for specific pigments within a layer. For example, azurite was often ground in a proteinaceous medium before being combined in an oil-based paint layer. This would have increased the brilliance of the colour as in oil the colour of azurite can appear less intense due to the similarity of the refractive indices of these materials. The use of protein would have had the additional effect of inhibiting the corrosion of this copper-based pigment. Gothic wall painting is also characterised by increasingly complex layering of colours with stratigraphies incorporating reflective metal leaf and translucent glazes to produce sumptuous effects, and the more frequent use of relief ornament. It is clear the medieval wall painters had a detailed understanding of the properties of the materials available to them and a mastery of their application.

It is also clear that the considerable developments in Gothic wall painting techniques were driven by artistic intent rather than any sudden introduction of new materials. All the materials intrinsically associated with Gothic wall painting – organic binding media, alkali-sensitive pigments such as lead white and red lead, and lake pigments – were all available and employed in the Romanesque period. Equally, it is inaccurate to suggest that painting technology became steadily more complex during the medieval period. It seems that Romanesque and Gothic wall paintings were equally sophisticated but employed different application methodologies of equal complexity to produce very different effects.

Pigment alteration in English medieval wall painting

Three pigment alterations previously unknown in English medieval wall painting have been detected in the present research. First, the lightening of red lead; second the alteration of vivianite to a yellow form, possibly its polymorph, metavivianite; and finally the alteration of verdigris to form a blue copper chloride-based alteration product.

In addition to these discoveries, the present work documents a number of other pigment alterations such as the darkening of vermilion, the darkening of lead white and red lead, the transformation of calcium carbonate to calcium sulphate, the oxidation of azurite and the fading of lake pigments. Each of these alterations has been observed and studied previously by other authors, so the information collated by the present research simply adds to the cumulated examples of these phenomena. Examples are given in the relevant pigment chapters.

The lightening of red lead

The lightening of red lead to form a white product has been observed by a number of authors and various factors have been implicated in the transformation process: high RH, light and carbon dioxide. Experiments have shown that lightening is more rapid in aqueous binding media than in linseed oil, and that leanly bound films alter more rapidly. This suggests that penetration of light and moisture into the paint layer is an important factor in the lightening process. Furthermore, the lightening of red lead in an oil medium has been attributed to a reaction between the red pigment and the oil to form lead-fatty acid salts and lead carbonate.

The phenomenon was first identified in English wall painting by the author in the early 12th-century scheme in the apse of St. Gabriel's Chapel, Canterbury Cathedral. Here, the alteration of the red pigment has resulted in the formation of a grey veil over many previously red areas and is particularly noticeable in zones of red drapery. Particles of the veil and of the underlying red paint layer were analysed separately by XRD, which confirmed the presence of cerussite and red lead respectively. Paint cross-sections from affected areas display an open, porous structure with a larger proportion of the transformation product towards the surface of the paint layer. This ratio is typical of an alteration product and is indicative of a process initiated by an external incident or by a treatment of the surface.

Paint cross-sections from the late 14th-century *Apocalypse* cycle in the chapter house of Westminster Abbey also display a partial alteration of red lead to form a white, lead-containing alteration product. Here again, considerable volume change associated with the transformation has caused disruption of the paint layers. However, features of the altered particles in this instance are comparable with examples of the lightening of red lead in panel painting: the presence of a yellow zone of incomplete alteration in the centre of the particles; radial cracks in the disrupted outer portion; and wavy striations perpendicular to the cracks. The transformation in this case is due to a reaction between red lead and the oil medium.

Recent experiments indicate that the lightening phenomenon will occur where RH exceeds 75%. Since high humidity is a causal factor in the lightening of red lead, consolidation by lime-watering – a process which involves multiple saturations of a wall painting – is likely to be most inadvisable where lead pigments are present. Consolidation involving hygroscopic milk proteins such as casein may also be inadvisable since their presence will increase the moisture content of the paint layer and thus favour the lightening process (see pp. 139–40).

Alteration of vivianite

The alteration of vivianite from its relatively stable blue form to yellow in the Romanesque and Gothic schemes in the Holy Sepulchre Chapel, Winchester Cathedral, represents a new finding in English medieval wall painting. It is of particular importance since the considerable alteration in the schemes' appearance at Winchester may also be the result of well-intentioned but misguided conservation treatment. In the only other instances where vivianite has been found in medieval wall paintings, the pigment also appears to have partially altered. The vivianite found in the 13th-century scheme at Mästerby, Gotland has a yellowish-green appearance (Nord and Tronner 2000: 97), while in the scheme of *c*1370 at Urschalling (Germany), a patchy blue-brown discoloration of the decoration is thought to be due to an alteration of the mineral.[2]

The mineral is often colourless when first excavated. Its initial colour change from colourless to blue is due to increased ferric ion concentration. Mechanical grinding of the colourless crystals, heating in air, storage in a vacuum or chemical treatment of samples can produce a more rapid conversion from ferrous to ferric ions and so to a blue colour (see p. 38). Careful observation of polished cross-sections has shown that the vivianite present in the Holy Sepulchre Chapel paintings has only retained its blue colour where it has apparently been protected from the external environment. This strongly suggests that the invasive conservation treatments to which these paintings have been subjected may have caused further oxidation to the ferric state and hence produced a further colour shift to yellow.

Identification of the yellow oxidation product of vivianite has proved challenging. However, XRD analysis indicates the presence of metavivianite – a polymorph of vivianite – and it is possible that this represents the yellow alteration product. The alteration of vivianite clearly requires more detailed and urgent investigation. If the oxidation hypothesis is correct, the next challenge is to determine what preventive measures might arrest further colour change in the important paintings at Winchester.

Alteration of verdigris

The alteration of verdigris to form an amorphous pale blue product has been detected by the author in a number of English Gothic wall paintings – at Brent Eleigh (Suffolk); in the Ante-Reliquary Chapel, Norwich Cathedral; at Ickleton (Cambridgeshire); and at Farleigh Hungerford Castle (Somerset) – and also in contemporary polychrome sculpture at Ingham (Norfolk), and in the Chapel of Our Lady Undercroft, Canterbury Cathedral.

The alteration appears to be associated with contact between verdigris and overlying layers of fresh limewash or lime plaster. For example, in the mid-14th-century paintings at Ickleton, the blue transformation product is not apparent where an original dark green glaze remains intact over the verdigris layer. Where the glaze is lost, however, subsequent limewash layers have come into direct contact with the verdigris layer and have effected the transformation. Verdigris is known to be unstable in alkaline conditions, forming copper hydroxides. Transformation in the presence of limewash or lime plaster layers is therefore unsurprising. Alteration appears to occur initially at the interface between the layers or where cracks in the underlying green paint have provided a passage for further penetration of the calcium hydroxide from the fresh limewash or plaster.

Analysis of the alteration product by SEM/EDX from the sites listed above indicated the presence of a copper chloride component, and further analysis by FTIR of a sample from the early 15th-century scheme at Farleigh Hungerford Castle subsequently

---

2. Analysis by H. Kühn; M. Alof, personal communication, 1988; see also Pursche 2000.

identified the copper chloride as atacamite. The origin of the chloride ions is uncertain. Although it is possible that salt was added to the limewash layers subsequently applied to the surface of these paintings, it seems more likely that chloride-containing salts were incorporated in the manufacture of the complex copper acetate (verdigris)-based green pigment.[3]

Further FTIR analysis of samples of the original green paint layer and its blue alteration product from one of the 14th-century tombs at Ingham suggests that the oil medium from the verdigris has been leached from the now-blue alteration product. Leaching of the binding medium results in a low pigment-to-binder ratio, and the consequent open porous structure is likely to encourage chemical reactions to occur.

Further instrumental analysis is needed to characterise the chlorine-rich alteration products of verdigris and to clarify the alteration mechanism, which at Ickleton, Ingham and Farleigh Hungerford is so clearly associated with the surface application of alkaline material (limewash or lime plaster).

Relationships with continental wall painting, polychrome sculpture and panel painting: implications for workshop practice

Relationship to continental wall painting

In several cases striking similarities have been established between English wall paintings and others elsewhere in northern Europe. Although different workshops were clearly responsible, the findings suggest diffusion of expertise and technique across national boundaries.

For example, the application of the pigment salt green at Hardham almost exactly parallels that found in the slightly later paintings at Idensen (Germany). Close parallels can also be drawn between the scheme of *c*1130 in St. Gabriel's Chapel, Canterbury Cathedral, and the contemporary scheme at Idensen. For example, ultramarine was employed extensively in both schemes, and the combination of charcoal black and white employed as a ground for ultramarine at Idensen is similar to that used both in St. Gabriel's Chapel at Canterbury and in the scheme of *c*1170 at Cormac's Chapel at Cashel (Ireland).

The results of the analysis of organic dyestuffs employed in the late 14th-century paintings in the chapter house of Westminster Abbey may also be significant in relation to investigations carried out elsewhere in northern Europe. Both kermes and brazil lake were detected in the chapter house, while brazil lake alone was identified as part of the original palette of the important paintings of *c*1332–49 on the inner face of the choir screens at Cologne Cathedral. Since these choir screen wall paintings are among the most important and splendid of their date in northern Europe, brazil lake – which is the most photosensitive of the red lake pigments and therefore particularly susceptible to fading – seems an unusual inclusion in an otherwise sumptuous palette. However, it is possible that, as at Westminster, the lake employed at Cologne was prepared from cloth clippings and that therefore a mixture of several dyestuffs may be present. Further analysis of the lake pigments used in the Cologne paintings would be particularly interesting in this respect.

3. See, for example, Theophilus 1979: 41.

More generally, patterns in the use of a particular pigment may be traced across northern Europe. Thus, the unique use of azurite in English Romanesque wall painting at Kempley (Gloucestershire) has continental parallels as in the Romanesque schemes at Mårslet (Denmark) and Saint Hilaire-le-Grand at Poitiers. Although ultramarine was the most frequently selected blue mineral pigment in the Romanesque period, these exceptional and geographically dispersed occurrences show that azurite was certainly available.

The use of green earth declines sharply at the end of the Romanesque period in English and other northern European wall painting. In Gothic wall painting it is much less common than copper green pigments such as verdigris. It is likely that the greater intensity of these copper-based pigments proved more compatible with the Gothic palette than the rather soft warm green of green earth.

In contrast with analytical evidence from England, verdigris has been detected in several Romanesque wall paintings on the continent; for example, in French wall paintings at Saint-Aignan, Brinay-sur-Cher, and Notre-Dame-la-Grande, Poitiers.

Relationship with polychrome sculpture and panel painting: implications for workshop practice

The relatively good condition of surviving polychromy on a number of sculptures, such as the late 14th-century de Bois tomb at Ingham (Norfolk), has provided valuable opportunities to gather contextual and comparative information for contemporary wall paintings. It is clear from these analyses that the complexity of painting technology on sculpture is often mirrored by that of contemporary wall painting. This further supports the contention that technologies crossed the boundaries of substrate. Thus, the use of particular binding media for specific pigments or layers within the stratigraphy of polychrome sculpture is comparable with similar results from analysis of wall paintings. For example, in a sample taken from the painted cushion beneath Lady Margaret de Bois' head, the binding medium of the azurite, applied over silver leaf, has both a proteinaceous and an oily component. This finding compares closely with analysis of the 13th-century wall paintings at Angers Cathedral, where azurite was ground in glue before being combined in an oil medium.[4] The complexity of painting techniques on Gothic sculpture is underscored by results from the examination of the de Ingham tomb (*c*1344), also at Ingham; for instance, while protein was employed to bind the vermilion paint layer, an oil medium was employed for the red lake glaze on the surface.

If similar painting technologies were used for sculptures and wall paintings, it seems likely that the same workshops may sometimes have undertaken both. Evidence from St. Albans Cathedral supports this view in the case of the shrine of St. Alban and the two adjacent early 14th-century wall paintings depicting standing saints situated at either end of the east side of the feretory.

Although these two figures are stylistically identical and certainly of the same date, analysis has established significant technical differences between them, showing that they are in fact by two different artists.

In the case of 'St William of York', a lime ground was applied to the tooled ashlar masonry. Over this are two types of secondary local ground, each applied to produce a

4. Analysis by Raymond White.

different effect. A red layer was applied under the face and hair to produce a warmer tone, and also under areas of blue drapery to achieve a more purple effect. An underpaint of yellow earth mixed with a little red was used below the green background and certain areas of the red chasuble. The technique of this red drapery is particularly interesting since three types of layer structure are identifiable, each associated with the different degree of shading that was required.

By contrast with the complex technique used in the St. William figure based on a multiple layer structure to produce the effects of modelling and rich translucent colour saturation, the artist of the other figure creates hues by means of single layers of mixed colours applied over a single lead white ground. In the cross-section from his purple chasuble, it is clear that a layer of red lake, combined with azurite to produce a purple effect, has been applied directly over a thick lead white ground.

Significantly, although these painting techniques are so different, remarkably close comparisons are provided between the shrine polychromy and the original techniques of both wall paintings. Orange and red paint samples from the shrine are directly comparable in their technique with those taken from the unidentified saint – comprising a lime substrate with a thick lead white ground above, and then a red lead or vermilion paint layer. Even the small red lead particles found in the lead white ground of the wall painting also occur in the similar ground on the shrine.

Similarly, a striking comparison can be made between the original green background of the 'St. William' figure and the green used on the shrine: in both a copper green (probably malachite) has been combined with lead white in very similar proportions. The technique of the gilding is also very similar; in a sample taken from a gilded lion on the shrine the gold leaf is applied with an oil mordant over a ground of yellow earth containing a few red particles, identical to that in the wall painting. Since the shrine has such close technical similarities with both wall paintings, it is difficult to escape the conclusion that a single workshop was responsible for the wall paintings and the shrine polychromy as part of a single, overall scheme at the time the shrine was created, which is known to have been between 1302 and 1308. It is, indeed, entirely possible that they were the work of one or more of the painters active at Westminster in the late 13th and early 14th centuries: the king's painters, Walter of Durham and Thomas of Westminster, may have lived in St. Albans, and the painter John of St. Albans is well known for dancing on a table for the entertainment of Edward II.[5]

A slightly different situation has been determined in the case of the two schemes of *c*1400 in the chapter house of Westminster Abbey – the *Judgement* which shows strong Italian influence in both style and iconography, and the *Apocalypse* cycle, *Old Testament* figures and *Angel Musicians* which are close in style to north German painting of the late 14th century, particularly to those by the workshop of Master Bertram of Hamburg. So close is the similarity in the latter case that it may be suggested that German painters were commissioned in this instance. Documentary evidence confirms that German painters were working in England at this time; for example, Herebright of Cologne was commissioned in 1398 to paint an image of St. Paul and its tabernacle in St. Paul's Cathedral (Harvey 1961: 11 n. 1).

At Westminster, scientific examination has shown that technical differences correlate exactly with the stylistic differences between the two schemes. Thus the *Judgement* was

5. For these painters, see Tristram 1955: 69–70, and Colvin 1963: 226–7.

executed on an overall lead white ground, and the *Apocalypse* and related paintings executed on a chalk ground. Whereas tin leaf is used extensively in the *Apocalypse* and *Angel Musicians*, it is absent from the *Judgement*. Bold raised motifs such as the crowns occur only in the *Judgement*, whereas parchment attachments are employed only in the *Apocalypse*. While at St. Albans the accumulated evidence points to the activity of two different artists within the same workshop, at Westminster two different workshops are indicated, the first executing the *Judgement* painting and the second undertaking the *Apocalypse* cycle and other paintings soon afterwards. The technical differences between the two schemes cannot elucidate their precise dating, although that little time intervened is perhaps suggested by the technical and stylistic break occurring within the *Last Judgement* to which iconographically the *Old Testament* figures belong (Binski 1995: 190–91).

It is clear from the accumulated evidence that wall painting and sculptural polychromy share a complex painting technology – the techniques employed are independent of substrate. In addition, analysis of Norwegian polychromy by Plahter and Wiik (1970) demonstrated that materials and techniques employed on sculpture and panel painting are very closely related. Researchers in Prague have also found close parallels between the methods employed for 14th-century wall painting and panel painting when they compared the Holy Rood Chapel in Karlštejn Castle, the paintings in the cloisters of Emmaus Abbey in Prague, the ambulatory chapels in St. Vitus' Cathedral and the panel paintings of Master Theodoric (Hamsík and Tomek 1983; Pasálek 1992). A recent study has now established a close technical connection between the Thornham Parva Retable and the wall paintings of *c*1300 in the Ante-Reliquary Chapel, Norwich Cathedral. Although the techniques employed for these paintings strongly reflect general early 14th-century painting practice, the technical similarities are striking – certainly more than coincidental – and it is clear that they were made within a well-established workshop tradition that could have had its centre in Norwich (Sauerberg *et al.* forthcoming). These studies demonstrate that medieval polychromy in northern Europe may be viewed as highly integrated. The techniques employed may vary considerably, even within a single workshop, but painting technology is independent of substrate.

Conservation implications of the present research

It would seem an obvious principle in the conservation of wall paintings that a detailed and accurate understanding of their technology is required to inform appropriate conservation treatment. Sadly this has not always been recognised by conservators in this field, and misunderstandings have resulted in sometimes irreparable damage. This research has begun to explore the complexity of English medieval wall painting. From it stems some important implications for the conservation of these paintings, however there is doubtless much yet to discover.

Perhaps the most important conclusions to be drawn are: first, that there exists enormous complexity in technique during the medieval period; second, that assumptions about technique without supporting analysis are often ill-founded; third, that techniques vary between schemes of similar date and even within the same scheme; fourth, that there is a distinct evolution of technique which approximately parallels the chronology of the Romanesque and Gothic periods; and finally, that inappropriate conservation and consequent damage may disguise the true quality of the original painting.

Until recently it was assumed that there was a fairly predictable homogeneity in the technology of English medieval wall painting. The presumed lack of organic binding media before the Gothic period meant that all Romanesque paintings were considered to have been painted in a lime-based technology, while glazes were not considered to be a feature of English medieval wall paintings of any period. However, a systematic study of such paintings reveals a startling complexity in the techniques employed. Particular findings are: the use of multiple binding media within a single scheme as early as the Romanesque period; the presence of lake pigments used in opaque paint layers and as glazes in Romanesque wall painting; multiple binding media within a single paint layer in the Gothic period; and complex stratigraphies involving both the mixing and layering of pigments throughout the medieval period. Thus careful examination and analysis is vital prior to any conservation to develop a clear understanding of the overall technology of a wall painting and to ensure appropriate materials and procedures are employed.

The seemingly authoritative nature of Professor Tristram's work in the 1940s and 1950s perhaps encouraged a relatively unquestioning approach to the original technique of wall paintings by later conservators. The assumption that the original technology of Romanesque wall painting was essentially lime-based and that these paintings would therefore require lime-based conservation treatments seemed reasonable enough. However, the research on which this book is based provides several examples where such an approach has proved detrimental to the original materials. Indeed, small differences in painting technique can make huge differences to the optimal approach to conservation. For example, in the early Romanesque scheme at Witley (Surrey) the majority of pigments would indeed have been bound by the carbonation of lime, and it might reasonably have been presumed that this was the exclusive method of application. However, much surface detail was added using lead pigments which required additional organic binding media. By treating this as an overall lime-based painting, lead pigments have altered and many details have been lost. The redesignation of the St. Gabriel's Chapel scheme at Canterbury Cathedral – regarded by Tristram as paradigmatic of northern *fresco* – as a multimedia painting is perhaps the most striking example where assumptions about technique of a work have had to be revised.

Such assumptions have frequently influenced the types of materials used in conservation treatments. For example, the pervasive use of wax to consolidate and provide an impervious and waterproof surface for wall paintings in the 19th and 20th centuries seems, at least in part, to stem from the revival of encaustic techniques. This revival coincided with the discovery of Roman wall paintings at Herculaneum and Pompeii and credited the durability of these Roman paintings to the use of wax media and varnishes.[6] The application of inappropriate materials such as wax in the absence of clear information regarding the original painting materials and the causes of deterioration has often resulted in severe and extensive damage. For example, in the early Romanesque scheme at Witley (Surrey), an accumulation of moisture behind the impervious wax coating resulted in significant losses in the paint layer. Moreover, it is likely that numerous details painted with lead pigments, which had subsequently darkened, were removed together with the severely darkened coating during cleaning. In addition, such materials have frequently compromised future conservation treatments and the study of aspects of

6. For an account of the rationale and use of wax and wax-resin preservatives in English wall painting, see Cather and Howard 1986.

original technique, in particular the analysis of original organic components such as lake pigments and binding media. The present research provides a significant body of information regarding some of the original materials of English medieval wall paintings, which should help in the selection of more suitable materials and processes for the conservation of these paintings.

Although this study has established some general patterns regarding the original palette and its application, it has also indicated that it is inadvisable to extrapolate information from one scheme to another. For example, although the phenomenon of dark flesh painting in English wall painting has generally been shown by Welford to be due to the alteration of lead pigments, this is not the case universally. In the Ante-Reliquary Chapel at Norwich the dark patchy appearance is due instead to the partial and selective cleaning of the surface. Moreover, careful study of the St. Albans feretory paintings has established that even with two stylistically identical paintings in a single scheme there may be fundamental differences in technology which have clear implications for the conservation of the overall scheme; for example, each may require different approaches to cleaning and consolidation.

The evolution of technique which approximately follows the chronology of the Romanesque and Gothic periods argues the need for an individual approach to the conservation of each particular scheme. It has been firmly established that organic binding media were used in conjunction with lime from much earlier than previously thought – indeed from as early as *c*1130. By the Gothic period, linseed oil was established as a major binding component. The parallel evolution of the palette is evidenced by increasing amounts of red lead, lead white and verdigris, all of which are unstable in alkaline conditions. Not only these pigments, but also the organic binding media, may be profoundly affected by lime-based conservation treatments.

Many late medieval wall paintings in particular have been subjected to inappropriate and disfiguring restoration/s, and this can lead to an undervaluing of the original scheme. This situation is typified by the 15th-century paintings in the chapel of Farleigh Hungerford Castle. It is hard to believe that this scheme, which now appears as an over-restored wreck, was once even more sumptuous and subtle in technique than the famous scheme of Eton College Chapel. The recent technical examination has revealed the astounding quality and richness of the original polychromy. Undoubtedly the most striking technical aspect of these paintings is the extensive use of gold and silver leaf with associated translucent glazes. For example, a gold effect was achieved for St. George's helm by a yellow glaze over silver leaf, and further embellishment provided by raised curvilinear details made from a second layer of silver leaf highlighted with a green glaze. Intricate relief ornament was used to imitate mail armour around the neck of the kneeling knight. Here, a thick paste of lead white combined with oil was applied and the design modelled *in situ*. The undulating surface was then covered with silver leaf, and though doubtless glazed to prevent tarnishing, no trace of the glaze survives. Red lake was employed extensively, most notably in conjunction with orpiment in the shield above St. George. Perhaps in view of the analytical results at Farleigh Hungerford, we need to revise our view of some other late medieval wall paintings which have been similarly over-restored, and which without technical examination may remain unjustly disregarded and undervalued. The complexity of the technology of English wall painting throughout the medieval period demands that each scheme be studied prior to conservation if appropriate conservation methods and materials are to be selected.

Avenues for further research

As further schemes of wall painting are subjected to detailed examination and analysis, important new data will be gleaned and the results incorporated into the accumulating body of evidence. The present research has highlighted a number of avenues which should prove particularly fruitful for further exploration.

First, the three pigment alterations detected by the present study – namely the alteration of red lead to cerussite, the transformation of vivianite, and the alteration of verdigris – all require further investigation as a matter of urgency. Until the mechanisms for these alterations are fully understood it will not be possible to prevent further occurrences and to arrest any changes which may be ongoing.

Second, the complexity of the relationship of pigments to their original binding media has been one of the most important findings of this work. These relationships are still poorly understood since only a small number of samples have been subjected to the detailed analysis required; more work is clearly needed. The most promising sites for investigation in this respect are those which have escaped conservation treatment, since the already extremely aged original binding materials are not further masked by other non-original materials. Success in this area requires a combination of sophisticated instrumental methods of analysis such as Fourier transform infrared microspectroscopy (FTIR mic.), gas chromatography-mass spectrometry (GC-MS), pyrolysis gas chromatography-mass spectrometry (PyGC-MS), thermally assisted hydrolysis and methylation-gas chromatography-mass spectrometry (THM-GC-MS), and high performance liquid chromatography (HPLC) undertaken by experienced researchers, as reliable interpretation of analytical results is vital.

Other instrumentation, particularly non-invasive instrumental equipment which can be used *in situ*, will be of particular importance for wall painting examination, analysis and conservation. For example, recent evaluation of a portable multispectral imaging system, MuSIS® 2007, for the examination, recording and monitoring of wall paintings *in situ* proved extremely successful.[7] The versatility of the MuSIS system was demonstrated in tests in the Chapter House of Westminster Abbey and at St. Albans Cathedral. In the latter, a multispectral survey of the medieval vault was undertaken from floor level, a distance of some 21 metres. Microscopy was also employed in conjunction with ultraviolet imaging to detect the remains of 19th-century surface coatings.[8] Other techniques have also been developed for the *in situ* analysis of pigments. For example, laser-induced breakdown spectroscopy (LIBS) has been shown to be capable of collecting elemental data from pigments and performing depth profile analysis. This technique is highly selective and sensitive and although microdestructive – ablating a sample area approximately 40μm in diameter and no more than 10μm in depth – it avoids the need for invasive sampling.[9] By contrast, laser-induced fluorescence (LIF) is a non-destructive technique which provides information that can be directly related to the molecular structure of the pigments or other

7. This joint research programme of the Foundation for Research and Technology, Hellas (FORTH) and the Courtauld Institute of Art, University of London was sponsored by the British Council and the General Secretariat for Research and Technology. The MuSIS® 2007, developed by Dr Costas Balas of FORTH, images from ultraviolet to near infrared wavelengths (320 to 1550 nm).
8. These surveys were undertaken by Sharon Cather of the Courtauld Institute of Art, Adrian Heritage of English Heritage and Dr Costas Balas of FORTH.
9. See Anglos *et al.* 1997 and Burgio *et al.* 2000.
10. See Anglos *et al.* 1996.

components of the painting, both organic and inorganic, providing a further valuable tool for *in situ* inspection.[10]

Third, the provenancing of pigments employed for wall painting is another area which merits further research. Lead isotope analysis undertaken by Stos-Gale confirmed that the lead isotope composition of lead white samples from the chapter house of Westminster Abbey are fully consistent with lead ores from Derbyshire. The lead white employed for the Thornham Parva Retable was also found to be made from lead ore from the same source. Lead isotope analysis of further samples from medieval wall paintings, panels and polychrome sculpture may provide information on the manufacture and distribution of these pigments.

Finally, the relatively good survival of medieval polychrome sculpture in England, and its clear relation to wall painting and other technologies, highlights this as an important area for further investigation. The results of such a study will almost certainly cast light on the technology of wall painting and other medieval polychromy.

❦ APPENDIX ONE

Glossary

attachment any optional (often prefabricated) material which has been applied to or incorporated in the polychromy. Attachments include parchment and relief decoration, but excludes metal leaf.

binding medium the binding agent for pigments in a paint layer. This may be inorganic – such as lime – or organic, including oil, egg, glue etc.

buon fresco technique of painting by which the pigment – which is combined with water – is held to the surface by the carbonation of the freshly applied *intonaco*. Fresh plaster includes slaked lime – calcium hydroxide – and this sets on contact with air.

carbonation carbonation (or setting) is an irreversible chemical process that binds pigments to the surface of the plaster – ($Ca(OH)_2 + CO_3 \rightarrow CaCO_3 + H_2O$).

giornate small sections of the *intonaco* which may be painted before the plaster is set.

glair egg white which has been whipped to a froth and then left to stand until the froth has subsided when it may be used as a paint medium.

glaze paint layer which is translucent because the pigment and the medium have similar or identical refractive indices. This layer is usually designed to interact optically with the underlying layers.

ground a layer used to prepare a surface for painting. This layer interacts topographically with the underlying structure and provides a compact, and often highly reflective surface for the application of paint layers.

imprimatura a thinly applied translucent paint layer applied over a white ground. The *imprimatura* tones the underlying white ground and reduces its absorbency.

intonaco uppermost plaster layer on which the painting is executed. This plaster layer may be applied directly to the support or over the levelling plaster.

lake a pigment produced by precipitation on to a base from a solution of dye.

levelling plaster the first coat of plaster or render which is applied directly to an uneven support to provide a uniform, level surface (may also be known as *arriccio*).

lime painting painting technique in which pigment – combined with limewater or lime milk – is applied directly to the dry plaster or on a layer limewash.

limewash (or whitewash) is a layer of lime (or gypsum) which is applied with a brush to the uppermost plaster layer, or directly on a stone surface.

pastiligia raised decoration usually made from gesso (calcium sulphate or calcium carbonate and animal glue) and applied freehand i.e. the pattern is made directly on the paint surface or ground and is not prefabricated.

preparatory drawing the preparatory drawing generally sets out the image within the visual field. By contrast with a *sinopia*, the preparatory drawing is undertaken on the uppermost plaster or ground layer.

sealant a layer applied over the support, and/or at the ground/paint interface to prevent absorption of binding media from the subsequent paint layers into the porous ground/support.

secco painting on dry (or set) plaster with an additional binding medium which may be organic or lime-based.

siccative any metallic salt or solution of metallic salts which are added to drying oils for the purpose of accelerating the rate of drying or oxidation. Driers are usually derived from lead, copper, cobalt or manganese.

sinopia or full-scale sketch – usually in red earth – on the penultimate layer of plaster, or sometimes directly to the wall. The presence of a *sinopia* has implications for painting technique and is usually associated with the application of at least the initial layers of paint in *buon fresco*.

size a term frequently applied to gelatin or to very pure glue. In the medieval period, the finest size was made from parchment clippings.

skin glue impure gelatin prepared from the skins of animals.

snapped line a snapped line may be employed to set out the vertical and horizontal divisions of a scheme. A piece of pigment-soaked twine is stretched across the wall and snapped to imprint a line in the desired position.

support material which supports all other components/strata of the polychromy – usually stone, rubble wall or brick.

tempera in its wider sense this term refers to any one of several paint media used to bind (temper) a pigment layer. However, the term is now generally taken to refer to a paint layer made with a whole egg medium.

tin-relief applied decoration (attachment) which has been prefabricated in a mould that incorporates a layer of tin leaf to act as a releasing agent.

underpaint (or local ground) a layer which has been applied in a specific area of the painting and which is designed to interact optically with the subsequently applied paint layer – for example, a grey or black underpaint may have been employed beneath a mineral blue pigment to increase its covering power.

❦ APPENDIX TWO

Examination proforma for use *in situ*

The Techniques of English Medieval Wall Painting: Proforma A Leverhulme Research Project, Courtauld Institute of Art	
1. **Site:** NORWICH CATHEDRAL	2. **Site Code:** NORLOS
3. **Town:** NORWICH	4. **County:** NORFOLK
5. **Contact (name, telephone no.):** Keith Darby Feilden & Mawson	6. **Permission granted:** October 1995

7. Location of painting within building:

Nave, south aisle, fourth bay from the east, vault and adjacent arch.The only substantial remains are on the north-west part of the groined vault and adjacent area of the arch soffit. It seems that originally each 1/2 segment contained a single subject and that there was a programme of 8 subjects on the vault. 3 roundels survive on the north side of the soffit (there were originally 6). On the corresponding arch of the east side of the bay there are remains of similar colouring and there was probably a similar set of roundels here.

Cuts in the masonry indicate the former presence of screens across the aisle at either end of the bay, and it has been suggested that this was a chapel of some importance. NB these cuts are not datable and may be later medieval insertions.

8. Subject/s of painting:

Story of the foundation of the cathedral by Bishop Herbert Losinga.
Upper roundel: Losinga's act of simony in 1091 – simony
Central roundel: Losinga is now Bishop and repenting his deed, hands clasped to denote grief – repentance
Lowest roundel: Losinga building Norwich Cathedral (shows central tower with wooden spire, destroyed in the riot of 1272) – building
Vault – scene misinterpreted by Tristram as the Virgin holding the Christ Child – however subjects are male and may be part of more extensive Losinga programme and may be something to do with the translation of St. William of Norwich's relics – his shrine was near to this bay in the nave.
Represented are one or more figures facing an enthroned king.

9. Date of painting:

c1190–1200 on stylistic grounds, see Park 1994: 161.

10. Historical references (technique):

E. Baker, 'Norwich Cathedral: some notes on the 12th century paintings on the vault in the south aisle', *Friends of Norwich Cathedral, Forty-sixth annual report, 1975*, 10–11. 'fresco, completed in secco – rich blues, greens & vermilion'
E. W. Tristram, 'The paintings of Norwich Cathedral', *Friends of Norwich Cathedral Annual Report, 1935* (1936) 10,11.
E. W. Tristram, *English Medieval Wall Painting: The Twelfth Century*, Oxford, 1944, 55–7, 138–9 pls. 84–5 sub pls. 5b,c.
D. Park and H. Howard, 'The medieval polychromy of Norwich Cathedral', forthcoming.
Watercolour by Fox, reproduced in The Victoria History of the Counties of England: Norfolk II, London 1906, Pl. 1, face to p. 532.

11. Bibliographic references:

W. H. St John Hope, in Proceedings of the Society of Antiquaries, 2nd series, 17 (1897–9), 308–9
W. H. St John Hope and W. T. Bensly, 'Recent discoveries in the cathedral church of Norwich', *Norfolk Archaeology*, 14 (1901) 111–12.
D. Park, *Simony & Sanctity: Herbert Losinga, St. Wulfstan of Worcester and Wall Paintings in Norwich Cathedral*, 1994.
E. Fernie, *An Architectural History of Norwich Cathedral*, Oxford, 1993.

12. Previous conservation treatments:

Uncovered during a programme of 'unflaking' limewash and cleaning the nave in 1898–9 (see W. H. St. John Hope (1897–9) and W. H. St. John Hope & Bensly (1901). Varnish mentioned by Eve Baker may have been applied at this time.

Tristram was active in the cathedral in the 1930s and may have applied the wax preservative mentioned by E. Baker.

Wax and varnish 'preservatives' were removed by Eve Baker in 1975. Limewash and old repairs were also removed and further small areas of painting found in exploration of the 'thick cement plaster' covering the rest of the vault. Saturated with limewater. Further painting may still be present.

13. Previous examination and sampling :

14. Comments:

Compare with Ely E bay of S aisle (c1150)
Vault of HSC, Winchester Cathedral (c1220)
Galilee Chapel, Durham Cathedral, Westminster Psalter and Sigena wall paintings
Life of St. Cuthbert illuminated at Durham c1200

Did the paintings in the adjoining bay to the west form part of this programme or relate to it? These featured St. Wulfstan of Worcester destroyed in 1860s. DP thinks that these paintings are c1250.

Fire damage "E. Baker, never saw a trace of pigment or plaster damaged by fire".

The Techniques of English Medieval Wall Painting: Proforma B Leverhulme Research Project, Courtauld Institute of Art	
1. Site: NORWICH CATHEDRAL, south aisle	**2. Site Code:** NORLOS
3. Date examined: 13–16 October 1995	**4. Examined by:** H. Howard

5. Relationship of architecture to painted decoration:

Founded by Losinga 1091
Fire 1171, rebuilding 1173
Riots in 1272 and second fire, repairs 1278
1362 spire falls
1463 fire caused by lightning

6. Areas examined and documentation of examination:

Nave south aisle, 4th bay from east. West segment, north side, single remaining scene from the 'Losinga' cycle.

7. Examination in normal and raking light:

7.1 Appearance of support:
The structure of the vault is presumably rubble with plaster infill and surface layers. Much of the evidence of original plastering is lost due to edging repairs and fills made during the 1960s. The surface is rather undulating though generally smooth.

7.2 Evidence for plastering techniques:
The edge of the plaster for the vault abuts the ashlar of the soffit, though it does not really appear to overlap it. The red paint layer does however overlap the plaster of the vault and the side of the soffit. The plaster itself is pale and rather coarse, the aggregate is as large as 7 × 5 × 4mm very close to the painted surface. Tool marks (for application and smoothing) are visible in raking light (re: Witley, Surrey c1110).

7.3 Appearance of ground:
Where the paint and preparatory layers survive well, it appears that the surface was well prepared and smoothed. A final lime (?) ground/fine plaster skim (?) was applied to the vault and carefully smoothed in preparation for painting. This layer is very well integrated with the plaster (applied wet-on wet?). The preparatory drawing has been applied on the surface of this layer. The integration of ground/final plaster skim with the underlying plaster is clearly visible in the cloak of the western figure.

7.4 Evidence of preparatory techniques:
Here, unlike on the soffit, the preparatory drawing is in yellow (see drapery) which has been refined and finalised in red in some areas (faces and hair).

7.5 Underpainting:
Drawing in yellow over the white ground/plaster skim followed by the basic background colours (underpaint), e.g. a scumble of blue/grey, rather cursorily applied, has been used as an underpaint for the blue of the central zone.

7.6 Paint layers:
The majority of the uppermost paint layers have been lost, but traces of thick blue and green layers are present in some places (drapery) and a small amount of the thick impasto in white which is the same as that found on the soffit. Much less of the thick uppermost paint layers survive in this area of the painting, just small traces of brilliant green, blue, white and black. Green and blue have been applied over white.

7.7 Apparent binding medium:
The drawing is well integrated with the ground/plaster skim and is presumably primarily bound by the carbonation of its lime binder. The thick uppermost paint layers however are clearly separate and would require an additional binder. This may be lime white in some cases. Some organic component may also be present.

7.8 Attachments:
None apparent.

7.9 Surface coatings/treatments:
The paintings have been treated with varnish and with wax. In UV light a lot of waxy material is still apparent on the surface. It seems likely that a large proportion of the uppermost paint layers were lost during the application and subsequent partial removal of the wax and varnish coating.

8. Other:

Repairs made by the Bakers in the 1960s have resulted in a deep step between painted zones of plaster adjacent to the rib and the painted rib itself. It is difficult to explain this feature – both areas of painting are clearly from the same scheme – but the vault must have been in a perilous state with much loss and badly detached zones of plaster before this conservation work.

The Techniques of English Medieval Wall Painting: Proforma B Leverhulme Research Project, Courtauld Institute of Art	
1. Site: NORWICH CATHEDRAL, south aisle	**2. Site Code:** NORLOS
3. Date examined: 13th–16th October 1995	**4. Examined by:** H. Howard

5. Relationship of architecture to painted decoration:

Founded by Losinga 1091
Fire 1171, rebuilding 1173
Riots in 1272 and second fire, repairs 1278
1362 spire falls
1463 fire caused by lightning

6. Areas examined and documentation of examination:

Nave south aisle, 4th bay from east.
Soffit of arch at west side, 'Losinga' cycle.
Three roundels remaining on the north side of the soffit.

7. Examination in normal and raking light:

7.1 Appearance of support:
Ashlar quoins – quite roughly tooled (this is particularly visible on the sides of the blocks) – with rubble infill for the centre of the soffit, over which plaster was applied to provide a support for the painting. No plaster has been applied over the side of the ashlar quoins. Here only a thick white ground was applied in preparation for painting.

7.2 Evidence for plastering techniques:
Edging repairs and fills made in the 1960s cover the evidence for plastering, and it is not now possible to find any evidence for plaster joins. The soffit itself has been plastered while the side of the ashlar quoins have only a lime (?) ground applied to their surfaces. The plaster is quite coarse with large pieces of aggregate 3 × 4 × 5mm quite near to the painted surface. In general, the plaster surface appears to have been well smoothed and it undulates gently.

7.3 Appearance of ground:
A fine lime ground or plaster skim has been applied over the rather coarse plaster to provide a smoother surface for painting. This layer has been well smoothed into the underlying plaster, and it is possible that the underdrawing and basic background colours have been slightly polished into the surface.

7.4 Evidence of preparatory techniques:
No incision or snapped lines are visible. Compasses must have been used to set out the roundels, but it is not possible to find any evidence for the centres (there are several losses in these areas). Initial setting out appears to have been in red earth, painted in a fine and assured hand.

7.5 Underpainting:
The concentric circles of the roundels have been set out in red paint. The outer band has then been painted green, the central band red (vermilion ?), and black for the central zone. The black functions as an underpaint for blue (ultramarine ?).

7.6 Paint layers:
The uppermost paint layers are quite distinct from the underpainting. These uppermost layers are approximately 0.5–1.0mm thick and clearly lie quite separately on the surface.

7.7 Apparent binding medium:
The underdrawing and underpainting appears to be bound by the carbonation of the plaster skim or lime ground layer. The uppermost layers are however separate from this and must have an additional binder. This may be lime, from lime white or lime milk, or may be organic or a mixture of these. Some of the thick white highlights appear to be lime white, and so are bound by carbonation.

7.8 Attachments:
No evidence for attachments survives.

7.9 Surface coatings/treatments:
The paintings were both varnished and waxed. These coatings were partially removed by the Bakers in the 1960s. A large quantity of wax remains on the surface and this is clearly visible in UV light. You can also see the remains of wax deep in the pores of the plaster.

8. Other:

There has been severe loss of the uppermost paint layers (perhaps during the application and subsequent removal of the surface coatings). The condition of what remains is reasonably sound – though there is much poorly slaked lime in the 1960s repairs. The retouchings undertaken in the 1960s have been coloured to fit the surrounding areas and toned to complete shapes e.g. the window of the cathedral.

<table>
<tr><td colspan="3">The Techniques of English Medieval Wall Painting: Proforma C
Conservation of Wall Painting Department, Courtauld Institute of Art</td></tr>
<tr><td colspan="2">SITE: Norwich Cathedral, 'Losinga' cycle, soffit</td><td>SITE CODE: NORLOS</td></tr>
<tr><td colspan="2">DATE SAMPLED: 15 November 1995</td><td>SAMPLE NO.: 1
ACCESSION NO.: 1509</td></tr>
<tr><td colspan="2">TYPE OF SAMPLE: Cross-section</td><td>SAMPLED BY: H. Howard</td></tr>
<tr><td></td><td>Appearance of:</td><td>Condition of:</td></tr>
<tr><td>Support</td><td>Coarse lime plaster with quite large particles (3 × 4 × 5mm) of aggregate near to the surface. These would actually penetrate the paint layer but for being rather carefully smoothed and aligned up to the surface.</td><td>Generally sound</td></tr>
<tr><td>Ground</td><td>There appears to be a lime ground or even a very fine plaster skim. However, it is also possible that lime was brought to the surface by smoothing the underlying plaster, and that this now looks like a separate ground. In areas of black, such as this sample, it is very difficult to see the white ground layer.</td><td>Generally sound</td></tr>
<tr><td>Medium</td><td>It seems likely that the underdrawing and underpaint are bound by the carbonation of the lime plaster and (?) ground. Subsequent paint layers are either mixed with lime white (or lime in some other form) or have an additional binding medium (or both).</td><td>Generally sound</td></tr>
<tr><td>Paint layer/s</td><td>Red over yellow on black underpaint. The red is quite a strong colour and does not appear to have been combined with lime. This may be red earth with an organic medium. The yellow layer has been applied in considerable impasto and brushstrokes are clearly visible in the surface. This layer is quite distinct and separate from the black layer beneath. NB the sample may well shear here. The black layer is very compact and intimately bound with the underlying plaster. Again, brushstrokes are clearly visible in this layer.</td><td>Generally sound</td></tr>
<tr><td>Attachments</td><td>None apparent.</td><td></td></tr>
<tr><td>Coatings / surface treatments</td><td>Varnish and wax coatings removed during the 1960s. Directly adjacent to the sample is a repair and retouching from the Bakers' treatment. Check that no trace of this later paint has contaminated the sample (retouching is grey/brown in colour).</td><td></td></tr>
<tr><td colspan="3">Examination of area sampled at magnification:
Two tiny zones of red pigment are visible on the surface, as is some watercolour retouching from the 1960s intervention (grey/brown). The yellow layer is thickly painted over the black local ground. Large particles of aggregate are clearly visible near the paint surface.</td></tr>
</table>

Examination of area sampled in ultraviolet and infrared light:

No additional information visible in infrared light. In UV, the remaining wax fluoresces a pale green colour and is clearly visible, particularly in the pores of the plaster. Quite a lot remains deeply embedded in both the paint and plaster layers.

Diagram of area sampled:

White

Black

retouching

retouching over original paint

Black

Area sampled

Red over yellow

White

AREA SAMPLED & DESCRIPTION:

Soffit, lowest roundel, cathedral, 2nd remaining window from east, just below the Bakers' repair: trace of red on yellow over black underpaint.

COMMENTS:

APPENDIX THREE

Tables

Table 1 Selected analytical data relating to the use of blue pigments in medieval polychromy.

Site and Location	Date	Pigment(s)	Technique and Analyst(s)	Reference(s)
Examples in English medieval wall painting				
Lewes Priory, Sussex: excavated wall painting fragments	*c*1090	Natural ultramarine	PLM, SEM/EDX, H. Howard (CIA)	Howard and Park 1997
St. Botolph's, Hardham, West Sussex	*c*1100	A false blue has been made by combining charcoal black and lime white	PLM, H. Howard (CIA)	Howard 1990; Sawdy *et al.* 1998
Coombes, West Sussex: nave, dado zone, fictive marbling	*c*1100	A false blue has been made by combining charcoal black and lime white	PLM, H. Howard (CIA)	Howard 1990
All Saints' Witley, Surrey: nave south and west walls	*c*1110	A false blue has been made by combining charcoal black and lime white	PLM, H. Howard (CIA)	Howard 1992a
Ickleton, Cambridgeshire: nave, north arcade	*c*1125	Natural ultramarine	PLM, MCT, H. Howard (CIA)	Howard 1988
St. Michael and All Angels, Copford, Essex: apse and chancel	*c*1125–30	Natural ultramarine (tentatively identified)	PLM, C. Babington (EH)	Organ *et al.* 1991
St. Mary's, Kempley, Gloucestershire: chancel and nave east wall	*c*1130	Azurite. There is some evidence of the alteration of azurite to malachite	PLM/MCT, S. Rickerby (CIA)	Rickerby 1990a
Canterbury Cathedral: St. Gabriel's Chapel: apse.	*c*1130	Natural ultramarine	PLM, G. Hedley (CIA); PLM, SEM/EDX, H. Howard (CIA)	Cather and Howard 1994
Chichester Cathedral: north transept	*c*1140–50	Ultramarine		Gärtner 1988
Canterbury Cathedral: St. Anselm's Chapel, apse, *St. Paul and the Viper*	Mid-12th century	Natural ultramarine	PLM, G. Hedley (CIA); H. Howard (CIA)	Howard 1988; Gärtner and Organ 1989
Winchester Cathedral: Holy Sepulchre Chapel	*c*1175	Natural ultramarine and vivianite	PLM, SEM/EDX, H. Howard (CIA); XRD, G. Cressy (NHM)	Howard 1995a; Howard 1996a
Sherborne Abbey, Dorset: chapter house, excavated painted wall plaster	*c*1180	A false blue has been made by combining carbon black and lime white	PLM, SEM/EDX, H. Howard (CIA)	Howard and Park forthcoming a
Durham Cathedral: Galilee Chapel, *St. Cuthbert*	*c*1180	Natural ultramarine	PLM, H. Howard (CIA)	Howard 1988
Guildford, Surrey: High Street, 'synagogue'	*c*1180	Indigo	PLM, SEM/EDX, FTIR, H. Howard (CIA)	Howard 1996c
Glastonbury Abbey, Somerset: Lady Chapel	*c*1187	Natural ultramarine	PLM, E. Sinclair	Sampson 1996; Sinclair pers. comm.
Norwich Cathedral: nave south aisle, vault, 4th bay from east and soffit of adjacent transverse arch, '*Losinga*' cycle	*c*1190	Lapis lazuli/ natural ultramarine	PLM, SEM/EDX, H. Howard (CIA)	Howard and Gasol 1996

Site and Location	Date	Pigment(s)	Technique and Analyst(s)	Reference(s)
Eastbridge Hospital, Canterbury, Kent: hall, north wall	*c*1220	Indigo	PLM, SEM/EDX, FTIR, H. Howard (CIA)	Howard 1994
Winchester Cathedral: Holy Sepulchre Chapel	*c*1220	Natural ultramarine and vivianite	PLM, SEM/EDX, H. Howard (CIA); XRD, G. Cressy (NHM)	Howard 1995a; Howard 1996a
Ely Cathedral: St. George's Chapel, entrance arch soffit	*c*1220–30	Azurite (this pigment has partially altered to form a pale blue/green copper chloride)	PLM, SEM/EDX, FTIR, H. Howard (CIA)	Howard and Pathak 1995
Norwich Cathedral: Ante-Reliquary Chapel, eastern arch	*c*1250	Natural ultramarine	PLM, SEM/EDX, H. Howard (CIA)	Howard 1995c; Park and Howard 1996
Horsham St. Faith Priory, Norfolk: refectory	*c*1250 with 15th-century additions	Azurite. The natural pigment has been adulterated with its synthetic equivalent	PLM, H. Howard (CIA); PLM, A. Roy (NG)	Babington and James 1988; Howard 1988
Winchester Cathedral: Guardian Angels' Chapel, vault	Early 13th century with later 13th-century additions	Natural ultramarine, azurite and vivianite	PLM, SEM/EDX, FTIR, H. Howard (CIA); XRD, G. Cressy (NHM)	
Rochester Cathedral: choir, *Wheel of Fortune*	Mid-13th century	Indigo	PLM, J. Darrah (V&A)	Darrah 1992
Clarendon Palace, Wiltshire: excavated wall painting fragments	13th century	Natural ultramarine	I. Bristow	Ashurst and James 1988
St. Augustine's Abbey, Canterbury, Kent: excavated fragments of wall painting from the ambulatory south wall and St. Thomas's Chapel	13th century	Azurite		Sherlock and Woods 1988
Hospital of St. Giles, Brompton Bridge, North Yorkshire: excavated wall painting fragments	13th century	A false blue has been made by combining carbon black and lime white	PLM, S. Stewart (EH)	Stewart 1995
Duxford, Cambridgeshire: west and south walls of the chancel	13th century	A false blue has been made by combining carbon black and lime white	PLM (CIA)	CIA 1987
York Minster: coffin lid of Archbishop Walter de Gray	13th century	Natural ultramarine and false blue made from combining charcoal black with lime white	J. Plesters (NG)	Ramm *et al.* 1971
Norwich Cathedral: Ante-Reliquary Chapel, vault and western arch	*c*1300	Azurite. The pigment has partially altered to form a copper chloride	PLM, SEM/EDX, H. Howard (CIA)	Howard 1995c; Park and Howard 1996
St. Albans Cathedral: feretory, south-east pier, *Unidentified archbishop*	*c*1302–08	Azurite	PLM, SEM/EDX, H. Howard (CIA)	Howard 1993a

Site and Location	Date	Pigment(s)	Technique and Analyst(s)	Reference(s)
St. Albans Cathedral: feretory, north-east pier, *St. William of York*	*c*1302–08	Azurite and a synthetic copper blue. Medieval repainting was undertaken with pigments including indigo	PLM, SEM/EDX, H. Howard (CIA)	Howard 1993a
Brent Eleigh, Suffolk: chancel, east wall	*c*1335–40	Synthetic copper blue. This blue pigment has a high chloride content	PLM/MCT, SEM/EDX H. Howard (CIA).	Howard 1988
Little Wenham, Suffolk: chancel east wall	1st half 14th century	Synthetic copper blue. This blue pigment has a high chloride content	PLM, FTIR, H. Howard (CIA); SEM/EDX, A. Roy (NG)	Howard 1988
St. Stephen's Chapel, Westminster Palace: fragments of wall painting now in British Museum	*c*1352–63	Natural ultramarine	J. Haslam; J. Plesters (NG)	Haslam 1807; van Geersdaele and Goldsworthy 1978; Plesters 1993
Westminster Abbey: chapter house, *Majesty*	*c*1375–1404	Azurite and indigo	PLM/MCT, SEM/EDX H. Howard (CIA)	Howard 1988; Howard 2000
Westminster Abbey: chapter house, *Apocalypse* cycle	*c*1375–1404	Azurite and indigo	PLM/MCT, SEM/EDX H. Howard (CIA)	Howard 1988; Howard 2000
Tower of London: the Byward Tower, south tower, 1st floor room, south wall	*c*1395	Azurite	PLM, H. Howard (CIA); PLM, C. Babington (EH)	Howard 1988
Ely Cathedral: Prior Crauden's Chapel, south tabernacle	14th century	Azurite (tentatively identified)	PLM, C. Babington (CIA).	Babington 1987
St. Nicholas' Hospital Church, Harbledown, Kent: chancel east wall and window embrasure	14th century	Natural ultramarine and false blue made from charcoal black combined with lime white	PLM, SEM/EDX, H. Howard (CIA)	Organ 1992
Farringdon, Hampshire: *Christ in Majesty*	14th century	Synthetic copper blue	PLM, J. Darrah (V&A)	Ballantyne 1988
Great Hockham, Norfolk: north aisle, north wall	14th century	Synthetic copper blue. This blue pigment has a high chloride content	PLM, FTIR, H. Howard (CIA); SEM/EDX, A. Roy (NG)	Howard 1988
Holy Trinity, Coventry, West Midlands: chancel arch, *Last Judgement*	*c*1435–60	A false blue has been made from carbon black combined with lime white	PLM, SEM/EDX, H. Howard (CIA)	Howard 1996b
Durham Cathedral: Deanery, entrance hall, north wall	*c*1470	Indigo	PLM/MCT, H. Howard (CIA)	Howard 1988
Eton College Chapel, Berkshire: north and south walls: subjects drawn from sources such as the *Golden Legend* and the *Speculum Vitae*	*c*1479–87	A blue pigment now faded (thought to be indigo) was identified		Plummer 1986
St. Margaret of Antioch, Cowlinge, Suffolk: *St Michael Weighing Souls*	15th century	Azurite	PLM/MCT, H. Howard (CIA)	Organ and Sawdy 1991

Site and Location	Date	Pigment(s)	Technique and Analyst(s)	Reference(s)
Winchester Cathedral: Lady Chapel, vaults	Late 15th century	Azurite	PLM/MCT, T. Curteis	Curteis and Paine 1995
Lord Mayor's Chapel (St. Mark's), Bristol: south aisle arcade: *Trinity, Nativity* and *Noli me Tangere*	*c*1510	Indigo	PLM, SEM/EDX, FTIR, H. Howard (CIA)	Gill and Howard 1997
York Minster: The Bedern, excavated wall plaster	Medieval	A false blue has been made from carbon black combined with lime white	PLM, L. Wong (CIA)	Wong forthcoming
Examples in northern European medieval wall painting				
Convent of St. John the Baptist, Müstair, Switzerland	*c*800	Egyptian blue	PLM/XRD, G. Banik	Mairinger and Schreiner 1986
Saint-Germain, Auxerre, France: crypt	9th–11th century	A false blue has been made by combining carbon black and calcium carbonate white	PLM, Raman spectrometry, C. Couprey	Sapin and Couprey 1992
St. Georg in Oberzell, Reichenau, Germany	Late 10th century	Azurite which has partially altered to form paratacamite	H. Kühn	Knoepfli and Emmenegger 1990
Lambach, Austria: west choir	*c*1089	Natural ultramarine		Knoepfli and Emmenegger 1990
Abbey of La Trinité, Vendome, France	*c*1100	False blue		Taralon 1986a
Idensen, Germany	*c*1130	Natural ultramarine	PLM/MCT, SEM/ EDS, FTIR, M. Matteini and A. Moles (OPD)	Geilmann 1938; Matteini and Moles 1994
Johanneskapelle, Pürgg, Austria	*c*1165	Natural ultramarine and false blue	PLM, SEM/EDX, FTIR, H. Howard (CIA); I. Hammer 1977	Howard 1995d
Cashel, Ireland: Cormac's Chapel	*c*1170	Natural ultramarine	PLM, McGrath; PLM, H. Howard (CIA)	McGrath 1987; Howard 1991
Convent of St. John the Baptist, Müstair, Switzerland: apses	*c*1190	Lapis lazuli	F. Mairinger	Emmenegger 1986
Mårslet, Denmark	Romanesque	Azurite		Graebe *et al.* 1986
Saint Hilaire-le-Grand, Poitiers, France: *Revelation* cycle	Romanesque	Azurite	LRMH	Taralon 1986b
Berzé-la-Ville, France: Monk's Chapel	Romanesque	Ultramarine	C. Couprey	Rollier-Hanselmann 1997
Allouis, France	Romanesque	Natural ultramarine		Kupfer 1982
Gurk Cathedral, Austria: west gallery of the Bishop's Chapel	*c*1263	Azurite which has partially altered to form paratacamite	F. Mairinger	Ellwanger-Eckel 1979

Site and Location	Date	Pigment(s)	Technique and Analyst(s)	Reference(s)
Angers Cathedral, France: choir, *Scenes of the Life of S. Maurille*	*c*1270–80	Azurite	LRMH	Subes-Picot 1992; Stirnemann 1995; Demailly *et al.* 1998
Mästerby, Gotland, Sweden	13th century	Natural ultramarine, azurite and iron phosphates (likely to be vivianite)	SEM/EDX, XRD	Nord and Tronner 2000
Källunge, Gotland, Sweden	13th century	Azurite	SEM/EDX, XRD	Nord and Tronner 2000
Garde, Gotland, Sweden	13th century	Natural ultramarine and azurite	SEM/EDX, XRD	Nord and Tronner 2000
Cologne Cathedral, Germany: choir screen wall paintings	*c*1332–49	Azurite	E. Jägers	Jägers and Schulze-Senger 1989; Heritage 1990; Maul 1992
Urschalling, Germany	*c*1370	Vivianite	H. Kühn	Pursche 2000
Härkeberga, Uppland, Sweden	*c*1500	Azurite	SEM/EDX, XRD	Nord and Tronner 2000
Täby, Uppland, Sweden	*c*1500	Azurite	SEM/EDX, XRD	Nord and Tronner 2000
Examples in English medieval panel painting				
Holy Innocent's, Adisham, Kent: nave, west wall, reredos	*c*1180	Natural ultramarine	PLM, SEM/EDX, H. Howard (CIA)	Howard 1993c
Peterborough Cathedral, nave ceiling	*c*1220	Azurite. FTIR analysis indicated a partial alteration of azurite to form copper oxalate	PLM, SEM/EDX, FTIR H. Howard (CIA)	Howard 1997b
Painted Chamber, Westminster Palace: *ceiling paterae*	*c*1263–66	Azurite and natural ultramarine	J. Lynn (HKI)	Liversidge *et al.* 1995; Lynn 1997
Thornham Parva, Suffolk: Thornham Parva Retable	*c*1335	Azurite	PLM, S. Bucklow (HKI)	Goldsmith and Bucklow 1998; Bucklow forthcoming
Kingston Lacy, Dorset: painted screen	Early 14th century	Azurite	PLM, L. Sheldon and C. Hassall (UCL)	Tracy 1997
Exeter Cathedral: pulpitum paintings	14th century	Azurite and indigo	PLM/MCT, E. Sinclair	Sinclair 1995b
Wilton Diptych (English or French), now in the National Gallery, London	Late 14th century	Natural ultramarine, azurite and indigo	PLM/ SEM/EDX, A. Roy (NG)	Roy 1997
St. Mary's, Worstead, Norfolk: screen	Early 16th century	Azurite	PLM/MCT, T. Curteis	Curteis 1993

Site and Location	Date	Pigment(s)	Technique and Analyst(s)	Reference(s)
Examples in northern European medieval panel painting				
Heddal altar frontal, Norway	*c*1250–60	Ultramarine and azurite. The ultramarine has been coated in protein before being bound in oil (U. Plahter, pers. comm.)	PLM, SEM/EDX, U. Plahter	Plahter 1995
Tingelstad III altarpiece	*c*1260–1300	Ultramarine ash, vivianite and azurite. Ultramarine is found on drapery while azurite is employed for architecture and the mandorla. Vivianite is an underpaint for azurite	SEM/EDX U. Plahter	Plahter 1995; U. Plahter pers. comm.
Regensburg, Germany	13th century	Vivianite	A. Burmester	Burmester and Krekel 2001
Kinsarvik altar frontal, Norway	Before *c*1300	Ultramarine	PLM, SEM/EDX, U. Plahter	Plahter 1995
Ulvik altar frontal, Norway	Before *c*1300	Ultramarine	PLM, SEM/EDX, U. Plahter	Plahter 1995
Trondheim altar frontal, Norway	After *c*1300	Mixture of azurite, indigo and lead white, glazed with azurite	PLM, SEM/EDX, U. Plahter	Plahter 1995
Nedstryn altar frontal, Norway	After *c*1300	Azurite and indigo. Azurite is employed for garments while indigo is used for the wave border on the inner side of the frame	PLM, SEM/EDX, U. Plahter	Plahter 1995
German triptych in the Gallery of the Courtauld Institute of Art, London (known as the Estouteville triptych)	Late 14th century	Azurite	PLM, G. van Heemstra; SEM/EDX, A. Burnstock (CIA)	Villers *et al.* 1997
Polychromed sculpture: examples in England				
St. Augustine's Abbey, Canterbury, Kent: excavated limestone capital	Early 7th or 9th century	Egyptian blue	PLM, H. Howard (CIA); SEM/EDX, A. Roy (NG)	Braine and Welford 1991
Lincoln Cathedral: west front, frieze	*c*1140	Natural ultramarine and a copper blue pigment	A. Roy (NG)	Larson 1992, 1993
St. Mary's Abbey, York: excavated stone sculpture, Moses	*c*1180–85	Natural ultramarine and indigo	A. Roy (NG)	Brodrick 1993
St. Albans Abbey: chapter house, excavated fragments of sculpture	12th century (?)	Egyptian blue	J. Plesters (NG)	M. Biddle, pers. comm.
Worcester Cathedral: crypt, west capital, below St. Thomas	Early13th century	A false blue has been made from carbon black combined with lime white	PLM, I. Kakoulli and H. Howard (CIA)	Park 1997
Lichfield Cathedral: polychromed stone column	13th century	A false blue has been made by combining charcoal black and lime white	PLM, J. Darrah (V & A)	Howard 1988
Salisbury Cathedral: west front, porch	13th century with numerous phases of repainting	Copper blue. Traces of ultramarine, azurite and indigo have also been found, though it is not clear to which phase of decoration they belong	PLM, SEM/EDX, E. Sinclair	Sinclair 1997

Site and Location	Date	Pigment(s)	Technique and Analyst(s)	Reference(s)
St. Albans Cathedral: feretory, Shrine of St. Alban	*c*1302–08	Azurite. The mineral blue pigment has partially altered to form copper chloride and copper oxide	PLM, C. Babington (EH); PLM, SEM/EDX, H. Howard (CIA)	
Abergavenny, Wales: wooden knight	*c*1325	Azurite. (The effigy may have been repainted, it is not clear whether the azurite is original)	A. Brodrick	Blair 1994
Ingham, Norfolk: chancel, north wall, tomb of Sir Oliver de Ingham	*c*1344	Synthetic copper blue which may be an alteration product of verdigris	PLM, SEM/EDX, H. Howard (CIA)	Howard 1993d
Canterbury Cathedral: Chapel of Our Lady Undercroft, screens enclosing chapel	1370s	Azurite, indigo and a false blue made by combining charcoal black and white. The azurite has partially altered to form a pale blue copper chloride	PLM, H. Howard (CIA); SEM/EDX, A. Roy (NG)	Howard 1993b
Ingham, Norfolk: nave south aisle, tomb of Sir Roger and Lady Margaret de Bois	*c*1380	Azurite	PLM, SEM/EDX, H. Howard (CIA)	Howard 1993d
Exeter Cathedral: roof bosses	Early 14th century	Azurite and indigo	A. Roy (NG); J. Darrah (V&A); U. Plahter.	Hulbert 1991
Exeter Cathedral: west front, polychromed sculpture	14th and 15th century	Azurite (15th-century polychromy) and indigo – or woad – (14th- and 15th-century polychromy)	PLM, E. Sinclair; SEM/EDX, A. Roy (NG)	Sinclair 1991, 1992
Tewkesbury Abbey: various painted stone fragments which may have originated from a late medieval high altarpiece and associated fittings	Early 15th century	Azurite	PLM, C. Hassel (UCL)	UCL 1996
Winchester Cathedral: Great Screen	*c*1475–90	Azurite and indigo	PLM, J. Darrah (V&A)	Brodrick and Darrah 1989
Fitzalan Chapel, Arundel: Fitzalan effigies	15th century	Azurite. Indigo was also tentatively identified	PLM, J. Darrah (V&A)	Brodrick and Darrah 1986
Cartmel Fell, Cumbria: Chapel of St. Anthony, wooden figure of Christ	Late 15th century	Indigo	J. Plesters and A. Roy (NG)	Behrens 1982

Site and Location	Date	Pigment(s)	Technique and Analyst(s)	Reference(s)
Miscellaneous British examples				
Antioch Chamber, Clarendon Palace, Wiltshire: excavated oyster-shell palette	*c*1235	Azurite	PLM, H. Hughes (EH)	Hughes and Lewis 1988
St. Augustine's Abbey, Kent: excavated oyster-shell palette (type of painting for which used unknown)	13th century	Azurite	XRF, XRD, (EH/AML)	Sherlock and Woods 1988
Greyfriars, Norwich: excavated shell palette	13th–14th century	Azurite. The natural pigment appears to have been adulterated with its synthetic equivalent	PLM, SEM/EDX, H. Howard (CIA)	Howard and Park forthcoming b
Ducklington Rectory, Oxfordshire: series of wooden beams with heraldic decoration which probably originated from Minster Lovell Manor House	15th century	Indigo		Steane 1985
Whitefriars, Carmelite Friary (Church of Our Lady of Mount Carmel), Coventry: excavated oyster-shell palette	Mid-14th–mid-15th centuries	Azurite	PLM, SEM/EDX, FTIR, H. Howard (CIA)	Howard 1997a
Linlithgow Carmelite Friary, Scotland: two pottery vessels retaining traces of pigments	Early post-Reformation	Azurite. A single lump of indigo or woad was also identified	PLM, H. Howard (CIA)	Howard 1989
Polychromed sculpture: northern European examples				
Chartres Cathedral, France: *Portail Royal*	*c*1140–50	Ultramarine and azurite (azurite was found only on the central portal, while ultramarine was identified only on the left portal, suggesting that they were painted at slightly different dates)	LRMH	Nonfarmale and Manaresi 1987
Crucifix of Hemse, Sweden	*c*1170–90	Natural ultramarine	U. Plahter	Plahter 1984
Notre-Dame Cathedral, Lausanne, Switzerland: south portal	*c*1216–20	Natural ultramarine		Furlano *et al.* 1981
Amiens Cathedral, France: *'Portail de la Mère Dieu'*	*c*1220–60	Azurite	LRMH	Weeks 1998
Enthroned Virgin Mary, South Germany	*c*1250	Vivianite		Westhoff 2002
Holy Cross Minster, Schwäbisch Gmünd, Germany: polychromed portals	14th century	Vivianite and smalt	PLM, C. Blauer; PLM/ MCT	C. Blauer pers. comm.; Hauff 1988
Herlin Altarpiece, Church of St. James, Rothenberg, Germany	*c*1466	Indigo	PLM, IR	Broekman-Bolstijn *et al.* 1970

Table 2 Selected analytical data relating to the use of green pigments in medieval polychromy.

Site and Location	Date	Pigment(s)	Technique and Analyst(s)	Reference(s)
Examples in English medieval wall painting				
Lewes Priory, Sussex: excavated wall painting fragments	*c*1090	Green earth	PLM, SEM/EDX, H. Howard (CIA)	Howard and Park 1997
St. Botolph's, Hardham, West Sussex	*c*1100	Copper chloride (salt) green	PLM, SEM/EDX, FTIR, H. Howard (CIA)	Sawdy *et al.* 1998
St. Mary's, Kempley, Gloucestershire: chancel and nave east wall	*c*1130	Malachite	PLM/MCT, S. Rickerby (CIA)	Rickerby 1990a
Canterbury Cathedral: St. Gabriel's Chapel, apse	*c*1130	Green earth and a mixed green of ultramarine combined with yellow earth. Verdigris has also been identified on the column capital at the entrance to the chapel, though this polychromy may be slightly later	Prof. Rees Jones (1936); and PLM, SEM/EDX, H. Howard (CIA)	Cather and Howard 1994
Canterbury Cathedral: St. Anselm's Chapel, apse, *St. Paul and the Viper*	Mid-12th century	Malachite and green made from combining yellow and black pigments	Examination with microscope	Gärtner and Organ 1989
Chichester Cathedral: north transept	*c*1140–50	Malachite		Gärtner 1988
Winchester Cathedral: Holy Sepulchre Chapel	*c*1175	Green earth	PLM, SEM/EDX, H. Howard (CIA)	Howard 1995a
Sherborne Abbey, Dorset: chapter house, excavated painted wall plaster	*c*1180	Paratacamite (copper-chloride green), which may be an alteration produce of malachite (also identified)	PLM, SEM/EDX, FTIR, H. Howard (CIA)	Howard and Park forthcoming a
Norwich Cathedral: nave south aisle, vault, 4th bay from east and soffit of adjacent transverse arch, *'Losinga'* cycle	*c*1190	Green earth and copper chloride (salt) green (atacamite)	PLM, SEM/EDX, FTIR, H. Howard (CIA)	Howard and Gasol 1996
Winchester Cathedral: Holy Sepulchre Chapel	*c*1220	Copper chloride (salt) green (which appears to be a mixture of atacamite and paratacamite, and may have been applied as an original salt (copper chloride) green or may be the result of a pigment alteration)	PLM, SEM/EDX, FTIR, H. Howard (CIA)	Howard 1995a
Eastbridge Hospital, Canterbury, Kent: hall, north wall	*c*1220	Verdigris and copper resinate	PLM, SEM/EDX, H. Howard (CIA)	Howard 1994
Ely Cathedral: St. George's Chapel, entrance arch soffit	*c*1220–30	Chrysocolla (tentatively identified)	PLM, SEM/EDX, FTIR, H. Howard (CIA)	Howard and Pathak 1995
Horsham St. Faith Priory, Norfolk: refectory	*c*1250 with 15th-century additions	Verdigris	PLM, C. Babington (CIA); PLM, J. Darrah (V&A)	Babington and James 1988

Site and Location	Date	Pigment(s)	Technique and Analyst(s)	Reference(s)
Worcester Cathedral: crypt, eastern bishops	Early 13th century	Green earth	PLM, SEM/EDX, H. Howard (CIA)	Park 1997
Winchester Cathedral: Guardian Angels' Chapel, vault	Early 13th century with later 13th-century additions	Malachite, verdigris and a synthetic copper chloride (salt) green	PLM, SEM/EDX, FTIR, H. Howard (CIA)	
Hospital of St. Giles, Brompton Bridge, North Yorkshire: excavated wall painting fragments	13th century	Green earth	PLM, S. Stewart (EH)	Stewart 1995
St. Mary's and All Saints' Willingham, Cambridgeshire: nave walls	13th century	Malachite	Prof. Wehlte	Wehlte 1954
Clarendon Palace, Wiltshire: excavated wall painting fragments	13th century	Copper green	I. Bristow	Ashurst and James 1988
Norwich Cathedral: Ante-Reliquary Chapel, vault and western arch	*c*1300	Verdigris	PLM, SEM/EDX, H. Howard (CIA)	Howard 1995c; Park and Howard 1996
St. Albans Cathedral: feretory, north-east pier, *St. William of York*	*c*1302–08	Malachite. Medieval repainting was undertaken in pigments including a synthetic copper green (verdigris)	PLM, SEM/EDX, H. Howard (CIA)	Howard 1993a
Merevale, Warwickshire: chancel arch	*c*1310–20	Synthetic copper green (containing a small quantity of copper chloride)	PLM, SEM/EDX, H. Howard (CIA)	
Brent Eleigh, Suffolk: chancel east wall	*c*1335–40	Verdigris	PLM, H. Howard (CIA)	Howard 1988
Rochester Cathedral: north-east transept, wall painting in tomb recess of William of Perth	*c*1340	Verdigris and copper resinate	PLM, S. Bradley (BM)	Rickerby 1988
St. Stephen's Chapel, Westminster Palace: fragments of wall painting now in British Museum	*c*1352–63	Verdigris and copper resinate	J. Haslam	Haslam 1807
Westminster Abbey: chapter house, *Majesty*	*c*1375–1404	Copper green	PLM, SEM/EDX, H. Howard (CIA)	Howard 2000
Westminster Abbey: chapter house, *Apocalypse scenes*	*c*1375–1404	Verdigris and copper green	PLM, SEM/EDX, FTIR, H. Howard (CIA)	Howard 2000
Canterbury Cathedral: Chapel of Our Lady Undercroft, eastern bay and screens enclosing chapel	*c*1370s	Malachite, verdigris and copper resinate	PLM, H. Howard (CIA); SEM/EDX, A. Roy (NG)	Howard 1993b

Site and Location	Date	Pigment(s)	Technique and Analyst(s)	Reference(s)
Tower of London: the Byward Tower, south tower, first floor room, south wall	*c*1395	Malachite, verdigris and copper resinate	PLM, C. Babington (EH).	
St. Nicholas' Hospital Church, Harbledown, Kent: chancel east wall and window embrasure	14th century	Malachite and a synthetic copper green	PLM, SEM/EDX, H. Howard (CIA)	
Ely Cathedral: Prior Crauden's Chapel, south tabernacle	14th century	Green earth (tentatively identified)	PLM, C. Babington (CIA)	Babington 1987
St. Margaret of Antioch, Cowlinge, Suffolk	14th century	Malachite	PLM, MCT, H. Howard (CIA)	
Ickleton, Cambridgeshire: chancel arch	Mid-14th century	Verdigris and perhaps copper resinate	PLM, MCT, H. Howard (CIA)	Howard 1988
Bradfield Combust, Suffolk	*c*1400	Verdigris	PLM	Rickerby 1990b
Lichfield Cathedral: south choir aisle, south wall, tomb recess with wall painting	*c*1400	Copper resinate	PLM, J. Darrah (V&A)	
Holy Trinity, Coventry, West Midlands: chancel arch, *Last Judgement*	*c*1435–60	Verdigris. A green copper resinate or copper oleate was also identified	PLM, SEM/EDX, FTIR, H. Howard (CIA)	Howard 1996b
Eton College Chapel, Berkshire: north and south walls, subjects drawn from sources such as the *Golden Legend* and the *Speculum Vitae*	*c*1479–87	Copper green		Plummer 1986
St. Leonard's Chapel, Farleigh Hungerford Castle, Somerset	Early 15th century	Verdigris	PLM, SEM/EDX, FTIR, H. Howard (CIA)	Howard *et al.* 1998
St. Mary's and All Saints' Willingham, Cambridgeshire: nave south wall	14th and 15th century	Verdigris, copper resinate and green verditer	PLM/MCT, T. Curteis	Curteis 1995a
Berry Pomeroy Castle, Devon: Gatehouse, east wall, *Adoration of the Magi*	15th century	Copper green	PLM	Gill *et al.* 1996
Winchester Cathedral: Lady Chapel, vaults	Late 15th century	Synthetic copper green and copper resinate	PLM/MCT, T. Curteis	Curteis and Paine 1995
St. Mary and St. Leodegarius, Ashby St. Ledgers, Northamptonshire: chancel arch and adjacent parts of north and south walls: *Passion* cycle	*c*1500	Copper green	PLM/MCT, T. Curteis	Curteis 1994a
Lord Mayor's Chapel (St. Mark's), Bristol: south aisle arcade, *Trinity, Nativity* and *Noli me Tangere*	*c*1510	Copper chloride (salt) green	PLM, SEM/EDX, FTIR, H. Howard (CIA)	Gill and Howard 1997
Westminster Abbey: chapter house, *Bestiary scenes*	*c*1510	Copper green	PLM, SEM/EDX, H. Howard (CIA)	Howard 2000

Site and Location	Date	Pigment(s)	Technique and Analyst(s)	Reference(s)
Examples in northern European medieval wall painting				
Cathedral of Verden, Germany	*c*1025–28	Malachite	H. Kühn	Gettens and Fitzhugh 1993b
St. Georg in Oberzell, Reichenau, Germany	*c*1060–80	Copper green (atacamite?)	H. Kühn	Knoepfli and Emmenegger 1990
Lambach, Austria: west choir	*c*1089	Malachite and green earth	H. Kortan	Hammer 1987
Idensen, Germany	*c*1130	Copper chloride (salt) green (thought to be in the form of copperhydroxychlorideacetate, M. Matteini, pers comm.)	PLM/MCT, SEM/ EDS, FTIR, M. Matteini and A. Moles (OPD)	Matteini and Moles 1994
Johanneskapelle, Pürgg, Austria	*c*1165	Malachite and a mixed green of yellow earth combined with charcoal black	PLM, SEM/EDX, FTIR [H. Howard (CIA)], and by I. Hammer 1977	Howard 1995d
Perschen Oberpfalz, Germany	*c*1160–70	Malachite	H. Kühn	Gettens and Fitzhugh 1993b
St. Bavo's Abbey, Ghent, Belgium	*c*1175	Green earth	MCT, Bontinck	Grissom 1986
Glasgow Cathedral: wall painting fragments excavated from the crypt	*c*1190	Green earth (celadonite)	PLM, SEM/EDX, FTIR, H. Howard (CIA)	Howard 1998a; Park 1998
Church of Kappel, Germany: choir	Early 12th century	Malachite	H. Kühn	Gettens and Fitzhugh 1993b
St. Peter and St. Paul in Niederzell, Reichenau, Germany: apse	12th century	Copper green		Knoepfli and Emmenegger 1990
Berzé-la-Ville, France: Monk's Chapel	Romanesque	Green earth and copper chloride green	PLM, C. Grissom; Raman spec. (C. Couprey).	Grissom 1986; Rollier-Hanselmann 1997
Auxerre Cathedral, France: *Crypt* cycle, apse	Romanesque	Green earth (glauconite)	PLM, XRF, FTIR, E. Cadet; Raman spec. (C. Couprey)	Cadet 1995
Saint-Aignan, Brinay-sur-Cher, France	Romanesque	Green earth, verdigris	LRMH	Rollier-Hanselmann 1997
Poitiers, France: Notre-Dame-la-Grande	Romanesque	Green earth, verdigris	LRMH	Rollier-Hanselmann 1997
Saint-Savin, France	Romanesque	Green earth	LRMH	Rollier-Hanselmann 1997
Abbey of la Trinité,Vedôme, France	Romanesque	Green earth	LRMH	Rollier-Hanselmann 1997

Site and Location	Date	Pigment(s)	Technique and Analyst(s)	Reference(s)
Church of St. Peter and St. Paul, Sussey-le-Maupas, Côte-d'Or, France	Romanesque	Yellow iron oxide and carbon black combined to make green. In addition, it appears that green earth was applied over a ground of calcium carbonate and lead white	LRMH	Rollier-Hanselmann 1997
Abbey of St. Philibert, Tournus, France	Romanesque	Green earth	LRMH	Rollier-Hanselmann 1997
Convent of St. John the Baptist, Müstair, Switzerland	*c*1200	Green earth and copper green	F. Mairinger	Emmenegger 1986
Fraugde Church, Insel Fünen, Denmark	*c*1200–25	Copper chloride (salt) green	(ITM)	Richter 1988
Angers Cathedral, France: choir, *Scenes of the Life of S. Maurille*	*c*1270–80	Copper greens (probably including copper resinate)	LRMH	Subes-Picot 1992; Stirnemann 1995; Demailly *et al.* 1998
Mästerby, Gotland, Sweden	13th century	Malachite and atacamite	SEM/EDX, XRD	Nord and Tronner 2000
Källunge, Gotland, Sweden	13th century	Atacamite	SEM/EDX, XRD	Nord and Tronner 2000
Garde, Gotland, Sweden	13th century	Atacamite	SEM/EDX, XRD	Nord and Tronner 2000
Cologne Cathedral, Germany: choir screen wall paintings	*c*1332–49	Verdigris	E. Jägers	Jägers and Schulze-Senger 1989; Heritage 1990; Maul 1992
Magdalenenkirche, Judenburg, Austria: choir, north wall, *Crucifixion*	14th century	Spherical malachite	PLM, FTIR, H. Howard (CIA)	Binder 1997
Aeschi, Switzerland	14th century	Malachite	H. Kühn	Gettens and Fitzhugh 1993b
Linden, Oberbayern, Germany	*c*1440	Malachite	H. Kühn	Gettens and Fitzhugh 1993b
Risinge, Östergötland, Sweden	Early 15th century	Green earth and malachite	SEM/EDX, XRD	Nord and Tronner 2000
Sachsen Cathedral, Freiberg, Germany	*c*1484–90	Malachite	H. Kühn	Gettens and Fitzhugh 1993b
Härkeberga, Uppland, Sweden	*c*1500	Atacamite, malachite and green earth	SEM/EDX, XRD	Nord and Tronner 2000
Härnevi, Uppland, Sweden	*c*1500	Atacamite and malachite	SEM/EDX, XRD	Nord and Tronner 2000
Täby, Uppland, Sweden	*c*1500	Atacamite, malachite and green earth	SEM/EDX, XRD	Nord and Tronner 2000
Katharinen Church, Lübeck, Germany		Copper chloride (salt) green	(ITM)	Richter 1988
Burgkloster, Lübeck, Germany		Copper chloride (salt) and copper sulphate greens	(ITM)	Richter 1988

Site and Location	Date	Pigment(s)	Technique and Analyst(s)	Reference(s)
Examples in English medieval panel painting				
Holy Innocent's, Adisham, Kent: nave, west wall, Adisham reredos	*c*1180	Verdigris	PLM, SEM/EDX, H. Howard (CIA)	Howard 1993c
Peterborough Cathedral: nave ceiling	*c*1220	Verdigris	PLM, SEM/EDX, FTIR, H. Howard (CIA)	Howard 1997b, 1998b.
Painted Chamber, Westminster Palace: *ceiling paterae*	*c*1263–66	Verdigris	J. Lynn (HKI)	Liversidge *et al.* 1995; Lynn 1997
Thornham Parva, Suffolk: Thornham Parva Retable	*c*1335	Copper greens (resinate or oleate)	S. Bucklow (HKI)	Goldsmith and Bucklow 1998
Kingston Lacy, Dorset: painted screen	Early 14th century	Verdigris and copper resinate	PLM, L. Sheldon and C. Hassall (UCL)	Tracy 1997
Exeter Cathedral, Devon: pulpitum paintings	14th century	Copper resinate	PLM/MCT, E. Sinclair	Sinclair 1995b
Wilton Diptych (English or French), now in the National Gallery, London	Late 14th century	Green earth and a copper green (perhaps copper resinate)	PLM/ SEM/EDX, A. Roy (NG)	Roy 1997
Examples in northern European medieval panel painting				
Heddal altar frontal, Norway	*c*1250–60	Copper green	U. Plahter	Plahter 1995
Alte Kapelle, Regensburg, Germany: *Christ Enthroned* on reverse of shrine	Early 13th century	Verdigris	Micr., ES, RS,(DI)	Kühn 1993a
Kinsarvik altar frontal, Norway	Before *c*1300	Mixed green of charcoal black and orpiment	PLM, SEM/EDX, U. Plahter	Plahter 1995
Dale II altar frontal, Norway	After *c*1300	Copper chloride green	PLM, SEM/EDX, U. Plahter	Plahter 1995
Odda altar frontal, Norway	After *c*1300	Copper chloride green and mixed green of azurite, ochre and orpiment	PLM, SEM/EDX, U. Plahter	Plahter 1995
Eid altar frontal, Norway	After *c*1300	Mixed green of indigo and orpiment	PLM, SEM/EDX, U. Plahter	Plahter 1995
German triptych in the Gallery of the Courtauld Institute of Art, London (known as the Estouteville triptych)	Late 14th century	Verdigris	PLM, G. van Heemstra; SEM/EDX, A. Burnstock (CIA)	Villers *et al.* 1997
Cathedral of St. Bavon, Ghent, Belgium: *The Mystic Lamb Altarpiece* (Jan van Eyck)	After *c*1432	Malachite	Micr, P. Coremans and J. Thissen	Gettens and Fitzhugh 1993b
Polychromed sculpture: examples in England				
Lincoln Cathedral: west front, frieze	*c*1140	Copper green	A. Roy (NG)	Larson 1992, 1993
Wells Cathedral: west front	*c*1225–40	Copper resinate	A. Roy and R. White (NG)	Roy and White 1982
St. Albans Cathedral: feretory, Shrine of St. Alban	*c*1302–08	Malachite and a copper chloride (salt) green	PLM, C. Babington (EH); PLM, SEM/EDX, H. Howard (CIA)	

Site and Location	Date	Pigment(s)	Technique and Analyst(s)	Reference(s)
Ingham, Norfolk: chancel, north wall, tomb of Sir Oliver de Ingham	*c*1344	Verdigris	PLM, SEM/EDX, H. Howard (CIA)	Howard 1993d
Ingham, Norfolk: nave south aisle, tomb of Sir Roger and Lady Margaret de Bois	*c*1380	Verdigris	PLM, SEM/EDX, H. Howard (CIA)	Howard 1993d
Exeter Cathedral: roof bosses	Early 14th century	Synthetic copper green	A. Roy (NG); J. Darrah (V&A); U. Plahter	Hulbert 1991
St. Mary's, Erwarton, Suffolk: polychromed monument of Sir Bartholomew Bacon and Joan de Heveningham	14th century	Malachite (or its synthetic equivalent, green verditer) and copper resinate	A. Roy (NG)	Roy 1984
Westminster Hall, London: polychromed kings	14th century	Copper green	PLM, C. Babington (EH)	Cherry and Stratford 1995
Exeter Cathedral: west front, polychromed sculpture	14th and 15th century	Verdigris or copper chloride (salt) green	PLM, E. Sinclair; SEM/EDX, A. Roy (NG)	Sinclair 1991, 1992
Winchester Cathedral: Great Screen	*c*1475–90	Verdigris and copper resinate	PLM, J. Darrah (V&A)	Brodrick and Darrah 1989
Fitzalan Chapel, Arundel: Fitzalan effigies	15th century	Verdigris and copper resinate	PLM, J. Darrah (V&A); GC-MS, R. White (NG)	Brodrick and Darrah 1986
Miscellaneous British examples				
St. Augustine's Abbey, Kent: excavated oyster-shell palette	13th-century	Copper green	XRF, XRD, (EH/AML)	Sherlock and Woods 1988
Linlithgow Carmelite Friary, Scotland: two pottery vessels retaining traces of pigments	Early post-Reformation	Synthetic copper green	PLM, H. Howard (CIA)	Howard 1989
Polychromed sculpture: northern European examples				
Monumental wooden doors from S. Maria im Kapitol, Cologne, Germany	*c*1050	Copper chloride green (which has partially altered to form tenorite)	E. Jägers	Jägers 1991
Hørning stave church, Denmark: Hørning plank	11th century	Copper chloride green	(ITM)	Richter 1988
Aby crucifix, Denmark	*c*1050–1100	Copper chloride green		Richter 1988
Chartres Cathedral, France: *Portail Royal*	*c*1140–50	Malachite	LRMH	Nonfarmale and Manaresi 1987
Vossestrand, Norway: lion capital	2nd half 12th century	Green earth	PLM, SEM/EDX, U. Plahter	Plahter 1990
Lectern at Freudenstädter, Germany	Romanesque	Copper chloride green		Richter 1988

Site and Location	Date	Pigment(s)	Technique and Analyst(s)	Reference(s)
Notre-Dame Cathedral, Lausanne, Switzerland: south portal	*c*1210–20	Malachite		Furlano *et al.* 1981
Cathedral of St. Mary, Freiberg, Germany: *Triumph of the Cross*	*c*1230	Verdigris	Micr., ES, IR (DI)	Kühn 1993a
Ruhpolding, Bavaria: *Madonna and Child*	*c*1220–40	Verdigris and green earth	Micr., ES, IR (DI)	Kühn 1993a
Amiens Cathedral, France: '*Portail de la Mère Dieu*'	*c*1220–60	Verdigris	LRMH	Weeks 1998
Maria Laach, Germany: Tomb of Pfalzgraf Heinrich II	13th century	Malachite and verdigris	E. Jägers	Oellermann 1993
Holy Cross Minster, Schwäbisch Gmünd, Germany	14th century	Verdigris	PLM/MCT	Hauff 1988
Bernt Notke, Stockholm, Sweden: statue of *St. George*	*c*1484–89	Verdigris	Micr., ES, IR (DI)	Kühn 1993a
Northern Netherlandish polychromed sculpture	Late 15th century		Malachite	Micr. K. Groen Kühn 1993a

Table 3 Selected analytical data relating to the use of red pigments in medieval polychromy.

Site and Location	Date	Pigment(s)	Technique and Analyst(s)	Reference(s)
Examples in English medieval wall painting				
Lewes Priory, Sussex: excavated wall painting fragments	*c*1090	Vermilion and red earth	PLM, SEM/EDX, H. Howard (CIA)	Howard and Park 1997
Worcester Cathedral: fragment from box WCCS 86 (crypt)	Late 11th century	Red earth	PLM/MCT, H. Howard and I. Kakoulli (CIA)	Park 1997
St. Botolph's, Hardham, West Sussex	*c*1100	Red earth	PLM, H. Howard and PLM, S. Hluvko (CIA)	Howard 1988; Hluvko 1991; Sawdy *et al.* 1998
All Saints' Witley, Surrey: nave south and west walls	*c*1110	Red earth and red lead	PLM, H. Howard (CIA)	Howard 1992a
St. Michael and All Angels, Copford, Essex: apse and chancel	*c*1125–30	Vermilion and red earth	PLM, C. Babington (EH)	Organ *et al.* 1991
St. Mary's, Kempley, Gloucestershire: chancel and nave east wall	*c*1130	Red earth	PLM/MCT, S. Rickerby (CIA)	Rickerby 1990a
Canterbury Cathedral: St. Gabriel's Chapel, apse	*c*1130	Vermilion, red lead and red earth	PLM, SEM/EDX, H. Howard (CIA)	Cather and Howard 1994
Canterbury Cathedral: St. Anselm's Chapel, apse, *St. Paul and the Viper*	Mid-12th century	Vermilion and red earth	Examined with microscope	Gärtner and Organ 1989
Winchester Cathedral: Holy Sepulchre Chapel	*c*1175	Red lake, vermilion, red lead and red earth	PLM, SEM/EDX, H. Howard (CIA)	Howard 1995a
Sherborne Abbey, Dorset: chapter house, excavated painted wall plaster	*c*1180	Vermilion, red lead and red earth	PLM, SEM/EDX, H. Howard (CIA)	Howard and Park forthcoming a
Guildford, Surrey: High Street, 'synagogue'	*c*1180	Red earth	PLM, SEM/EDX, H. Howard (CIA)	Howard 1996c
Norwich Cathedral: nave south aisle, vault, 4th bay from east and soffit of adjacent transverse arch, '*Losinga*' cycle	*c*1190	Vermilion and red earth	PLM, SEM/EDX, H. Howard (CIA)	Howard and Gasol 1996
Keynsham Abbey, Somerset: south chapel, excavated fragments with masonry pattern and fleur-de-lys	12th century	Red earth	J. Plesters (NG)	Lowe and Mason 1987
St. Andrew's Priory, York: excavated plaster fragments	12th century	Red earth		Coppack 1988
Sherborne Abbey, Dorset: Novice's Room, excavated fragments	Late 12th or early 13th century	Red earth	PLM, SEM/EDX, H. Howard (CIA)	Howard and Park forthcoming a
Eastbridge Hospital, Canterbury, Kent: hall, north wall	*c*1220	Red earth	PLM, SEM/EDX, H. Howard (CIA)	Howard 1994

Site and Location	Date	Pigment(s)	Technique and Analyst(s)	Reference(s)
Winchester Cathedral: Holy Sepulchre Chapel	*c*1220	Vermilion, red lead and red earth	PLM, SEM/EDX, H. Howard (CIA)	Howard 1995a
Ely Cathedral: St. George's Chapel, entrance arch soffit	*c*1220–30	Red lead and red earth	PLM, SEM/EDX, H. Howard (CIA)	Howard and Pathak 1995
Ely Cathedral: St. Edmund's Chapel, north wall, *Martyrdom of St. Edmund*	*c*1250	Red earth	PLM/MCT, S. Hluvko	Hluvko 1991
Norwich Cathedral: Ante-Reliquary Chapel, eastern arch	*c*1250	Vermilion and red lead	PLM, SEM/EDX, H. Howard (CIA)	Howard 1995c; Park and Howard 1996
Horsham St. Faith Priory, Norfolk: refectory	*c*1250 with 15th-century additions	Red lake (perhaps madder), vermilion, red lead and red earth	PLM, C. Babington (CIA); PLM J. Darrah (V&A)	Babington and James 1988
Winchester Cathedral: south transept, south-east corner	*c*1270	Red earth	PLM/MCT, T. Curteis	Curteis *et al.* 1994
Worcester Cathedral: crypt, eastern bishops	Early 13th century	Red earth	PLM, SEM/EDX, H. Howard and I. Kakoulli (CIA)	Park 1997
Worcester Cathedral: crypt, excavated wall painting fragments from box 'painted wall plaster from 6 stone blocks'	Early 13th century	Vermilion, red lead and red earth	PLM/MCT, H. Howard and I. Kakoulli (CIA)	Park 1997
Worcester Cathedral: crypt, rosette of masonry pattern	Early 13th century	Red lead	PLM/MCT, H. Howard and I. Kakoulli (CIA)	Park 1997
Winchester Cathedral: Guardian Angels' Chapel, vault	Early 13th century with later 13th-century additions	Vermilion, red lead and red earth	PLM, SEM/EDX, H. Howard (CIA)	
Rochester Cathedral: choir, *Wheel of Fortune*	Mid-13th century	Vermilion and red lead	PLM, J. Darrah (V&A)	Darrah 1992
St. Albans Cathedral: nave, north arcade, western painted pier, *Crucifixion*	Mid-13th century	Vermilion	PLM/MCT, P. Welford (CIA)	Welford 1991
Clarendon Palace, Wiltshire: excavated wall painting fragments	13th century	Vermilion	I. Bristow	Ashurst and James 1988
St. Augustine's Abbey, Canterbury, Kent: excavated fragments of wall painting from the ambulatory south wall and St. Thomas's Chapel	13th century	Red lead and red iron oxide (haematite)	EH/AML	Sherlock and Woods 1988
York Minster: coffin lid of Archbishop Walter de Gray	13th century	Vermilion	J. Plesters (NG)	Ramm *et al.* 1971

Site and Location	Date	Pigment(s)	Technique and Analyst(s)	Reference(s)
Hospital of St. Giles, Brompton Bridge, North Yorkshire: excavated wall painting fragments	13th century	Vermilion, red lead and red earth	PLM, S. Stewart (EH)	Stewart 1995
Duxford, Cambridgeshire: west and south walls of the chancel	13th century	Red earth	PLM (CIA)	CIA 1987
Westminster Abbey: nave, shields	13th century	Vermilion	PLM/MCT, A. Gardner (CIA)	
St. Mary's and All Saints' Willingham, Cambridgeshire: nave walls	13th century	Red earth	Prof. Wehlte	Wehlte 1954
St. Mary's, Wissington, Suffolk	13th century	Vermilion, red earth (and perhaps red lead)	PLM/MCT, T. Curteis	Curteis 1996a
Cleeve Abbey, Somerset	13th century	Red earth	PLM/MCT, S. Hluvko (CIA)	Hluvko 1991
Norwich Cathedral: Ante-Reliquary Chapel, vault and western arch	*c*1300	Madder lake and vermilion	PLM, SEM/EDX, H. Howard (CIA); UV spectroscopy A. Wallert (GCI)	Howard 1995c; Park and Howard 1996
St. Albans Cathedral: feretory, south-east pier, *Unidentified archbishop*	*c*1302–08	Madder lake, vermilion, red lead and red earth	PLM, SEM/EDX, H. Howard (CIA); DT-MS A. Wallert (RM)	Howard 1993a
St. Albans Cathedral; feretory, north-east pier, St. *William of York*	*c*1302–08	Red lake, vermilion and red earth	PLM, SEM/EDX, H. Howard (CIA)	Howard 1993a
Bushmead Priory, Bedfordshire: refectory	*c*1310	Vermilion		Park 1986
Merevale, Warwickshire: chancel arch	*c*1310–20	Red lead and red earth	PLM, SEM/EDX, H. Howard (CIA)	
Longthorpe Tower, Cambridgeshire: first floor solar	*c*1330	Vermilion, red lead and red earth	PLM, SEM/EDX, H. Howard (CIA)	
Holy Trinity, Bradwell-juxta-Coggeshall, Essex	*c*1330–40	Vermilion, red lead and red earth	PLM, H. Howard (CIA)	Cather and Park 1998
Rochester Cathedral: north-east transept, wall painting in tomb recess of William of Perth	*c*1340	Red lead and red earth (in addition, a red lake glaze may be present on the poppinjays)	PLM, S. Bradley (BM)	Rickerby 1988
St. Stephen's Chapel, Westminster Palace: fragments of wall painting now in British Museum	*c*1352–63	Red lake, vermilion, red lead and red earth	J. Haslam; J. Plesters and J. Kirby (NG)	Haslam 1807; van Geersdaele and Goldsworthy 1978
Ely, Cambridgeshire: 9 Silver Street, vinescroll ornament	*c*1350–75	Vermilion and red lead	PLM/MCT, S. Hluvko (CIA)	Hluvko 1991
Canterbury Cathedral: Chapel of Our Lady Undercroft, vault of eastern bay and screens enclosing chapel	1370s	Vermilion, red lead and red earth	PLM, H. Howard (CIA)	Howard 1993b

Site and Location	Date	Pigment(s)	Technique and Analyst(s)	Reference(s)
Westminster Abbey: chapter house, *Majesty*	*c*1375–1404	Red lake, vermilion, red lead and red earth	PLM/MCT, S. Hluvko (CIA); PLM, SEM/EDX, H. Howard (CIA)	Hluvko 1991; Howard 2000
Westminster Abbey: chapter house, *Apocalypse* cycle	*c*1375–1404	Red lake, vermilion, red lead and red earth	PLM/MCT, S. Hluvko (CIA); PLM, SEM/EDX, H. Howard (CIA)	Hluvko 1991; Howard 2000
Tower of London: the Byward Tower, south tower, first floor room, south wall	*c*1395	Red lake, vermilion and red lead	PLM, S. Hluvko (CIA); PLM, C. Babington (EH)	Hluvko 1991
Slapton, Northamptonshire: nave south aisle	Early 14th century	Vermilion and perhaps red lead	PLM, P. Welford (CIA)	Welford 1991
Fiddleford, Dorset: Manor House, solar, *Annunciation*	14th century	Vermilion	PLM/MCT, S. Hluvko (CIA)	Hluvko 1991
Ely Cathedral: Prior Crauden's Chapel, south tabernacle	14th century	Vermilion and red lead (tentatively identified)	PLM, C. Babington (CIA)	Babington 1987
St. Nicholas' Hospital Church, Harbledown, Kent: chancel east wall and window embrasure	14th century	Red lead and red earth	PLM, SEM/EDX, H. Howard (CIA)	
Farringdon, Hampshire: *Christ in Majesty*	14th century	Red lead	PLM, J. Darrah (V&A)	Ballantyne 1988
Canterbury Cathedral: Jesus Chapel, vault	14th century	Vermilion		Langslow *et al.* 1984
Great Hockham, Norfolk: north aisle, north wall	14th century	Red earth	PLM, H. Howard (CIA)	Howard 1988
Worcester Cathedral: nave, north arcade, fourth pier from east, north-west side	14th century	Red lead	PLM/MCT, H. Howard and I. Kakoulli (CIA)	Park 1997
St. Nicholas', Little Horwood, Buckinghamshire	14th and 15th century	Vermilion and red earth	PLM/MCT, T. Curteis	Curteis 1994b
Bradfield Combust, Suffolk	*c*1400	Vermilion and red earth	PLM	Rickerby 1990b
Lichfield Cathedral: south choir aisle, south wall, tomb recess with wall painting	*c*1400	Vermilion, red lead and red earth	PLM, J. Darrah (V&A)	
Holy Trinity, Coventry, West Midlands: chancel arch, *Last Judgement*	*c*1435–60	Red lake, vermilion, red lead and red earth	PLM, SEM/EDX, H. Howard (CIA)	Howard 1996b
Durham Cathedral: Deanery, entrance hall, north wall	*c*1470	Red earth	PLM/MCT, H. Howard (CIA)	Howard 1988
Eton College Chapel, Berkshire: north and south walls, subjects drawn from sources such as the *Golden Legend* and the *Speculum Vitae*	*c*1479–87	Red lake, vermilion and red lead	National Gallery, London; PLM/MCT, S. Hluvko (CIA)	Plummer 1986; Hluvko 1991

Site and Location	Date	Pigment(s)	Technique and Analyst(s)	Reference(s)
Canterbury Cathedral: north choir aisle, *Legend of St. Eustace*	*c*1480	Vermilion, red lead and red earth	PLM/MCT, S. Hluvko (CIA)	Hluvko 1991
St. Leonard's Chapel, Farleigh Hungerford Castle, Somerset	Early 15th century	Red lake, vermilion, red lead and red earth	PLM, SEM/EDX, FTIR, H. Howard (CIA)	Howard *et al.* 1998
St. Margaret of Antioch, Cowlinge, Suffolk: *St Michael Weighing Souls*	15th century	Vermilion and red lead	PLM/MCT, H. Howard (CIA)	
Berry Pomeroy Castle, Devon: Gatehouse, east wall, *Adoration of the Magi*	15th century	Vermilion, red lead and red earth	PLM	Gill *et al.* 1996
All Saints', Warlingham, Surrey: *St. Christopher*	15th century	Vermilion		Granville and Burbidge 1994
Cleeve Abbey, Somerset	15th century	Red earth	PLM/MCT, S. Hluvko (CIA)	Hluvko 1991
St. Martin's, Colchester, Essex: *Doom*	Mid-15th century	Vermilion and red earth	PLM/MCT, T. Curteis	Curteis 1996b
All Saints' and St. Andrew, Kingston, Cambridgeshire	Late 15th century	Red lead and red earth	PLM/MCT, T. Curteis	Curteis 1995b
Winchester Cathedral: Lady Chapel, vaults	Late15th century	Vermilion, red lead and red earth	PLM/MCT, T. Curteis	Curteis and Paine 1995
St. Mary and St. Leodegarius, Ashby St. Ledgers, Northamptonshire: chancel arch and adjacent parts of north and south walls, *Passion* cycle	*c*1500	Vermilion and red lead	PLM/MCT, T. Curteis	Curteis 1994a
Bramall Hall, Cheshire	*c*1500	Red lead and red earth		
Lord Mayor's Chapel (St. Mark's), Bristol: south aisle arcade: *Trinity, Nativity* and *Noli me Tangere*	*c*1510	Vermilion	PLM, SEM/EDX, H. Howard (CIA)	Gill and Howard 1997
Westminster Abbey: chapter house, Bestiary scenes	*c*1510	Red lead	PLM/MCT, S. Hluvko (CIA)	Hluvko 1991
Westminster Abbey: chapter house, dado rose decoration beneath Bestiary	Before 1510	Vermilion and red lead	PLM/MCT, S. Hluvko (CIA)	Hluvko 1991
Ely, Cambridgeshire: 7 Silver Street, first floor paintings	*c*1530	Red lead and red earth	PLM/MCT, S. Hluvko (CIA)	Hluvko 1991
Haughmond Abbey, Shropshire: chapter house	Medieval	Red iron oxide (haematite)	XRF	Heyworth 1991
York Minster: The Bedern, excavated wall plaster	Medieval	Vermilion, red lead and red earth	PLM, L. Wong (CIA)	Wong forthcoming

Site and Location	Date	Pigment(s)	Technique and Analyst(s)	Reference(s)
Examples in northern European medieval wall painting				
Convent of St. John the Baptist, Müstair, Switzerland	*c*800	Red/purple lake pigment likely to be madder, red lead and red earth	F. Mairinger	Mairinger and Schreiner 1986
Idensen, Germany	*c*1130	Cinnabar and red earth	PLM/MCT, SEM/ EDS, FTIR, M. Matteini and A. Moles (OPD)	Geilmann 1938; Matteini and Moles 1994
Johanneskapelle, Pürgg, Austria	*c*1165	Red earth	PLM, SEM/EDX, H. Howard (CIA); I. Hammer 1977	Howard 1995d
Cashel, Ireland: Cormac's Chapel	*c*1170	Vermilion, red lead and red earth	EDX, ES, XRD, M. McGrath; PLM, H. Howard (CIA)	McGrath 1987; Howard 1991
Glasgow Cathedral: wall painting fragments excavated from the crypt	*c*1190	Red lake, vermilion, red lead and red earth	PLM, SEM/EDX, H. Howard (CIA)	Howard 1998a; Park 1998
Convent of St. John the Baptist, Müstair, Switzerland	*c*1200	Vermilion, red lead and red earth	F. Mairinger	Emmenegger 1986
Berzé-la-Ville, France: Monk's Chapel	Romanesque	Vermilion, red lead and red earth	C. Couprey (LRMH)	Rollier-Hanselmann 1997
Auxerre Cathedral, France: *Crypt* cycle	Romanesque	Red earth and red lead (though the latter may represent a later retouching)	PLM, XRF, FTIR, E. Cadet	Cadet 1995
Angers Cathedral, France: choir, *Scenes of the Life of S. Maurille*	*c*1270–80	Vermilion, red lake, red lead and red earth. Vermilion was applied over a layer of red lead	LRMH	Subes-Picot 1992; Stirnemann 1995; Demailly *et al.* 1998
Cologne Cathedral, Germany: choir screen wall paintings	*c*1332–49	Brasil-wood lake and vermilion	E. Jägers	Jägers and Schulze-Senger 1989; Heritage 1990; Maul 1992
Master of Waltensburg, Waltensburg, Kanton Graubünden, Germany	*c*1330–50	Red lead	XRDH. Kühn	Fitzhugh 1986
Landkartengalerie, Salzburg, Austria	Late 16th century	Cochineal lake		
Kirche St. Marien, Briefkapelle, Lübeck, Germany	Medieval	Red lead	MCTE. Richter	Fitzhugh 1986

Site and Location	Date	Pigment(s)	Technique and Analyst(s)	Reference(s)
Examples in English medieval panel painting				
Holy Innocent's, Adisham, Kent: nave, west wall, reredos	*c*1180	Vermilion and red lead	PLM, SEM/EDX, H. Howard (CIA)	Howard 1993c
Peterborough Cathedral: nave ceiling	*c*1220	Vermilion, red lead and red earth	PLM, SEM/EDX, H. Howard (CIA)	Howard 1997b
Painted Chamber, Westminster Palace: *ceiling paterae*	*c*1263–66	Vermilion, red lead, an organic red lake and red earth	J. Lynn (HKI)	Liversidge *et al.* 1995; Lynn 1997
Westminster Abbey: Westminster Retable	*c*1270	Red lake on a hydrated alumina substrate		
Thornham Parva, Suffolk: Thornham Parva Retable.	*c*1335	Polish cochineal lake, vermilion, red lead and red earth	HPLC, J. Kirby (NG); PLM, S. Bucklow (HKI)	Bucklow pers. comm.; Goldsmith and Bucklow 1998
Kingston Lacy, Dorset: painted screen	Early 14th century	Red lake, vermilion, red lead and haematite	PLM, L. Sheldon and C. Hassall (UCL)	Tracy 1997
Exeter Cathedral: pulpitum paintings	14th century	Vermilion and red lead	PLM/MCT, E. Sinclair	Sinclair 1995b
Wilton Diptych (English or French), now in the National Gallery, London	Late 14th century	Red lake and vermilion	PLM, SEM/EDX, A. Roy (NG)	Roy 1997
St. Mary's, Worstead, Norfolk: screen	Early 16th century	Red lake, vermilion and red lead	PLM/MCT, T. Curteis	Curteis 1993
Examples in northern European medieval panel painting				
Heddal altar frontal, Norway	*c*1250–60	Red lake, vermilion, red lead and red earth	PLM, SEM/EDX, U. Plahter	Plahter 1995
German triptych in the Gallery of the Courtauld Institute of Art, London (known as the Estouteville triptych)	Late 14th century	Madder lake, vermilion, haematite	PLM, G. van Heemstra; SEM/EDX, A. Burnstock (CIA)	Villers *et al.* 1997
Part of an altar, Wallref-Richartz Museum, Cologne (No. 60-62-389, 755)	*c*1420–30	Red lead	XRD, H. Kühn	Fitzhugh 1986
Master of the Heisterbacher Altars, Altar, Wallref-Richartz Museum, Cologne (No. 73)	*c*1440–50	Red lead	XRD, H. Kühn	Fitzhugh 1986
Herlin altarpiece, Church of St. James, Rothenburg, Germany	*c*1466	Red lead and madder lake	PLM, IR spectroscopy, Broekman-Bolstijn	Broekman-Bolstijn *et al.* 1970; Fitzhugh 1986
High altar, Århus Cathedral, Denmark	*c*1479	Red lead	PLM/MCTLarsen	Fitzhugh 1986
Blaubeuren High Altarpiece, Württemberg, Germany	*c*1490	Red lead	XRD, E. Richter	Fitzhugh 1986
Altarpiece wing from Zittau, Germany	*c*1510–20	Madder lake		Heydenreich 1994

Site and Location	Date	Pigment(s)	Technique and Analyst(s)	Reference(s)
Polychromed sculpture: examples in England				
St. Augustine's Abbey, Canterbury, Kent: excavated limestone capital	Early 7th or 9th century	Red earth	PLM, P. Welford (CIA)	Brain and Welford 1991
Lincoln Cathedral: west front, frieze	*c*1140	Vermilion and red lead	A. Roy (NG)	Larson 1992, 1993
St. Mary's Abbey, York: excavated stone sculpture	*c*1180–85	Vermilion, red lead and red earth	A. Roy (NG)	Brodrick 1993
Wells Cathedral: west front	*c*1225–40	Vermilion and red earth	A. Roy and R. White (NG)	Roy and White 1982
Ely Cathedral: Lady Chapel	*c*1231–49	Vermilion and red lead	PLM/MCT, S. Hluvko (CIA)	Hluvko 1991
Canterbury Cathedral: cloister, *Martyrdom Doorway*	*c*1236–39	Vermilion and red lead	PLM/MCT, S. Hluvko (CIA)	Hluvko 1991
Canterbury Cathedral: cloister, niche in wall within the blind arcade to the south of the *Martyrdom Doorway*	*c*1236	Vermilion, red lead and red earth	PLM/MCT, S. Hluvko (CIA)	Hluvko 1991
All Saints', Sawley, Derbyshire: polychromed angel which originally formed part of a canopied tomb	*c*1270–80	Vermilion		Alexander and Binski 1987
York Minster: chapter house, excavated stone pendant	Late 13th century	Vermilion and red earth	PLM, J. Darrah (V&A)	Heywood and Oosterwijk 1993
Salisbury Cathedral: west front, porch	13th century with numerous phases of repainting	Red earth. Red earth appears to have been used in the earliest priming layers. Vermilion and red lead have also been found, though it is not clear to which phase of decoration they belong	PLM, E. Sinclair	Sinclair 1997
St. Albans Cathedral: feretory, Shrine of St. Alban	*c*1302–08	Red lake, vermilion and red lead	PLM, C. Babington (EH); PLM, SEM/EDX, H. Howard (CIA)	
Abergavenny, Wales: wooden knight	*c*1325	Red lead	A. Brodrick	Blair 1994
Ingham, Norfolk: chancel, north wall, tomb of Sir Oliver de Ingham	*c*1344	Red lake, vermilion, red lead and red earth	PLM, SEM/EDX, H. Howard (CIA)	Howard 1993d
Ingham, Norfolk: nave south aisle, tomb of Sir Roger and Lady Margaret de Bois	*c*1380	Vermilion, red lead and red earth	PLM, SEM/EDX, H. Howard (CIA)	Howard 1993d
Salisbury Cathedral: west front	Mid-13th–mid-14th century	Vermilion and red earth	PLM, SEM/EDX, H. Howard (CIA)	Howard 1995e, 1996d
Exeter Cathedral: roof bosses	Early 14th century	Red lake, vermilion, red lead and red earth	A. Roy (NG); J. Darrah (V&A); U. Plahter	Hulbert 1991

Site and Location	Date	Pigment(s)	Technique and Analyst(s)	Reference(s)
Worcester Cathedral: Edgar Tower, polychromed foot of a sculpture	14th century	Vermilion and red lead	PLM/MCT, I. Kakoulli and H. Howard (CIA)	Park 1997
St. Mary's, Erwarton, Suffolk: polychromed monument of Sir Bartholomew Bacon and Joan de Heveningham	14th century	Vermilion, red lead and red earth	A. Roy (NG)	Roy 1984
Effigy of Geoffry de Lucy, now in the Victoria and Albert Museum	14th century	Red lead	PLM/MCT, S. Boulter (V&A)	Boulter 1993
Westminster Hall: polychromed kings	14th century	Vermilion, red lead and red earth	PLM, C. Babington (EH)	Cherry and Stratford 1995
Exeter Cathedral: west front	14th and 15th century	Red lake, vermilion, red lead and red earth	PLM (E. Sinclair); SEM/EDX, A. Roy (NG)	Sinclair 1991, 1992
Winchester Cathedral: Great Screen	*c*1475–90	Red lake, vermilion, red lead and red earth	PLM, J. Darrah (V&A)	Brodrick and Darrah 1989
Fitzalan Chapel, Arundel: Fitzalan effigies	15th century	Red lake, vermilion, red lead and red earth	PLM, J. Darrah (V&A)	Brodrick and Darrah 1986
Cartmel Fell, Cumbria: Chapel of St. Anthony, wooden figure of Christ	Late 15th century	Red earth (a red lake pigment was also suggested)	J. Plesters and A. Roy (NG)	Behrens 1982
Worcester Cathedral: wooden pilgrim's staff		Kermes lake	A. Burnstock and R. White (NG)	Lubin *et al.* 1990
Miscellaneous British examples				
Antioch Chamber, Clarendon Palace, Wiltshire: excavated medieval artists' palette	*c*1235	Red lead	PLM, H. Hughes (EH)	Hughes and Lewis 1988
Westminster Abbey: Coronation Chair	*c*1297–1300	Vermilion	Prof. Rees Jones (CIA)	Percival-Prescott 1957; Rees Jones 1954
St. Augustine's Abbey, Kent: excavated oyster-shell palette	13th century	Vermilion	XRF, XRD (EH/AML)	Sherlock and Woods 1988
Greyfriars, Norwich: excavated oyster-shell palette	13th–14th century	Vermilion	PLM, SEM/EDX, H. Howard (CIA)	Howard and Park forthcoming b
Whitefriars Carmelite Friary, Coventry: excavated oyster-shell palette	Mid-14th–mid-16th centuries	Red lake	PLM, SEM/EDX, FTIR, H. Howard (CIA)	Howard 1997a
Acton Court, Avon: oyster-shell palette excavated from fill of south arm of moat	Mid–late 16th century	Vermilion	PLM, H. Hughes and S. Strong (EH)	Hughes and Strong 1990
Linlithgow Carmelite Friary, Scotland: three shells, two of which appear to have been used as palettes or containers	Early post-Reformation	Vermilion and red lead	PLM, H. Howard (CIA)	Howard 1989

Site and Location	Date	Pigment(s)	Technique and Analyst(s)	Reference(s)
Polychromed sculpture: northern European examples				
Grave of the King of Gorm at Jelling, Jutland, Denmark: fragments of carved wood	*c*960	Red earth	M. Christensen and J. Glastrup (NM)	Poster at ICOM '96
Crucifix from Lichtenstein Castle (now in Württembergisches Landesmuseum, Stuttgart)	*c*1050	Red lead	MCT, Westhoff *et al.*	Fitzhugh 1986
Hørning stave church, Denmark: Hørning plank	11th century	Vermilion and red lead	M. Christensen and J. Glastrup (NM)	Poster at ICOM '96
Chartres Cathedral, France: *Portail Royal*	*c*1140–50	Red earth	LRMH	Nonfarmale and Manaresi 1987
Freudenstädter Lesepult, town church of Freudenstädter (Schwarzwald) Germany	*c*1150	Red lead	XRD, E. Richter	Fitzhugh 1986
Virgin and Child, Hovland Stave Church, Eggedal, Norway	12th century	Red lead	MCT, Simonson and U. Plahter	Fitzhugh 1986
Vossestrand, Norway: lion capital	2nd half 12th century	Vermilion, red lead and red earth	PLM, SEM/EDX, U. Plahter	Plahter 1990
Notre-Dame Cathedral, Lausanne, Switzerland: south portal	*c*1216–20	Vermilion, red lead and red earth		Furlano *et al.* 1981
Amiens Cathedral, France: '*Portail de la Mère Dieu*'	*c*1220–60	Red lake, vermilion, red lead and red earth	LRMH	Weeks 1998
Maria Laach, Germany: Tomb of Pfalzgraf Heinrich II	13th century	Madder lake, vermilion and red lead		Oellermann 1993
Dal Church, *Virgin and Child*, now in the University Museum of National Antiquities, Oslo, Norway	Mid-13th century	Red lead	PLM, U. Plahter	Plahter and Wiik 1970
Holy Cross Minster, Schwäbisch Gmünd, Germany: polychromed portals	14th century	Red lead and red earth	PLM/MCT	Hauff 1988

Table 4 Selected analytical data relating to the use of yellow pigments in medieval polychromy.

Site and Location	Date	Pigment(s)	Technique and Analyst(s)	Reference(s)
Examples in English medieval wall painting				
Lewes Priory, Sussex: excavated wall painting fragments	*c*1090	Yellow earth	PLM, SEM/EDX, H. Howard (CIA)	Howard and Park 1997
All Saints' Witley, Surrey: nave south and west walls	*c*1110	Yellow earth	PLM, SEM/EDX, H. Howard (CIA)	Howard 1992a
St. Michael and All Angels, Copford, Essex: apse and chancel	*c*1125–30	Yellow earth (tentatively identified)	PLM, C. Babington (EH)	Organ *et al.* 1991
St. Mary's, Kempley, Gloucestershire: chancel and nave east wall	*c*1130	Yellow earth	PLM/MCT, S. Rickerby (CIA)	Rickerby 1990a
Canterbury Cathedral: St. Gabriel's Chapel, apse	*c*1130	Yellow earth	PLM, SEM/EDX, H. Howard (CIA)	Cather and Howard 1994
Canterbury Cathedral: St. Anselm's Chapel, apse, *St. Paul and the Viper*	Mid-12th century	Yellow earth	Examination with microscope	Gärtner and Organ 1989
Chichester Cathedral: north transept	*c*1140–50	Yellow earth		Gärtner 1988
Winchester Cathedral: Holy Sepulchre Chapel	*c*1175	Yellow earth and umber	PLM, SEM/EDX, H. Howard (CIA)	Howard 1995a
Sherborne Abbey, Dorset: chapter house, excavated painted wall plaster	*c*1180	Yellow earth	PLM, SEM/EDX, H. Howard (CIA)	Howard and Park forthcoming a
Guildford, Surrey: High Street, 'synagogue'	*c*1180	Yellow earth	PLM, SEM/EDX, H. Howard (CIA)	Howard 1996c
Norwich Cathedral: nave south aisle, vault, 4th bay from east and soffit of adjacent transverse arch, '*Losinga*' cycle	*c*1190	Yellow earth	PLM, SEM/EDX, H. Howard (CIA)	Howard and Gasol 1996
Winchester Cathedral: Holy Sepulchre Chapel	*c*1220	Yellow earth	PLM, SEM/EDX, H. Howard (CIA)	Howard 1995a
Eastbridge Hospital, Canterbury, Kent: hall, north wall	*c*1220	Yellow earth	PLM, SEM/EDX, H. Howard (CIA)	Howard 1994
Worcester Cathedral: crypt, excavated plaster fragment with head	*c*1220	Orpiment (artificial, probably formed by sublimation)	PLM, SEM/EDX, A. Roy (NG)	Park 1997
Norwich Cathedral: Ante-Reliquary Chapel, eastern arch	*c*1250	Yellow earth	PLM, SEM/EDX, H. Howard (CIA)	Howard 1995c; Park and Howard 1996
Horsham St. Faith Priory, Norfolk: refectory	*c*1250 with 15th-century additions	Yellow earth	PLM, C. Babington (CIA) and J. Darrah (V&A)	Babington and James 1988

Site and Location	Date	Pigment(s)	Technique and Analyst(s)	Reference(s)
Winchester Cathedral: south transept, south-east corner	*c*1270	Yellow earth	PLM/MCT, T. Curteis	Curteis *et al.* 1994
Worcester Cathedral: excavated wall plaster fragments from box (painted wall plaster from 6 stone blocks)	Early 13th century	Yellow earth	PLM, H. Howard and I. Kakoulli (CIA)	Park 1997
Worcester Cathedral: excavated wall plaster fragments from unlabelled box with two pieces with 13th-century foliage	Early 13th century	Yellow earth	PLM, H. Howard and I. Kakoulli (CIA)	Park 1997
Winchester Cathedral: Guardian Angels' Chapel, vault	Early 13th century with later 13th-century additions	Yellow earth	PLM, SEM/EDX, H. Howard (CIA)	
Rochester Cathedral: choir, *Wheel of Fortune*	Mid-13th century	Orpiment	PLM, J. Darrah (V&A)	Darrah 1992
Clarendon Palace, Wiltshire: excavated wall painting fragments	13th century	Yellow earth	I. Bristow	Ashurst and James 1988
Hospital of St. Giles, Brompton Bridge, North Yorkshire: excavated wall painting fragments	13th century	Yellow earth	PLM, S. Stewart (EH)	Stewart 1995
St. Mary's and All Saints' Willingham, Cambridgeshire: nave walls	13th century	Yellow earth	Prof. Wehlte	Wehlte 1954
Norwich Cathedral: Ante-Reliquary Chapel, vault and western arch	*c*1300	Yellow earth	PLM, SEM/EDX, H. Howard (CIA)	Howard 1995c; Park and Howard 1996
St. Albans Cathedral: feretory, north-east pier, *St. William of York*	*c*1302–08	Yellow earth	PLM, SEM/EDX, H. Howard (CIA)	Howard 1993a
Merevale, Warwickshire: chancel arch	*c*1310–20	Yellow earth	PLM, SEM/EDX, H. Howard (CIA)	
Longthorpe Tower, Cambridgeshire: first floor solar	*c*1330	Yellow earth	PLM, SEM/EDX, H. Howard (CIA)	
Holy Trnity, Bradwell-juxta-Coggeshall, Essex	*c*1330–40	Yellow earth	PLM, H. Howard (CIA)	Cather and Park 1998
Rochester Cathedral: north-east transept, wall painting in tomb recess of William of Perth	*c*1340	Yellow earth	PLM, S. Bradley (BM)	Rickerby 1988
Canterbury Cathedral: Chapel of Our Lady Undercroft, vault of eastern bay and screens enclosing chapel	*c*1370s	Yellow earth	PLM, H. Howard (CIA); SEM/EDX, A. Roy (NG)	Howard 1993b
Westminster Abbey: chapter house, *Apocalypse* cycle	*c*1375–1404	Orpiment and yellow earth	PLM, SEM/EDX, H. Howard (CIA)	unpublished analysis H. Howard 1998

Site and Location	Date	Pigment(s)	Technique and Analyst(s)	Reference(s)
Tower of London: the Byward Tower, south tower, first floor room, south wall	*c*1395	Yellow earth	PLM, C. Babington (EH)	
Ely Cathedral: Prior Crauden's Chapel, south tabernacle	14th century	Yellow earth (tentatively identified)	PLM, C. Babington (CIA)	Babington 1987
Farringdon, Hampshire: *Christ in Majesty*	14th century	Orpiment	PLM, J. Darrah (V&A)	Ballantyne 1988
Lichfield Cathedral: south choir aisle, south wall, tomb recess with wall painting	*c*1400	Lead-tin yellow (tentatively identified)	PLM, J. Darrah (V&A)	
Bradfield Combust, Suffolk	*c*1400	Yellow earth	PLM,	Rickerby 1990b
Holy Trinity, Coventry, West Midlands: chancel arch, *Last Judgement*	*c*1435–60	Yellow earth (and massicot in earlier scheme on SW face of S. pier of chancel arch)	PLM, SEM/EDX, H. Howard (CIA)	Howard 1996b
Eton College Chapel, Berkshire: north and south walls, subjects drawn from sources such as the *Golden Legend* and the *Speculum Vitae*	*c*1479–87	Lead-tin yellow		Plummer 1986
St. Leonard's Chapel, Farleigh Hungerford Castle, Somerset	Early 15th century	Orpiment and lead-tin yellow (preliminary XRD analysis suggests that lead-tin yellow type I may be present)	PLM, SEM/EDX, FTIR, H. Howard (CIA); XRD, M. Ward (EH/AML)	Howard *et al.* 1998
St. Nicholas', Little Horwood, Buckinghamshire	14th and 15th century	Yellow earth	PLM/MCT, T. Curteis	Curteis 1994b
St. Mary's and All Saints' Willingham, Cambridgeshire: nave south wall	14th and 15th century	Orpiment	PLM/MCT, T. Curteis	Curteis 1995a
St. Martin's, Colchester, Essex: *Doom*	Mid-15th century	Lead-tin yellow	PLM/MCT, T. Curteis	Curteis 1996b
Berry Pomeroy Castle, Devon: Gatehouse, east wall, *Adoration of the Magi*	15th century	Yellow earth	PLM	Gill *et al.* 1996
All Saints', Warlingham, Surrey: *St. Christopher*	15th century	Yellow earth		Granville and Burbidge 1994
All Saints' and St. Andrew, Kingston, Cambridgeshire	Late 15th century	Yellow earth	PLM/MCT, T. Curteis	Curteis 1995b
St. Mary and St. Leodegarius, Ashby St. Ledgers, Northamptonshire: chancel arch and adjacent parts of north and south walls, *Passion* cycle	*c*1500	Orpiment	PLM/MCT, T. Curteis	Curteis 1994a
Lord Mayor's Chapel (St. Mark's), Bristol: south aisle arcade, *Trinity, Nativity* and *Noli me Tangere*	*c*1510	Lead-tin yellow	PLM, SEM/EDX, H. Howard (CIA)	Gill and Howard 1997
York Minster: The Bedern, excavated wall plaster	Medieval	Yellow earth and lead-tin yellow	PLM, (Wong); SEM/EDX, H. Howard (CIA)	Wong forthcoming

Site and Location	Date	Pigment(s)	Technique and Analyst(s)	Reference(s)
Examples in northern European medieval wall painting				
Convent of St. John the Baptist, Müstair, Switzerland	*c*800	Massicot (applied in a medium containing protein and oil) and yellow earth	F. Mairinger	Mairinger and Schreiner 1986
Idensen, Germany	*c*1130	Yellow earth	PLM/MCT, SEM/ EDS, M. Matteini and A. Moles (OPD)	Matteini and Moles 1994
Johanneskapelle, Pürgg, Austria	*c*1165	Yellow earth	PLM, SEM/EDX, H. Howard (CIA)	Howard 1995d
Cashel, Ireland: Cormac's Chapel	*c*1170	Yellow earth	PLM, H. Howard (CIA)	McGrath 1987; Howard 1991
Glasgow Cathedral: wall painting fragments excavated from the crypt	*c*1190	Yellow and brown earths	PLM, SEM/EDX, H. Howard (CIA)	Howard 1998a; Park 1998
Convent of St. John the Baptist, Müstair, Switzerland	*c*1200	Yellow earth and raw sienna	F. Mairinger	Emmenegger 1986
Angers Cathedral, France: choir, *Scenes of the Life of S. Maurille*	*c*1270–80	Orpiment, lead-tin yellow and yellow earth		Subes-Picot 1992; Demailly *et al.* 1998
Cologne Cathedral, Germany: choir screen wall paintings	*c*1332–49	Yellow earth	E. Jägers	Jägers and Schulze-Senger 1989; Heritage 1990; Maul 1992
Cathedral of St. Marien, Freiberg, Germany: ceiling painting	*c*1484–90	Lead-tin yellow	ES, H. Kühn	Kühn 1993b
Cathedral of St. Marien, Freiberg, Germany: upper sacristy	*c*1499	Lead-tin yellow	ES, H. Kühn	Kühn 1993b
Vleeshuis, Ghent: Nabur Martinus wall painting, *Nativity*	15th century	Lead-tin yellow	MCT (IRPA)	Kühn 1993b
Collégiale de Nivelles, Ghent	15th century	Lead-tin yellow	MCT (IRPA)	Kühn 1993b
Härkeberga, Uppland, Sweden	*c*1500	Lead-tin yellow and massicot	SEM/EDX, XRD	Nord and Tronner 2000
Täby, Uppland, Sweden	*c*1500	Lead-tin yellow and massicot	SEM/EDX, XRD	Nord and Tronner 2000
Examples in English medieval panel painting				
Peterborough Cathedral: nave ceiling	*c*1220	Yellow earth and massicot	PLM, SEM/EDX, H. Howard (CIA)	Howard 1997b
Painted Chamber, Westminster Palace: *ceiling paterae*	*c*1263–66	Yellow earth	J. Lynn (HKI)	Liversidge *et al.* 1995; Lynn 1997
Kingston Lacy, Dorset: painted screen	Early 14th century	Yellow earth	PLM, L. Sheldon and C. Hassall (UCL)	Tracy 1997

Site and Location	Date	Pigment(s)	Technique and Analyst(s)	Reference(s)
Exeter Cathedral: pulpitum paintings	14th century	Yellow earth	PLM/MCT, E. Sinclair	Sinclair 1995b
Wilton Diptych (English or French), now in the National Gallery, London	Late 14th century	Yellow lake, orpiment (combined with indigo), lead-tin yellow (type II) and yellow earth	LM/ SEM/EDX, A. Roy (NG)	Roy 1997; Fitzhugh 1997
St. Mary's, Worstead, Norfolk: screen	Early 16th century	Lead-based yellow pigment suggested	PLM/MCT, T. Curteis	Curteis 1993
Examples in northern European medieval panel painting				
Heddal altar frontal, Norway	*c*1250–60	Yellow earth	U. Plahter	Plahter 1995
St. Michel's Church, Hildesheim, Germany: painted wooden ceiling	Mid-13th century	Orpiment	PLM, E. Denninger	Denninger 1969
Tingelstad I, II and III altarpieces, Norway	*c*1250–1350	Orpiment	PLM, XRD, U. Plahter	Plahter *et al.* 1974
Kinsarvik altar frontal, Norway	Before *c*1300	Orpiment (mixed with charcoal black)	U. Plahter	Plahter 1995
Eid altar frontal, Norway	After *c*1300	Orpiment (mixed with indigo)	U. Plahter	Plahter 1995
Nârodni Gallery, Prague: *Wittingauer Altar*	Late 14th century	Lead-tin yellow II	ES, XRD, H. Kühn	Kühn 1993b
Master of the Wittingauer Altar workshop, Nârodni Gallery, Prague: *Martyrdom of St. Barbara*	Late 14th century	Lead-tin yellow II	ES, XRD, H. Kühn	Kühn 1993b
German triptych in the Gallery of the Courtauld Institute of Art, London (known as the Estouteville triptych)	Late 14th century	Lead-tin yellow, yellow earth	PLM, G. van Heemstra; SEM/EDX, A. Burnstock (CIA)	Villers *et al.* 1997
Polychromed sculpture: examples in England				
St. Augustine's Abbey, Canterbury, Kent: excavated limestone capital	Early 7th or 9th century	Yellow earth	PLM, P. Welford (CIA)	Braine and Welford 1991
St. Mary's Abbey, York: excavated stone sculpture	*c*1180-85	A bright 'saffron' colour is mentioned	A. Roy (NG)	Brodrick 1993
York Minster: chapter house, excavated stone pendant	Late 13th century	Yellow earth	PLM, J. Darrah (V&A)	Heywood and Oosterwijk 1993
St. Albans Cathedral: feretory, Shrine of St. Alban	*c*1302–08	Yellow earth	PLM, C. Babington (EH) and H. Howard (CIA)	
Salisbury Cathedral: west front	13th century with numerous phases of repainting	Yellow earth	PLM, SEM/EDX, H. Howard (CIA)	Howard 1995e, 1996d
Ingham, Norfolk: chancel, north wall, Tomb of Sir Oliver de Ingham	*c*1344	Yellow earth	PLM, SEM/EDX, H. Howard (CIA)	Howard 1993d
Ingham, Norfolk: nave south aisle, Tomb of	*c*1380	Yellow earth	PLM, SEM/EDX,	Howard 1993d

Site and Location	Date	Pigment(s)	Technique and Analyst(s)	Reference(s)
Westminster Hall, London: polychromed kings	14th century	Yellow earth	PLM, C. Babington (EH)	Cherry and Stratford 1995
Exeter Cathedral: west front, polychromed sculpture	14th and 15th century	Yellow earth, orpiment and lead-tin yellow	PLM (E. Sinclair); SEM/EDX, A. Roy (NG)	Sinclair 1991, 1992; Fitzhugh 1997
St. Leonard's Church, Apethorpe, Northamptonshire: Monument to Sir Richard Dalton	*d*1442	Massicot?	PLM, MCT, T. Curteis	Curteis 1998
Winchester Cathedral: Great Screen	*c*1475–90	Yellow earth	PLM, J. Darrah (V&A)	Brodrick and Darrah 1989
Fitzalan Chapel, Arundel: Fitzalan effigies	15th century	Yellow earth and lead-tin yellow	PLM, XRF, J. Darrah (V&A)	Brodrick and Darrah 1986
Cartmel Fell, Cumbria: Chapel of St. Anthony, wooden figure of Christ	Late 15th century	Yellow earth	J. Plesters and A. Roy (NG)	Behrens 1982
Miscellaneous British examples				
St. Augustine's Abbey, Canterbury, Kent: excavated oyster-shell palette	13th century	Lead-tin yellow	XRF, XRD (EH/AML)	Sherlock and Woods 1988
Polychromed sculpture: northern European examples				
Grave of the King of Gorm at Jelling, Jutland, Denmark: fragments of carved wood	*c* 960	Orpiment	M. Christensen and J. Glastrup (NM)	Poster at ICOM '96
Hørning stave church, Denmark: Hørning plank	11th century	Orpiment	M. Christensen and J. Glastrup (NM)	Poster at ICOM '96
Vossestrand, Norway: lion capital	2nd half 12th century	Orpiment	PLM, SEM/EDX, U. Plahter	Plahter 1990
Notre-Dame Cathedral, Lausanne, Switzerland: south portal	*c*1216–20	Yellow earth		Furlano *et al.* 1981
Amiens Cathedral, France: *'Portail de la Mère Dieu'*	*c*1220–60	Yellow earth	LRMH	Weeks 1998
Dal Church, *Virgin and Child*, now in the University Museum of National Antiquities, Oslo, Norway	Mid-13th century	Orpiment	PLM, U. Plahter	Plahter and Wiik 1970
Diocesan Museum, Freising, Germany: *Virgin*	*c*1350	Lead-tin yellow	ES, H. Kühn	Kühn 1993b
Holy Cross Minster, Schwäbisch Gmünd, Germany	14th century	Yellow earth	PLM/MCT	Hauff 1988
Moerlin Epitaph, now in the Städtische Kunstsammlungen (No. 1340), Germany	*c*1497	Lead-tin yellow	PLM, ES, IR (DI)	Kühn 1993b
Berne Minster, Switzerland: excavated polychromed sculptures	Late Gothic	Lead-tin yellow		Gutscher *et al.* 1989

Table 5 Selected analytical data relating to the use of white pigments in medieval polychromy.

Site and Location	Date	Pigment(s)	Technique and Analyst(s)	Reference(s)
Examples in English medieval wall painting				
Lewes Priory, Sussex: excavated wall painting fragments	*c*1090	Lime white	PLM, SEM/EDX, H. Howard (CIA)	Howard and Park 1997
Worcester Cathedral: crypt, fragments of wall painting from box WCCS 86	Late 11th century	Lime white	PLM/MCT, H. Howard and I. Kakoulli (CIA)	Park 1997
Coombes, West Sussex: nave, dado zone, fictive marbling	*c*1100	Lime white	PLM, H. Howard (CIA)	Howard 1990
St. Botolph's, Hardham, West Sussex	*c*1100	Lime white	PLM, H. Howard and S. Hluvko (CIA)	Howard 1990; Sawdy *et al.* 1998; Hluvko 1991
All Saints' Witley, Surrey: nave south and west walls	*c*1110	Lead white and lime white	PLM, H. Howard (CIA)	Howard 1992a
Ickleton, Cambridgeshire: nave, north arcade	*c*1125	Lime white	PLM, MCT, H. Howard (CIA)	Howard 1988
St. Michael and All Angels, Copford, Essex: apse and chancel	*c*1125–30	Lime white	PLM, C. Babington (EH)	Organ *et al.* 1991
St. Mary's, Kempley, Gloucestershire: chancel and nave east wall	*c*1130	Lime white	PLM/MCT, S. Rickerby (CIA)	Rickerby 1990a
Canterbury Cathedral: St. Gabriel's Chapel, apse	*c*1130	Lead white and lime white	PLM, SEM/EDX, H. Howard (CIA)	Cather and Howard 1994
Binstead, West Sussex: chancel, north wall, *St Margaret of Antioch*	*c*1135	Lead white	PLM/MCT, P. Welford (CIA)	Welford 1991
Little Easton, Essex: nave, apostle	*c*1130–40	Lead white	PLM/MCT, P. Welford (CIA)	Welford 1991
Canterbury Cathedral: St. Anselm's Chapel, apse, *St. Paul and the Viper*	Mid-12th century	Lime white	PLM	Gärtner and Organ 1989
Winchester Cathedral: Holy Sepulchre Chapel	*c*1175	Chalk and lime white	PLM, SEM/EDX, H. Howard (CIA)	Howard 1995a
Sherborne Abbey, Dorset: chapter house: excavated painted wall plaster	*c*1180	Chalk and lime white	PLM, SEM/EDX, H. Howard (CIA)	Howard and Park forthcoming a
Guildford, Surrey: High Street, 'synagogue'	*c*1180	Lime white	PLM, SEM/EDX, FTIR, H. Howard (CIA)	Howard 1996c
Norwich Cathedral: nave south aisle, vault, 4th bay from east and soffit of adjacent transverse arch, '*Losinga*' cycle	*c*1190	Lime white	PLM, SEM/EDX, H. Howard (CIA)	Howard and Gasol 1996

Site and Location	Date	Pigment(s)	Technique and Analyst(s)	Reference(s)
Sherborne Abbey, Dorset: Novice's Room, excavated fragments	Late 12th or early 13th century	Lime white	PLM, SEM/EDX, H. Howard (CIA)	Howard and Park forthcoming a
Eastbridge Hospital, Canterbury, Kent: hall, north wall	*c*1220	Lead white and lime white	PLM, SEM/EDX, FTIR, H. Howard (CIA)	Howard 1994
Winchester Cathedral: Holy Sepulchre Chapel	*c*1220	Lead white and lime white	PLM, SEM/EDX, H. Howard (CIA)	Howard 1995a
Ely Cathedral: St. George's Chapel, entrance arch soffit	*c*1220–30	Lime white (which has partially transformed to form calcium sulphate)	PLM, SEM/EDX, H. Howard (CIA)	Howard and Pathak 1995
Norwich Cathedral: Ante-Reliquary Chapel, eastern arch	*c*1250	Lead white and lime white	PLM, SEM/EDX, H. Howard (CIA)	Howard 1995c; Park and Howard 1996
Horsham St. Faith Priory, Norfolk: refectory	*c*1250 with 15th-century additions	Lead white and lime white	PLM, C. Babington (CIA); PLM, J. Darrah (V&A)	Babington and James 1988
Winchester Cathedral: south transept, south-east corner	*c*1270	Lime white	PLM/MCT, T. Curteis	Curteis *et al.* 1994
Worcester Cathedral: crypt, eastern bishops	Early 13th century	Lime white	PLM, SEM/EDX, H. Howard and I. Kakoulli (CIA)	Park 1997
Worcester Cathedral: crypt, masonry pattern with rosettes	Early 13th century	Lime white (which has partially transformed to form calcium sulphate)	PLM/MCT, H. Howard and I. Kakoulli (CIA)	Park 1997
Worcester Cathedral: fragments of wall painting from box 'painted wall plaster from 6 stone blocks'	Early 13th century	Lime white (which has partially transformed to form calcium sulphate)	PLM/MCT, H. Howard and I. Kakoulli (CIA)	Park 1997
Winchester Cathedral: Guardian Angels' Chapel, vault	Early 13th century with later 13th-century additions	Lead white and gypsum which may be an alteration product	PLM, SEM/EDX, H. Howard (CIA)	
Rochester Cathedral: choir, *Wheel of Fortune*	Mid-13th century	Lead white and chalk	PLM, J. Darrah (V&A)	Darrah 1992
St. Albans Cathedral: nave, north arcade, western painted pier, *Crucifixion*	Mid-13th century	Lead white	PLM/MCT, P. Welford (CIA)	Welford 1991
St. Augustine's Abbey, Canterbury, Kent: excavated fragments of wall painting from the ambulatory south wall and St. Thomas's Chapel	13th century	Lime white		Sherlock and Woods 1988

Site and Location	Date	Pigment(s)	Technique and Analyst(s)	Reference(s)
Hospital of St. Giles, Brompton Bridge, North Yorkshire: excavated wall painting fragments	13th century	Lead white and lime white	PLM, S. Stewart (EH)	Stewart 1995
Duxford, Cambridgeshire: west and south walls of the chancel	13th century	Lime white	PLM (CIA)	CIA 1987
Westminster Abbey: nave, shields	13th century	Lead white	PLM/MCT, A. Gardner (CIA)	
St. Mary's, Wissington, Suffolk	13th century	Lime white and a lead-based pigment which has now altered but may originally have been lead white or red lead	PLM/MCT, T. Curteis	Curteis 1996a
York Minster: coffin lid of Archbishop Walter de Gray	13th century	Lime white	J. Plesters (NG)	Ramm *et al.* 1971
Norwich Cathedral: Ante-Reliquary Chapel, vault and western arch	*c*1300	Lead white, lime white and gypsum which may be an alteration product	PLM, SEM/EDX, H. Howard (CIA)	Howard 1995c; Park and Howard 1996
Durham Cathedral: Galilee Chapel, nave, south arcade, *Crucifixion and Martyrdom of St. Peter*	*c*1300	Lead white	PLM/MCT, P. Welford (CIA); SEM/EDX, A. Roy (NG)	Welford 1991
St. Albans Cathedral: feretory, south-east pier, *Unidentified archbishop*	*c*1302–08	Lead white	PLM, SEM/EDX, H. Howard (CIA)	Howard 1993a
St. Albans Cathedral: feretory, north-east pier, *St. William of York*	*c*1302–08	Lead white and lime white	PLM, SEM/EDX, H. Howard (CIA)	Howard 1993a
Little Wenham, Suffolk: chancel east wall	1st half 14th century	Lead white	PLM, FTIR, H. Howard (CIA)	Howard 1988
Merevale, Warwickshire: chancel arch	*c*1310–20	Lead white and perhaps bone white	PLM, SEM/EDX, H. Howard (CIA)	
Longthorpe Tower, Cambridgeshire: first floor solar	*c*1330	Lead white	PLM/MCT, P. Welford (CIA); SEM/EDX, A. Roy (NG)	Welford 1991
Hailes, Gloucestershire: choir, south and north walls	*c*1330	Lead white	PLM/MCT, P. Welford (CIA)	Welford 1991
Holy Trinity, Bradwell-juxta-Coggeshall, Essex	*c*1330–40	A lead-based pigment which has now altered and may have originally been lead white or red lead	PLM, H. Howard (CIA)	Cather and Park 1998
Brent Eleigh, Suffolk: chancel, east wall	*c*1335–40	Lime white	PLM/MCT, H. Howard (CIA)	Howard 1988
Rochester Cathedral: north-east transept, wall painting in tomb recess of William of Perth	*c*1340	Lead white	PLM, S. Bradley (BM)	Rickerby 1988
St. Stephen's Chapel, Westminster Palace: fragments of wall painting now in British Museum	*c*1352–63	Lead white and lime white	J. Haslam; J. Plesters and J. Kirby (NG)	Haslam 1807; van Geersdaele and Goldsworthy 1978

Site and Location	Date	Pigment(s)	Technique and Analyst(s)	Reference(s)
Canterbury Cathedral: Chapel of Our Lady Undercroft, vault of eastern bay and screens enclosing chapel	*c*1370s	Lead white, lime white and gypsum	PLM, H. Howard (CIA); SEM/EDX, A. Roy (NG)	Howard 1993b
Westminster Abbey: chapter house, *Majesty*	*c*1375–1404	Lead white, bone white	PLM/MCT, S. Hluvko (CIA); PLM, SEM/EDX, FTIR, H. Howard (CIA)	Howard 1988; Hluvko 1991, Howard 2000
Westminster Abbey: chapter house, *Apocalypse* cycle	*c*1375–1404	Lead white, bone white and chalk (ground)	PLM/MCT, S. Hluvko (CIA); PLM, SEM/EDX, FTIR, H. Howard (CIA)	Hluvko 1991, Howard 2000
Tower of London: the Byward Tower, south tower, first floor room, south wall	*c*1395	Lead white	PLM, H. Howard and S. Hluvko (CIA); PLM, C. Babington (EH).	Howard 1988; Hluvko 1991
Slapton, Northamptonshire: nave, south aisle	Early 14th century	A lead-based pigment which has now altered and may have originally been lead white or red lead	PLM, P. Welford (CIA)	Welford 1991
Ely Cathedral: Prior Crauden's Chapel, south tabernacle	14th century	Lead white	PLM, C. Babington (CIA)	Babington 1987
St. Nicholas' Hospital Church, Harbledown, Kent: chancel east wall and window embrasure	14th century	Lime white	PLM, SEM/EDX, H. Howard (CIA)	
Great Hockham, Norfolk: north aisle, north wall	14th century	Lime white	PLM, H. Howard (CIA)	Howard 1988
Worcester Cathedral: nave, north arcade, 4th pier from east, north-west side	14th century	Lead white	PLM/MCT, H. Howard and I. Kakoulli (CIA)	Park 1997
Ickleton, Cambridgeshire: chancel arch	Mid-14th century	Lead white	PLM, MCT, H. Howard (CIA)	Howard 1988
Little Easton, Essex: nave, south wall, *Passion* cycle	Late 14th century	Lead white	PLM/MCT, P. Welford (CIA)	Welford 1991
St. Nicholas', Little Horwood, Buckinghamshire	14th and 15th century	Lead white and lime white	PLM/MCT, T. Curteis	Curteis 1994b
Lichfield Cathedral: south choir aisle, south wall, tomb recess with wall painting	*c*1400	Lead white	PLM, J. Darrah (V&A)	
Holy Trinity, Coventry, West Midlands: chancel arch, *Last Judgement*	*c*1435–60	Lead white and lime white	PLM, SEM/EDX H. Howard (CIA)	Howard 1996b
Barnby, Suffolk: nave, south wall, *Crucifixion*	*c*1450	Lead white	PLM, P. Welford (CIA); SEM/EDX, A. Roy (NG)	Welford 1991
Durham Cathedral: Deanery, entrance hall, north wall	*c*1470	Lead white	PLM/MCT, H. Howard (CIA)	Howard 1988

Site and Location	Date	Pigment(s)	Technique and Analyst(s)	Reference(s)
Eton College Chapel, Berkshire: north and south walls: subjects drawn from sources such as the *Golden Legend* and the *Speculum Vitae*	*c*1479–87	Lead white	PLM/MCT, S. Hluvko (CIA)	Plummer 1986; Hluvko 1991
Canterbury Cathedral: north choir aisle, *Legend of St. Eustace*	*c*1480	Lead white and chalk	PLM/MCT, S. Hluvko (CIA)	Hluvko 1991
St. Leonard's Chapel, Farleigh Hungerford Castle, Somerset	Early 15th century	Lead white and calcium carbonate white	PLM, SEM/EDX, FTIR, H. Howard (CIA)	Howard *et al.* 1998
St. Margaret of Antioch, Cowlinge, Suffolk: *St Michael Weighing Souls*	15th century	Lime white and a lead-based pigment which has now altered and may have originally been lead white or red lead	PLM/MCT, H. Howard (CIA)	
Berry Pomeroy Castle, Devon: Gatehouse, east wall, *Adoration of the Magi*	15th century	Lead white and lime white	PLM	Gill *et al.* 1996
St. Martin's, Colchester, Essex: *Doom*	Mid-15th century	Lead white	PLM/MCT, T. Curteis	Curteis 1996b
All Saints' and St. Andrew, Kingston, Cambridgeshire	Late 15th century	Lead white and lime white	PLM/MCT, T. Curteis	Curteis 1995b
Winchester Cathedral: Lady Chapel, vaults	Late15th century	Lead white	PLM/MCT, T. Curteis	Curteis and Paine 1995
Lord Mayor's Chapel (St. Mark's), Bristol: south aisle arcade, *Trinity, Nativity and Noli me Tangere*	*c*1510	Lead white, lime white and gypsum	PLM, SEM/EDX, FTIR, H. Howard (CIA)	Gill and Howard 1997
Westminster Abbey: chapter house, *Bestiary scenes*	*c*1510	Lead white, chalk (ground)	PLM/MCT, S. Hluvko (CIA); PLM, SEM/EDX, H. Howard (CIA)	Hluvko 1991
St. Augustine's, Broxbourne, Hertfordshire: *Doom*	Early 16th century	Lead white may be present	PLM/MCT, T. Curteis	
York Minster: The Bedern, excavated wall plaster	Medieval	Lime white	PLM, L. Wong (CIA)	Wong forthcoming
Examples in northern European medieval wall painting				
Convent of St. John the Baptist, Müstair, Switzerland	*c* 800	Chalk and *bianco di San Giovanni*	F. Mairinger	Mairinger and Schreiner 1986
Idensen, Germany	*c*1130	Lead white and lime white	PLM/MCT, SEM/ EDS, FTIR, M. Matteini and A. Moles (OPD)	Geilmann 1938; Matteini and Moles 1994
Johanneskapelle, Pürgg, Austria	*c*1165	Lime white	PLM, SEM/EDX, FTIR, H. Howard (CIA); I. Hammer 1977	Howard 1995d

Site and Location	Date	Pigment(s)	Technique and Analyst(s)	Reference(s)
Cashel, Ireland: Cormac's Chapel	*c*1170	Lead white and lime white	EDS, M. McGrath; PLM, H. Howard (CIA)	McGrath 1987; Howard 1991
Glasgow Cathedral: wall painting fragments excavated from the crypt	*c*1190	Lime white	PLM, SEM/EDX, FTIR, H. Howard (CIA)	Howard 1998a; Park 1998
Convent of St. John the Baptist, Müstair, Switzerland	*c*1200	Lime white	F. Mairinger	Emmenegger 1986
Angers Cathedral, France: choir, *Scenes of the Life of S. Maurille*	*c*1270–80	Lead white. A thin lead white or lime ground was used to prepare the stone surface	LRMH	Subes-Picot 1992; Stirnemann 1995; Demailly *et al.* 1998
Cologne Cathedral, Germany: choir screen wall paintings	*c*1332–49	Lead white	E. Jägers.	Jägers and Schulze-Senger 1989; Heritage 1990; Maul 1992
Examples in English medieval panel painting				
Holy Innocent's, Adisham, Kent: nave, west wall, Adisham reredos	*c*1180	Lead white and gypsum	PLM, SEM/EDX, H. Howard (CIA)	Howard 1993c
Peterborough Cathedral: nave ceiling	*c*1220	Lead white, calcium carbonate and calcium sulphate which may represent an alteration of calcium carbonate	PLM, SEM/EDX, H. Howard (CIA)	Howard 1997b, 1998b
Painted Chamber, Westminster Palace: *ceiling paterae*	*c*1263–66	Calcium carbonate (employed as a ground) and lead white	J. Lynn (HKI)	Liversidge *et al.* 1995; Lynn 1997
Westminster Abbey: Westminster retable	*c*1270	Calcium sulphate combined with size was employed for the ground, which was sometimes applied over a lead white priming	PLM, SEM/EDX, A Roy (NG)	Binski and Freestone 1995
Thornham Parva, Suffolk: Thornham Parva Retable	*c*1335	Lead white (which is likely to have come from Derbyshire) and chalk (ground)	S. Bucklow (HKI); Lead-isotope analysis S. Stos-Gale (OU)	Goldsmith and Bucklow 1998
Kingston Lacy, Dorset: painted screen	Early 14th century	Lead white	PLM, L. Sheldon and C. Hassall (UCL)	Tracy 1997
Exeter Cathedral, Devon: pulpitum paintings	14th century	Lead white	PLM/MCT, E. Sinclair	Sinclair 1995b
Norwich Cathedral: *Retable of Five Scenes from the Passion*	Late 14th century	Chalk (ground)	PLM/MCT, J. Plesters (NG)	Plummer 1959; Gettens *et al.* 1993b
Wilton Diptych (English or French), now in the National Gallery, London	Late 14th century	Lead white and chalk (ground)	PLM/ SEM/EDX, A. Roy (NG)	Roy 1997
Chichester Cathedral: series of English School portraits on oak panels	*c*1530–35	Chalk (ground)	PLM/MCT, J. Plesters (NG)	Gettens *et al.* 1993b
St. Mary's, Worstead, Norfolk: screen	Early 15th century	Lead white and chalk	PLM/MCT, T. Curteis	Curteis 1993

Site and Location	Date	Pigment(s)	Technique and Analyst(s)	Reference(s)
Examples in northern European medieval panel painting				
Heddal altar frontal, Norway	*c*1250–60	Lead white	PLM, SEM/EDX, U. Plahter	Plahter 1995
Tingelstad I, II and III altarpieces, Norway	*c*1250–1350	Chalk (ground)	PLM, K. Perch-Nielsen	Plahter and Plahter 1972; Gettens *et al.* 1993b
German triptych in the Gallery of the Courtauld Institute of Art, London (known as the Estouteville triptych)	Late 14th century	Lead white, chalk (ground and constituent of mordant)	PLM, G. van Heemstra; SEM/EDX, A. Burnstock (CIA)	Villers *et al.* 1997
Polychromed sculpture: examples in England				
Lincoln Cathedral: west front, frieze	*c*1140	Gypsum which may be an alteration product		Larson 1992
St. Mary's Abbey, York: excavated stone sculpture	*c*1180–85	Lead white	A. Roy (NG)	Brodrick 1993
Wells Cathedral: west front	*c*1225–40	Lead white and gypsum, which may be an alteration product	MCT and spectrographic analysis, A. Roy (NG)	Roy and White 1982; Gettens *et al.* 1993a
Ely Cathedral: Lady Chapel	*c*1231–49	Lead white	PLM/MCT, S. Hluvko (CIA)	Hluvko 1991
Canterbury Cathedral: Great Cloister, *Martyrdom Doorway*	*c*1236–39	Lead white and chalk	PLM/MCT, S. Hluvko (CIA)	Hluvko 1991
Worcester Cathedral: crypt, west capital, below St. Thomas	Early 13th century	Lime white which has partially altered to form calcium sulphate	PLM, I. Kakoulli and H. Howard (CIA)	Park 1997
York Minster: chapter house, excavated stone pendant	Late 13th century	Lead white	PLM, J. Darrah (V&A)	Heywood and Oosterwijk 1993
St. Albans Cathedral: feretory, *Shrine of St. Alban*	*c*1302–08	Lead white	PLM, Babington (EH); PLM, SEM/EDX, Howard (CIA)	
Abergavenny, Wales: wooden knight	*c*1325	Gypsum (this may not be original since the effigy appears to have been repainted)	A. Brodrick	Blair 1994
Ingham, Norfolk: chancel, north wall, tomb of Sir Oliver de Ingham	*c*1344	Lead white and gypsum	PLM, SEM/EDX, H. Howard (CIA)	Howard 1993d
Ingham, Norfolk: nave south aisle, tomb of Sir Roger and Lady Margaret de Bois	*c*1380	Lead white, lime white, chalk and gypsum	PLM, SEM/EDX, H. Howard (CIA)	Howard 1993d
Salisbury Cathedral: west front	Mid-13th–mid-14th century	Lime white (which has partially transformed to form calcium sulphate and calcium oxalate)	PLM, SEM/EDX, H. Howard (CIA)	Howard 1995e, 1996d
Exeter Cathedral: roof bosses	Early 14th century	Lead white	A. Roy (NG); J. Darrah (V&A); U. Plahter	Hulbert 1991

Site and Location	Date	Pigment(s)	Technique and Analyst(s)	Reference(s)
Worcester Cathedral: Edgar Tower, polychromed foot of a sculpture	14th century	Lead white	PLM/MCT, I. Kakoulli and H. Howard (CIA)	Park 1997
St. Mary's, Erwarton, Suffolk: polychromed monument of Sir Bartholomew Bacon and Joan de Heveningham	14th century	Lead white	A. Roy (NG)	Roy 1984
Effigy of Geoffry de Lucy, now in the Victoria and Albert Museum	14th century	Calcium carbonate	PLM/MCT, S. Boulter (V&A)	Boulter 1993
Westminster Hall, London: polychromed kings	14th century	Lead white	PLM, C. Babington (EH)	Cherry and Stratford 1955
Exeter Cathedral: west front, polychromed sculpture	14th and 15th century	Lead white, lime white and chalk	PLM, E. Sinclair; SEM/EDX, A. Roy (NG)	Sinclair 1991, 1992
Winchester Cathedral: Great Screen	*c*1475–90	Lead white and chalk	PLM, J. Darrah (V&A)	Brodrick and Darrah 1989
Fitzalan Chapel, Arundel: Fitzalan effigies	15th century	Lead white and chalk	PLM, J. Darrah (V&A)	Brodrick and Darrah 1986
Cartmel Fell, Cumbria: Chapel of St. Anthony, wooden figure of Christ	Late 15th century	Lead white and chalk	J. Plesters and A. Roy (NG)	Behrens 1982
Miscellaneous British examples				
Westminster Abbey: Coronation Chair	*c*1297–1300	Lead white was identified as the ground of the original decoration. Chalk was identified as the ground for the early 14th-century redecoration	Prof. Rees Jones (CIA)	Percival-Prescott 1957; Rees Jones 1954; Gettens *et al.* 1993b
St. Augustine's Abbey, Kent: excavated oyster-shell palette	13th century	Lead white	XRF, XRD (EH/AML)	Sherlock and Woods 1988
Whitefriars Carmelite Friary, Coventry: excavated oyster-shell palette with paint	Mid-14th–mid-16th centuries	Lead white	PLM, SEM/EDX, H. Howard (CIA)	Howard 1997a
Polychromed sculpture: northern European examples				
Grave of the King of Gorm at Jelling, Jutland, Denmark: fragments of carved wood	*c* 960	Lead white	M. Christensen and J. Glastrup (NM)	Poster at ICOM '96
Chartres Cathedral, France: *Portail Royal*	*c*1140–50	Lead white	LRMH	Nonfarmale and Manaresi 1987
Crucifix of Hemse, Sweden	*c*1170–90	Lead white and chalk (from Denmark or Sweden)	PLM, SEM/EDX, U. Plahter	Plahter 1984
Vossestrand, Norway: lion capital	2nd half 12th century	Lead white	PLM, SEM/EDX, U. Plahter	Plahter 1990

Site and Location	Date	Pigment(s)	Technique and Analyst(s)	Reference(s)
Notre-Dame Cathedral, Lausanne, Switzerland: south portal	*c*1216–20	Lead white		Furlano *et al.* 1981
Amiens Cathedral, France: *'Portail de la Mère Dieu'*	*c*1220–60	Lead white, bone white and gypsum	LRMH	Weeks 1998
Maria Laach, Germany: Tomb of Pfalzgraf Heinrich II	13th century	Lead white and chalk (ground)		Oellermann 1993
Dal Church, *Virgin and Child*, now in the University Museum of National Antiquities, Oslo, Norway	Mid-13th century	Chalk (ground)	PLM, U. Plahter	Plahter and Wiik 1970; Gettens *et al.* 1993b
Holy Cross Minster, Schwäbisch Gmünd, Germany: polychromed portals	14th century	Lead white, chalk and calcium sulphate	PLM/MCT	Hauff 1988
Herlin altarpiece, Church of St. James, Rothenburg, Germany	*c*1466	Chalk (ground)	PLM, IR spectroscopy, M. Broekman-Bolstijn	Broekman-Bolstijn *et al.* 1970
St. Wolfgang, Pfarrkirche, Austria: altarpiece, sculpture and panels (Michael Pacher)	1471–81	Chalk (ground)	XRD, P. Hallebeek	Gettens *et al.* 1993b

Table 6 Selected analytical data relating to the use of black pigments in medieval polychromy.

Site and Location	Date	Pigment(s)	Technique and Analyst(s)	Reference(s)
Examples in English medieval wall painting				
Lewes Priory, Sussex: excavated wall painting fragments	*c*1090	Carbon black	PLM, SEM/EDX, H. Howard (CIA)	Howard and Park 1997
St. Botolph's, Hardham, West Sussex	*c*1100	Charcoal black	PLM, H. Howard (CIA)	Howard 1990
Coombes, West Sussex: nave, dado zone, fictive marbling	*c*1100	Charcoal black	PLM, H. Howard (CIA)	Howard 1990
All Saints' Witley, Surrey: nave south and west walls	*c*1110	Charcoal black	PLM, H. Howard (CIA)	Howard 1992a
St. Michael and All Angels, Copford, Essex: apse and chancel	*c*1125–30	Carbon black	PLM, C. Babington (EH)	Organ *et al.* 1991
St. Mary's, Kempley. Gloucestershire: chancel and nave east wall	*c*1130	Carbon black	PLM/MCT, S. Rickerby (CIA)	Rickerby 1990a
Canterbury Cathedral: St. Gabriel's Chapel, apse	*c*1130	Carbon black	PLM, SEM/EDX, H. Howard (CIA)	Cather and Howard 1994
Canterbury Cathedral: St. Anselm's Chapel, apse, *St. Paul and the Viper*	Mid-12th century	Brown iron oxide and charcoal black	PLM	Gärtner and Organ 1989
Winchester Cathedral: Holy Sepulchre Chapel	*c*1175	Brown iron oxide and charcoal black	PLM, SEM/EDX, H. Howard (CIA)	Howard 1995a
Sherborne Abbey, Dorset: chapter house, excavated painted wall plaster	*c*1180	Carbon black	PLM, SEM/EDX, H. Howard (CIA)	Howard and Park forthcoming a
Guildford, Surrey: High Street, 'synagogue'	*c*1180	Charcoal black	PLM, SEM/EDX, H. Howard (CIA)	Howard 1996c
Glastonbury Abbey, Somerset: Lady Chapel	*c*1187	Carbon black	PLM, E. Sinclair	Sampson 1996; Sinclair pers. comm.
Norwich Cathedral: nave south aisle, vault, 4th bay from east and soffit of adjacent transverse arch, '*Losinga*' cycle	*c*1190	Charcoal black	PLM, SEM/EDX, H. Howard (CIA)	Howard and Gasol 1996
Keynsham Abbey, Somerset: south chapel, excavated fragments with masonry pattern and fleur-de-lys	12th century	Carbon black	J. Plesters (NG)	Lowe and Mason 1987
Eastbridge Hospital, Canterbury, Kent: hall, north wall	*c*1220	Brown iron oxide and charcoal black	PLM, SEM/EDX, H. Howard (CIA)	Howard 1994
Winchester Cathedral: Holy Sepulchre Chapel	*c*1220	Charcoal black and carbon black	PLM, SEM/EDX, H. Howard (CIA)	Howard 1995a

Site and Location	Date	Pigment(s)	Technique and Analyst(s)	Reference(s)
Ely Cathedral: St. George's Chapel, entrance arch soffit	*c*1220–30	Carbon black	PLM, SEM/EDX, H. Howard (CIA)	Howard and Pathak 1995
Norwich Cathedral: Ante-Reliquary Chapel, eastern arch	*c*1250	Carbon black	PLM, SEM/EDX, H. Howard (CIA)	Howard 1995c; Park and Howard 1996
Horsham St. Faith Priory, Norfolk: refectory	*c*1250 with 15th-century additions	Carbon black	PLM, C. Babington (CIA); PLM J. Darrah (V&A)	Babington and James 1988
Winchester Cathedral: south transept, south-east corner	*c*1270	Carbon black	PLM/MCT, T. Curteis	Curteis *et al.* 1994
Worcester Cathedral: crypt, eastern bishops	Early 13th century	Carbon black	PLM, SEM/EDX, H. Howard and I. Kakoulli (CIA)	Park 1997
Worcester Cathedral: excavated wall painting fragments from box 'painted wall plaster from 6 stone blocks'	Early 13th century	Charcoal black and carbon black	PLM/MCT, H. Howard and I. Kakoulli (CIA)	Park 1997
Winchester Cathedral: Guardian Angels' Chapel, vault	Early 13th century with later 13th-century additions	Charcoal black	PLM, SEM/EDX, H. Howard (CIA)	
Rochester Cathedral: choir, *Wheel of Fortune*	Mid-13th century	Brown iron oxide and bone black	PLM, J. Darrah (V&A)	Darrah 1992
Clarendon Palace, Wiltshire: excavated wall painting fragments	13th century	Carbon black	I. Bristow	Ashurst and James 1988
St. Augustine's Abbey, Canterbury, Kent: excavated fragments of wall painting from the ambulatory south wall and St. Thomas's Chapel	13th century	Carbon black		Sherlock and Woods 1988
Hospital of St. Giles, Brompton Bridge, North Yorkshire: excavated wall painting fragments	13th century	Carbon black	PLM, S. Stewart (EH)	Stewart 1995
Duxford, Cambridgeshire: west and south walls of the chancel	13th century	Carbon black	PLM (CIA)	CIA 1987
Westminster Abbey: nave, shields	13th century	Charcoal black	PLM/MCT, A. Gardner (CIA)	
York Minster: coffin lid of Archbishop Walter de Gray	13th century	Charcoal black	J. Plesters (NG)	Ramm *et al.* 1971
Norwich Cathedral: Ante-Reliquary Chapel, vault and western arch	*c*1300	Charcoal black	PLM, SEM/EDX, H. Howard (CIA)	Howard 1995c; Park and Howard 1996
St. Albans Cathedral: feretory, south-east pier, *Unidentified archbishop*	*c*1302–08	Charcoal black	PLM, SEM/EDX, H. Howard (CIA)	Howard 1993a

Site and Location	Date	Pigment(s)	Technique and Analyst(s)	Reference(s)
St. Albans Cathedral: feretory, north-east pier, *St. William of York*	*c*1302–08	Charcoal black	PLM, SEM/EDX, H. Howard (CIA)	Howard 1993a
Longthorpe Tower, Cambridgeshire: first floor solar	*c*1330	Carbon black	PLM/MCT, P. Welford (CIA)	Welford 1991
Rochester Cathedral: north-east transept, wall painting in tomb recess of William of Perth	*c*1340	Carbon black	PLM, S. Bradley (BM)	Rickerby 1988
Canterbury Cathedral: Chapel of Our Lady Undercroft, vault of eastern bay and screens enclosing chapel	*c*1370s	Charcoal black	PLM, H. Howard (CIA)	Howard 1993b
Westminster Abbey: chapter house, *Apocalypse* cycle: lateral group	*c*1375–1404	Bone black	PLM/MCT, H. Howard (CIA)	Howard 1988
Tower of London: the Byward Tower, south tower, first floor room, south wall	*c*1395	Charcoal black	PLM, H. Howard PLM C. Babington (EH)	Howard 1988
Ely Cathedral: Prior Crauden's Chapel, south tabernacle	14th century	Carbon black	PLM, C. Babington (CIA)	Babington 1987
St. Nicholas' Hospital Church, Harbledown, Kent: chancel east wall and window embrasure	14th century	Charcoal black	PLM, SEM/EDX, H. Howard (CIA)	
Great Hockham, Norfolk: north aisle, north wall	14th century	Charcoal black	PLM, H. Howard (CIA)	Howard 1988
Worcester Cathedral: nave, north arcade, 4th pier from east, north-west side	14th century	Charcoal black	PLM/MCT, H. Howard and I. Kakoulli (CIA)	Park 1997
St. Nicholas', Little Horwood, Buckinghamshire	14th and 15th century	Carbon black	PLM/MCT, T. Curteis	Curteis 1994b
Lichfield Cathedral: south choir aisle, south wall, tomb recess with wall painting.	*c*1400	Manganese black	PLM, J. Darrah (V&A)	
Bradfield Combust, Suffolk	*c*1400	Carbon black	PLM	Rickerby 1990b
Holy Trinity, Coventry, West Midlands: chancel arch, *Last Judgement*	*c*1435–60	Carbon black	PLM, SEM/EDX, H. Howard (CIA)	Howard 1996b
Durham Cathedral: Deanery, entrance hall, north wall	*c*1470	Bone black	PLM/MCT, H. Howard (CIA)	Howard 1988
Eton College Chapel, Berkshire: north and south walls, subjects drawn from sources such as the *Golden Legend* and the *Speculum Vitae*	*c*1479–87	Carbon black		Plummer 1986
St. Leonard's Chapel, Farleigh Hungerford Castle, Somerset	Early 15th century	Charcoal black	PLM, SEM/EDX, FTIR, H. Howard (CIA)	Howard *et al.* 1998
St. Margaret of Antioch, Cowlinge, Suffolk: *St Michael Weighing Souls*	15th century	Charcoal black	PLM/MCT, H. Howard (CIA)	

Site and Location	Date	Pigment(s)	Technique and Analyst(s)	Reference(s)
Berry Pomeroy Castle, Devon: Gatehouse, east wall, *Adoration of the Magi*	15th century	Carbon black	PLM	Gill *et al.* 1996
All Saints' and St. Andrew, Kingston, Cambridgeshire	Late 15th century	Carbon black	PLM/MCT, T. Curteis	Curteis 1995b
Winchester Cathedral: Lady Chapel, vaults	Late 15th century	Carbon black	PLM/MCT, T. Curteis	Curteis and Paine 1995
Lord Mayor's Chapel (St. Mark's), Bristol: south aisle arcade, *Trinity, Nativity* and *Noli me Tangere*	*c*1510	Carbon black	PLM, SEM/EDX, H. Howard (CIA)	Gill and Howard 1997
York Minster: The Bedern, excavated wall plaster	Medieval	Carbon black	PLM, L. Wong (CIA)	Wong forthcoming
Examples in northern European medieval wall painting				
Convent of St. John the Baptist, Müstair, Switzerland	*c*800	Charcoal black	PLM, F. Mairinger	Mairinger and Schreiner 1986
Idensen, Germany	*c*1130	Carbon black	PLM/MCT, SEM/EDS, M. Matteini and A. Moles (OPD)	Matteini and Moles 1994
Johanneskapelle, Pürgg, Austria	*c*1165	Brown iron oxide and charcoal black	PLM, SEM/EDX, H. Howard (CIA); I. Hammer 1977	Howard 1995d
Cashel, Ireland: Cormac's Chapel	*c*1170	Charcoal black	PLM, H. Howard (CIA)	Howard 1991
Glasgow Cathedral: wall painting fragments excavated from the crypt	*c*1190	Charcoal black	PLM, SEM/EDX, H. Howard (CIA)	Howard 1998a; Park 1998
Convent of St. John the Baptist, Müstair, Switzerland	*c*1200	Charcoal black	F. Mairinger	Emmenegger 1986
Angers Cathedral, France: choir, *Scenes of the Life of S. Maurille*	*c*1270–80	Carbon black and 'animal black'	XRF, SEM/EDX	Subes-Picot 1992; Demailly *et al.* 1998
Cologne Cathedral, Germany: choir screen wall paintings	*c*1332–49	Carbon black (vine black?)	E. Jägers	Jägers and Schulze-Senger 1989; Heritage 1990; Maul 1992
Examples in English medieval panel painting				
Holy Innocent's, Adisham, Kent: nave, west wall, Adisham reredos	*c*1180	Charcoal black	PLM, SEM/EDX, H. Howard (CIA)	Howard 1993c
Peterborough Cathedral: nave ceiling	*c*1220	Carbon black	PLM, SEM/EDX, H. Howard (CIA)	Howard 1997b

Site and Location	Date	Pigment(s)	Technique and Analyst(s)	Reference(s)
Painted Chamber, Westminster Palace: *ceiling paterae*	*c*1263–66	Carbon black	J. Lynn (HKI)	Liversidge *et al.* 1995; Lynn 1997
Thornham Parva, Suffolk: Thornham Parva Retable	*c*1335	Carbon black	PLM, S. Bucklow (HKI)	Goldsmith and Bucklow 1998
Kingston Lacy, Dorset: painted screen	Early 14th century	Carbon black	PLM, L. Sheldon and C. Hassall (UCL)	Tracy 1997
Wilton Diptych (English or French), now in the National Gallery, London	Late 14th century	Carbon black and charcoal black	PLM/ SEM/EDX, A. Roy (NG)	Roy 1997
St. Mary's, Worstead, Norfolk: screen	Early16th century	Carbon black	PLM/MCT, T. Curteis	Curteis 1993
Examples in northern European medieval panel painting				
Heddal altar frontal, Norway	*c*1250–60	Charcoal black	U. Plahter	Plahter 1995
German triptych in the Gallery of the Courtauld Institute of Art, London (known as the Estouteville triptych)	Late 14th century	Bone black	PLM, G. van Heemstra; SEM/EDX, A. Burnstock (CIA)	Villers *et al.* 1997
Polychromed sculpture: examples in England				
Lincoln Cathedral: west front, frieze	*c*1140	Brown iron oxide	A. Roy (NG)	Larson 1992, 1993
St. Mary's Abbey, York: excavated stone sculpture	*c*1180–85	Charcoal black	A. Roy (NG)	Brodrick 1993
Wells Cathedral: west front	*c*1225–40	Charcoal black	A. Roy and R. White (NG)	Roy and White 1982
Canterbury Cathedral: Great Cloister, *Martyrdom Doorway*	*c*1236–39	Brown iron oxides	PLM/MCT, S. Hluvko (CIA)	Hluvko 1991
Worcester Cathedral: crypt, west capital, below St. Thomas	Early 13th century	Charcoal black	PLM, I. Kakoulli and H. Howard (CIA)	Park 1997
St. Albans Cathedral: feretory, Shrine of St. Alban	*c*1302–08	Charcoal black	PLM, Babington (EH); PLM, SEM/EDX, Howard (CIA)	
St. Mary's, Erwarton, Suffolk: polychromed monument of Sir Bartholomew Bacon and Joan de Heveningham	14th century	Bone black and charcoal black	A. Roy (NG)	Roy 1984
Exeter Cathedral: west front, polychromed sculpture	14th and 15th century	Carbon black	PLM, E. Sinclair; SEM/EDX, A. Roy (NG)	Sinclair 1991, 1992
Winchester Cathedral: Great Screen	*c*1475–90	Black ochre and brown iron oxide	PLM, J. Darrah (V&A)	Brodrick and Darrah 1989
Fitzalan Chapel, Arundel: Fitzalan effigies	15th century	Carbon black	PLM, J. Darrah (V&A)	Brodrick and Darrah 1986

Site and Location	Date	Pigment(s)	Technique and Analyst(s)	Reference(s)
Cartmel Fell, Cumbria: Chapel of St. Anthony, wooden figure of Christ	Late 15th century	Charcoal black	J. Plesters and A. Roy (NG)	Behrens 1982
Worcester Cathedral: wooden pilgrim's staff		Bone black	A. Burnstock and R. White (NG)	Lubin *et al.* 1990
Polychromed sculpture: northern European examples				
Grave of the King of Gorm at Jelling, Jutland, Denmark: fragments of carved wood	*c.*960	Charcoal black	M. Christensen and J. Glastrup (NM)	Poster at ICOM '96
Vossestrand, Norway: lion capital	2nd half 12th century	Charcoal black	PLM, SEM/EDX, U. Plahter	Plahter 1990
Amiens Cathedral, France: '*Portail de la Mère Dieu*'	*c*1220–60	Carbon black	LRMH	Weeks 1998
Maria Laach, Germany: Tomb of Pfalzgraf Heinrich II	13th century	Bone black	E. Jägers	Oellermann 1993
Holy Cross Minster, Schwäbisch Gmünd, Germany: polychromed portals	14th century	Charcoal black	PLM/MCT	Hauff 1988

Table 7 Selected analytical data relating to the use of oil in medieval polychromy.

Site and Location	Date	Oil type (if known) and application	Technique and Analyst(s)	Reference(s)
Examples in English medieval wall painting				
Canterbury Cathedral: St Gabriel's Chapel, apse	*c*1130	Both linseed oil and a proteinaceous component were detected in a sample from the ultramarine background	GC-MS, R. White (NG)	Cather and Howard 1994
Winchester Cathedral: Holy Sepulchre Chapel	*c*1175	Linseed oil – which was partially prepolymerised (heat-bodied) to promote drying – was employed as a mordant for the gold leaf of the censing angel's halo in the *Entombment*	FTIR, J. Pilc (NG); GC-MS, R. White (NG)	Howard 1995a; White 1998
Sherborne Abbey, Dorset: chapter house, excavated painted wall plaster	*c*1180	An oil-based preparation for gilding, applied over a layer of chalk and haematite has been identified in wall painting fragments of type II	FTIR, H. Howard (CIA)	Howard and Park forthcoming
Eastbridge Hospital, Canterbury, Kent: hall, north wall	*c*1220	Analysis suggests the presence of verdigris in an oil medium in the border of the mandorla and in the green background beneath the inscription	FTIR, H. Howard (CIA)	Howard 1994
Winchester Cathedral: Guardian Angels' Chapel, vault	Early 13th century with later 13th-century additions	A little drying oil was detected in a sample taken from the lowest rosette on the south side of the west segment of the vault	GC-MS, R. White (NG)	White 1998
Horsham St. Faith Priory, Norfolk: refectory	*c*1250	Linseed oil was identified both in rich green and reddish brown paint	GC-MS, R. White (NG)	White 1989, 1995
Norwich Cathedral: Ante-Reliquary Chapel, eastern arch	*c*1250	Both oil and protein were identified in the ground of the border above the bishop on the south-east pier. The oil was identified as partially prepolymerised linseed oil	FTIR, H. Howard (CIA); GC-MS, R. White (NG)	Park and Howard 1996; Howard 1995c; White 1998
Norwich Cathedral: Ante-Reliquary Chapel, vault and western arch	*c*1300	Analysis of a sample from an area of green drapery on the western arch indicates that oil was used as the original binding medium	FTIR, H. Howard (CIA); GC-MS, R. White (NG)	Park and Howard 1996; Howard 1995c; White 1998
St. Albans Cathedral: feretory, north-east pier, *St. William of York*	*c*1302–08	Linseed oil – which was prepolymerised (heat-bodied) to promote drying – with the addition of a very small amount of a conifer resin has been identified in *St. William of York*	GC-MS, R. White (NG)	Howard 1993a
Rochester Cathedral: north-east transept, wall painting in tomb recess of St. William of Perth	*c*1340	Walnut oil and a resinous material have been identified in the paint layers	GLC, K. Groen (HKI)	Rickerby 1988

Site and Location	Date	Oil type (if known) and application	Technique and Analyst(s)	Reference(s)
St. Stephen's Chapel, Westminster Palace: fragments of wall painting now in the British Museum	*c*1352–63	A drying oil – perhaps linseed – was identified in the two uppermost paint layers	J. Plesters and J. Mills (NG)	van Geersdaele and Goldworthy 1978
Canterbury Cathedral: Chapel of Our Lady Undercroft, vault of eastern bay and screens enclosing the chapel	1370s	A rich translucent effect was achieved on the raised decorative elements of the canopy by glazing silver leaf with verdigris in an oil medium	FTIR, H. Howard (CIA)	Howard 1993b
Westminster Abbey: chapter house, *Apocalypse* cycle	*c*1375–1404	FTIR analysis confirmed the presence of oil combined with lead white and with verdigris. In addition, an oil-based mordant was employed beneath gold leaf. GC-MS analysis of the red lead in a sample taken from the tiny angels above the Old Testament figures in the *Apocalypse* wall painting cycle confirmed the presence of linseed oil and perhaps a trace of pine resin	FTIR, H. Howard (CIA) GC-MS (NG)	Howard 2000
Tower of London, the Byward Tower, south tower, first floor room, south wall	*c*1395	The original binding medium of the scheme is generally linseed oil	GC-MS, R. White (NG)	
Holy Trinity, Coventry, West Midlands: chancel arch, *Last Judgement*	*c*1435–60	Analysis confirmed the presence of oil combined with lead white, chalk and verdigris	FTIR, H. Howard (CIA)	Howard 1996b
Horsham St. Faith Priory, Norfolk: refectory	15th-century additions	Prepolymerised linseed oil has been identified in both green and reddish brown paint layers from the 15th-century repainting	GC-MS, R. White (NG)	White 1989, 1995
Eton College Chapel, Berkshire: north and south walls, subjects drawn from sources such as the *Golden Legend* and the *Speculum Vitae*	*c*1479–87	The stone support has been primed with oil and the paint medium itself has been identified as linseed oil	GC-MS, J. Mills (NG)	Plummer 1986
St. Leonard's Chapel, Farleigh Hungerford Castle, Somerset	Early 15th century	A drying oil was identified combined with both verdigris and lead white	FTIR, H. Howard (CIA)	Howard *et al.* 1998
The Lord Mayor's Chapel (St. Mark's), Bristol: south aisle arcade, *Trinity, Nativity* and *Noli me Tangere*	*c*1510	Oil was used both as a sealant (combined with lead white added as a drier) and as the binding medium for the paint	FTIR, H. Howard (CIA)	Gill and Howard 1997
Examples in northern European medieval wall painting				
Convent of St. John the Baptist, Müstair, Switzerland	*c*800	Traces of oil and protein were identified in the binding media of lime-sensitive pigments such as red and yellow lead		Mairinger and Schreiner 1986
Idensen, Germany	*c*1130	Lead white combined with siccative oil – likely to be linseed oil – was identified in a sample taken from Christ's mantle and in the lead white-based preparation for gilding	M. Matteini and A. Moles (OPD)	Matteini and Moles 1994

Site and Location	Date	Oil type (if known) and application	Technique and Analyst(s)	Reference(s)
Angers Cathedral, France: choir, *Scenes of the Life of S. Maurille*	*c*1270–80	The presence of linseed oil was confirmed in the red paint from bay 3, and linseed oil with a small amount of pine resin was identified in a green copper resinate-containing layer. The blue paint was first surrounded by a layer of glue before being bound in linseed oil	GC-MS and FTIR, R. White (NG)	Subes-Picot 1992; White 1995; Demailly *et al.* 1998
Cologne Cathedral, Germany: choir-screen wall paintings	*c*1332–49	The binding medium is egg tempera with an addition of oil	E. Jägers	Jägers and Schulze-Senger 1989; Heritage 1990; Maul 1992
Examples in English medieval panel painting				
Peterborough Cathedral: nave ceiling	*c*1220	Oil was identified combined with lead white	FTIR, H. Howard (CIA)	Howard 1997b, 1998b
Painted Chamber, Westminster Palace: *ceiling paterae*	*c*1263–66	Linseed oil has been confirmed as the principal binding medium, though walnut oil was employed for areas of lead white paint	GC-MS, R. White (NG)	Liversidge *et al.* 1995; Lynn 1997
Westminster Abbey: Westminster Retable	*c*1270	Samples from the Virgin's robe and green paints, though leanly bound, were found to contain linseed oil	GC-MS, R. White (NG)	White 1995
Westminster Abbey: Sedilia panels	*c*1310	The red, white and mustard coloured ground of the Sedilia panels were all found to contain linseed oil	GC-MS, R. White (NG)	White 1995
Thornham Parva Retable	*c*1330–40	Linseed oil was identified as the predominant binding medium. In some instances the oil was partially prepolymerised (heat-bodied) while for other paint layers it was strongly heat-bodied. The chalk ground is bound with animal glue.		Goldsmith and Bucklow 1998; Bucklow 2003 forthcoming
St. Cuthbert's, Wells: *Tree of Jesse* reredos and north reredos panel	*c*1470	Both have underpaint in egg tempera with the main paint layers bound in linseed oil	GC-MS, R. White (NG)	R. White lecture at CIA 11.2.92
Examples in northern European medieval panel painting				
Heddal altar frontal, Norway	*c*1250–60	Lead white was combined with unprepolymerised linseed oil, and a pine resin was employed for the green glazes. For red lake glazes over silver leaf, prepolymerised linseed oil mixed with pine resin has been used. Ultramarine has been coated with a proteinaceous material before being bound in oil (pers. comm. U. Plahter)	GC-MS, R. White (NG)	White 1995

Site and Location	Date	Oil type (if known) and application	Technique and Analyst(s)	Reference(s)
Tingelstad I Altarpiece, Norway	*c*1260–1300	Rich translucent copper resinate greens, consisting of verdigris complexed with pine resin acids and mixed with a drying oil, have been identified. Lead white is combined with linseed oil. For red lake glazes over silver leaf, prepolymerised linseed oil mixed with pine resin has been used. Azurite is bound in a mixture of egg tempera and linseed oil	GC-MS, R. White (NG)	Plahter and Plahter 1972; White 1995
Tingelstad II Altarpiece, Norway	*c*1250–1350	Azurite is bound in a mixture of egg tempera and linseed oil (there is more oil present than in Tingelstad I)	GC-MS, R. White (NG)	Plahter and Plahter 1972; White 1995
St. Paul, Gausdal, Norway	*c*1260–70	Walnut oil (perhaps combined with linseed oil)	GC-MS, R. White (NG)	White 1995
Trondheim altarpiece, Norway	*c*1310–20	Lead white is bound in linseed oil and verdigris appears to have been combined with oil	GC-MS, R. White (NG)	White 1995
German triptych in the Gallery of the Courtauld Institute of Art, London (known as the Estouteville triptych)	Late 14th century	Heat-bodied walnut oil with the addition of pine resin was employed for the white of an angel's robe and for the green of the angel's wing; walnut oil with pine resin was also used for the black outlines, but here the mixture was subjected to a greater degree of prepolymerisation. Partially prepolymerised linseed oil combined with pine resin was employed in a green sample taken from the architecture. The chalk ground was sealed with an unpigmented oil film	GC-MS and FTIR, R. White (NG)	Villers *et al.* 1997
Jan van Eyck, *Giovanni Arnolfini and his wife* (Netherlandish)	1434	Heat-bodied linseed oil with pine resin was employed for an area of green drapery. For a red lake paint layer, linseed oil was used	GC-MS, R. White (NG)	Campbell *et al.* 1997
Stephen Lochner, *Saints Matthew, Catherine of Alexandria and John the Evangelist* (German)	*c*1445	Strongly heat-bodied linseed oil with pine resin was used for the green of St. Matthew's robe. Heat-bodied linseed oil with some pine resin was used for green foliage and zones of red lake pigment, while for areas of white drapery, the white background and pale blue-grey drapery, linseed oil alone was employed	GC-MS, R. White (NG)	Campbell *et al.* 1997
Simon Marmion, *The Soul of Saint Bertin carried up to God* (Netherlandish)	1459	Linseed oil was employed for areas of the greenish-blue background	GC-MS, R. White (NG)	Campbell *et al.* 1997

Site and Location	Date	Oil type (if known) and application	Technique and Analyst(s)	Reference(s)
Joos van Wassenhove (attrib.) *Music* (Netherlandish)	*c*1480	Linseed oil was employed for an opaque green paint layer, while walnut oil was used for a white highlight	GC-MS, R. White (NG)	Campbell *et al.* 1997
Gerard David, *Christ Nailed to the Cross* (Netherlandish)	*c*1480s	Walnut oil was used for the pale flesh colour of Christ and for the blue sky	GC-MS, R. White (NG)	Campbell *et al.* 1997
Workshop of the Master of the Life of the Virgin, *The Mass of Saint Hubert* (German)	1485–90?	Heat-bodied walnut oil was employed for areas of thick white paint	GC-MS, R. White (NG)	Campbell *et al.* 1997
Polychromed sculpture: examples in England				
Lincoln Cathedral: west front, frieze	*c*1140	Traces of linseed oil have been identified in the original paint	GC-MS, R. White (NG)	Larson 1992, 1993
Canterbury Cathedral: cloister, Martyrdom Doorway	*c*1236–39	Linseed oil with fatty contaminants was identified as the medium	GC-MS, R. White (NG)	Hluvko 1991
Westminster Abbey: Tomb of Edmund Crouchback	*d*1296	The binding medium has been identified as linseed oil	GC-MS, J. Mills and R. White (NG)	R. White lecture at CIA 11.2.92
St. Albans Cathedral: feretory, Shrine of St. Alban	*c*1302–08	The presence of linseed oil was also confirmed in samples of both green and red paint	GC-MS, R. White (NG)	Howard 1993a
Ingham, Norfolk: chancel, north wall, tomb of Sir Oliver de Ingham	*c*1344	Oil was combined with red lead and lead white in the red layer beneath the *pastiglia*. Further analysis confirmed the presence of prepolymerised linseed oil combined with a little pine resin in a sample taken from the raised decoration of the camail	FTIR, J. Pilc (NG); GC-MS, R. White (NG)	Howard 1993d; White 1998
Ingham, Norfolk: nave south aisle, tomb of Sir Roger and Lady Margaret de Bois	*c*1380	The ground and the majority of the paint layers have a predominantly oily binder. The oil in a sample taken from the decoration on the cushion under the head of Lady Margaret was identified as linseed oil with no evidence of prepolymerisation, while heat-bodied linseed oil was identified in a sample taken from the impasto decoration on Sir Roger's helm	FTIR, J. Pilc (NG); GC-MS, R. White (NG)	Howard 1993d; White 1998
St. Mary's, Erwarton, Suffolk: polychromed monument to Sir Bartholomew Bacon and Joan de Heveningham	14th century	Binding medium of the paint is a drying oil which may have had a little resin added to it	J. Plesters and A. Roy (NG)	Roy 1984
Exeter Cathedral: west front	14th and 15th century	The presence of both linseed oil and egg tempera has been confirmed, either in separate paint layers or, more usually, combined to form an emulsion	GC-MS, R. White (NG)	Sinclair 1991, 1992

Site and Location	Date	Oil type (if known) and application	Technique and Analyst(s)	Reference(s)
Winchester Cathedral: Great Screen, sculptural polychromy	*c*1475–90	The binding medium has been identified as egg was combined with a little oil	GC-MS, R. White (NG)	Brodrick and Darrah 1989
Fitzalan Chapel, Arundel: Fitzalan effigies	15th century	Analysis indicated that the medium of all paint layers was linseed oil with a little protein, likely to be egg	GC-MS, R. White (NG)	Brodrick and Darrah 1986
Cartmel Fell, Cumbria: Chapel of St. Anthony, wooden figure of Christ	Late 15th century	Oil is employed as the binding medium of the paint layers, and also – possibly with the addition of a resinous component – as the mordant for the gold leaf	J. Plesters and A. Roy (NG)	Behrens 1982
Miscellaneous British examples				
Westminster Abbey: Coronation Chair	*c*1297–1300	Lead white combined with a drying oil was identified as the ground of the original decoration. In addition, a drying oil was also a component of the adhesive putty for the glass appliqué of the second decorative phase	Prof. Rees Jones (CIA)	Percival-Prescott 1957; Rees Jones 1954
Polychromed sculpture: northern European examples				
Grave of the King of Gorm at Jelling, Jutland, Denmark: fragments of carved wood	*c*960	The medium of the paint was identified as a drying oil	M. Christensen and J. Glastrup (NM)	Poster at ICOM '96
Hørning stave church, Denmark: Hørning plank	11th century	Linseed oil was identified as the binding medium for the paint layers	M. Christensen and J. Glastrup (NM)	Poster at ICOM '96
Crucifix of Hemse, Sweden	*c*1170–90	A ground of lead white in a drying oil was applied to the head, crown, nape of neck and legs in preparation for the paint layers	J. Mills (NG)	Plahter 1984
Crucifix from Forest, Brussels	Late 12th century	The majority of the polychromy is bound in oil, but the gilding on the loin cloth is aqueous. Various binding media can also be found on the crucifixes at Väte, Alskog and Endre		Plahter 1984
Crucifix from Cappenberg, Westphalia	*c* 1220	The preparation layer consists of lead white and oil	U. Plahter	Plahter 1984

Site and Location	Date	Oil type (if known) and application	Technique and Analyst(s)	Reference(s)
Hedal Madonna, Norway	*c*1250–60	Linseed oil with small amounts of resin and traces of egg tempera were employed for the oil gilding of Christ's hair. The white lining of the mantle is bound in linseed oil over a chalk ground containing fish or animal glue. A glue content was also identified in the blue pigment which had a medium principally consisting of oil and pine resin. Heat-bodied linseed oil was also employed for black paint layers on the throne, and nut oil has been identified in opaque red paint. For red lake: egg tempera; egg tempera combined with linseed oil; and linseed oil combined with pine resin have been identified	GC-MS, R. White (NG)	White 1995
Gausdal, *St. Paul*, Norway	1260–70s	The white robe of St. Paul was found to contain nut oil. A pine resin was also identified. Red lead is bound in non-heat bodied linseed oil	GC-MS, R. White (NG)	White 1995
Dal Church, *Virgin and Child*, now in the University Museum of National Antiquities, Oslo, Norway	Mid-13th century	The polychromy includes animal glue, chalk, drying oil and protein, and analysis has indicated that there is a drying oil base to which a proteinaceous material such as egg has been added. Green paint has been identified as verdigris in oil		Plahter and Wiik 1970; White 1995
St. Nicholas, Eidsborg stavechurch: polychrome sculpture	Late 13th century	The binding medium has been identified as a drying oil		Skaug 1963–4
Holy Cross Minster, Schwäbisch Gmünd, Germany: polychromed portals	14th century	Oil gilding and both a proteinaceous and an oily binder for the pigments were identified	MCT, FTIR, chromatography	Hauff 1988
Berne Minster, Switzerland: excavated polychrome sculpture	Late Gothic	Oil gilding has been identified on the polychromy sculpture		Gutscher *et al.* 1989; Zumbrunn 1994.

Table 8 Selected analytical data relating to the use of resin and oil/resin in medieval polychromy.

Site and Location	Date	Resin or oil/resin type (if known) and application	Technique and Analyst(s)	Reference(s)
Examples in England				
Wells Cathedral: west front, sculptural polychromy	*c*1225–40	The presence of natural resin has been confirmed in a sample of green paint	GC-MS, R. White (NG)	Roy and White 1982
St. Albans Cathedral: feretory wall paintings	*c*1302–08	Linseed oil – which was prepolymerised (heat-bodied) to promote drying – with the addition of a very small amount of a conifer resin has been identified in *St. William of York*	GC-MS, R. White (NG)	Howard 1993a
Ingham, Norfolk: chancel, north wall, Tomb of Sir Oliver de Ingham	*c*1344	Oil was combined with red lead and lead white in the red layer beneath the *pastiglia*. Further analysis confirmed the presence of prepolymerised linseed oil combined with a little pine resin in a sample taken from the raised decoration of the camail	FTIR, J. Pilc (NG); GC-MS, R. White (NG)	Howard 1993d; White 1998
Rochester Cathedral: north-east transept, wall painting in tomb recess of St. William of Perth	*c*1340	Walnut oil and a resinous material have been identified in the paint layers	GLC, K. Groen (HKI)	Rickerby 1988
St. Mary's, Erwarton, Suffolk: monument to Sir Bartholomew Bacon and Joan de Heveningham	14th century	Binding medium of the paint is a drying oil which may have had a little resin added to it	J. Plesters and A. Roy (NG)	Roy 1984
Fitzalan chapel, Arundel: Fitzalan effigies	15th century	The presence of pine resin has been confirmed	GC-MS, R. White (NG)	Brodrick and Darrah 1986
Cartmel Fell, Cumbria: Chapel of St. Anthony, wooden figure of Christ	Late 15th century	Oil is employed as the binding medium of the paint layers, and also – possibly with the addition of a resinous component – as the mordant for the gold leaf	J. Plesters and A. Roy (NG)	Behrens 1982
Examples in northern Europe				
Notre-Dame Cathedral, Lausanne, Switzerland: south portal	1216–1220	Interestingly, a brown resin combined with crushed tile was employed by the 13th-century sculptor for repairs in the portal		Furlano *et al.* 1981
Heddal altar frontal, Norway	*c*1250–60	Linseed oil was identified and a pine resin was employed for the green glazes. For red lake glazes over silver leaf, prepolymerised linseed oil mixed with pine resin has been used	GC-MS, R. White (NG)	White 1995

Site and Location	Date	Resin or oil/resin type (if known) and application	Technique and Analyst(s)	Reference(s)
Hedal Madonna, Norway	*c*1250–60	Linseed oil with small amounts of resin and traces of egg tempera were employed for the oil gilding of Christ's hair. The white lining of the mantle is bound in linseed oil over a chalk ground containing fish or animal glue. A glue content was also identified in the blue pigment which had a medium principally consisting of oil and pine resin. Heat-bodied linseed oil was also employed for black paint layers on the throne, and nut oil has been identified in opaque red paint. For red lake: egg tempera; egg tempera combined with linseed oil; and linseed oil combined with pine resin have been identified	GC-MS, R. White (NG)	White 1995
Gausdal, *St. Paul*, Norway	1260–70s	The white robe of St. Paul was found to contain nut oil. A pine resin was also identified. Red lead is bound in non-heat-bodied linseed oil	GC-MS, R. White (NG)	White 1995
Tingelstad I Altarpiece, Norway	*c*1260–1300	Rich translucent copper resinate greens, consisting of verdigris complexed with pine resin acids and mixed with a drying oil, have been identified. Lead white is combined with linseed oil. For red lake glazes over silver leaf, prepolymerised linseed oil mixed with pine resin has been used. Azurite is bound in a mixture of egg tempera and linseed oil	GC-MS, R. White (NG)	White 1995
Angers Cathedral, France: choir, wall paintings depicting *Scenes of the Life of S. Maurille*	*c*1270–80	The presence of linseed oil was confirmed in the red paint from bay 3, and linseed oil with a small amount of pine resin was identified in a green copper resinate-containing layer. The blue paint was first surrounded by a layer of glue before being bound in linseed oil	GC-MS and FTIR, R. White (NG)	Subes-Picot 1992; White 1995; Demailly *et al.* 1998
Odda altar frontal, Norway	After *c*1300	Larch resin was employed for the yellow glaze applied over silver to produce gold effects	GC-MS, R. White (NG)	White 1995
Dale II altar frontal, Norway	After *c*1300	Yellow glaze over silver is bound in pine resin and heat-bodied linseed oil		White 1995
Altar frontal of Möðruvellir, Iceland		Yellow glaze over silver has been applied with a medium of predominantly pine resin with a component of linseed oil. Glair varnish has been identified on the surface of the altar frontal	GC-MS, R. White (NG)	White 1995

Site and Location	Date	Resin or oil/resin type (if known) and application	Technique and Analyst(s)	Reference(s)
German triptych in the Gallery of the Courtauld Institute of Art, London (known as the Estouteville triptych)	Late 14th century	Heat-bodied walnut oil with the addition of pine resin was employed for the white of an angel's robe and for the green of the angel's wing; walnut oil with pine resin was also used for the black outlines, but here the mixture was subjected to a greater degree of prepolymerisation. Partially prepolymerised linseed oil combined with pine resin was employed in a green sample taken from the architecture. The chalk ground was sealed with an unpigmented oil film	GC-MS and FTIR, R. White (NG)	Villers *et al.* 1997
Jan van Eyck, *Giovanni Arnolfini and his wife* (Netherlandish)	1434	Heat-bodied linseed oil with pine resin was employed for an area of green drapery. For a red lake paint layer, linseed oil was used	GC-MS, R. White (NG)	Campbell *et al.* 1997
Stephen Lochner, *Saints Matthew, Catherine of Alexandria and John the Evangelist* (German)	*c*1445	Strongly heat-bodied linseed oil with pine resin was used for the green of St. Matthew's robe. Heat-bodied linseed oil with some pine resin was used for green foliage and zones of red lake pigment, while for areas of white drapery, the white background and pale blue-grey drapery, linseed oil alone was employed	GC-MS, R. White (NG)	Campbell *et al.* 1997
Herlin altarpiece, Church of St. James, Rothenburg, Germany	*c*1466	Although the binding medium of matt areas has been identified as glue, other zones are bound with a resinous material combined with varying quantities of oil. In addition, a translucent red glaze is thought to be madder in a resinous medium		Broekman-Bolstijn *et al.* 1970

Table 9 Selected analytical data relating to the use of protein in medieval polychromy.

Site and Location	Date	Protein type (if known) and application	Technique and Analyst(s)	Reference(s)
Examples in English medieval wall painting				
Canterbury Cathedral: St. Gabriel's Chapel, apse	*c*1130	Small quantities of protein are combined with ultramarine and in red and charcoal black underpainting for ultramarine	Ninhydrin test, R. White (NG)	Cather and Howard 1994
Winchester Cathedral: Holy Sepulchre Chapel	*c*1175	Protein was identified as a component (combined with calcium carbonate) of the raised decorative border on the Magdalene's drapery, and combined with red lead in the raised decorative elements on the geometric border dividing the narrative tiers. Further analysis indicated the presence of a glue distemper or casein in the vermilion and red lead highlight on Nicodemus' leg	FTIR, H. Howard (CIA); Ninhydrin test, R. White (NG)	Howard 1995a; White 1998
Norwich Cathedral: Ante-Reliquary Chapel, eastern arch	*c*1250	Both protein and oil were identified in the ground of the border above the bishop on the south-east pier	FTIR, H. Howard (CIA)	Park and Howard 1996; Howard. 1995c
Horsham St. Faith Priory, Norfolk: refectory	*c*1250 with 15th-century additions	Analysis has indicated that rabbit-skin glue was employed to bind azurite in the first Foundation scene in the narrative cycle. The pigment is leanly bound. Egg tempera was also identified in the gold from St. Faith's crown	R. White (NG)	White 1989, 1995
Keynsham Abbey, Somerset: south chapel, excavated wall painting fragments	12th century	Protein – thought to be in the form of animal glue – was identified in paint layers from these excavated fragments	J. Plesters (NG)	Lowe and Mason 1987
Norwich Cathedral: Ante-Reliquary Chapel, vault and western arch	*c*1300	A proteinaceous component was found in the azurite of Virgin's robe	FTIR, H. Howard (CIA)	Park and Howard 1996; Howard 1995c
Rochester Cathedral: north-east transept, wall painting in tomb recess of William of Perth	*c*1340	Protein was identified combined with red pigments	K. Groen (HKI)	Rickerby 1988
St Stephen's Chapel, Westminster Palace: fragments of wall painting now in the British Museum	*c*1352–63	The ground of consists of red lead bound in egg tempera (see p. 131, note 118)	J. Mills and R. White (NG)	van Geersdaele and Goldworthy 1978; White 1995
Canterbury Cathedral: Chapel of Our Lady Undercroft, vault of eastern bay and screens enclosing the chapel	1370s	A proteinaceous material, likely to be glue, appears to have been employed as the medium for the gesso (calcium sulphate) ground. There is also a tentative identification of whole egg in the polychromy	H. Howard (CIA)	Howard 1993b

Site and Location	Date	Protein type (if known) and application	Technique and Analyst(s)	Reference(s)
Tower of London: the Byward Tower, south tower, first floor room, south wall	*c*1395	Although the paint layers are generally bound in linseed oil, the azurite of St. John's robe appears to have been bound in a glue distemper	R. White (NG)	
Holy Trinity, Coventry, West Midlands: chancel arch, *Last Judgement*	*c*1435–60	Analysis indicated that different media were employed for specific pigments or layers. For example, in the layer of red lead combined with a little red lake, FTIR analysis suggested the presence of egg, while oil was indicated in other parts of the stratigraphy	FTIR, H. Howard (CIA)	Howard 1996b
St. Cuthbert's, Wells: *Tree of Jesse*	*c*1470	The main paint layers, bound in linseed oil, have been applied over an underpaint with an egg tempera medium	R. White (NG)	R. White, pers. comm.
Examples in northern European medieval wall painting				
Convent of St. John the Baptist, Müstair, Switzerland	*c*800	Traces of oil and protein were identified in the binding media of lime-sensitive pigments such as red and yellow lead		Mairinger and Schreiner 1986
Idensen, Germany	*c*1130	The use of egg has been established as the binder for blue and green pigments	M. Matteini and A. Moles (OPD)	Matteini and Moles 1994
Glasgow Cathedral: wall painting fragments excavated from the crypt	*c*1190	Instrumental analysis undertaken using GC-MS confirmed the presence of protein, and suggested that casein or casein with animal glue may have been employed as the original binding medium for the red lake pigment in one of the paint samples	GC-MS, P. Colombini (DCCI/UP)	Howard 1998a; Park 1998
Udabno Monastery, Georgia	Late 12th–early 13th century	The binding medium of these wall paintings has been confirmed as glue		Dneprovskaya 1995
Bertubani and Natlis-Mtsemeli Monasteries, Georgia	12th–13th century	The presence of egg has been confirmed in both schemes of wall painting		Dneprovskaya 1995
St. Georgen, Austria	*c*1240	The binding medium is casein combined with lime	Antibody staining, K. Peterson	H. Leitner, pers. comm.
Angers Cathedral, France: choir, *Scenes of the Life of S. Maurille*	*c*1270–80	Analysis of the organic component of these wall paintings demonstrated that within a blue layer painted in linseed oil the individual azurite particles were coated in glue. Casein was also detected in the ground layers	R. White (NG)	Subes-Picot 1992; White 1995; Demailly *et al.* 1998

Site and Location	Date	Protein type (if known) and application	Technique and Analyst(s)	Reference(s)
Cologne Cathedral. Germany: choir screen wall paintings	*c*1332–49	The binding medium is egg tempera with an addition of oil	E. Jägers	Jägers and Schulze-Senger 1989; Heritage 1990; Maul 1992
St. Mary's, Danzig, Poland: Chapel of St. George, polychromed wooden figures and associated mural paintings	16th century	Casein		Kozarzewski and Zankowski 1988
Salzburg, Austria: Landkartengalerie	Late 16th century	Casein was identified as the binding medium in the preparatory layer		H. Leitner, pers. comm.
Examples in English medieval panel painting				
Adisham, Kent: nave, west wall, Adisham reredos	*c*1180	Preliminary analytical evidence suggests that both the calcium sulphate and lead white grounds were bound with a proteinaceous material, likely to be glue. Further analysis by GC-MS suggested the presence of a mixed proteinaceous binder – perhaps a combination of egg/glue or milk/glue	FITC, H. Howard (CIA). GC-MS, P. Colombini (DCCI/UP)	Howard 1993c
Painted Chamber, Westminster Palace: *ceiling paterae*	*c*1263–66	Analysis indicated that for the blue glaze of the background, the ultramarine may have been ground in egg before being bound in oil	GC-MS, R. White (NG)	Liversidge *et al.* 1995; Lynn 1997
Norwich Cathedral: Despenser Retable	*c*1380	The binding medium is generally glue with the addition of a few oil-based glazes. The frame has also been painted in size-bound pigments	J. Plesters (NG)	Plummer 1959
Wilton Diptych (French or English), now in the National Gallery, London	Late 14th century	Egg tempera has been established as the binding medium of the Wilton Diptych, with the exception of the ultramarine on the shield which is bound in animal glue	R. White (NG)	Roy 1997
Examples in northern European panel painting				
Heddal Altar frontal, Norway	*c*1250–60	An unpigmented glair (egg white) varnish has been identified on this altarpiece. Ultramarine has been coated with a proteinaceous material before being bound in oil (U. Plahter, pers. comm.)	R. White (NG)	White 1995; Plahter 1995
Tingelstad I altar frontal, Norway	*c*1260–1300	An unpigmented glair (egg white) varnish has been identified on this altarpiece. Azurite has been bound in egg tempera combined with a little linseed oil	R. White (NG)	Plahter and Plahter 1974; White 1995

Site and Location	Date	Protein type (if known) and application	Technique and Analyst(s)	Reference(s)
Tingelstad II altar frontal, Norway	*c*1250–1350	Azurite has been bound in egg tempera combined with a little linseed oil. There is more oil present here than in Tingelstad I	R. White (NG)	Plahter and Plahter 1974; White 1995
St. Michel's Church, Hildesheim, Germany: painted wooden ceiling	Mid-13th century	A lime casein medium was identified, and in addition, the priming of the ceiling is chalk combined with animal glue		Denninger 1969
Trondheim altar frontal, Norway	After *c*1300	An unpigmented glair (egg white) varnish has been identified on this altarpiece	R. White (NG)	White 1995
Icelandic Altarfrontal of Möðruvellir		Traces of an unpigmented glair (egg white) varnish has been identified on this altarpiece	R. White (NG)	White 1995
Polychromed sculpture: examples in England				
St Mary's Abbey, York: excavated stone sculptures	*c*1180	An egg tempera medium has been identified in these column figures, and glue was employed as the mordant for the gilding on the hair of St John	R. White (NG)	Brodrick 1993
Wells Cathedral: west front	*c*1225–40	An egg tempera medium has been identified in the sculpted figures. In addition, a gum medium was also suggested	R. White (NG)	Roy and White 1982
All Saints', Sawley, Derbyshire: polychromed angel which originally formed part of a canopied tomb	*c*1270–80	A glue distemper (fish or animal) has been identified in this polychromy	R. White (NG)	Alexander and Binski 1987
York Minster: chapter house, excavated stone pendant	Late 13th century	Glue size combined with a little ochre and lead white was identified as the adhesive for gold leaf	J. Darrah (V&A)	Heywood and Oosterwijk 1993
St. Albans Cathedral: feretory, Shrine of St. Alban	*c*1302–08	Glue was detected in samples of blue and black pigment. The white ground layer also has a proteinaceous binder	R. White (NG)	White 1993
Ingham, Norfolk: chancel, north wall, tomb of Sir Oliver de Ingham	*c*1344	Red lead combined with a glue binder was identified by GC-MS. In addition, vermilion combined with protein was identified by MCT on a carved stone fragment, likely to relate to the de Ingham tomb	GC-MS, P. Colombini (DCCI/UP); FITC, H. Howard (CIA)	Howard 1993d

Site and Location	Date	Protein type (if known) and application	Technique and Analyst(s)	Reference(s)
Ingham, Norfolk: nave south aisle, tomb of Sir Roger and Lady Margaret de Bois	*c*1380	Detailed examination indicated that different media were employed for specific pigments or layers. In a sample taken from the west end of the tomb chest, the uppermost layer of red earth combined with lead white is bound in protein, while all other layers in the sample have a predominantly oily binder. Samples taken from both the lead white layers and the *pastiglia* decoration indicated the presence of both protein and oil, and in addition, a proteinaceous adhesive was employed for the silver leaf	FTIR, H. Howard (CIA)	Howard 1993d
Exeter Cathedral: west front, sculptural polychromy	14th and 15th century	Analysis indicated the presence of both egg tempera and linseed oil. It appears that generally the ground may have an egg-based binder, while for subsequent layers, egg and oil were either mixed to form an emulsion or were used for separate specific layers within the stratigraphy. In one sample however, a proteinaceous binding medium (which may be animal glue) was identified in the pink ground	R. White (NG)	Sinclair 1991, 1992
Winchester Cathedral: Great Screen, sculptural polychromy	*c*1475–90	The presence of egg with a little oil is indicated in samples taken from the white cloaks of two of the kings	R. White (NG)	Brodrick and Darrah 1989
Fitzalan Chapel, Arundel: Fitzalan effigies	15th century	A little egg appears to have been added to the linseed oil which forms the principal component of the binding medium. In addition, glue size was used as an adhesive for the effigies	J. Darrah (V&A); R. White (NG)	Brodrick and Darrah 1986
Cartmel Fell, Cumbria, Chapel of St. Anthony: wooden figure of Christ	Late 15th century	Glue has been identified as a component of the chalk ground and was also used to seal the ground layer	J. Plesters and A. Roy (NG)	Behrens 1982
Worcester Cathedral: wooden pilgrim's staff	Medieval	Glue (fish or animal) was detected in the polychromy of this staff	R. White (NG)	Lubin *et al.* 1990
Miscellaneous British examples				
Westminster Abbey: Coronation Chair	*c*1297–1300	Protein, likely to be in the form of size, was identified as a component of the original chalk ground	Professor Rees Jones (CIA)	Percival-Prescott 1957; Rees-Jones 1954

Site and Location	Date	Protein type (if known) and application	Technique and Analyst(s)	Reference(s)
Westminster Abbey: Coronation Chair	2nd quarter 14th century	The elaborately decorated gilding of the second decoration of the Coronation Chair is likely to have been undertaken using a glair (egg white) mordant. For this decorative phase calcium carbonate was mixed with a proteinaceous binder to form a lean paste and generally applied over a ground of lead white and oil	Prof. Rees Jones (CIA)	Percival-Prescott 1957; Rees-Jones 1954
Polychromed sculpture: northern European examples				
Church of S. Michele, Val Susa, Italy: polychromed figures in the apse	12th century	The stone sculpture was coated with a layer of red iron oxide combined with calcium caseinate in preparation for painting		S. Pulga, pers. comm.
Notre-Dame Cathedral, Lausanne, Switzerland: south portal	*c*1216–20	A medium likely to be either egg white or casein was used for the paint layers of the cathedral's south portal		Furlano *et al.* 1981
Amiens Cathedral, France: '*Portail de la Mère Dieu*'	*c*1220–60	A proteinaceous material (probably rabbit-skin glue) was identified in the early strata of the polychromy	LRMH	Weeks 1998
Hedal Madonna, Norway	*c*1250–60	The ground consists of chalk bound in fish or animal glue. The red gemstones in the crown are painted in red lake bound in an egg tempera medium and applied over 'gold', which consists of silver glazed with yellow. The red lake pigment is also bound in linseed oil combined with egg tempera and in linseed oil mixed with a pine resin. A glue content has been identified in blue pigment principally bound in oil and pine resin	R. White (NG)	White 1995
Maria Laach, Germany: Tomb of Pfalzgraf Heinrich II	13th century	The ground is composed of chalk bound with animal glue. In addition, lead white and bone black pigments are also in a water-soluble medium, likely to be protein		Oellermann 1993
Dal Church, *Virgin and Child*, now in the University Museum of National Antiquities, Oslo, Norway	Mid-13th century	The binding medium is a protein-based material such as egg that has been combined with a drying oil		Plahter and Wiik 1970
Holy Cross Minster, Schwäbisch Gmünd, Germany: polychromed portals	14th century	Both a proteinaceous and an oily binder for the pigments were identified	MCT, FTIR, chromatography	Hauff 1988

❦ APPENDIX FOUR

Abbreviations used in the tables

Institutions

BM	British Museum
CIA	Courtauld Institute of Art
DI	Doerner Institute
DCCI(UP)	Dipartimento di Chimica e Chimica Industriale (Università di Pisa)
EH	English Heritage
EH/AML	English Heritage/Ancient Monuments Laboratory
GCI	Getty Conservation Institute
HKI	Hamilton Kerr Institute
IRPA	Institut Royal du Patrimoine Artistique
ITM	Institut für Technologie d. Malerei
LRMH	Laboratoire de Recherches des Monuments Historiques
NG	National Gallery
NHM	Natural History Museum
NM	Nationalmuseet
OPD	Opificio delle Pietre Dure
OU	Oxford University
RM	Rijksmuseum
UCL	University College
V&A	Victoria and Albert Museum

Analytical techniques

DT-MS	Direct temperature-resolved mass spectrometry
ES	Emission spectrography
FITC	Fluorescein isothiocyanate
FTIR	Fourier transform infrared spectroscopy
GC-MS	Gas chromatography-mass spectrometry

HPLC	High-performance liquid chromatography
IR	Infrared spectroscopy
MCT	Microchemical tests
Micr.	Microscopy
PLM	Polarised light microscopy
Raman spec.	Raman spectroscopy
SEM/EDX	Scanning electron microscopy/energy-dispersive X-ray microanalysis
XRF	X-ray fluorescence
XRD	X-ray diffraction

References

van Acker, L. (1972) 'Petri Pictoris carmina nec non Petri de Sancto Audemaro Librum de coloribus faciendis', *Corpus Christianorum, Continuatio Medievalis* **25**, 45–98.

Agrawal, O. P. (1969–70) 'A study of the technique of Indian wall paintings', *Journal of Indian Museums* **25–26**, 99–118.

Alexander, J. and P. Binski (eds) (1987) *The Age of Chivalry: Art in Plantagenet England 1200–1400* (catalogue of an exhibition at the Royal Academy), London.

Allardyce, F. (1984) 'The painting of the legend of St. Eustace in Canterbury', *Archaeologia Cantiana* **101**, 115–30.

Allardyce, F., L. Medhurst and T. Organ (1983) 'The Deanery Wall Painting, Durham' [unpublished report].

Anglos, D., S. Couris and C. Fotakis (1997) 'Laser diagnostics of painted artworks: laser-induced breakdown spectroscopy in pigment identification', *Applied Spectroscopy* **51**, 1025–30.

Anglos, D., M. Solomidou, I. Zergioti, V. Zafiropulos, T. Papazoglou and C. Fotakis (1996) 'Laser-induced fluorescence in artwork diagnostics: an application in pigment analysis', *Applied Spectroscopy* **50**, 1331–4.

Apted, M. R. and S. Hannabuss (eds) (1978) *Painters in Scotland 1301–1700: A Biographical Dictionary* (Scottish Record Society, new ser., 7), Edinburgh.

Arnold, A., A. Kung and K. Zehnder (1986) 'Deterioration and preservation of Carolingian and medieval mural paintings in the Müstair Convent (Switzerland). Part I: decay mechanisms and preservation', in *Case Studies in the Conservation of Stone and Wall Paintings* (Preprints of the IIC 11th International Congress, Bologna), 190–94, London.

Arts Council (1969) *Frescoes from Florence* (catalogue of an exhibition at the Hayward Gallery, London), London.

Ashurst, J. and T. B. James (1988) ' Stonework and plasterwork', in T. B. James and A. M. Robinson (eds), *Clarendon Palace: The History and Archaeology of a Medieval Palace and Hunting Lodge near Salisbury, Wiltshire*, 234–58, London.

van Asperen de Boer, J. R. J. (1974) 'An examination of particle size distributions of azurite and natural ultramarine in some early Netherlandish paintings', *Studies in Conservation* **19**, 233–43.

Augusti, S. (1965) 'Analysis of the material and technique of ancient mural paintings', in S. Rees Jones (ed.), *Application of Science in Examination of Works of Art*, 67–70, Boston.

Aze, S., J. Vallet and O. Grauby (2002) 'Chromatic degradation processes of red lead pigment', in *Preprints of the ICOM-CC 13th Triennial Meeting, Rio de Janeiro, September 2002*, 549–55, London.

Babington, C. (1987) 'Microscopic analysis of paint samples taken from the south tabernacle, Prior Crauden's Chapel, Ely', Conservation of Wall Painting Department, Courtauld Institute of Art, London [unpublished report].

Babington, C. and J. James (1988) 'The wall paintings of Horsham St. Faith Priory: their technique, discovery and conservation', unpublished lecture for the exhibition, *Age of Chivalry: Art in Plantagenet England 1200–1400*, Royal Academy of Arts, 18 February 1988.

Baker, A. (1998) 'Adam and Eve and the Lord God: the Adam and Eve cycle of wall paintings in the church of Hardham, Sussex', *Archaeological Journal* **155**, 207–25.

Baker, E. (1965) 'The Ante-Reliquary Chapel, Norwich Cathedral', unpublished report for the Council for the Care of Churches [undated, but probably 1965].

Ballantyne, A. (1988) 'All Saints' Church, Farringdon, Hampshire' [unpublished report, June 1988].

Ballantyne, A. (1989) 'Church of Saint Mary the Virgin, Henham, Essex' [unpublished report, March 1989].

Ballantyne, A. (1992) 'White Notley', in *The Colchester Area* (Supplement to *Archaeological Journal* **149**), 38–40.

van den Berg, K. J., M. van Eikema Hommes, K. Groen, J. Boon and B. Berrie (2000) 'On copper green glazes in paintings', in J. Goupy and J. P. Mohen (eds) *Art et chimie, la couleur*, actes du congrés, 18–21, Paris.

Behrens, G. A. (1982) 'Conservation work on the Cartmel Fell figure of Christ', *Transactions of the Cumberland and Westmorland Antiquarian and Archaeological Society* **82**, 125–34.

Bensi, P. (1990) 'La pellicola pittorica nella pittura murale in Italia: materiali e tecniche esecutive dall'alto medioevo al XIX secolo', in C. Danti, M. Matteini and A. Moles (eds), *Le Pitture Murali: Techniche, Problemi, Conservazione*, 73–102, Florence.

Bertelli, C. (1988) 'Castelseprio e Milano', *Bisanzio, Roma e l'Italia nel'alto medioevo: XXXIV settimana di studio del CISAM,* Spoleto, Vol. 2, 869–917.

Best, S. P., R. J. Clark, M. Daniels, C. Porter and R. Withnall (1995) 'Identification by Raman microscopy and visible reflectance spectroscopy of pigments on an Icelandic manuscript', *Studies in Conservation* **40**, 31–40.

Biddle, M. and B. Kjølbye-Biddle (1980) 'England's premier abbey: the medieval chapter house of St. Albans Abbey, and its excavation in 1978', *Expedition* **22**, 17–32.

Binder, C. (1997) 'The wall paintings in the choir of the Magdalenenkirche, Judenburg: a comparative study of their techniques', unpublished diploma dissertation, Conservation of Wall Painting Department, Courtauld Institute of Art, London.

Binski, P. (1986) *The Painted Chamber at Westminster*, Society of Antiquaries Occasional Paper No. 9, London.

Binski, P. (1995) *Westminster Abbey and the Plantegenets: Kingship and the Representation of Power, 1200–1400*, New Haven and London.

Binski, P. and I. Freestone (1995) 'The technique of the Westminster Retable: a preliminary report with an analysis of the glass component', in M. Malmanger, L. Berczelly and S. Fuglesang (eds), *Norwegian Medieval Altar Frontals and Related Material* (Acta ad archaeologiam et artivm historiam pertinentia, Vol. 11), 59–71, Rome.

Binski, P. and D. Park (1986) 'A Ducciesque episode at Ely: the mural decorations of Prior Crauden's Chapel', in W. M. Ormrod (ed.), *England in the Fourteenth Century* (Proceedings of the 1985 Harlaxton Symposium), 28–41, Woodbridge.

Birelli, G. B. (1601) *Opere di Giovambatista Birelli* ...Florence.

Birley, R. and B. Young (1993) 'Vivianite of post-Roman origin from Vindolanda, Northumberland', *Transactions of the Natural History Society of Northumbria* **65**, 65–7.

Blair, C. (1994) 'The Wooden Knight at Abergavenny', *Church Monuments* **9**, 33–52.

Boas, J., M. Alexander and K. Fryer (1996) 'A 12th century room found in Guildford', *Surrey Archaeological Society Bulletin* **300**, 13–14.

Bomford, D., J. Brough and A. Roy (1980) 'Three panels from Perugino's Certosa di Pavia Altarpiece', *National Gallery Technical Bulletin* **4**, 3–32.

Bomford, D., J. Dunkerton, D. Gordon and A. Roy (1990) *Art in the Making: Italian Painting before 1400*, London.

Bomford, D., A. Roy and A. Smith (1986) 'The technique of Dieric Bouts: two paintings contrasted', *National Gallery Technical Bulletin* **4**, 39–58.

Borsook, E. (1980) *The Mural Painters of Tuscany*, 2nd edn, Oxford.

Boulter, S. (1993) 'Polychrome and petrographical analysis of the De Lucy effigy', *V&A Conservation Journal* **8**, 4–7.

Braine, D. and P. Welford (1991) 'Composite capital', in *The Making of England: Anglo-Saxon Art and Culture AD 600–900* (catalogue of an exhibition at the British Museum), 34–5, London.

Brill, T. B. (1980) *Light: Its Interaction with Art and Antiquities*, New York.

Brindle, S. and B. Kerr (1997) *Windsor Revealed: New Light on the History of the Castle*, English Heritage.

Brodrick, A. (1993) 'Painting techniques of early medieval sculpture', in *Romanesque: Stone Sculpture from Medieval England* (catalogue of an exhibition at the Henry Moore Institute, Leeds), 18–27, Leeds.

Brodrick, A. and J. Darrah (1986) 'The 15th century polychromed limestone effigies of William Fitzalan, 9th Earl of Arundel, and his wife, Joan Nevill, in the Fitzalan Chapel, Arundel', *Church Monuments* **1**, 65–94.

Brodrick, A. and J. Darrah (1989) 'A description of the polychromy on the fragments of limestone figure-sculpture from the Great Screen at Winchester Cathedral', *Burlington Magazine* **131**, 615–17.

Broekman-Bolstijn, M., J. R. J. van Asperen de Boer, E. H. van't Hul-Ehrnreich and C. M. Verduyn-Groen (1970) 'The scientific examination of the polychromed sculpture in the Herlin Altarpiece', *Studies in Conservation* **15**, 370–400.

Bucklow, S. (forthcoming) 'The Thornham Parva – a summary of the technique', in A. Massing (ed.), *Paintings and Practice, the Thornham Parva Retable: Technique, Conservation and Context of an English Medieval Painting*, Cambridge.

van Buren, A. H. (1986) 'Thoughts, old and new, on the sources of early Netherlandish painting', *Simiolus* **16**, 93–112.

Burgio, L., R. Clark, T. Stratoudaki, D. Anglos and M. Doulgeridis (2000) 'Pigment identification in painted artworks: a dual analytical approach employing laser-induced breakdown spectroscopy and Raman microscopy,' *Applied Spectroscopy* **54**, 1–7.

Burmester, A. and C. Krekel (1998) 'The relationship between Albrecht Dürer's palette and fifteenth/sixteenth-century pharmacy price lists: the use of azurite and ultramarine', in A. Roy and P. Smith (eds), *Painting Techniques: History, Materials and Studio Practice* (Preprints of the IIC 17th International Congress, Dublin, 1998), 101–5, London.

Burmester, A. and C. Krekel (2001) 'Das Blaupigment Vivianit als Schlüssel für Datierung und die Herkunft des Regensburger Gnadenbildes?' in *Alte Kapelle Regensburg* (Arbeitschefte des Bayerischen Landesamtes für Denkmalpflege, 114), 83–6, München.

Cadet, E. (1995) 'Les peintures murales romanes de Saint-Etienne, Auxerre: approches historique, technique, stylistique et iconographique', unpublished MA thesis, University of Paris, Pantheon-Sorbonne.

Caiger-Smith, A. (1963) *English Medieval Mural Painting*, Cambridge.

Campbell, L., S. Foister and A. Roy (eds) (1997) 'The methods and materials of northern European painting 1400–1550', *National Gallery Technical Bulletin* **18**, 6–55.

Cardon, D. (1990) *Guide des teintures naturelles: plantes, lichens, champignons, mollusques et insectes*, Paris.

Cather, S. and H. Howard (1986) 'The use of wax and wax-resin preservatives on English mediaeval wall paintings: rationale and consequences', in N. S. Brommelle and P. Smith (eds), *Cases Studies in the Conservation of Stone and Wall Paintings* (Preprints of the IIC 11th International Congress, Bologna, 1986), 48–53, London.

Cather, S. and H. Howard (1994) 'Romanesque wall paintings in the apse of St Gabriel's Chapel, Canterbury Cathedral: their technique, condition and environment reassessed', in *Forschungsprojekt Wandmalerei-Schäden* (Arbeitshefte zur Denkmalpflege in Niedersachsen, 11), 141–56, Hannover.

Cather, S. and D. Park (1998) 'Wall paintings', in W. J. Rodwell, 'Holy Trinity Church, Bradwell-juxta-Coggeshall: a survey of the fabric and appraisal of the Norman brickwork', *Essex Archaeology and History* 29, 105–10.

Cather, S., D. Park and P. Williamson (eds) (1990) *Early Medieval Wall Painting and Painted Sculpture in England* (BAR British Series 216), Oxford.

Cennini, Cennino (1960) *Il Libro dell'arte, the Craftsman's Handbook of Cennino d'Andrea Cennini*, transl. D. V. Thompson, New Haven, 1933 (repr. New York 1960).

Cennino, Cennini (1991) '*Il Libro dell'arte*', pref., comm. and notes Mario Serchi, Felice de Monnier, Florence.

Chapman, F. R. (ed) (1907) *Sacrist Rolls of Ely*, 2 vols, Cambridge.

Cheetham, F. (1984) *English Medieval Alabasters*, Oxford.

Cherry, J. and N. Stratford (1995) *Westminster Kings and the Medieval Palace of Westminster* (British Museum Occasional Paper 115), London.

Childs, W. R. (1978) *Anglo-Castilian Trade in the Later Middle Ages*, Manchester.

Church, A. H. (1915) *The Chemistry of Paints and Painting*, 4th edn, London.

Clarke, M. (2001) *The Art of All Colours*, London.

Collins, J. H. (1892) *A Handbook to the Mineralogy of Cornwall and Devon* (second edition) facsimile reprint 1969, Truro.

Colombini, M. P., R. Fuoco, A. Giacomelli and B. Muscatello (1998) 'Characterization of proteinaceous

binders in wall painting samples by microwave assisted acid hydrolysis and GC-MS determination of amino acids', *Studies in Conservation* 43, 33–41.
Colvin, H. M. (ed.) (1963) *The History of the King's Works: The Middle Ages*, 2 vols, HMSO.
Colvin, H. M. (ed.) (1971) *Building Accounts of King Henry III*, Oxford.
Coomaraswamy, A. K. (1934) 'The technique and theory of Indian painting', *Technical Studies* 3, 59–89.
Coppack, G. (1988) *Abbeys: Yorkshire's Monastic Heritage* (exhibition catalogue, Yorkshire Museum), York.
Courtauld Institute of Art, Department of Wall Painting Conservation (1987) 'The wall paintings of St. John's Church, Duxford, Cambridgeshire' [unpublished report, January 1987], London.
Courtauld Institute of Art, Conservation of Wall Painting Department (1990) 'Secular wall paintings in Ely: recently discovered 14th- and 16th-century schemes and their architectural context. Preliminary report on the conservation programme 1989,' in H. Davis (ed.), *Medieval Wall Paintings: Silver Street, Ely*, 6–15, Cambridge.
Craig, G. Y. (ed.) (1991) *Geology of Scotland*, London.
Cramp, R. J. and J. Cronyn (1990) 'Anglo-Saxon polychrome plaster and other materials from the excavations of Monkwearmouth and Jarrow: an interim report', in S. Cather, D. Park and P. Williamson (eds), *Early Medieval Wall Painting and Painted Sculpture in England* (BAR British Series 216), 17–30, Oxford.
Crews, P. C. (1987) 'The fading rates of some natural dyes', *Studies in Conservation* 32, 65–72.
Curteis Paine Associates (1993) 'Analytical examination and solvent gel development for the 16th century screen at St. Mary's, Worstead' [unpublished report for English Heritage, March 1993].
T. Curteis Associates (1994a) 'Analytical investigation of the wall paintings, St. Mary's & St. Leodegarius' Church, Ashby St. Ledgers' [unpublished report, November 1994].
T. Curteis Associates (1994b) 'Analytical investigation of the wall paintings, St. Nicholas' Church, Little Horwood' [unpublished report, August 1994].
T. Curteis Associates (1995a) 'Condition survey and proposals for the conservation of the wall paintings, St. Mary and All Saints, Willingham, Cambridgeshire' [unpublished report, December 1995].
T. Curteis Associates (1995b) 'All Saints and St. Andrew's Church, Kingston, Cambridgeshire: analytical examination of the wall paintings' [unpublished report, August 1995].
T. Curteis Associates (1996a) 'Condition survey and proposals for the conservation of the wall paintings, St. Mary's Church, Wissington, Suffolk' [unpublished report, January 1996].
T. Curteis Associates (1996b) 'Condition survey and proposals for the conservation of the wall paintings, St. Martin's Church, Colchester, Essex' [unpublished report, January 1996].
T. Curteis Associates (1998) 'Technical survey and proposals for the conservation of the poychromy on the monument to Sir Roger Dalton, St. Leonard's Church, Apethorpe, Northamptonshire' [unpublished report, March 1998].
Curteis, T. and S. Paine (1995) 'Conservation of the polychromy in the vaults, Winchester Cathedral, Lady Chapel' [unpublished report, August 1995].
Curteis, T., S. Paine and D. Park (1994) 'Recently conserved wall paintings in the south transept', *Winchester Cathedral Record* 63, 27–32.
Daniels, V. (1987) 'The blackening of vermilion by light', in J. Black (ed.), *Recent Advances in the Conservation and Analysis of Artifacts* (Proceedings of the Institute of Archaeology Jubilee Conservation Conference), 280–82, London.
Darrah, J. (1992) 'The Wheel of Fortune, Rochester Cathedral wall painting, *c.* 1250: an examination of the paint layers' [unpublished report].
Degering, H. (1928) 'Theophilus Presbiter qui et Rugerus', *Westfälische Studien ... Alois Bömer gewidmet*, 248–62, Leipzig.
Delamare, F., L. Delamare, B. Guineau and G. S. Odin (1990) 'Couleur, nature et origine des pigments verts employés en peinture murale gallo-romaine', in *Pigments et Colorants de l'Antiquité et du Moyen Age: teinture, peinture, enluminure études historiques et physico-chimiques,* Colloque international du CNRS, 103–16, Paris.
Delaval, E. H. (1777) *An Experimental Inquiry into the Cause of the Changes of Colour in Opake and Coloured Bodies with an Historical Preface*, London.
Demailly, S., P. Hugon, M. Stefanaggi and W. Nowick (1998) 'The technique of the mural paintings in the choir of Angers Cathedral', in A. Roy and P. Smith (eds) *Painting Techniques: History, Materials and Studio Practice* (Preprints of the IIC 17th International Congress, Dublin, 1998), 10–15, London.
Demus, O. (1970) *Romanesque Mural Painting*, trans. M. Whittall, London.
Denninger, E. (1969) 'The examination of pigments and media from the painted wooden ceiling of St. Michael's Church at Hildesheim, W. Germany', *Studies in Conservation* 14, 91–5.

Denninger, E. (1974) 'What is "Bianco di San Giovanni" of Cennino Cennini?', *Studies in Conservation* **19**, 185–7.
Dionysius of Fourna (1974) *The Painter's Manual*, trans. P. Hetherington, London.
Dneprovskaya, M. (1995) 'Medieval pigment and plaster technology in the XII–XIII century mural paintings at David-Garedji, Georgia', in P. Vandiver *et al.* (eds), *Material Issues in Art and Archaeology IV* (Materials Research Society Symposium Proceedings 352), 727–32, Pittsburgh.
Dodwell, C. R. (1993) *The Pictorial Arts of the West 800–1200*, Yale.
Drescher, D., O. Emmenegger, R. Möller and J. Pursche (1990) 'Maltechnische Befunde', *Schäden an Wandmalereien und ihre Ursachen* (Arbeitshefte zur Denkmalpflege in Niedersachsen, 8), 75–93, Hannover.
Drury, P. (1990) 'Anglo-Saxon painted plaster fragments excavated at Colchester Castle, Essex', in S. Cather. D. Park and P. Williamson (eds), *Early Medieval Wall Painting and Painted Sculpture in England* (BAR British Series 216), 111–22, Oxford.
Dumay, M. (1874) 'De la peinture à l'huile en France au commencement du XIVe siècle', *Mémoires de la Société Nationale des Antiquaires de France*, 237–46.
Dunkerton, J., A. Roy and A. Smith (1987) 'The unmasking of Tura's "Allegorical Figure": a painting and its concealed image', *National Gallery Technical Bulletin* **11**, 5–35.
Duval, A. R. (1992) 'Les préparations colorées des tableaux de l'ecole Française des dix-septième et dix-huitième siècles', *Studies in Conservation* **37**, 239–62.
Duxson, W. E. (1933) *Maddermarket*, Norwich.
Eastlake, C. L. (1960) *Methods and Materials of Painting of the Great Schools and Masters*, 2 vols, New York.
Eibner, A. (1914) *Über lichtechte Zinnober*, Munich.
Eibner, A. (1917a–f) 'Über lichtwirkungen auf Malerfarbstoffe, III. Erhöhung der Lichtechtheit der auf nassem Wege hergestellten Zinnober', *Chemiker-Zeitung* (a) **41/55** (1917) 385–6; (b) **41/58** (1917) 402–4; (c) **41/61** (1917) 423–5; (d) **41/65-6** (1917) 447–50; (e) **41/70** (1914) 482–3; (f) **41/74-5** (1917) 510–13.
Eibner, A. (1918) 'Lichtwirkungen auf Malerfarbstoffe, IV. Darstellung von Zinnobern mit erhöhter Lichtechtheit aus den Erzen durch Aus laugen', *Chemiker Zeitung* **42/116-7**, 469–70.
van Eikema Hommes, M. (2001) 'Verdigris glazes in historical oil paintings: recipes and techniques', *Zeitschrift für Kunsttechnologie und Konservierung* **15**, 163–95.
Eikhoff, A. K. (1973) 'Pigmentation of coatings', in T. C. Patton (ed.), *Pigments Handbook*, Vol. 2, 49–55, New York.
Ellwanger-Eckel, F. (1979) 'Herstellung und verwendung künstlicher grüund blauer kupferpigmente in der malerei', unpublished diploma dissertation, Institut für technologie der Malerei, Staatliche Akademie der Bildenden Künste, Stuttgart, May 1979.
Embrey, P. G. and R. F. Symes (1987) *Minerals of Cornwall and Devon*, London.
Emmenegger, O. (1986) 'Deterioration and preservation of Carolingian and medieval mural paintings in the Müstair Convent (Switzerland). Part III: Techniques and materials used and past restorations', in N. S. Brommelle and P. Smith (eds), *Case Studies in the Conservation of Stone and Wall Paintings* (Preprints of the IIC 11th International Congress, Bologna, 1986), 197–9, London.
Englefield, W. D. A. (1950) *History of the Painter-Stainers' Company of London*, London.
Eraclius (1967) '*De Coloribus et artibus romanorum*', in M. P. Merrifield, *Original Treatises on the Arts of Painting*, London 1849 [repr. New York, 1967] Vol. 1, 166–257.
Eraclius (1996) *I Colori e le arti dei Romani: e la compilazione pseudo eracliana*, ed. and trans., C. Garzya Romano, Napoli.
Erskine, A. (ed.) (1981) *The Accounts of the Fabric of Exeter Cathedral*, 1279–1353 (Devon & Cornwall Record Society, new ser., Vol. 24), Torquay.
Erskine, A. (ed.) (1983) *The Accounts of the Fabric of Exeter Cathedral*, 1279–1353 (Devon & Cornwall Record Society, new ser., Vol. 26), Torquay.
Evans, J. (1948) *Art in Mediaeval France 987–1498*, Oxford.
Faye, G. H., P. G. Manning and E. H. Nickel (1968) 'The polarized optical absorption spectra of tourmaline, cordierite, chloritoid and vivianite: ferrous-ferric electronic interaction as a source of pleochroism', *American Mineralogist* **53**, 1174–1201.
Federspiel, B. (1995) 'Questions about medieval gesso grounds', in A. Wallert, E. Hermens and M. Peek (eds), *Historical Painting Techniques, Materials, and Studio Practice* (Preprints of a symposium at the University of Leiden, June 1995), 58–64, Los Angeles.
Feller, R. J. (1967) 'Studies on the darkening of vermilion by light', in *Report and Studies in the History of Art*, 99–11, National Gallery of Art, Washington DC.

Fitzhugh, E. W. (1986) 'Red lead and minium', in R. L. Feller (ed.), *Artists' Pigments: A Handbook of their History and Characteristics*, Vol. 1, 109–39, Cambridge.

Fitzhugh, E. W. (1997) 'Orpiment and realgar', in E. W. Fitzhugh (ed.), *Artists' Pigments: A Handbook of their History and Characteristics*, Vol. 3, 47–79, Oxford.

Fuchs, R. and D. Oltrogge (1994) 'Colour material and painting technique in the Book of Kells', in F. O'Mahony (ed.), *The Book of Kells* (Proceedings of a conference at Trinity College, Dublin, September 1992), 133–71, Dublin.

Furlano, V., R. Pancella and T. Hermanès (1981) 'Portail peint de la cathédrale de Lausanne: analyses pour une restauration', *Chantier* **12**, 13–20.

Gärtner, W. (1987) 'Church of the Blessed Virgin, Eastry, Kent' [unpublished report, July 1987].

Gärtner, W. (1988) 'Chichester Cathedral, Treasury: conservation of the 12th century wall paintings' [unpublished report, November 1988].

Gärtner, W. and T. Organ (1989) 'Monitoring and conservation of the Romanesque wall painting of St. Anselm's Chapel, Canterbury Cathedral', *The Conservator* **13**, 7–14.

Gasol Fargas, R. M. (1999) 'Study of the original technique of the wall paintings of the chapter house of Santa Maria de Sigena, Spain (1190–1200)', in *Preprints of the ICOM-CC 12th Triennial Meeting, Lyon, 29th August–3 September 1999, II*, 467–72, London.

Gauthier, J. (1958) 'Etude de quelques propriétés des sels neutres et basiques de cuivre des acides formique, acétique, propionique' (1st thesis); 'Propositions données par la Faculté' (2nd thesis). PhD theses, Procédé Sertic, l'Université de Paris, Lyon.

Gay, M. (1970) 'Essais d'identification et de localisation des liants picturaux par des colourations spécifiques sur coupes minces', *Laboratoire de Recherche des Musées de France, Annales*, 8–24.

Gee, E. (1984) *A Glossary of Building Terms used in England from the Conquest to c. 1550*, Frome.

van Geersdaele, P. C. and L. J. Goldsworthy (1978) 'The restoration of wall painting fragments from St. Stephen's Chapel, Westminster', *The Conservator* **2**, 9–12.

Geilmann, W. (1938) 'Chemische-technische Untersuchung der Wand und Gewölbemalereien in der romanischen Kirche zu Idensen', *Nachrichten von Gesellschaft der Wissenschaften zu Göttingen* **2**, 71–84.

Gettens, R. J. (1938a) 'The materials in the wall paintings of Bāmiyān, Afghanistan', *Technical Studies in the Field of the Fine Arts* **6**, 186–93.

Gettens, R. J. (1938b) 'The materials from the wall paintings at Kizil in Chinese Turkestan', *Technical Studies in the Field of the Fine Arts* **6**, 281–94.

Gettens, R. J. and E. W. Fitzhugh (1993a) 'Azurite and blue verditer', in A. Roy (ed.), *Artists' Pigments: A Handbook of their History and Characteristics*, Vol. 2, 23–35, Oxford.

Gettens, R. J. and E. W. Fitzhugh (1993b) 'Malachite and green verditer', in A. Roy (ed.), *Artists' Pigments: A Handbook of their History and Characteristics*, Vol. 2, 183–202, Oxford.

Gettens, R. J. and M. E. Mrose (1954) 'Calcium sulphate minerals in the grounds of Italian paintings', *Studies in Conservation* **1**, 174–89.

Gettens, R. J. and F. W. Sterner (1941) 'The compatibility of pigments in artists' oil paints', *Technical Studies in the Field of the Fine Arts* **10**, 18–28.

Gettens, R. J. and G. L. Stout (1958) 'A monument of Byzantine wall painting – the method of construction', *Studies in Conservation* **3**, 107–19.

Gettens, R. J. and G. L. Stout (1966) *Painting Materials: A Short Encyclopaedia*, New York.

Gettens, R. J., H. Kühn and W. T. Chase (1993a) 'Lead white', in A. Roy (ed.), *Artists' Pigments: A Handbook of their History and Characteristics*, Vol. 2, 67–81, Oxford.

Gettens, R. J., E. W. Fitzhugh and R. L. Feller (1993b) 'Calcium carbonate whites', in A. Roy (ed.), *Artists' Pigments: A Handbook of their History and Characteristics*, Vol. 2, 203–26, Oxford.

Gettens, R. J., R. L. Feller and W. T. Chase (1993c) 'Vermilion and cinnabar', in A. Roy (ed.), *Artists' Pigments: A Handbook of their History and Characteristics*, Vol. 2, 159–82, Oxford.

Gill, M. and H. Howard (1997) 'Glimpses of Glory: paintings from St Mark's Hospital, Bristol', in L. Keen (ed.), *'Almost the Richest City': Bristol in the Middle Ages* (British Archaeological Association Conference Transactions 19), 97–106, London.

Gill, M., T. Manning, D. Park and S. Stewart (1996) 'The gatehouse wall painting: stylistic analysis', in *Berry Pomeroy Castle* (Proceedings of the Devon Archaeological Society **54** [1996]), 317–23.

Giovannoni, S., M. Matteini and A. Moles (1990) 'Studies and developments concerning the problem of altered lead pigments in wall painting', *Studies in Conservation* **35**, 21–5.

Goldsmith, L. and S. Bucklow (1998) 'The Thornham Parva retable: workshop practice in fourteenth-century English panel painting', in A. Roy and P. Smith (eds), *Painting Techniques: History,*

Materials and Studio Practice (Preprints of the IIC 17th International Congress, Dublin, 1998), 40–44, London.

Goldstein, J. I., D. E. Newbury, P. Echlin, D. C. Joy, A. D. Romig, C. E. Lyman, C. Fiori and E. Lifshin (1992) *Scanning Electron Microscopy and X-ray Microanalysis: A Text for Biologists, Materials Scientists, and Geologists*, 2nd edn, New York.

Gowing, R. (1997) 'Monitoring solar radiation on painted architectural surfaces: an investigation into the use of a CCD camera system', unpublished diploma dissertation, Conservation of Wall Painting Department, Courtauld Institute of Art, London.

Graebe, H., K. Trampedach and M. Jensen (1986) 'Kalkmalerierne i Mårslet Kirke', *National Museete Arbejdsmark* 23, 164–82.

Granger, F. (ed. and trans.) (1962) *Vitruvius de Architectura*, 2 vols, London.

Granville, J. and J. Burbidge (1994) 'Conservation of wall paintings: All Saints' Church, Warlingham, Surrey' [unpublished report, July 1994].

Gribble, C. D. and A. J. Hall (1992) *Optical Mineralogy: Principles and Practice*, London.

Grissom, C. A. (1986) 'Green earth', in R. L. Feller (ed.), *Artists' Pigments: A Handbook of their History and Characteristics*, Vol. 1, 141–67, Cambridge.

Grout, R. (1999) 'Investigations into the causes and nature of blackened vermilion', unpublished diploma dissertation, Department of Conservation and Technology, Courtauld Institute of Art, London.

Grout, R. and A. Burnstock (2000) 'A study of the blackening of vemilion', *Zeitschrift für Kunsttechnologie und Konservierung* 14, 15–22.

Gutscher, D., B. Mühlethaler, A. Portmann and A. Reller (1989) 'Conversion of azurite to tenorite', *Studies in Conservation* 34, 117–22.

Halliwell, J. O. (1848) 'Early English artistical receipts', *Journal of the British Archaeological Association* 3, 107–11.

Halliwell, J. O. (ed.) (1855) *Early English Miscellanies in Prose and Verse, Selected from an Inedited Manuscript of the 15th Century*, Wharton Club Publication 2, 72–91, London.

Hammer, I. (1987) 'The conservation in situ of the Romanesque wall paintings of Lambach', in S. Cather (ed.), *The Conservation of Wall Paintings* (Proceedings of a symposium organised by the Courtauld Institute of Art and the Getty Conservation Institute, London, July 1987), 43–55, Los Angeles.

Hamsík, M. and Tomek, J. (1983) 'Technické paralely deskové a nástênné malby 14, století', *Umêní* 31, 308–16.

Hanzel, D., W. Meisel, D. Hanzel, and P. Gütlich (1990) 'Mössbauer effect study of the oxidation of vivianite', *Solid State Communications* 76, 307–10.

Harley, R. (1982) *Artists' Pigments c.1600–1835*, London.

Harrison, R. K., P. Stone, I. B. Cameron, R. W. Elliot and R. R. Harding (1987) 'Geology, petrology and geochemistry of Ailsa Craig, Ayrshire', *British Geological Survey Report* Vol. 16, No. 9, HMSO.

Harvey, J. H. (1961) 'The Wilton Diptych – a re-examination', *Archaeologia* 98, 1–28.

Haslam, J. (1807) in *Antiquities of Westminster*, 223–5, London.

Hauff, G. (1988) 'Study and conservation of the polychrome portal sculptures of the Holy Cross Minster, Schwäbisch Gmünd', in J. Ciabach (ed.), *6th International Congress on Deterioration and Conservation of Stone, Toruń, 12–14 September 1988*, 712–21, Toruñ.

Haydocke, R. (1598) *A Tracte Containing the Artes of Curious Paintings, Carvinge and Buildinge*, Oxford.

Hedley, G. (1979) 'Analysis of pigments in St. Gabriel's and St. Anselm's Chapels, Canterbury Cathedral', unpublished notes, Department of Conservation and Technology, Courtauld Institute of Art, London.

Heritage, A. (1990) 'Historical and technical report on the wall paintings of the inner choir screen of Cologne Cathedral ...' [unpublished report].

Heydenreich, G. (1994) 'A painted wing of a late Gothic altarpiece from Zittau: iconography, painting technique and restoration', *Restauro* 1, 32–40.

Heywood, B. and S. Oosterwijk (1993) 'Catalogue', *Romanesque: Stone Sculpture from Medieval England* (catalogue of an exhibition at the Henry Moore Institute, Leeds), 35–73, Leeds.

Heyworth, M. (1991) 'Examination of pigment samples from Haughmond Abbey, Shropshire', Ancient Monuments Laboratory Report 12/91.

Higgitt, C., M. Spring and D. Saunders (2002) 'Pigment–medium interactions in oil paint films containing red lead or lead-tin yellow', *Deterioration of Artists' Paints: Effects and Analysis.* Extended abstracts of presentations given at a joint meeting of ICOM-CC Working Groups Paintings 1 & 2 and The Paintings Section, UKIC, British Museum, London, 10–11 September 2001, 21–4.

Hluvko, S. (1991) 'Red pigments in English medieval wall painting', unpublished dissertation, Conserva-

tion of Wall Painting Department, Courtauld Institute of Art, London.
Hofenk-de Graaff, J. H. and W. G. Th. Roelofs (1972) 'On the occurrence of red dyestuffs in textile materials from the period 1450–1600', *ICOM Committee for Conservation, Plenary Meeting*, October 2–5 1972, Madrid, 7–27.
Howard, H. (1988) 'Blue pigments in English medieval wall painting', unpublished dissertation, Conservation of Wall Painting Department, Courtauld Institute of Art, London.
Howard, H. (1989) 'Wall plaster, pigment containers and pigments', in J. A. Stones (ed.), *Three Scottish Carmelite Friaries: Excavations at Aberdeen, Linlithgow and Perth 1980–1986*, 153–4, Edinburgh.
Howard, H. (1990) '"Blue" in the Lewes Group', in S. Cather. D. Park and P. Williamson (eds), *Early Medieval Wall Painting and Painted Sculpture in England* (BAR British Series 216), 195–99, Oxford.
Howard, H. (1991) 'Cormac's Chapel, Cashel: analysis of paint samples', Conservation of Wall Painting Department, Courtauld Institute of Art, London [unpublished report, August 1991].
Howard, H. (1992a) 'All Saints' Church Witley, Surrey: scientific examination of the Romanesque wall paintings', Conservation of Wall Painting Department, Courtauld Institute of Art, London [unpublished report, May 1992].
Howard, H. (1992b) 'Scientific examination of the wall paintings in St. Gabriel's Chapel, Canterbury Cathedral', Conservation of Wall Painting Department, Courtauld Institute of Art, London [unpublished report, December 1992].
Howard, H. (1993a) 'Workshop practices and the identification of hands: Gothic wall paintings at St Albans', *The Conservator* **17**, 34–45.
Howard, H. (1993b) 'The Chapel of Our Lady Undercroft, Canterbury Cathedral, and the relationship of English and Bohemian painting techniques in the second half of the 14th century', *Technologia Artis* 3, 31–4.
Howard, H. (1993c) 'Scientific examination of the polychromy on the early 13th-century reredos in Holy Innocents' Church, Adisham', Conservation of Wall Painting Department, Courtauld Institute of Art, London [unpublished report, April 1993].
Howard, H. (1993d) 'Scientific examination of the polychromy of the de Bois and de Ingham tombs at Ingham, Norfolk', Conservation of Wall Painting Department, Courtauld Institute of Art, London [unpublished report, December 1993].
Howard, H. (1994) 'Eastbridge Hospital, Canterbury: scientific examination of the early thirteenth-century wall paintings in the hall', Conservation of Wall Painting Department, Courtauld Institute of Art, London [unpublished report, July 1994].
Howard, H. (1995a) 'Techniques of the Romanesque and Gothic wall paintings in the Holy Sepulchre Chapel, Winchester Cathedral', in A. Wallert, E. Hermens and M. Peek (eds), *Historical Painting Techniques, Materials, and Studio Practice* (Preprints of a symposium at the University of Leiden, June 1995), 91–104, Los Angeles.
Howard, H. (1995b) 'The Lord Mayor's Chapel (St. Mark's), Bristol: scientific examination of the paintings on stone of *c.* 1500', Conservation of Wall Painting Department, Courtauld Institute of Art, London [unpublished report, January 1995].
Howard, H. (1995c) 'Scientific examination of the wall paintings in the Ante-Reliquary Chapel, Norwich Cathedral,' Conservation of Wall Painting Department, Courtauld Institute of Art, London [unpublished report, March 1995].
Howard, H. (1995d) 'Scientific examination of the wall paintings in the Johanneskapelle, Pürgg, Styria', Conservation of Wall Painting Department, Courtauld Institute of Art, London [unpublished report, March].
Howard, H. (1995e) 'Samples from the west front of Salisbury Cathedral', Conservation of Wall Painting Department, Courtauld Institute of Art, London [unpublished report, July 1995].
Howard, H. (1996a) 'Romanesque and Gothic wall paintings in Winchester Cathedral: an unusual use and alteration of vivianite', *Journal of the Russell Society* 6, 93–6.
Howard, H. (1996b) 'Scientific examination of the "Doom" wall painting in Holy Trinity Church, Coventry', Conservation of Wall Painting Department, Courtauld Institute of Art, London [unpublished report, May 1996].
Howard, H. (1996c) 'Scientific examination of the paintings in the "synagogue" of *c.* 1180, High Street, Guildford', Conservation of Wall Painting Department, Courtauld Institute of Art, London [unpublished report, April 1996].
Howard, H. (1996d) 'Salisbury Cathedral: scientific examination of samples taken from the west front', Conservation of Wall Painting Department, Courtauld Institute of Art, London [unpublished report, June 1996].

Howard, H. (1996e) 'The techniques of English medieval wall painting', Leverhulme Research Fellowship, Conservation of Wall Painting Department, Courtauld Institute of Art, London, compilation of reports and articles, 1991–1996.
Howard, H. (1997a) 'Scientific examination of an oyster-shell palette from the Carmelite Friary, Coventry', Conservation of Wall Painting Department, Courtauld Institute of Art, London [unpublished report, February 1997].
Howard, H. (1997b) 'Peterborough Cathedral, nave ceiling: scientific examination of the original decoration', Conservation of Wall Painting Department, Courtauld Institute of Art, London [unpublished report, September 1997].
Howard, H. (1998a) 'Glasgow Cathedral: scientific examination of Romanesque wall painting fragments excavated from the crypt', Conservation of Wall Painting Department, Courtauld Institute of Art, London [unpublished report, August 1998].
Howard, H. (1998b) 'Peterborough Cathedral: scientific examination of the nave ceiling, bays 36–39', Conservation of Wall Painting Department, Courtauld Institute of Art, London [unpublished report, September 1998].
Howard, H. (1998c) 'Guildford Castle: scientific examination of traces of paint remaining in the antechamber of the medieval chapel', Conservation of Wall Painting Department, Courtauld Institute of Art, London [unpublished report, December 1998].
Howard, H. (2000) 'Westminster Abbey, Chapter House: scientific examination of the wall paintings', Conservation of Wall Painting Department, Courtauld Institute of Art, London [unpublished report, October 2000].
Howard, H. (2001) 'Analysis of the painted plaster', in J. Mitchell and I. Hansen (eds), *San Vincenzo Al Volturno 3: The Finds from 1980–86 Excavations*, 287–995, Spoleto.
Howard, H. (forthcoming) 'Technology of the painted past: recent scientific examination of the medieval wall paintings of the Chapter House in Westminster Abbey', in J. Fidler, R. Gowing and A. Heritage (eds), *Conserving the Painted Past: Developing Approaches to Wall Painting Conservation*, Proceedings of the Conference, London, 2–4 December 1999.
Howard, H. and R. Gasol (1996) 'Scientific examination of the "Losinga Cycle" of Romanesque wall paintings in Norwich Cathedral', *Technologia Artis* 4, 19–24.
Howard, H. and D. Park (1997) 'The painted wall plaster', in M. Lyne, *Lewes Priory: Excavations by Richard Lewis, 1969–82* ed. M. Gardiner, 124–31, Lewes.
Howard, H. and D. Park (forthcoming a) 'The painted wall plaster' in forthcoming report on the excavations of Sherborne Abbey.
Howard, H. and D. Park (forthcoming b) 'The medieval shell palette', in forthcoming volume on the Greyfriars Norwich (East Anglian Archaeological Report series) (in press).
Howard, H. and R. Pathak (1995) 'St. George's Chapel, Ely Cathedral: scientific examination of the newly discovered 13th-century paintings on the soffit of the entrance arch', Conservation of Wall Painting Department, Courtauld Institute of Art, London [unpublished report, March 1995].
Howard, H., T. Manning and S. Stewart (1998) 'Late medieval wall painting techniques at Farleigh Hungerford Castle and their context', in A. Roy and P. Smith (eds), *Painting Techniques: History, Materials and Studio Practice* (Preprints of the IIC 17th International Congress, Dublin, 1998), 59–64, London.
Hughes, H. and M. Heyworth (1990) 'Examination of pigments from Wroxeter, Shropshire', Ancient Monuments Laboratory Report, 91/90.
Hughes, H. and P. Lewis (1988) 'Paint palettes', in T. B. James and A. M. Robinson (eds), *Clarendon Palace: The History and Archaeology of a Medieval Palace and Hunting Lodge near Salisbury, Wiltshire*, 258–60, London.
Hughes, H. and S. Strong (1990) 'Examination of a shell containing pigment from Acton Court, Avon', Ancient Monuments Laboratory Report 82/90.
Hulbert, A. (1986) 'Holy Trinity Church, Coventry, Warwickshire' [unpublished report, February 1986].
Hulbert, A. (1991) 'An examination of the polychromy of Exeter Cathedral roof bosses, and its documentation', in F. Kelly (ed.) *Medieval Art and Architecture at Exeter Cathedral* (British Archaeological Association Conference Transactions XI, 1985), 188–98, London.
Hulbert, A. (1992) 'New discoveries', in A. Hulbert, J. Marsden and V. Todd (eds), *The Conservator as Art Historian*, 33–4, London.
Hulbert, A. (1997) 'Two unusual late fifteenth-century plasters' *Conservation News* 63, 57.
Hulbert, A. (1998) 'English fourteenth-century interior polychromy: manuscript sources and workshop practice at Exeter Cathedral', in A. Roy and P. Smith (eds), *Painting Techniques: History, Materials and*

Studio Practice (Preprints of the IIC 17th International Congress, Dublin, 1998) 35–9, London.

van't Hul-Ehrnreich, E. H. and P. B. Hallebeek (1972) 'A new kind of old green copper pigment found', *ICOM Committee for Conservation, Plenary Meeting,* 2–5 October 1972, Madrid.

Hunt, T. (1995) 'Early Anglo-Norman receipts for colours', *Journal of the Warburg and Courtauld Institutes* 58, 203–9.

Ilg, A. (1874) 'Theophilus Presbyter Schedula diversarum artium. 1. Band. Revidierter Text, Übersetzung und Appendix', *Quellenschriften für Kunstgeschichte und Kunsttechnik des Mittelalters und der Renaissance* 7, 1–374.

Jägers, E. (1991) 'Farbmaterial und technischer Aufbau der Polychromie – Zusammenfassung der naturwissenschaftlichen Untersuchungen', *Zeitschrift für Kunsttechnologie und Konservierung* 5, 214–17.

Jägers, E. and C. Schulze-Senger (1989) 'Zur maltechnik der chorschrankenmalereien im Kölner Dom', *Kölner Domblatt*, 187–98.

James, M. R. and E. W. Tristram (1928–9) 'The wall paintings in Eton College Chapel and in the Lady Chapel of Winchester Cathedral', *Walpole Society* 17, 1–43.

Johnston-Feller, R. (1986) 'Reflections on the phenomenon of fading', *Journal of Coatings Technology* 58, 39–40.

Johnston-Feller, R. and C. W. Bailie (1982) 'An analysis of the optics of paint glazes: fading', in N. S. Brommelle and G. Thompson (eds) *Science and Technology in the Service of Conservation* (Preprints of the IIC 9th International Congress, Washington, 1982), 180–85, London.

Kahn, D. (1986) 'Recent discoveries of Romanesque sculpture at St. Albans', in S. Macready and F. H. Thompson (eds), *Art and Patronage in the English Romanesque* (Society of Antiquaries of London Occasional Paper, new ser. 8), 71–89, London.

Kakoulli, I. (1995) 'Hellenistic and Roman wall paintings in Cyprus: a scientific examination of their technology', MA dissertation (September 1955) Conservation of Wall Painting Department, Courtauld Institute of Art, London.

Kargère, L. (2002) 'The *Montvianex Madonna*: materials and techniques in 12th-century Auvergne', in *Preprints of the ICOM-CC 13th Triennial Meeting, Rio de Janeiro, September 2002*, 507–12, London.

Katkov, M. and C. Oldenbourg (1982) 'St. Mary le Crypt, Gloucester: conservation of wall paintings, chancel, north wall' [unpublished report, November 1982].

Kauffman, A. J. and E. D. Dilling (1950) 'Differential thermal curves of certain hydrous and anhydrous minerals', *Economic Geology* 45, 222–44.

Kerber, G., M. Koller and F. Mairinger (1972) 'Studies of blue-green alterations in Austrian medieval wall paintings', *ICOM Committee for Conservation, Plenary Meeting,* 2–5 October 1972, Madrid.

Keyser, C. E. (1883) *List of Buildings ... having Mural ... Decorations*, HMSO.

Kirby, J. (1977) 'A spectrophotometric method for the identification of lake pigment dyestuffs', *National Gallery Technical Bulletin* 1, 35–44.

Kirby, J. (1999) 'Westminster Abbey Chapter House, Apocalypse scenes: HPLC analysis of dyestuffs', unpublished report, National Gallery Scientific Department, 22 September 1999.

Kirby, J. and R. White (1996) 'The identification of red lake pigment dyestuffs and a discussion of their use', *National Gallery Technical Bulletin* 17, 56–80.

Kirby, J., D. Saunders and J. Cupitt (1997) 'Colorants and colour change', in T. Bakkenist, R. Hoppenbrouwers and H. Limburg (eds), *Early Italian Paintings, Techniques and Analysis* (Postprints of a symposium at Maastricht, 9–10 October 1996), 65–71.

Klein, C. and C. S. Hurlbut (1985) *Manual of Mineralogy* (after J. D. Dana), 20th edn, New York.

Knoepfli, A. and O. Emmenegger (1990) 'Wandmalerei bis zum Ende des Mittelalters', *Reclams Handbuch der künstlerischen Techniken*, II, 7–212, Stuttgart.

Knoepfli, A., O. Emmenegger, M. Koller and A. Meyer (1990) *Reclams Handbuch der künstlerischen Techniken: Wandmalerei Mosaik*, II, Stuttgart.

Koller, M., H. Leitner and H. Paschinger (1990) 'Reconversion of altered lead pigments in alpine mural painting', *Studies in Conservation* 35, 15–20.

Kozarzewski, M. and R. Zankowski (1988) 'Zwei Wandgemälde in der Marienkirche zu Danzig', *Restauro* 94/2, 79–94.

Krautheimer, R. (1927) *Mittelalterliche Synagogen*, Berlin.

Kühn, H. (1993a) 'Verdigris and copper resinate', in A. Roy (ed.), *Artists' Pigments: A Handbook of their History and Characteristics*, Vol. 2, 131–58, Oxford.

Kühn, H. (1993b) 'Lead tin yellow', in A. Roy (ed.), *Artists' Pigments: A Handbook of their History and Characteristics*, Vol. 2, 83–112, Oxford.

Kunz, G. F. (1915) *The Magic of Jewels and Charms*, London.

Kupfer, M. (1982) 'The Romanesque frescoes in the church of Saint-Martin at Nohant-Vicq', unpublished PhD dissertation, Yale University.

Kyi, C. P. and R. E. Drew (1998) 'Initial investigations of the potential for the bioremediation of soluble nylon coatings', unpublished report for English Heritage, Conservation of Wall Painting Department, Courtauld Institute of Art, London.

Langslow, D., F. Allardyce, A. Worrall and D. Winfield (1984) 'The restoration of the painted ceiling, the Jesus Chapel, Canterbury Cathedral', *The Conservator* 8, 15–21.

Larson, J. (1992) 'An outline proposal for the conservation of the Romanesque frieze at Lincoln Cathedral', in D. Kahn (ed.), *The Romanesque Frieze and its Spectator*, 183–92, London.

Larson, J. (1993) 'The Lincoln frieze: a problem of conservation and historical investigation', in *Romanesque: Stone Sculpture from Medieval England* (catalogue of an exhibition at the Henry Moore Institute, Leeds), 28–34, Leeds.

Laurie, A. P. (1913) 'Ancient pigments and their identification in works of art', *Archaeologia* **64**, 316–33.

Laurie, A. P. (1914) *The Pigments and Mediums of the Old Masters*, London.

Laurie, A. P. (1967) *The Painters' Methods and Materials* (abr. repub.), New York 1967.

Lazzarini, L. (1982) 'The discovery of Egyptian blue in a Roman fresco of the medieval period: ninth century', *Studies in Conservation* **27**, 84–7.

Leitner, H. (1989) 'Bericht über die Restaurierung', in *Pfarrkirche St. Georgen ob Judenburg*, 109–13, Graz.

Leveto, P. D. (1985) 'Castel Seprio: architecture and painting', unpublished PhD thesis, Indiana University, Michigan.

Leveto-Jabr, P. D. (1987) 'Carbon-14 dating of wood from the east apse of Santa Maria at Castel Seprio', *Gesta* **26**, 17–18.

Lindley, P. and C. Galvin (1995) 'Pietro Torrigiano's portrait bust of King Henry VII', in P. Lindley (ed.), *Gothic to Renaissance: Essays on Sculpture*, 170–87, Stamford.

Liversidge, M., P. Binski and J. Lynn (1995) 'Two ceiling fragments from the Painted Chamber at Westminster Palace', *Burlington Magazine* **137**, 491–501.

Lombardi, C., M. Mariconda, S. Salvatori, F. Mainieri, U. Santamaria and R. Varoli-Piazza (2000) 'Experimental study on the production and use of artificial blue and green pigments (copper acetates and carbonate) from antiquity to the sixteenth century', in J. Goupy and J. P. Mohen (eds), *Art et chimie, la couleur*, actes du congrès, 31–7, Paris.

Lowe, B. and E. Mason (1987) 'Keynsham Abbey: excavations 1961–1985', *Somerset Archaeology and Natural History* (Proceedings of the Somerset Archaeological and Natural History Society **131**), 81–127.

Lubin, H., J. Spriggs, S. Kingsley, J. L. Wilkinson, J. Swann and C. Kightley (1990) *The Worcester Pilgrim*, ed. P. Barker, Worcester.

Lynn, J. (1997) 'Two thirteenth century panels from the Painted Chamber, Westminster Palace: a discussion of the materials and techniques', *Zeitschrift für Kunsttechnologie und Konservierung* **11**, 15–28.

Mairinger, F. and M. Schreiner (1986) 'Deterioration and preservation of Carolingian and medieval mural paintings in the Müstair Convent (Switzerland). Part I: materials and renderings of the Carolingian wall paintings', in N. S. Brommelle and P. Smith (eds), *Case Studies in the Conservation of Stone and Wall Paintings* (Preprints of the IIC 11th International Congress, Bologna, 1986), 195–6, London.

Malaguzzi-Valerji, V. (1973) 'Ancient fresco technique in the light of scientific examination', in W. J. Young (ed.), *Application of Science in Examination of Works of Art* (Proceedings of the seminar, Boston, June 1970), 164–9, Boston.

'*Mappae Clavicula*' (1974), transl. G. S. Smith and J. G. Hawthorne, *Transactions of the American Philosophical Society* (new ser.) **64**, 3–123.

Martin, E. (1977) 'Some improvements in techniques of analysis of paint media', *Studies in Conservation* **22**, 63–7.

Martin, E. and A. Duval (1990) 'Les deux variétés de jaune de plomb et d'étain: étude chronologique', *Studies in Conservation* **35**, 117–36.

Martin, E. and M. Eveno (1992) 'Contribution to the study of old green copper pigments in easel painting', in *Non-destructive Testing, Microanalytical Methods and Environmental Evaluation for Study and Conservation of Works of Art* (Preprints of the 3rd International Conference on Non-destructive Testing, Microanalytical Methods and Environmental Evaluation for Study and Conservation of Works of Art, Viterbo, October 1992), 781–91, Rome.

Matteini, M. and A. Moles (1981) 'The reconversion of oxidised white lead in mural painting: a control after a five year period', in *ICOM Committee for Conservation 6th Triennial Meeting, Ottawa*, 1–8.

Matteini, M. and A. Moles (1986) 'Elaborazione di un test per il riconoscimento e la localizzazione stratigrafica dei fenomeni di solfatazione nei marmi', *Restauro* **1**, 74–83.

Matteini, M. and A. Moles (1994) 'Mural paintings in Wunstorf-Idensen: chemical investigations on paintings, materials and techniques', in *Forschungsprojekt Wandmalerei-Schäden* (Arbeitshefte zur Denkmalpflege in Niedersachsen, 11), 82–6, Hannover.

Matteini, M., A. Moles and M. Seracini (1989) 'Indagini diagnostiche', in *Un progetto per Piero della Francesca*, 232–55, Florence.

Maul, G. (1992) 'Geschichte, Konservierung und Technologie der Chorschrankenmalereien', in *Kölner Domblatt* (Jahrbuch des zentral-dombau-vereins), 239–60, Sonderdruck.

McCormack, J. K. (2000) 'The darkening of cinnabar in sunlight', *Mineralium Deposita* **35**, 796–8.

McMcrone, W., J. Delly and S. Palenik (1979) *The Particle Atlas, Vol. V: Light Microscopy Atlas and Techniques*, 2nd edn., Michigan.

McGrath, M. (1987) 'The materials and techniques of Irish medieval wall painting', *Journal of the Royal Society of Antiquaries of Ireland* **117**, 96–124.

McLean, T. (1981) *Medieval English Gardens*, London.

Medhurst, L. (1993) 'Hospital of St. Thomas, Eastbridge: condition report and conservation proposals for wall paintings in the refectory' [unpublished reports, August and December 1993].

Merrifield, M. P. (1967) *Original Treatises on the Arts of Painting*, 2 vols, London (1849), repr. edn, New York 1967.

Mills, J. and R. White (1994) *The Organic Chemistry of Museum Objects*, 2nd edn, London.

Miskimin, H. A. (1975) *The Economy of Early Renaissance Europe 1300–1460*, Englewood Cliffs (NJ) 1969 (repr. 1975, Cambridge, Mass.).

Mitchell, J. (1985) 'The painted decoration of the early medieval monastery', in R. Hodges and J. Mitchell (eds), *San Vincenzo al Volturno: The Archaeology, Art and Territory of an Early Medieval Monastery* (BAR International Series 252), 125–76, Oxford.

Moffatt, E. A., N. T. Adair and G. S. Young (1985) 'The occurrence of oxalates on three Chinese wall paintings', in *Application of Science in Examination of Works of Art* (Proceedings of the Seminar, Boston, 7–9 September 1983), 235–8, Boston.

Mora, P., L. Mora and P. Philippot (1984) *Conservation of Wall Paintings*, London.

Morgan, N. J. (1982) *Early Gothic Manuscripts (I) 1190–1250*, London.

Morgan, N. J. (1988) *Early Gothic Manuscripts (II) 1250–1285*, London.

Nadolny, J. (2001) 'The techniques and use of gilded relief decoration by northern European painters, *c*1200–1500', unpublished PhD dissertation, Courtauld Institute of Art, University of London.

Naumova, M. M. and S. A. Pisareva (1994) 'A note on the use of blue and green copper compounds in paintings', *Studies in Conservation* **39**, 277–83.

Naumova, M. M., S. A. Pisareva and G. O. Nechiporenko (1990) 'Green copper pigments in old Russian frescoes', *Studies in Conservation* **35**, 81–8.

Nonfarmale, O. and R. Rossi Maneresi (1987) 'Il restauro del "Portail Royal" della cattedrale di Chartres', *Arte Medievale* (2nd ser.) **I**, 259–75.

Noppen, J. G. (1931) 'The Westminster Apocalypse and its source', *Burlington Magazine* **61**, 146–59.

Nord, A. G. and K. Tronner (2000) 'Chemical analysis of medieval mural paintings in Sweden', in J. Goupy and J. P. Mohen (eds), *Art et chimie, la couleur*, actes du congrès, 97–101, Paris.

Norton, C., D. Park and P. Binski (1987) *Dominican Painting in East Anglia: The Thornham Parva Retable and the Musée de Cluny Frontal*, Woodbridge.

Oakeshott, W. (1972) *Sigena: Romanesque Paintings in Spain and the Winchester Bible Artists*, London.

Odin, G. S. (1988) 'Approche analytique des pigments verts des peintures murales romaines', in *Pigments and Colorants* (Preprints of the International Colloquium at Orléans, December 1988), 75–6, CNRS.

Oellermann, E. (1993) 'Das Grabmal des Pfalzgrafen Heinrich II. Maltechnische Bemerkungen zur Fassung einer Holzskulptur aus dem 13. Jahrhundert', *Restauro* **4**, 234–42.

Organ, T. (1992) 'St. Nicholas' Hospital Church, Harbledown, chancel east wall and window enbrasure: report on the conservation of the mural paintings' [unpublished report, April–May 1992].

Organ, T. and A. Sawdy (1991) 'The Church of St. Margaret of Antioch, Cowlinge, Suffolk' [unpublished report, June–July 1991].

Organ, T., A. Sawdy and T. Smith (1991) 'St. Michael and All Angels, Copford' [unpublished report].

Orna, M. V., M. J. D. Low and N. S. Baer (1980) 'Synthetic blue pigments: ninth to sixteenth centuries: I. Literature', *Studies in Conservation* **25**, 53–63.

Orna, M. V., M. J. D. Low and M. M. Julian, 'Synthetic blue pigments: ninth to sixteenth centuries: II "Silver Blue"', *Studies in Conservation* **30**, 155–60.

Osborne, J. (1984) 'Early medieval wall painting in the lower Church of San Clemente, Rome', unpublished PhD thesis, University of London.

Pächt, O. (1961) 'A cycle of English frescoes in Spain', *The Burlington Magazine* **103**, 166–75.

Paine, S. (1996) 'Conservation of the painted arcades, Winchester Cathedral, Lady Chapel' [unpublished report, March 1996].

Paint Technology Manual 6: Pigments, Dyestuffs and Lakes (1966) eds C. J. A. Taylor and S. Marks, New York.

Palache, C., H. Berman and C. Frondel (1951) *Dana's system of mineralogy,* Vol. 2, New York.

Park, D. (1983) 'The wall paintings of the Holy Sepulchre Chapel', in *Medieval Art and Architecture at Winchester Cathedral* (British Archaeological Association Conference Transactions **6**, 1980), 38–62, London.

Park, D. (1984) 'The "Lewes Group" of wall paintings in Sussex', in R. Allen Brown (ed.), *Anglo Norman Studies VI* (Proceedings of the Battle Conference 1983), 200–237, Woodbridge.

Park, D. (1986) 'The Creation, marginalia and ornament in the refectory paintings of Bushmead Priory', *Bedfordshire Archaeology* **17**, 72–6.

Park, D. (1990) 'The paintings in St. Gabriel's Chapel, Canterbury Cathedral', in *Schäden an Wandmalerei unde ihre Ursachen* (Arbeitshefte zur Denkmalpflege in Niedersachsen, 8), 59–66, Hannover.

Park, D. (1997) 'Survey of the medieval and later polychromy of Worcester Cathedral', unpublished report for the Dean and Chapter.

Park, D. (1998) 'Late twelfth-century polychromy from Glasgow Cathedral', in R. Fawcett (ed.), *Medieval Art and Architecture in the Diocese of Glasgow* (British Archaeological Association Conference Transactions **23**), 35–41, London.

Park, D. and H. Howard (1996) 'The medieval polychromy of Norwich Cathedral Priory', in I. Atherton, E. Fernie, C. Harper-Bill and H. Smith (eds), *Norwich Cathedral: Church, City and Diocese, 1096–1996*, 378–409, London.

Park, D. and H. Howard (forthcoming) 'Examination of Romanesque wall painting fragments excavated from the crypt of Glasgow Cathedral', in S. T. Driscoll (ed.), *Excavations at Glasgow Cathedral 1988–1997*, Society for Medieval Archaeology Monograph 18, Leeds.

Park, D. and P. Welford (1993) 'The medieval polychromy of Winchester Cathedral', in J. Crook (ed.), *Winchester Cathedral: Nine Hundred Years, 1093–1993*, 123–38, Chichester and Winchester.

Pasálek, J. (1992) 'Master Theodoric, mural paintings in the Holy Rood Chapel at Karlštejn Castle', *Technologia Artis* **2**, 25–32.

Perch-Nielsen, K. von Salis and U. Plahter (1995) 'Analysis of fossil coccoliths in chalk grounds of medieval art in Norway', in M. Malmanger, L. Berczelly and S. Fuglesang (eds), *Norwegian Medieval Altar Frontals and Related Material* (Acta ad archaeologiam et artivm historiam pertinentia, Vol. 11), 145–55, Rome.

Percival-Prescott, W. (1957) *The Coronation Chair: An Investigation into the History and Present Condition of the Chair*, London.

Perry, M. (1998) 'St. Albans Cathedral: the upper west wall of the presbytery' [unpublished report].

Petushkova, J. P. and N. Lyalikova (1986) 'Microbiological degradation of lead-containing pigments in mural paintings', *Studies in Conservation* **31**, 65–9.

Petzold, A. (1992) '"His face like lightening": colour as signifier in representations of the holy women at the tomb', *Arte Medievale* (2nd ser.) **2**, 149–55.

Petzold, A. (1995) '*De coloribus et mixtionibus*: The earliest manuscripts of a Romanesque illuminator's handbook', in L. Brownrigg (ed.), *Making the Medieval Book: Techniques of Production*, Proceedings of the 4th Conference of The Seminar in the History of the Book to 1500, Oxford, July 1992, 59–66, Vermont.

Pilc, J. and R. White (1995) 'The application of FTIR-microscopy to the analysis of paint binders in easel paintings', *National Gallery Technical Bulletin* **16**, 73–84.

Plahter, L. E. and U. Plahter (1972) 'The technique of a group of Norwegian Gothic oil paintings', in *Conservation of Paintings and the Graphic Arts* (Preprints of the IIC Lisbon Congress, 9–14 October 1972), 131–8, London.

Plahter, L. E., E. Skaug and U. Plahter (1974) *Gothic Painted Altar Frontals from the Church of Tingelstad: Materials, Technique and Restoration*, Oslo.

Plahter, U. (1984) 'The crucifix from Hemse: analyses of the painting technique', *Maltechnik* **90**, 35–44.

Plahter, U. (1990) 'The capital-lion from Vossestrand: an investigation of the technique of its painting', in *Pigments et Colorants de l'Antiquité et du Moyen Âge: teinture, peinture, enluminure études historiques et physico-chimiques*, Colloque international du CNRS, 273–81, Paris.

Plahter, U. (1995) 'Colour and pigments used in Norwegian altar frontals', in M. Malmanger, L. Berczelly and S. Fuglesang (eds), *Norwegian Medieval Altar Frontals and Related Material* (Acta ad archaeologiam et artivm historiam pertinentia, Vol. 11), 112–26, Rome.

Plahter, U. (2002) 'A survey on pigments and pigment application on thirty-one Norwegian medieval oil paintings dated to 1250–1350', in *Preprints of the ICOM-CC 13th Triennial Meeting, Rio de Janeiro, September 2002*, 446–54, London.

Plahter, U. and S. A. Wiik (1970) 'The Virgin and Child from the church of Dal (Norway): examination and restoration', *Studies in Conservation* **15**, 304–15.

Plender, S. and R. Marchant (2000) 'The Kingston Lacy medieval screenwork', ABPR Meetings Notes, 27 November 2000, *The Picture Restorer* (2000), 23–5.

Plesters, J. (1956) 'Cross-sections and chemical analysis of paint samples', *Studies in Conservation* **2**, 110–55.

Plesters, J. (1963) 'Sancta Sophia, Trebizond: a note on the materials and technique', *Studies in Conservation* **8**, 131–5.

Plesters, J. (1993) 'Ultramarine blue, natural and artificial', in A. Roy (ed.), *Artists' Pigments: A Handbook of their History and Characteristics*, Vol. 2, 37–65, Oxford.

Pliny (1984) C. Plinius Secundus, *Naturalis Historia*, trans. H. Rackham, Cambridge, Mass.

Plummer, P. (1959) 'Restoration of a retable in Norwich Cathedral', *Studies in Conservation* **4**, 106–15.

Plummer, P. (1986) 'The wall paintings in Eton College Chapel', in P. Burman (ed.), *Conservation of Wall Paintings*, 36–40, London.

Pollard, A. M., R. G. Thomas and P. A. Williams, 'Synthesis and stabilities of the basic copper (II) chlorides atacamite, paratacamite and botallackite', *Mineralogical Magazine* **53**, 557–63.

Pursche, J. (2000) 'Konservierung und Restaurierung der Wandmalerein in Urschalling', in R. Gobiet (ed.), *Salzburger Beiträge zur Kunst- und Denkmalpflege*, Vol. 1: *Die Spätgotische Wandmalerei der Michaelskapelle in Piesendorf – Zur Erhaltung und Erforschung mittelalterlicher Wandmalerei im Ostalpenraum*, 143–62, Salzburg.

Rahn-Koltermann, G., D. H. Buss, R. Fuchs and O. Glemser (1991) 'Zur Kenntnis basischer Kupferacetate', *Zeitschrift für Naturforschung* **46b**, 1020–24.

Ramm, H. G. *et al.* (1971) 'The tombs of Archbishops Walter de Gray (1216–55) and Godfrey de Ludham (1258–65) in York Minster, and their contents', *Archaeologia* **103**, 101–47.

Rees Jones, S. (1936) 'Samples of pigment from certain English medieval wall paintings', unpublished report, Courtauld Institute of Art, London.

Rees Jones, S. (1954) 'The Coronation Chair', *Studies in Conservation* **1**, 103–14.

Rees-Jones, S. G. (1990) 'Early experiments in pigment analysis', *Studies in Conservation* **25**, 93–101.

Richard, J. M. (1887) *Mahaut, comtesse d'Artois et de Bourgogne, 1302–1329...*, Paris.

Richter, E. L. (1988) 'Seltene pigmente im Mittelalter', *Zietschrift für Kunsttechnologie und Konservierung* **2**, 171–7.

Rickerby, S. (1988) 'Conservation of medieval wall paintings in Rochester Cathedral', *Friends of Rochester Cathedral Report for 1988*, 9–12.

Rickerby, S (1990a) 'Kempley: a technical examination of the Romanesque wall paintings', in S. Cather, D. Park and P. Williamson (eds), *Early Medieval Wall Painting and Painted Sculpture in England* (BAR British Series 216), 249–61, Oxford.

Rickerby, S. (1990b) 'Bradfield Combust: report on the medieval wall painting' [unpublished report, June 1990].

Rickerby, S. (1991) 'Heat alterations to pigments painted in the fresco technique', *The Conservator* **15**, 39–44.

Rickerby, S. (1996) 'St. Mary the Virgin, Houghton-on-the-Hill, Norfolk: a preliminary assessment of the wall paintings' [unpublished report, May 1996].

Riederer, J. (1974) 'Recently identified Egyptian pigments', *Archaeometry* **16**, 102–9.

Riederer, J. (1997) 'Egyptian blue', in E. W. Fitzhugh (ed.), *Artists' Pigments: A Handbook of their History and Characteristics*, Vol. 3, 23–45, Oxford.

Rohl, B. M. (1996) 'Lead isotope data from the Isotrace Laboratory, Oxford: *Archaeometry* data base 2, galena from Britain and Ireland', *Archaeometry* **38**, 165–80.

Rollier-Hanselmann, J. (1997) 'D'Auxerre à Cluny: technique de la peinture murale entre le XIIIe et le XIIe siècle en Bourgogne', *Cahiers de civilisation médiévale* **40**, 57–90.

Rossi Maneresi, R. (1987) 'Considerazione tecniche sulla scultura monumentale policromata, romanica e gotica', *Bollettino d'Arte* **41**, 173–86.

Rouse, E. C. and A. Baker (1967) 'Wall paintings in Stoke Orchard Church, Gloucestershire', *Archaeological Journal* **123**, 79–117.

Roy, A. (1984) 'St. Mary's Church, Erwarton, Suffolk: polychromed monument to Sir Bartholomew Bacon and Joan de Heveningham', unpublished analytical report, Scientific Department, National Gallery, London.

Roy, A. (1997) 'The technique of the Wilton Diptych', in D. Gordon, L. Monnas and C. Elam (eds), *The Regal Image of Richard II and the Wilton Diptych*, 125–35, London.

Roy, A. (1998) 'Forbes Prize Lecture', *IIC Bulletin* **6**, 1–5.

Roy, A. and R. White (1982) 'Wells Cathedral: west front polychromy', unpublished report, Scientific Department, National Gallery, London.

Rudler, F. W. (1905) *A Handbook to a Collection of Minerals of the British Islands Mostly Selected from the Ludlam Collection in the Museum of Practical Geology*, HMSO.

Salzman, L. F. (1952) *Building in England Down to 1540: A Documentary History* (repr. 1992), Oxford.

Sampson, J. (1996) 'Glastonbury Abbey: report on the surviving polychromy of the Lady Chapel' [unpublished report, 2 vols, March 1996].

Sampson, J. (1998) *Wells Cathedral West Front: Construction, Sculpture and Conservation*, Stroud.

Sapin, C. and C. Couprey (1992) 'Les Peintures de Saint-Germain d'Auxerre', in *Edifices & Peintures aux IV–XI siècles* (Actes du Colloque CNRS, November 1992), 81–98, Auxerre.

Sauerberg, M. L., H. Howard and A. Tavares da Silva (forthcoming) 'The wall paintings of *c.* 1300 in the Ante-Reliquary Chapel, Norwich Cathedral and the Thornham Parva Retable: a technical comparison' in A. Massing (ed.), *Paintings and Practice, the Thornham Parva Retable: Technique, Conservation and Context of an English Medieval Painting*, Hamilton Kerr Institute, University of Cambridge.

Saunders, D. and J. Kirby (1994) 'Light-induced colour changes in red and yellow lake pigments', *National Gallery Technical Bulletin* **15**, 79–97.

Saunders, D. and J. Kirby (1996) 'Light-induced damage: investigating the reciprocity principle', in *11th Triennial Meeting Edinburgh, Scotland*, Vol. 1 (Preprints of the ICOM Committee for Conservation Meeting, September 1996), 87–90, London.

Saunders, D., M. Spring and C. Higgitt (2002) 'Colour change in red lead-containing paint films', *Preprints of the ICOM-CC 13th Triennial Meeting, Rio de Janeiro, September 2002*, 455–63, London.

Saunders, H. W. (1930) *An Introduction to the Obedientiary and Manor Rolls of Norwich Cathedral Priory*, Norwich.

Sawdy, A. (1995) 'An examination of the kinetics of deliquescence and recrystallisation of some soluble salts found in wall paintings', unpublished MSc dissertation, Conservation of Wall Painting Department, Courtauld Institute of Art, London.

Sawdy, A., A. Heritage and S. Cather (1998) 'St Botolph's Church, Hardham: conservation of the wall paintings. Phase 1: investigations, recording, treatment testing and assessment', unpublished report for English Heritage (February 1998), Conservation of Wall Painting Department, Courtauld Institute of Art, London.

Schäfer, G. (1855) *Das Handbuch der Malerei vom Berge Athos*, Trier.

Schuler, F. (1995) 'The pictorial program of the chapterhouse of Sigena', unpublished PhD thesis, Institute of Fine Arts, New York University.

Schweppe, H. (1997) 'Indigo and woad', in E. W. Fitzhugh (ed.), *Artists' Pigments: A Handbook of their History and Characteristics*, Vol. 3, 81–107, Oxford.

Schweppe, H. and H. Roosen-Runge (1986) 'Carmine – cochineal carmine and kermes carmine', in R. L. Feller (ed.), *Artists' Pigments: A Handbook of their History and Characteristics*, Vol. 1, 255–83, Cambridge.

Schweppe, H. and J. Winter (1997) 'Madder and alizarin', in E. W. Fitzhugh (ed.), *Artists' Pigments: A Handbook of their History and Characteristics*, Vol. 3, 109–42, Oxford.

Scott, D. A. (2000) 'A review of copper chorides and related salts in bronze corrosion and as painting pigments', *Studies in Conservation* **45**, 39–53.

Scott, D. A., Y. Taniguchi and E. Koseto (2001) 'The verisimilitude of verdigris: a review of the copper carboxylates', *Reviews in Conservation* **2**, 73–91.

Scott Robertson, W. A. (1880) 'The crypt of Canterbury Cathedral', *Archaeologia Cantiana* **13**, 48–80.

Setton, K. M. (1977) 'A history of the Crusades', in K. M. Setton (ed.), *The Art and Architecture of the Crusader States* IV, Wisconsin.

Shekede, L. (1998) 'English domestic wall painting on earthen supports: materials and deterioration', unpublished MA dissertation, Conservation for the Conservation of Wall Painting Department, Courtauld Institute of Art, London.

Sherlock, D. and H. Woods (1988) *St. Augustine's Abbey: Report on Excavations, 1960–78*, Maidstone.

Sibilia, E. *et al.* (1988) 'Santa Maria Foris Portas. Analisi di datazione con termoluminescenza', *Notiziario della Soprintendenza Archeologica della Lombardia, 1987*, 81, Milan.

Sinclair, E. (1991) 'The west front: II. The west front polychromy', in F. Kelly (ed.), *Medieval Art and Architecture at Exeter Cathedral* (British Archaeological Association Conference Transactions XI, 1985), 116–33, Leeds.

Sinclair, E. (1992) 'Exeter Cathedral: exterior polychromy', in *The Conservator as Art Historian* (Proceedings of the UKIC Wall Paintings Section Conference, June 1992), 7–14, London.

Sinclair, E. (1995a) 'The polychromy of Exeter and Salisbury Cathedrals: a preliminary comparison', in A. Wallert, E. Hermens and M. Peek (eds), *Historical Painting Techniques, Materials, and Studio Practice* (Preprints of a symposium at the University of Leiden, June 1995), 105–10, Los Angeles.

Sinclair, E. (1995b) 'Pulpitum paintings', *Friends of Exeter Cathedral Annual Report* 65, 15–19.

Sinclair, E. (1997) 'Salisbury Cathedral west front central porch: polychromy report' [unpublished report].

Sinkankas, J. (1972) *Gemstone and Mineral Data Book*, Winchester.

Skaug, E. (1963–4) *St. Nicholas fra Eidsborg Stavkirke (U. 733): malerteknisk undersøkelse og kopiering*, Universitetets Oldsaksamlings Årbok.

Smith, J. T. (1807) *Antiquities of Westminster*, London.

Smith, M. E. (1994) 'Vivianite from south Humberside and Yorkshire, England', *Journal of the Russell Society* 5, 119–20.

Spence, N. and A. Kent (1970) *The Old Churches of Norwich*, Norwich.

Spring, M. and R. Grout (2002) 'The blackening of vemilion: an analytical study of the process in paintings', *National Gallery Technical Bulletin* **23**, 50–61.

Staniland, K. (1986) 'Court style, painters and the Great Wardrobe', in W. M. Ormrod (ed.), *England in the Fourteenth Century* (Proceedings of the 1985 Harlaxton Symposium), 236–46, Woodbridge.

Steane, J. M. (1985) *The Archaeology of Medieval England and Wales*, Beckenham.

Stewart, S. (1995) 'Painted wall plaster', in 'Excavation of the hospital of St. Giles by Brompton Bridge, North Yorkshire', *Archaeological Journal* **152**, 109–245.

Stirnemann, P. (1995) 'Current research in France concerning colour and technique in manuscripts and wall paintings of the 13th and 14th centuries', in M. Malmanger, L. Berczelly and S. Fuglesang (eds), *Norwegian Medieval Altar Frontals and Related Material* (Acta ad archaeologiam et artivm historiam pertinentia, Vol. 11), 137–43, Rome.

Stolow, N. (1957) 'The action of solvents on drying oil films', *Journal of the Oil and Colour Chemists' Association*, Part I **40/5**, 377–402; Part II **40/6**, 488–500.

Stos-Gale, S. (1999) 'The lead isotope compositions and the origin of the lead used for production of pigments in the wall paintings in Westminster Abbey', unpublished report (21 September, 1999), Isotrace Laboratory, Oxford.

Stothard, C. A. (1876) *The Monumental Effigies of Great Britain*, London.

Stowe, B. B., E. Epstein and M. W. Nabors (1967) 'Origin of indigo of woad', *Nature* **216**, 547–9.

The Strasbourg Manuscript: A Medieval Painter's Handbook (1966) eds V. and R. Borradaile, London.

Subes-Picot, M. P. (1992) 'Peinture sur pierre: note sur la technique des peintures du XIIIe siècle découvertes à la cathédrale d'Angers', *Revue de l'Art* **97**, 85–93.

Tångeberg, P. (1984) 'The crucifix from Hemse', *Maltechnik* **90**, 24–34.

Taralon, J. (1986a) 'Les peintures murales romanes de la salle capitulaire de l'ancienne abbaye de Vedôme (I)', Congrès Archéologique de France, 139th session, *Blésois et Vedômois* (1981) (1986) 405–36.

Taralon, J. (1986b) 'Montoire. Chapelle Saint-Gilles', Congrès Archéologique de France, 139th session, *Blésois et Vedômois* (1981) (1986) 259–89.

Tatton-Brown, T. (1989) *Great Cathedrals of Britain*, London.

Tennent, N. H. and K. M. Antonio (1981) 'Bronze disease: synthesis and characterisation of botallackite, paratacamite and atacamite by infra red spectroscopy', in *ICOM Committee for Conservation 6th Triennial Meeting, Ottawa*, 18/23/3, 1–11.

Theophilus (1979) *On Divers Arts. The Treatise of Theophilus*, eds J. G. Hawthorne and C. S. Smith, Chicago 1963 (repr. 1979, New York).

Theophilus (1986) *Theophilus: The Various Arts, De Diversis Artibus*, ed. and transl. C. R. Dodwell, Oxford.

Thompson, D. V. (1926) 'Liber de coloribus illuminatorum siue pictorum from Sloane MS no. 1754', *Speculum* **1**, 280–307.

Thompson, D. V. (1933) 'Artificial vermilion in the Middle Ages', in *Technical Studies in the Field of the*

Fine Arts **2**, 62–70.

Thompson, D. V. (1935a) 'Medieval color-making: Tractatus qualiter quilibet artificialis color fieri possit from Paris, B.N., MS latin 6749[b]', *ISIS* **64**, 456–68.

Thompson, D. V. (ed.) (1935b) 'De Coloribus, Naturalia Exscripta et Collecta, from Erfurt, Stadt-Bücherei, MS. Amplonius Quarto 189 (XIII–XIV century)', *Technical Studies* **3**, 133–45.

Thompson, D. V. (1935c) 'Trial index to some unpublished sources for the history of medieval craftsmanship', *Speculum* **10**, 410–31.

Thompson, D. V. (1935d) 'The Liber Magistri Petri de Sancto Audemaro de coloribus faciendis', *Technical Studies in the Field of the Fine Arts* **4**, 28–33.

Thompson, D. V. (1956) *The Materials and Techniques of Medieval Painting*, New York.

Thomson, G. (1986) *The Museum Environment*, London.

Thornton, G. A. (1928) *A History of Clare, Suffolk*, Cambridge.

Tite, M., M. Bimson and M. Cowell (1982) 'Technical examination of Egyptian blue', *Archaeological Chemistry* **3**, 215–37.

Tomlinson, P. (1985) 'Use of vegetative remains in the identification of dyeplants from waterlogged ninth-tenth century A.D. deposits at York', *Journal of Archaeological Science* **12**, 269–83.

Torraca, G. (1988) *Porous Building Materials*, 3rd edn, Rome.

Tracy, C. (1997) 'An English painted screen at Kingston Lacy', *Apollo* **146**, 20–28.

Tristram, E. W. (1944) *English Medieval Wall Painting: The Twelfth Century*, Oxford.

Tristram, E. W. (1950) *English Medieval Wall Painting: The Thirteenth Century*, Oxford.

Tristram, E. W. (1955) *English Wall Painting of the Fourteenth Century*, London.

University College London (UCL) (1996) UCL Painting Analysis, 'Tewkesbury Abbey: paint samples from various stone fragments' [unpublished report].

Vale, M. (1990) *The Angevin Legacy and the Hundred Years War 1250–1340*, Oxford.

Veen, van der M., A. R. Hall and J. May (1993) 'Woad and the Britons painted blue', *Oxford Journal of Archaeology* **12**, 367–71.

Véliz, Z., C. Babington, M. González and A. Illán (1992) 'Technique and treatment of the St. Michael Weighing Souls in the Royal Convent of Santa Clara, Toledo, Spain', in H. Hodges *et al.* (eds), *Conservation of the Iberian and Latin American Cultural Heritage* (Preprints of the IIC 14th International Congress, Madrid, 1992), 168–74, London.

Vetterli, W. A. (1950) 'Zur Geschichte des Indigos', *Ciba-Rundschau* **93**, 3416–21.

Villers, C., G. van Heemstra and C. Reynolds (1997) 'A fourteenth-century German triptych in the Courtauld Gallery', *Burlington Magazine* **139**, 668–75.

Wallert, A. (1984) 'Orpiment and realgar', *Maltechnik* **4**, 45–55.

Wallert, A. (1986a) 'Verzino and roseta colours in 15th century Italian manuscripts', *Maltechnik (Restauro)* **3**, 52–68.

Wallert, A. (1986b) 'Fluorescent assay of quinone, lichen and redwood dyestuffs', *Studies in Conservation* **31**, 145–55.

Wallert, A. (1991a) 'Wie man im Mittelalter Blaupigmente herstellte', *Restauro* **97**, 13–17.

Wallert, A. (1991b) 'Cimatura de Grana: identification of natural organic colorants and binding media in medieval manuscript illumination', *Zeitschrift für Kunsttechnologie und Konservierung* **5**, 74–83.

Watson, T. L. (1918) 'The color change in vivianite and its effect on the optical properties', *The American Mineralogist* **3**, 159–61.

Weber, F. W. (1923) *Artists' Pigments, their Chemical and Physical Properties*, New York.

Weeks, C. (1998) 'The "Portail de la Mere Dieu" of Amiens Cathedral: its polychromy and conservation', *Studies in Conservation* **43**, 101–8.

Wehlte, K. (1954) 'St. Mary and All Saints', Willingham, Cambridge' [unpublished analytical report, July 1954].

Wehlte, K. (1975) *The Materials and Techniques of Painting*, New York.

Welford, P. (1991) 'An investigation into the phenomenon of dark flesh areas in English medieval wall paintings', unpublished dissertation, Conservation of Wall Painting Department, Courtauld Institute of Art, London.

Westhoff, H. (2002) 'Ein kleinod romanischer fassmalerei: maltechnik und restaurierung einer thronenden Muttergottes', *Restauro* **1**, 34–9.

White, R. (1989) 'Wall paintings, Horsham St. Faith Priory, Norfolk: organic analysis', unpublished report, Scientific Department, National Gallery, London.

White, R. (1993) 'Shrine of St. Albans: polychromed fragments', unpublished report (January 1993), Scientific Department, National Gallery, London.

White, R. (1995) 'Analyses of Norwegian medieval paint media: a preliminary report', in M. Malmanger, L. Berczelly and S. Fuglesang (eds), *Norwegian Medieval Altar Frontals and Related Material* (Acta ad archaeologiam et artivm historiam pertinentia, Vol. 11), 127–35, Rome.

White, R. (1998) 'Analysis of samples from wall paintings', unpublished report (March 1998), Scientific Department, National Gallery, London.

White, R. and J. Pilc (1996) 'Media analysis', *National Gallery Technical Bulletin* **17**, 91–103.

Carus Wilson, E. M. (1937) *The Overseas-trade of Bristol in the Later Middle Ages*, Bristol.

Winfield, D. (1968) 'Middle and later Byzantine wall painting methods: a comparative study', *Dumbarton Oaks Papers* **22**, 62–139.

Winfield, D. (1977) 'The Cathedral of Christchurch at Canterbury: a report on the wall paintings' [unpublished report, July 1977].

Winter, J. (1983) 'The characterization of pigments based on carbon', *Studies in Conservation* **28**, 49–66.

Wong, L. (forthcoming) 'The medieval painted wall plaster', in J. D. Richards, *et al.* (eds), *The College of the Vicars Chapel in the Bedern* (The Archaeology of York, 10: The Medieval City North-east of the Ouse), York (in press).

Woudhuysen-Keller, R. and P. Woudhuysen (1998) 'Thoughts on the use of the green glaze called "copper resinate" and its colour changes', in E. Hermens (ed.), *Looking Through Paintings: The Study of Painting Techniques and Materials in Support of Art Historical Research*, 133–46, London.

Wunderlich, C. H. (1998) 'Zur diskussion gestellt: Kupfercalciumacetat – wichtiges blaupigment in der wandmalerei des mittelalters?' *Restauro* **1**, 22–5.

Zuixiong, L. (1989) 'Effects of light and humidity on the three pigments of haematite, vermilion and red lead', in *Compilation of the Dissertation on Dunhuang Magao Grottoes Conservation*, Dunhuang Research Academy, 239–62.

Zuixiong, L. (1990) 'Research on the discolouration of red pigments of Dunhuang wall paintings', unpublished technical paper, Dunhuang Research Academy.

Zumbrunn, U. (1994) 'Der Skulpturenfund der Münsterplattform in Bern', *Zeitschrift für Kunsttechnologie und Konservierung* **8**, 112–29.

Index

Abbey of La Trinité, Vendôme, France, chapter house 191
Aby crucifix, Denmark 77
Adisham reredos, Kent *see* Holy Innocent's, Adisham
Aeschi, Switzerland 83
alkali-sensitive pigments 5, 6, 11, 13, 68, 166, 170, 201, 202
All Saints', Witley, Surrey 4, 5, 127, 135, 144–7, 168, 177–9, 183, 183, 198, 199, 209
All Saints' and St. Andrew, Kingston, Cambridgeshire 128, 130
alumina (hydrated) 111, 113, 114, 120, 122
Amiens Cathedral, France
  Portail de la Mère Dieu 23, 107, 122, 123, 133, 151, 175, 183
Angers Cathedral, France 2, 6, 7
  choir 22, 46, 47, 83, 94, 106, 120, 132, 150, 152, 157, 158, 162, 169, 181, 93, 206
Anglo-Saxon 5, 9, 37, 39, 40, 57, 196
Arezzo, Italy *see* S. Francesco, *Legend of the True Cross*
arriccio 143
Ashby St. Ledgers, Northamptonshire *see* St. Mary and St. Leodegarius
Auxerre Cathedral, France *see* Saint-Etienne, Auxerre
azure 27, 40, 41, 46, 50–52, 54, 59, 154, 176, 189
azurite 7, 11, 27, 28, 30, 31, 32, 34–6, **40–50**, 62, 63, 78, 80, 83, 84, 101, 104, 119, 121, 122, 125, 132, 179, 182, 190, 196, 199, 202, 203, 206, 207
  alteration 42, 49–50, 55, 74, 77

Badākshān, Afganistan 27, 33
Bāmiyān, Afganistan 33
Berne Minster, Switzerland,
  excavated polychromed sculpture 163
Berry Pomeroy Castle, Devon
  Gatehouse, east wall 128, 135, 152, 192
Berzé-la-Ville, France
  Monk's Chapel 70, 76, 106, 131, 136, 145, 150
Beverley, Humberside, Dominican Friary 175
bice 50
Binsted, West Sussex
  chancel, north wall 177, 184
black chalk 187
black earth pigments **194**
Bradfield Combust, Suffolk 88, 94, 101, 102, 104
Bradwell-juxta-Coggeshall, Essex *see* Holy Trinity
Brent Eleigh, Suffolk
  chancel, east wall 53, 56, 91, 168, 204
Brinay-sur-Cher, France *see* St.-Aignan
Bristol *see* St. Mark's, The Lord Mayor's Chapel, Bristol
British Museum (St. Stephen's Chapel wall painting fragments) *see* Westminster Palace
Brompton Bridge, N. Yorkshire *see* St. Giles Hospital, Brompton Bridge
brushes 21, 143
bole 142, 174
bone black 124, 187, 188, **193**
bone white **185**
Bourges Cathedral, France
  external polychromy 183
Bouts, Dieric and workshop *see* National Gallery, London
buon fresco 3–6, 8, 13, 16, 22, 28, 42, 46, 62, 87, 101, 105, 114, 126, 136, 143, 149, 155, 161, 167, 198, 201, 202, 209
Bushmead Priory, Bedfordshire, refectory 101, 110

calcite 28, 29, 182
calcium
  carbonate 2, 61, 63, 74, 75, 82, 105, 111, 113, 114, 116–18, 120, 124, 125, 129, 131–3, 149, 165, 171, 181, 185, 197, 203
  hydroxide 3, 4, 8, 114, 149, 155, 165, 166, 167, 194, 201, 204
  oxalate 170
  sulphate dihydrate (gypsum) 33, 60, 74, 75, 77, 107, 111, 113, 116, 120, 133, 166, 169, 170, 172, **173–6**, 197

anhydrite 173
calcium sulphate grounds 175
Canterbury *see also* Eastbridge Hospital, St. Augustine's Abbey and shell palettes
Canterbury Cathedral 3, 6
Chapel of Our Lady Undercroft 8, 60, 89, 101, 103, 110, 128, 130, 134, 151, 175, 176, 204
polychromed screens 47, 49, 54, 55, 83, 92
cloister 5, 148
Martyrdom Doorway 107, 133, 182
Legend of St. Eustace 103, 128, 129, 131, 146
St. Anselm's Chapel 29, 68, 80, 144
St. Gabriel's Chapel 2, 4, 5, 11–13, 21, 29, 30, 31, 32, 33, 34, 67, 72, 80, 87, 88, 100, 101, 103, 105, 111, 127, 130, 134, 137, 139, 140, 145, 147, 148, 150, 168, 177–9, 183, 191, 198, 199–203, 205, 209
carbon black 2, 76, 91, 100, 102, 105, 117, 118, 124, 129, 130, 136, 147, 151, 180, 183, **187–94**, 197, 201
charcoal black 22, 31, 32, 33, 34, 37, 47, 50, 54, 61, 68, 69, 75, 77, 80, 81, 88, 89, 104, 105, 107, 108, 116, 118, 125, 130, 133, 137, 143, 148, 149, 157, 167, 168, 179, 180, 182, 187, 188, **189–92**, 193, 199, 205
carbonation 3–6, 91, 131, 135, 148, 149, 158, 166, 167, 168, 169, 184, 199, 201, 209
Cartmel Fell, Cumbria *see* Chapel of St. Anthony
casein 6, 44, 46, 105, 119, 131, 139, 140, 166, 169, 203
Cashel, Ireland
Cormac's Chapel 10, 12, 13, 22, 29, 30, 31, 33, 34, 101, 102, 104, 127, 148, 167, 177, 190, 192, 198, 199, 205
Castelseprio, Italy 39, 40
Cennini, *see* manuscripts, Cennino Cennini
ceruse *see* lead white
cerussite 137–40, 178, 183, 203, 211
chalk 94, 105, 114, 118, 119, 123, 129, 131, 148, 149, 156, 169, **171–3**, 175
grounds 108, 172, 173, 183
Chapel of St. Anthony, Cartmel Fell, Cumbria
figure of Christ 63, 172
Chaptal, Jean 1
Chartres Cathedral, France
Portail Royal 83, 183
Chichester Cathedral
north transept 68, 80, 81, 146
Chippenham, Cambridgeshire 192
chrysocolla *see* copper silicate
cinnabar *see* vermilion
Clarendon Palace, Wiltshire
Antioch Chamber, shell palette 44, 128
excavated wall painting fragments 30, 88
Cluny Abbey, France 132
coal black **193–4**
Colchester Castle, Essex
excavated wall plaster 189
Cologne Cathedral, Germany 2
choir screen wall paintings 2, 7, 22, 46, 90, 106, 120, 121, 124, 150, 151, 182, 188, 205
Cologne, Germany *see* S. Maria im Kapitol
Coombes, W. Sussex, nave, dado zone 4, 168, 189, 190
Copford, Essex *see* St. Michael and All Saints, Copford
copper hydroxides 91, 204
copper oxalate 42, 75
copper pigments 7, 24, 154, 158, 159, 181, 200
acetates (*see also* verdigris) 50
alteration 14
carbonates *see* azurite and malachite
chlorides *see* salt greens
copper blue 120, 158
oleates 87, 93–5
phosphate 84
resinate 44, 63, 87, 90, **93–5**
silicates 45
sulphates 80, 200
antlerite 84
posnjakite 84
synthetic copper blues 11, 29, 35, 41, 44, 45, **50–57**, 168, 190, 199, 200
Courtauld Institute of Art, London 2, 23
Gallery, Estouteville Triptych 23, 48, 90, 91, 122, 163
Coventry, West Midlands *see* Holy Trinity, Coventry and shell palettes
Cowlinge, Suffolk *see* St. Margaret of Antioch

Danish wall painting 76
Davy, Sir Humphrey 1
Ducklington Rectory, Oxfordshire 63
Durham Cathedral 103
Deanery, entrance hall, north wall 24, 60, 61, 64, 146, 193
Galilee Chapel 29, 30, 33, 149, 184
Duxford, Cambridge *see* St. John's

earth pigments 22
Eastbridge Hospital, Canterbury, Kent
hall, north wall 60–64, 88, 93, 146, 148, 149, 167, 170, 179–81
Edward II 207
Edward III 21, 154
egg shells 51, 60, 115
egg tempera
egg white 46, 106, 115, 130, 166
whole egg and egg yolk 5–7, 28, 33, 46, 47, 82, 87, 106, 109, 114, 125, 131, 156, 157, 158, 160, 182, 188, 199
Egyptian blue 27, **39–40**
Ely Cathedral 20
accounts 41, 54, 98, 126, 154, 157
Lady Chapel 107, 132
Prior Crauden's Chapel 67, 69, 104
St. George's Chapel 6, 45, 146, 167
Ely, Cambridgeshire
Silver Street
No. 7 128, 130
No. 9 103
Emmaus Abbey, Prague, Czech Republic 208
Erwarton, Suffolk *see* St. Mary's
Estouteville Triptych *see* Courtauld Institute of Art, Gallery
Eton College Chapel, Berkshire

north and south walls 7, 22, 64, 117, 120, 125, 128, 161, 162, 178, 210
Exeter Cathedral 9, 20
accounts 9, 20, 41, 54, 58, 59, 86, 98, 113, 165, 176
external polychromy 23
roof bosses 20, 23, 41, 63, 121, 150
west front 47, 63, 77, 90, 92, 95, 107, 121, 123, 132, 150, 152, 157, 162, 170, 172, 182, 193

'false blue' 31, 50, 167, **189–91**, 199
Farleigh Hungerford Castle, Somerset, St. Leonard's Chapel 7, 88, 89, 91, 92, 117, 118, 120, 123, 125, 128, 131, 155, 156, 159, 161–3, 177, 179, 180, 182, 204, 205, 210
Farringdon, Hampshire
Christ in Majesty 53, 56, 135, 155, 158, 159
Ferrara Cathedral
facade portals 107
Fiddleford Manor, Dorset
solar 101, 102
Fitzalan effgies, Fitzalan Chapel, Arundel 23, 90, 121, 162, 172, 182, 191
folium 58, 155
Freiberg, Germany, St. Marien Cathedral
15th-century wall paintings 83, 163
Freising, Germany
Diocesan Museum, *Virgin* 163
French treasury accounts
Countess of Artois 20, 58, 59, 113, 121, 142, 154, 182
fresco see buon fresco
Freudenstädter, Schwarzwald, Germany
Romanesque lectern 37, 77, 196

Garde, Gotland, Sweden 77
gelatin 187
gesso 174, 175, 183
giallorino see also lead-tin yellow 160, 161
Glasgow Cathedral, Scotland 10
crypt, excavated wall painting fragments 13, 65, 67, 68, 69, 100–102, 104, 105, 116, 127, 129, 144, 149, 169, 191, 198
glass 17, 18, 154, 161
glaze 6–8, 89, 93, 94, 106, 109, 111, 113, 116–19, 122, 124, 134, 156, 159, 180, 198, 202, 204, 206, 209, 210
Glastonbury Abbey, Somerset
Lady Chapel 23, 29, 30, 31
Ghent, Belgium *see* St. Bavo's Abbey
Gloucester *see* St. Mary le Crypt
glue (animal) 5, 13, 28, 43, 46, 105, 106, 119, 122, 131, 132, 135, 156, 169, 171, 174, 206
rabbit skin 7
goethite *see also* yellow earth 141, 143, 194
gold 27
gold application (gilding, gold leaf) 5, 7, 20, 21, 39, 41, 107, 119, 121, 125, 129, 133, 148, 151, 156, 172, 174, 180, 181, 194, 202, 207, 210
Gothic (English wall painting technique) 6–8

Great Hockham, Norfolk
north aisle, north wall 53, 54, 55
green earth **65–72**, 75, 76, 82, 116, 119, 143, 168, 169, 200, 206
celadonite 65, 66, 67
glauconite 65, 66, 70
green, mixed 32, 34, 62, 68, 105, 161
Guildford Castle, keep, medieval chapel 150
Guildford, High Street,
12th-century 'synagogue' 60–63
gum 28, 109, 130, 131, 139, 158
gum arabic 43, 158
Gurk Cathedral, Austria
Bishop's Chapel, west gallery 49, 77, 144

haematite *see also* red earth 141, 142, 148, 152
Harbledown, Kent *see* St. Nicholas' Hospital Church, Harbledown
Hardham, West Sussex *see* St. Botolph's
Härkeberga, Uppland, Sweden 77, 83
Härnevi, Uppland, Sweden 77, 83
Haslam, J. 1, 2
heat alteration of pigments 22, 42, 50, 66, 72, 86, 126, 143, 152, 154, 166, 171
Hemse crucifix, Sweden 107, 182, 183
Henham, Essex
chancel, north wall 144
Henry III 21, 128
Herlin altarpiece, St. James, Rothenburg, Germany 63, 122
high performance liquid chromatography 11
Hildesheim, Germany *see* St. Michael
Holy Cross Minster, Schwäbisch Gmünd, Germany
polychromed portals 37, 90, 133, 183, 196
Holy Innocent's, Adisham, Kent
reredos 34, 90, 108, 133, 175, 183
Holy Trinity, Bradwell-juxta-Coggeshall, Essex 127, 135, 138, 146
Holy Trinity, Coventry, West Midlands
chancel arch, *Last Judgement* 101, 105, 111, 117, 119, 120, 123, 125, 146, 160, 175, 193
Hørning stave church, Denmark,
Hørning plank 77, 132, 157, 182, 199
Horsham St. Faith Priory, Norfolk,
refectory 6, 7, 43, 44, 53, 88, 89, 117, 118, 120, 127, 178, 195
Houghton-on-the-Hill, Norfolk 4
House of Livia, Rome 39

Ickleton, Cambridgeshire
chancel arch 53, 55, 88, 91, 94, 205
nave, north arcade 30, 33
Idensen, Germany 2, 5, 12, 22, 34, 74, 76, 77, 106, 145, 150, 169, 180, 181, 183, 191, 197–9, 205
impasto highlights 4, 67, 131, 132, 146, 147, 149, 168, 170, 180–82, 201
imprimatura 108, 183
indigo 11, 14, 35, 41, 47, 50, **57–64**, 98, 104, 154–157, 160, 172, 176, 180, 190, 193, 199, 200
fading 59, 60, 63, 64
Ingham, Norfolk 10, 204, 205

de Bois tomb 8, 47, 90, 132, 133, 150, 172, 182, 206
de Ingham tomb 22, 54, 90–92, 106, 107, 121, 132, 133, 150, 206
intonaco 4, 143, 144, 145
Irish medieval wall painting 22
iron oxides and oxyhydroxides *see* red and yellow earth pigments
ivory black 187, 188

jars (for pigments) 20
Jelling, Jutland, Denmark
Grave of the King of Gorm 157
Johanneskapelle, Pürgg, Austria 34, 68, 82, 150, 169, 191, 192, 199
Judenburg, Austria *see* Magdalenenkirche and St Georgen ob Judenburg

Källunge, Gotland, Sweden 77
kaolin 111, 113, 141
Kappel Church, Oberschwaben, Germany
choir 82
Karlštejn Castle, Czech Republic
Holy Rood Chapel 208
Kempley, Gloucestershire *see* St. Mary's, Kempley
kermes lake *see* lake pigments, red
Kingston, Cambridgeshire *see* All Saints' and St. Andrew
Kingston Lacy, Dorset
14th-century painted screen 23, 48, 91, 133, 151, 183
Kirby, J. 11, 155

lake pigments 6, 7, 17, 209
fading 14, 114, 117, 120, 123, **124–5**, 203
red lake pigments 7, 11, 14, 21, 32, 45, 47, 98, 99, 103, 105–7, **111–25**, 126, 129, 133, 150–52, 156, 162, 169, 179, 181, 192, 197,198, 200, 202, 206, 207, 210
brazil wood lake 34, 59, 111, 112, 114–16, 119–21, 123, 124, 196, 200, 205
'cimatura di grana' 112, 113
kermes lake 11, 34, 97, 111–16, 118–20, 123, 124, 195, 196, 198, 200, 205
lac lake 112, 115, 123
madder lake 11, 33, 45, 111–17, 119–23, 151, 195, 196, 198, 200
Polish cochineal 112, 121, 122
yellow lake pigments 71, 93, 94, 151
Lambach, Austria
west choir 70–72, 82
lamp black 188, 193
Landkartengalerie, Salzburg, Austria 121
lapis lazuli 27, 28, 29, 30, 32, 33
Laurie, A. P. 1, 13
Lausanne cathedral, Switzerland
south portal 23, 34, 122, 123, 133, 151, 183
lead oxalate 137
lead pigments 4, 5, 13, 73, 99, 140, 154, 158, 159, 161, 181, 198, 199, 209
alteration 13, 14, 210
lead-tin yellow 7, 46, 86, 87, 88, 139, 150–152, 156, 157, **159–63**, 179, 201
lead white 5, 7, 11, 20, 34, 37, 44, 47, 48, 59, 60–63, 73, 76, 80–82, 86–91, 93, 94, 102–8, 111, 113, 117–20, 122, 125–34, 138, 139, 148, 150, 151, 154–6, 158, 160, 162, 166, 167, 169, 170, 172, 175, **176–85**, 192–4, 197, 201, 202, 210
lead white alteration 22, 177–9, 184–5, 203
lead white ground 6, 7, 34, 61, 95, 102, 105–8, 117, 120, 130, 144, 177–9, 181–4, 193, 201, 202, 207
levelling plaster *see also arriccio* 144
Lewes Priory, Sussex
excavated wall painting fragments 10, 29, 30, 67, 68, 100, 104, 144
Lichfield Cathedral, south choir aisle,
south wall, tomb recess 94, 102, 161, 191
lime ground 4, 6, 37, 88, 101, 118, 129, 136, 144, 167, 177
lime plaster 4, 30–32, 37, 42, 66–9, 88, 91, 92, 101, 105, 108, 130, 137, 148, 149, 167, 170, 204, 205
limewash 63, 91, 126, 204, 205
lime-water 13, 14, 38, 44, 78, 139, 169
lime white (*bianco di San Giovanni*) 4, 22, 31, 32, 33, 37, 45, 46, 50, 52, 54, 61, 67–9, 75–7, 81, 99, 101, 104–6, 108, 126, 146–50, 155, **165–70**, 171, 173, 185, 189, 190, 191, 192, 194, 201
Lincoln Cathedral
west front, frieze 5, 23, 34, 43, 62, 107, 132, 149, 176, 198, 199
Linden Church, Oberbayern, Germany 83
linseed oil *see* oil, linseed
lithage 126, 160, 177
Little Easton, Essex
nave 177, 184
Little Wenham, Suffolk
chancel, east wall 53, 54, 55, 56, 168, 184
local ground 4, 5, 31, 32, 33, 44, 45, 69, 80, 91, 102, 105, 147, 148, 167, 178, 180, 191–3, 201, 206
London *see* British Museum; Courtauld Institute of Art; National Gallery; St. Paul's Cathedral; Tower of London; Victoria and Albert Museum; Westminster Abbey; Westminster Palace
Longthorpe Tower, Cambridgeshire
first floor solar 13, 101, 110, 128, 172, 178, 184
Lübeck, Germany
Katharinen Church 77
St. Marien, Briefkapelle 132

madder lake *see* lake pigments, red
Magdalenenkirche, Judenburg, Austria
choir, north wall 79, 83
magnetite 152
malachite 42, 44, 49, 68, 71–4, 77, **78–84**, 146, 200, 207
alteration 74, 77, 83, 84
manuscripts
Azurium transmarinum hoc modo per pastillum afinatur, Gonville and Caius College, Cambridge *MS 181* 28
Book of Kells, Trinity College Library, Dublin

71, 157, 159
BL Canterbury Psalter (Vesp. A. I.) 122
BL Canterbury Psalter (Arundel 155) 122
BL Egerton MS 840 18
BL MS Egerton 1139 (Queen Melisend's Psalter) 33
BL Sloane 342 28
BL Sloane 1754 18
BL Sloane 2584 16
BL Cotton Titus D. XXIV 16, 42, 71, 106, 114
BL Winchester Psalter (Nero C.IV) 122
Brogyntyn 2.1 16
Cambridge, St John's College MS79 60
Cambridge, Trinity College MS1451 51
Compositiones Variae, Lucca MS 158 *(*MS Codex *490)* 17, 50, 99, 100, 108
*De Architectura,*Vitruvius 60, 65, 141
De Coloribus et Artibus Romanorum, Eraclius 16, 17, 18, 71, 73, 74, 85, 97, 98, 111, 114, 126, 155, 156, 181, 198
De Coloribus et mixtionibus 71
De Diversis Artibus, Theophilus 16, 18, 19, 46, 51, 58, 70, 71, 74, 85, 97, 100, 104, 106, 114, 127, 130, 143, 150, 155, 156, 168, 175, 177, 189, 198
Il Libro dell'arte, Cennino Cennini 16, 19, 28, 34, 46, 62, 66, 68, 70, 71, 78, 80, 82, 85, 87, 100, 101, 105, 114, 117, 126, 130, 137, 143, 155, 156, 161, 166, 167, 174, 177, 180, 181, 185, 187, 188
Jehan Le Bègue 18, 19, 85, 113, 155, 156, 160, 189
Liber de coloribus illuminatorum siue pictorum 126
Liber claritatis totius alkimicae Bologna Cod. Lat. *164 (153)* 28
Liber diversarum arcium (Montpellier, Ecole de Médecine MS 277*)* 16
Liber Magistri Petri de Sancto Audemaro de coloribus faciendis (Peter of St. Omer) 16, 18, 19, 46, 97, 100, 111, 115, 130, 177, 180, 188
Likneskjusmith 3025 16
Lindisfarne Gospels 136
Mappae Clavicula 16, 17, 50, 51, 85, 111, 127, 154, 155
Phillipps' Corning MS3175 17
Sélestat MS 17
Naturalis Historia, C. Plinius Secundus (Pliny) 40, 60, 70, 71, 73, 74, 78, 79, 97, 98, 100, 125, 134, 153, 176
Strasbourg MS 16, 19
The Painter's Manual, Dionysius of Fourna 70, 180
Winchester Bible 37, 68
Maria Laach, Germany
Tomb of Pfalzgraf Heinrich II 23, 83, 90, 122, 133, 183, 193
Mårslet, Denmark 2, 43, 206
massicot *see also* red lead and lead-tin yellow 160, 177
Mästerby, Gotland, Sweden 38, 77, 83, 196, 204
Matteini, M. 2
metacinnabar 99, 106, 110
metavivianite *see* vivianite, alteration
minium *see* red lead
mixed media 6, 7
Moles, A. 2
Monkwearmouth, Northumbria 5
Montvianex Madonna 6, 34, 122, 157, 199
mordant 5, 98, 129, 132, 151, 156, 172, 173, 181, 202, 207
mosaic gold 153
Müstair, Switzerland, Convent of St. John the Baptist 2, 5, 22, 40, 70, 71, 106, 120, 131, 132, 136, 144, 145, 150, 169, 192, 198, 199

National Gallery, London,
Bouts, Dieric and workshop
The Virgin and Child with Saints Peter and Paul 122
Christ Crowned with Thorns 122
van de Weyden, Rogier and workshop
The Exhumation of Saint Hubert 122
Wilton Diptych 23, 35, 63, 70, 157
Nevers Cathedral, France 70
Nohant-Vicq, France *see* Saint-Martin
Norwegian altar frontals 23, 133, 173
Dale I 133
Dale II 76
Eid 157
Hauge 133
Kinsarvik 157, 159
Nedstryn 63
Odda 76
Skaun 133
Tingelstad I 157
Tingelstad II 157
Tingelstad III 29, 38, 48, 157, 196
Trondheim 63
Norwich Cathedral
Ante-Reliquary Chapel 6, 7, 12, 13, 30, 31, 35, 43, 45, 46, 49, 53, 55, 74, 88, 89, 91, 101, 102, 105, 110, 116, 117, 119, 120, 123, 128, 129, 134, 149, 151, 178, 179, 181, 184, 192, 196, 204, 208, 210
'Losinga' cycle 29, 31, 34, 67, 69, 74–6, 100, 101, 105, 145, 168, 197, 201
Priory, accounts 9, 20, 49, 111, 126, 157, 181
Norwich, Greyfriars, shell palette 44
Notre-Dame-la-Grande, Poitiers, France 90, 206

oil 5, 7, 8, 13, 16, 20, 28, 41, 46, 47, 56, 62, 76, 87–9, 91–4, 98, 103, 106, 109, 110, 113, 114, 117–22, 124, 126, 128, 129, 131, 132, 134–7, 139, 148, 154, 156, 158–62, 167, 172, 173, 175, 177–81, 184, 193, 194, 199–203, 205–7, 210
heat-bodied (prepolymerised) 5, 7, 82, 87, 89, 94, 132, 170, 180–82, 202
linseed oil 5–7, 33, 62, 82, 89, 90, 94, 109, 113, 128, 132, 139, 162, 180–83, 198, 199, 202, 210
walnut oil 89, 90, 131
Opificio delle Pietre Dure, Florence 2
orange (mixed) 127
organic binding media 5–7, 11, 13, 91, 105, 106,

108, 124, 128, 149, 151, 157, 163, 169, 173, 180, 183, 198, 201, 202, 209–11
organic green colorants 71
moss green 71
saffron green 71
sap green 71
organic yellow colorants 71, 74
orpiment 7, 20, 34, 62, 73, 110, 118, 125–7, 150–52, **153–9**, 162, 181, 182, 201, 210
alteration 154, 157–9

painters and colourmen/women etc.
Agnes the limeburner 165
Edmund de Hemenhall 20
Herebright of Cologne 207
Jehan Coste 41
John Crepelgate 171
John Oxford 21
John of St. Albans 207
Master Bertram of Hamburg 207
Master Hugh of St. Alban 21
Master Theodoric 208
Odo the Goldsmith 21
Peter Pictor 19
Richard Norwich 21
Thomas of Westminster 207
Walter of Durham 207
William Page 41, 154
Park, D. 9
pastiligia 172, 173
pentimenti 145
Perschen Oberpfalz, Germany 82
Peterborough Cathedral
nave ceiling 47, 91, 108, 133, 151, 160, 183
pigment alterations 14
plaster joins 4, 201
plaster of Paris 173–5
plattnerite (lead dioxide) 135, 184
Plesters, Joyce 1
Poitiers, France *see* St. Hilaire-le-Grand and Notre-Dame-la-Grande
pontate 4
Prague, Czech Republic *see* Emmaus Abbey and St. Vitus' Cathedral
prasinus 143
preliminary drawing 4, 6, 12, 22, 143–7, 149, 150, 168, 201
primary sources 15–21
protein 5–7, 33, 46, 47, 87, 93, 106, 119, 124, 131, 134, 136, 169, 180, 198, 199, 202, 206
Pürgg, Austria *see* Johanneskapelle
purple, mixed 31, 32, 34, 45, 46, 61, 104, 116, 119, 121, 124, 125, 148, 151, 190, 192, 199, 207

realgar 127, 153, 154, 155, 159
red earth 31, 39, 45, 47, 54, 61, 67, 69, 80, 88, 91, 98, 100, 102, 104–8, 118, 126, 127, 129, 130, 132, 135–7, **141–52**, 167, 168, 169, 172, 176, 178, 179, 182, 189, 191–3, 200
red lake *see* lake pigments, red
red lead 5, 11, 21, 45, 53, 88, 97, 98, 99, 102, 103, 105–7, 110, 111, 118, 122, **125–40**, 151, 152, 154, 155, 158, 160, 166, 172, 175–9, 181, 182, 184, 191, 192, 198, 199, 200, 202, 203, 207, 210
alteration 126, 127, **134–40**, 183, 202, 211
red ochre *see* red earth
Rees-Jones, S.1, 2
Reformation 10
Regensburg, Germany 38, 196
Reichenau, Germany *see* St Georg in Oberzell
relief decoration 6–8, 121, 150, 172, 173, 175, 180, 202, 210
resin 87, 93, 94
conifer 82
pine 87, 93, 139
Reymerswaal regulations 111
Risinge, Östergötland, Sweden 83
Rochester Cathedral
choir (*Wheel of Fortune*) 24, 60, 61, 101, 102, 103, 127, 149, 155, 156, 172, 193
north-east transept, tomb recess of St. William of Perth 88, 89, 93, 117, 118, 120, 127, 129, 131
Roger of Helmarshausen 18
Romanesque (English wall painting technique) 3–6
Rome, Italy *see* House of Livia, S. Clemente and S. Maria Antiqua
Rouen green 73
Roy, A. 1, 155, 161, 182
Royal Wardrobe accounts 124, 154

saffron 154
St.-Aignan, Brinay-sur-Cher, France 90, 206
St. Albans Cathedral
chapter house, excavated sculpture fragments 39, 40
feretory, wall paintings 12, 14, 45, 53, 60, 62, 81, 89, 101, 102, 103, 116–19, 123, 125, 128, 130, 134, 144, 148, 178, 179, 192, 196, 206–8, 210
feretory, Shrine of St. Alban 10, 47, 49, 74, 78, 118, 121, 179, 206, 207
presbytery 161
St. Augustine's Abbey, Canterbury 39, 44, 138
St. Bavo's Abbey, Ghent, Belgium 70
St. Botolph's, Hardham, West Sussex 23, 74, 76, 77, 78, 144, 146–8, 167, 170, 190, 197, 205
S. Clemente, Rome, Italy 39
Saint-Etienne, Auxerre, France
crypt 24, 70, 136, 150, 191
S. Francesco, Arezzo, Italy
Legend of the True Cross 136
St. Georg in Oberzell, Reichenau, Germany 43
St. Georgen ob Judenburg, Austria 144–5, 169
St. Giles Hospital, Brompton Bridge, N. Yorkshire 67, 103
St. Hilaire-le-Grand, Poitiers, France
Revelation cycle 43, 206
St. John's, Duxford, Cambridgeshire chancel, west and south walls 147
St. Margaret of Antioch, Cowlinge, Suffolk 81, 82, 192
Santa Maria de Sigena, Spain

chapter house 22, 50, 152
S. Maria im Kapitol, Cologne, Germany 6, 77
S. Maria Antiqua, Rome, Italy 40
St. Marien Cathedral, Freiberg, Germany, 15th-century wall paintings 83, 163
St. Marien, Lübeck, Germany
Briefkapelle 132
St. Mark's (The Lord Mayor's Chapel), Bristol 10, 60, 61, 62, 64, 74, 76, 101, 102, 104, 146, 161, 162, 172, 175, 179–81, 193, 197
Saint-Martin, Nohant-Vicq, France 24
St. Mary and St. Leodegarius, Ashby St. Ledgers, Northamptonshire
Passion cycle 155, 156
St. Mary's Abbey, York 5, 23, 34, 62, 107, 132, 182, 192, 199
St. Mary's, Erwarton, Suffolk, 14th-century monument 95, 107, 193
St. Mary's, Kempley, Gloucestershire 4, 22, 34, 35, 43, 49, 80, 199, 206
St. Mary's and All Saints', Willingham, Cambridge 81, 88, 128, 130
St. Mary le Crypt, Gloucester 43
St. Michael, Hildesheim, Germany
painted wooden ceiling 157, 169
St. Michael and All Saints, Copford, Essex 29, 100, 104, 145
S. Michele, Val Susa, Italy 169
St. Nicholas' Hospital Church, Harbledown, Kent 30, 81, 128, 130
St. Paul's Cathedral, London 207
St. Philibert Abbey, Tournus, France(crypt) 70
St. Vitus' Cathedral, Prague, Czech Republic 208
S. Vincenzo al Volturno, Molise, Italy 39, 190
Sal ammoniac (ammonium chloride) 51, 52
Salisbury Cathedral
external polychromy, west front 23, 150, 170, 185
salt greens (copper chloride greens) 18, 22, 42, 69, **72–8**, 80, 84, 147, 180, 195, 197, 200, 204–5
alteration 73, 78
atacamite 69, 72–7, 92, 167, 169, 197, 205
botallackite 72, 73
clinoatacamite 72, 73
paratacamite 72–5, 81, 82
Salzburg, Austria (see Landkartengalerie)
Schwäbisch Gmünd, Germany *see* Holy Cross Minster
Scottish Treasury accounts 20, 59, 121, 154, 171
sealant 129, 131, 172, 175, 181
secco 4, 68–71, 82, 108, 131, 136, 155, 156, 161
shell palettes
Clarendon Palace, Wiltshire
Antioch Chamber 44, 128
Greyfriars, Norwich 44
St Augustine's Abbey, Canterbury 44, 162–3
Whitefriars, Carmelite Friary, Coventry 44, 117
Sherborne Abbey, Dorset
chapter house, excavated wall plaster 74, 80–82, 84, 100, 148, 172, 190, 200
Sigena, Spain *see* Santa Maria de Sigena, chapter house
silver leaf 7, 8, 20, 41, 47, 89, 90, 121, 129, 132, 133, 165, 180, 182, 206, 210
silver blue 50
Silver Street, Ely (see Ely, Cambridgeshire)
sinopia 4, 104, 141, 143, 144
size *see* glue, animal
snapped cord lines 4, 143
Spanish green (Theophilus) 51
stencils 61
Stoke Orchard, Gloucestershire
chancel arch 147
Stow, Lincolnshire,
north transept 149
Swedish medieval wallpainting 22
synthethic copper blues *see* copper pigments

Täby, Uppland, Sweden 77, 83
tenorite (cupric oxide) 49, 50, 78
terre-verte *see* green earth
Teweskbury Abbey, Gloucestershire
stone fragments 23, 47, 95, 172, 182
Thornham Parva, Suffolk,
Thornham Parva Retable 23, 48, 108, 122, 133, 151, 173, 180, 183, 208
tin leaf 21, 46, 102, 119, 156, 180, 185, 208
shears to cut 21
Tournus, France *see* St. Philibert Abbey
Tower of London, Byward Tower
south tower, 1st floor room 43, 44, 101, 103, 111, 117, 119, 128, 130, 134, 192, 195
'transitional' period 6, 7, 33
translucency 6–8, 11, 13, 28, 33, 45, 66, 68, 82, 88, 89, 93, 94, 111, 113, 114, 116–19, 121, 129, 131, 146, 148, 149, 154, 159, 165, 167, 170, 178, 179, 184, 194, 198, 201, 202, 207, 210
Tristram, E. W. 2, 13, 22, 89

ultramarine 10, 11, **27–35**, 36, 37, 41, 42, 43, 48, 49, 50, 62, 63, 81, 84, 86, 104, 116, 133, 148, 167, 168, 169, 179, 182, 190–92, 196, 197, 199, 205, 206
ultramarine ash 29
ultraviolet fluorescence 32, 33, 123, 137
umber **141, 149**, 150
Urschalling, Germany 37, 196, 204

Val Susa, Italy *see* S. Michele
varnish 13, 20, 93, 98
Vaudreuil Castle, Normandy 41
Vendôme, France *see* Abbey of La Trinité
veneda 189
Verden Cathedral, Germany 82
verdigris 8, 11, 13, 44, 50, 51, 52, 56, 62, 63, 67, 73, 80, 82, **85–92**, 93, 94, 98, 155, 158, 160, 162, 163, 165, 176, 179–81, 193, 194, 200, 205, 206, 210
alteration 14, 22, 54, 55, 56, 91–2, 202, 204, 211
verditer 50
vermilion 2, 30, 47, 48, 49, 69, 86, **96–111**, 113, 116, 118, 122, 125–7, 130–34, 137, 138, 144, 148, 150–52, 156, 160, 167, 169, 175–7, 179–81, 190, 192, 193, 198, 200, 202, 206, 207
darkening 24, 99, **109–11**, 203

Victoria and Albert Museum
bust of Henry VII 124, 196
viride salsum see salt greens
vivianite 11, **35–9**, 48, 168, 192, 195, 196
alteration 14, 36–9, 196, 202–4, 211
Vossestrand, Norway (lion-capital) 70, 133, 151, 159, 182

Waltensburg, Kanton Graubünden, Germany 132
Wells Cathedral
West Front, external polychromy 23, 71, 94, 106, 150, 176, 182
Westminster Abbey 21
accounts 187
chapter house 1, 43, 44, 45, 49, 60, 61, 64, 101–5, 108, 111, 116–21, 123, 128, 130, 136, 138, 139, 155, 156, 172, 173, 178–80, 185, 193, 196, 200, 203, 205, 207, 208
Retable 23, 35, 108, 122, 133, 151, 175
Westminster Palace 9, 20, 21
accounts 9, 20, 41, 52, 57, 86, 98, 113, 121, 126, 142, 153, 154, 158, 160, 171, 174, 187, 195
The Painted Chamber 21, 48
ceiling paterae 23, 35, 91, 108, 122, 133, 173
The queen's chamber 21
St. Stephen's Chapel 1, 2, 8, 20, 21, 22, 27, 30, 35, 41, 48, 88, 117, 118, 120, 128, 131, 178, 195, 199
van de Weyden, Rogier and workshop *see* National Gallery, London
White, R. 5, 33, 82, 140
White Notley, Essex 138
Willingham, Cambridge *see* St. Mary's and All Saints'
Wilton Diptych *see* National Gallery
Winchester Cathedral 3
Great Screen 23, 47, 63, 90, 95, 107, 121, 132, 133, 151, 172, 182, 194
south transept 147
Guardian Angels' Chapel 30, 31, 36, 38, 48, 75, 106, 128, 129, 135, 196
Holy Sepulchre Chapel 4, 5, 11, 29–33, 35–38, 67–71, 75, 100–6, 111, 116, 120, 129, 130, 131, 135, 138, 143–5, 148, 149, 167, 168, 172, 177, 180, 185, 192, 196–8, 202, 204
Lady Chapel, vaults 24, 105, 107
south transept 22
Windsor Castle
Rose Tower 20
St John's Tower
accounts 52
Witley, Surrey *see* All Saints, Witley
woad 57, 58, 59, 60
Worcester Cathedral
crypt 67, 103
Edgar Tower 107
pilgrim's staff 124, 196
13th-century excavated fragments 104, 155, 156, 158, 159
workshop practice 11, 12, 14, 145, 207, 208

yellow earth 2, 32, 39, 44, 47, 67, 68, 69, 81, 88, 90, 102, 104, 105, 118, 127, **141–52**, 154, 157, 167, 168, 178, 185, 189, 190, 194, 207
yellow ochre *see* yellow earth
York *see also* St. Mary's Abbey
York Minster 30
Bedern, excavated wall plaster 161, 201
central tower vault, fifteenth-century painting 58, 59, 176
chapter house 106
coffin-lid of Archbishop Walter de Gray 30, 168, 190
fabric rolls 160, 176

Zittau altarpiece, Germany 122